S0-BWX-177

ADVANCES IN RADIOBIOLOGY

H. J. MULLER

ADVANCES IN RADIOBIOLOGY

PROCEEDINGS OF THE FIFTH INTERNATIONAL CONFERENCE ON RADIOBIOLOGY HELD IN STOCKHOLM ON 15th-19th AUGUST, 1956

EDITED BY

GEORGE CARL de HEVESY

ARNE GUNNAR FORSSBERG

AND

JOHN D. ABBATT

CHARLES C THOMAS · PUBLISHER

Springfield · Illinois · U.S.A.

FIRST PUBLISHED . . . 1957

PRINTED IN GREAT BRITAIN BY
OLIVER AND BOYD LTD., EDINBURGH

This book is respectfully dedicated to

PROFESSOR H. J. MULLER

in appreciation of his pioneer
work in genetics

PREFACE

IN THE PREFACE to the Proceedings of the Fourth International Conference on Radiobiology, Professor Mitchell and his co-editors state that at that meeting more than ever before, the value of collaboration between investigators trained in a wide range of scientific disciplines was demonstrated as an obvious necessity for progress in the very difficult field of scientific inquiry represented by radiobiology. The truth of this statement was fully brought out in the course of the Fifth International Conference on Radiobiology held in Stockholm in August 1956. A great variety of subjects were discussed by scientists experienced in a wide range of scientific disciplines. The effect of irradiation on the hæmopoietic system being the first and that of the application of induced mutation in plant breeding the last items of the vast field covered. In the single year after the Cambridge meeting, marked progress has been achieved in the whole field covered by the Stockholm Conference. To mention one example only, our knowledge was markedly advanced both of the technique and the mechanism of replacement of damaged marrow cells by healthy ones. Though these and other investigations in the realm of radiobiology were carried out without regard to practical applications, these applications are in the future, but the immense difficulties which are involved in cellular replacement without damage to the organism will have to be overcome.

At the Stockholm meeting about 180 delegates attended from 16 different countries ; these included, for the first time, members of the Chinese Medical Association representing China. The appreciation by this Association of the importance of radiobiology is reflected in the statement of their President, Dr. Fu Lien-Chang, in a letter to the Chairman of the Fifth International Conference on Radiobiology, when he said: "Radiobiology is closely connected with man's health; hence the holding of your conference at this juncture is of great significance and importance to human welfare."

It is planned that the 1958 meeting shall be held in the United States of America.

G. HEVESY
ARNE FORSSBERG
JOHN D. ABBATT

Stockholm and London

March 1957

SWEDISH COMMITTEE

Chairman:
Professor George de Hevesy

Vice-Chairman:
Professor Åke Gustafsson

Secretaries:
Dr. Arne Forssberg
Dr. Arne Nelson

Treasurer:
Dr. Matts Helde

Members:
Dr. Lars Ehrenberg
Dr. Karl-Gustav Lüning

The Committee acknowledges the receipt of generous grants from the following authorities, foundations and corporations:

Swedish Government
Swedish Cancer Society
Swedish Defence Medical Research Committee
Swedish Atomic Energy Company
Swedish Atomic Energy Commission
Almqvist & Wiksell, Publishers
Elema Company
Hugo Tillquist & Company
Swedish Philips Company
Swedish Seed Company
Swedish Sugar Company
Weibullsholm Plant Breeding Institute

CONTENTS

PAGE

Preface vii

Swedish Committee ix

SECTION I

INITIAL OR PRIMARY CHEMICAL EFFECTS OF IRRADIATION

F. Hutchinson — Radiation sensitivity of molecules in intact cells 3

P. Alexander — The relative importance of direct and indirect radiochemical processes in radiobiology 8

G. Stein and A. J. Swallow — The biological action of ionising radiations from the point of view of radiation chemistry 16

SECTION II

BIOCHEMICAL IRRADIATION EFFECTS—ON ENZYMES AND OTHER CELLULAR CONSTITUENTS

B. Rajewsky, G. Gerber and H. Pauly — X-ray inactivation of the components of the succinic acid dehydrogenase-cytochrome - cytochrome oxidase - system 25

J. A. V. Butler, P. Cohn and A. R. Crathorn — Effects of ionising radiation on the *in vivo* incorporation of amino acids into proteins 33

M. L. Mendelsohn — The combined action of X-rays and chemical inhibitors on *in vitro* kidney transport and respiration 38

G. Höhne, H. A. Künkel, H. Maas, G. H. Rathgen — Primary biochemical effects in the X-irradiated *Yoshida* ascites sarcoma 43

F. G. Sherman and A. B. Almeida — The incorporation of ^{32}P into liver phospholipids and RNA mononucleotides of irradiated and non-irradiated mice 49

L. G. Lajtha, R. Oliver and F. Ellis — Effects of irradiation on DNA synthesis by human bone marrow cells *in vitro* ... 54

P. Mandel, C. M. Gros, J. Rodesch, C. Jaudel and P. Chambon — Effect of various doses of X-rays—whole body and local irradiation—on the nucleic acids of the bone marrow ... 59

M. G. Ord and L. A. Stocken — The effect of X-radiation on rat thymus nucleic acids at short intervals after exposure *in vivo* 65

Section III

PHYSIOLOGICAL AND MORPHOLOGICAL IRRADIATION CHANGES

		PAGE
D. E. Smith and Y. S. Lewis	Prolongation of the clotting time of peritoneal fluid after X-irradiation ...	73
V. Slouka	The participation of the peripheral nervous system in the reaction to irradiation	76
L. Révész	The effects of lethally damaged cells upon survivors in X-irradiated experimental tumours	80
W. Dittrich	Induction of chromosome breaks in the *Yoshida* sarcoma ascites cells of the rat by X-rays and fast electrons at different oxygen partial pressures ...	86
T. Alper	Observations on bacterial growth and morphology shortly after irradiation and some remarks on the oxygen effect	90

Section IV

ALPHA PARTICLE IRRADIATION OF SINGLE CELLS

R. J. Munson	The 'shooting' of bacteria one at a time by single alpha particles ...	105
R. Munro	Alpha irradiation of parts of single metaphase cells in chick tissue cultures	108
M. I. Davis, I. Simon-Reuss and C. L. Smith	The irradiation of single cells and parts of single cells in tissue culture with microbeams of alpha-particles ...	114

Section V

MODIFICATION OF SYSTEMIC IRRADIATION EFFECTS

I. General Principles

A. Hollaender	The effects of pre- and post-treatment on the radiation sensitivity of micro-organisms	123
J. A. Cohen, O. Vos and D. W. van Bekkum	The present status of radiation protection by chemicals and biological agents in mammals	134

II. Chemical Methods: effects of cysteamine cystein and related compounds

		PAGE
A. Pihl and L. Eldjarn	Studies on the mechanism of protection against ionising radiation by compounds of the cysteamine-cysteine group	147
Z. M. Bacq	Recent research on the chemical protectors and particularly on cysteamine-cystamine	160
R. Koch	The problem and constitution of radiation-sensitising agents	170
H. A. Künkel, G. Höhne and H. Maass	Radiobiological investigations on the hibernating loir (*glis glis*)	176
A. Catsch	The dose reduction factor for cysteamine and isothiuronium in the case of whole body X-irradiated mice and rats	181
U. Hagen	Chemical and biological action of protective SH-compounds	187
H. Marcovich and J. F. Duplan	A bacteriological test for the study of radioprotection problems in mammals	192

III. Transfer of Cellular Material—Irradiation and Immune Reactions

C. E. Ford, J. L. Hamerton, D. W. H. Barnes and J. F. Loutit	Studies of radiation chimæras by the use of chromosome markers	197
J. Soška, V. Drášil and Z. Karpfel	The cell factor in spleen homogenates after irradiation	204
D. W. H. Barnes, M. P. Esnouf and L. A. Stocken	Some experiments in favour of the cellular hypothesis for the spleen curative factor	211
E. L. Simmons, L. O. Jacobson and J. Denko	Studies on the mechanism of post-irradiation protection	214
V. L. Troitsky, M. A. Tumanjan and A. J. Friedenstein	The influence of ionising radiation upon the natural immunity	221

IV. Other Approaches to the Modification of Irradiation Effects

F. Devik	Modification of the X-ray reaction in the skin of mice by shielding of minute areas of the skin	226
O. Bogomolets, A. Boico, G. Diadiucha, Z. Zekhova, V. Lavrick and G. Levtchouk	Changes in the reactivity of the organism under the influence of ionising radiation	231

PAGE

Z. M. Bacq, P. Martinović, P. Fischer, M. Pavlović and G. Sladić — Irradiation and adrenal and pituitary function 237

A. Brohult — Alkoxyglycerol esters in irradiation treatment 241

S. Hornsey — The protective effect of the reduction of the body temperature on adult and young mice after whole-body irradiation 248

Section VI

THE TIME FACTOR IN RADIOBIOLOGY

H. and M. Langendorff — The effect of repeated small doses on the fertility of the white mouse ... 257

H. J. Curtis and R. Healey — Effects of radiation on ageing ... 261

B. Rajewsky, K. Aurand and I. Wolf — Studies on the time-intensity factor after whole-body X-irradiation ... 267

P. Desaive — Restoration of primordial follicles in the irradiated ovary 274

H. Marcovich and R. Latarjet — Effects of long-term irradiation on lysogenic bacteria 281

Section VII

STUDIES ON THE DISTRIBUTION OF ISOTOPES IN TISSUES

M. Owen and J. Vaughan — Changes in the rabbit tibia and an estimation of the dose received by the bone tissue following a single injection of strontium 287

F. Björnerstedt, C.-J. Clemedson, A. Engström, and A. Nelson — Bone and radiostrontium—an attempt to evaluate the dose distribution ... 294

R. Lewin, B. Rosoff, H. E. Hart, K. G. Stern and D. Laszlo — Decontamination studies. (A simple *in vitro* system to study the interaction of radioactive metals with proteins from body fluids and tissues) ... 298

J. D. Abbatt and H. E. A. Farran — Iodinated tyrosines and radiosensitivity in thyrotoxicosis 305

W. Jacobi, K. Aurand and A. Schraub — The radiation exposure of the organism by inhalation of naturally radioactive aerosols 310

SECTION VIII

IRRADIATION EFFECTS ON THE HÆMOPOIETIC SYSTEM

I. ERYTHROPOIESIS

		PAGE
L. F. LAMERTON and E. H. BELCHER	Effect of whole-body irradiation and various drugs on erythropoietic function in the rat. Studies with radioactive iron	321
E. B. HARRISS	*In vivo* uptake of radioactive iron by the erythroid cells of rat bone marrow	333
H. MAISIN, A. DUNJIC, P. MALDAGUE and J. MAISIN	Erythropoietic activity in irradiated rats injected with homologous and heterologous bone marrow study with ^{59}Fe	341
E. V. HULSE	Quantitative changes in the erythropoietic cells of the rat bone marrow during the first forty-eight hours after whole-body X-irradiation	349
C. W. GILBERT, E. PATERSON and M. V. HAIGH	The life span of red cells of the *rhesus* monkey following whole-body X-radiation	357

II. OTHER HÆMOPOIETIC FUNCTIONS

M. HELDE	Read-off methods in radiohæmatological control	361
L. A. ELSON	Comparison of the physiological response to radiation and radiomimetic chemicals. Patterns of blood response	372
B. LINDELL and J. ZAJICEK	The effect of whole body X-irradiation on the megakaryocytic system in rat femur	376
E. M. LEDLIE	The immediate effects of large doses of radioactive phosphorus on the peripheral blood compared with those of external irradiation in patients with malignant disease	382

III. RADIATION LEUKÆMIA

J. F. LOUTIT	Induction of leukæmia by radiation ...	388
M. FABER	Radiation induced leukæmia in Denmark	397

Section IX

RADIATION GENETICS

		PAGE
H. J. Muller and I. I. Oster	Principles of back mutation as observed in *Drosophila* and other organisms ...	407
T. C. Carter	Genetic implications of irradiation in man	416
K. G. Lüning and S. Jonsson	The induction of detrimental mutations in *Drosophila* by X-rays ...	425
G. Bonnier	Rate of development of X-ray induced detrimentals and the influence of selection pressure	433
O. G. Fahmy and M. J. Fahmy	Comparison of chemically and X-ray induced mutations in *Drosophila melanogaster*	437
F. H. Sobels	The possible role of peroxides in radiation and chemical mutagenesis in *Drosophila*	449
P. Oftedal and J. C. Mossige	Incorporation and mutagenicity of ^{32}P in *Drosophila* sperm	457
S. Wolff	Recent studies on chromosome breakage and rejoining	463
I. I. Oster	Modification of X-ray mutagenesis in *Drosophila*	475
K. Nordback and C. Auerbach	Recovery of chromosomes from X-ray damage	481
List of Participants ...	... ...	487
Subject Index	... ...	494
Index to Contributors	... ...	501

SECTION I

INITIAL OR PRIMARY CHEMICAL EFFECTS OF IRRADIATION

RADIATION SENSITIVITY OF MOLECULES IN INTACT CELLS

Franklin Hutchinson
Department of Biophysics, Gibbs Laboratories,
Yale University, New Haven, Conn., U.S.A.

Any reasonably satisfactory understanding of radiobiology will require a way of estimating quantitatively, even in a crude way, the effects of ionising radiation on the molecules of which the cell is composed. The direct effect may be calculated, under some conditions at least, by methods previously discussed by Pollard *et al.* (1955). The indirect effect in cells is caused by the migration of chemically active intermediates (such as OH, HO_2, H_2O_2) created by the ionising radiation. The magnitude of the indirect effect on a given molecule is specified by two parameters. One of these is the sensitivity of the molecule in question to the intermediates, and is frequently given in terms of the number Y of molecules inactivated per ionisation in a dilute water solution, under such conditions that all the intermediates formed react with the molecules. The value of Y is different for each kind of molecule. The other parameter is the mean distance P that the intermediates diffuse before they react with another molecule. The magnitude of P can be different for different parts of the cell.

Zirkle and Tobias (1953) developed a 'migration model' in which the dose and the biological effects were related through the 'diffusion distance' P. They applied it to the survival of yeast cells, but since so little is known about the targets involved, they had to calculate a value of P from other data and see if it was consistent.

In the present experiments yeast cells were irradiated both in a wet state and dry in vacuum, then assays carried out on the enzymes invertase and alcohol dehydrogenase (ADH) and on coenzyme A. The difference between the wet and the dry irradiations was assumed to measure the indirect effect only. By working with comparatively simple enzyme systems about which a good deal is known, it was then possible to interpret the results in terms of a numerical value of the parameter P.

Methods

Because of the very high doses needed (1–1,000 million rads) most of the irradiations were carried out on the Yale cyclotron, using 4 MeV deuterons and 8 MeV alpha particles, with a few points using the 40 MeV alpha particle beam of the Brookhaven cyclotron. The dry irradiations were carried out in vacuum by techniques previously described (Pollard *et al.*, 1955). Briefly, 1–3 mg. (dry weight) of yeast cells in 0·5 ml. water were pipetted on a $\frac{1}{2}$-inch round microscope glass cover slip and dried by slow pumping in a vacuum desiccator. After irradiation, the samples were re-suspended in 1 ml. of water.

For the wet irradiations, 1–3 mg. (dry weight) of yeast cells suspended in 0·5 ml. of 0·01 M phosphate buffer (*p*H of 7) were pipetted on to a $\frac{1}{2}$-inch diameter disk of Millipore 'molecular sieve' type filter which was placed on an absorbent pad saturated with buffer.

Invertase and ADH activities were measured on a haploid strain SC–7 obtained from Zirkle's laboratory. The Co–A activity of these cells was quite low, so a commercial brand (Fleischman's) of dried yeast was used.

Results

For invertase and ADH, if the logarithm of the activity surviving a given irradiation were plotted against the dose expressed in incident particles per square centimeter, the resultant curves were all straight lines. For Co–A, the dry curves were also straight lines. The wet curves indicate logarithmic inactivation down to 30% survival, with this percentage activity surviving doses about ten times the 50% dose. Thus the survival curves were of the form

$$f = e^{-\,SB}$$

where f is the fraction surviving a dose of B particles per square centimeter, and the parameter S, having the dimensions of an area, is usually referred to as a cross section. A large cross section denotes a high sensitivity to radiation.

The experimental values are collected in Table I. It is seen that the invertase in wet cells is about twice as sensitive to radiation as in dry cells, or the direct and indirect effects are about of the same order of magnitude. ADH, which is a sulphhydryl enzyme, is about twenty times as sensitive wet. Co–A is of the order of one hundred times as sensitive wet. Its dry cross section is in reasonable agreement with that expected on the basis of it low molecular weight of about 750. The direct effect on invertase is in excellent agreement with earlier measurements by Pollard, Powell and Reaume (1952) on purified invertase and by Powell and Pollard (1955) on invertase in dried cells.

Discussion

To obtain quantitative information on the movement in the cell of the intermediates which are active in the case of indirect action, use will be made of Zirkle and Tobias' migration model (1953). Under the assumption that the probability of an intermediate reacting with the surface of a molecule is large compared with the ratio of the mean free path of the intermediate in water (the order of an Ångstrom or so) to the radius r_0 of the molecule, Wijsman (1952) has shown that the probability that an intermediate formed a distance r from the molecule will react with that molecule is given by

$$\frac{r_0}{r} e^{-\frac{r-r_0}{P}} \qquad \text{......(1)}$$

where P is a convenient measure of the distance that an intermediate can travel, and is equal to $\sqrt{Dt}$, where D is the diffusion constant of the intermediate, and t is the time it takes a given concentration of intermediates to drop to $1/e$ of its initial value by collision with cell constituents. By integrating this expression, it was shown by Zirkle and Tobias that the fraction f of activity which survives a dose of beta particles per square centimeter is

$$f = e^{-(S'+S'')B} \qquad \text{......(2)}$$

where S′ is the cross section for direct action, S″ the cross section for the indirect effect

$$S'' = 4\pi Yi(P^2 r_0 + P r_0^2) \qquad \text{......(3)}$$

where i is the number of ion pairs per unit path length along the particle track, and Y is the ionic yield.

From the data in Table I, Table II can be constructed, adding values of Y determined from previous work. The values of the radii of the target molecules used are those calculated from the molecular weight under the assumption of a spherical molecule.

TABLE I

Particle	Rate of energy loss ev/100 Å in protein	Cross section A²	
		Wet	Dry
Invertase			
4 MeV deuterons	230	4,600	2,300
8 MeV alphas	1,000	23,000	11,000
ADH			
4 MeV deuterons	230	25,000	1,000
8 MeV alphas	1,000	23,000	—
Co-A			
4 MeV deuterons ...	230	10,000	150
40 MeV alphas	250	>3,000 (one run)	180 (one run)

TABLE II

Substance	ev/100 Å in protein	S″, Å²	Y	Assumed r_0, Å	Corresponding molecular wt.	*P*, Å
Co-A ...	230	10,000	2·3*	6	750	29
Invertase ...	230	2,300	0·053†	33	120,000	26
	1,000	12,000	0·053	33	120,000	30
ADH ...	230	24,000	1*	22	36,000‡	26

* E. S. G. Barron (1954).

† Unpublished measurements of F. DeFilippes.

‡ A previously reported molecular weight for ADH (Hayes and Velick, 1954) of 150,000 is completely out of line with the dry cross sections tabulated in Table I, so a molecular weight for the yeast ADH has been calculated from the radiation data.

The extremely good agreement among the four calculated values of *P* must be regarded as fortuitous, since the experimental errors in either S″ or Y alone are large enough to lead one to expect a much wider spread. Nonetheless, it seems clear that they point to an action radius much smaller than seems to be widely assumed in the radiobiological literature. Whether the order of magnitude would be true for other parts of the cell, particularly the nucleus, cannot be answered at present. The 30% of the Co–A irradiated in wet cells which was much more resistent to radiation, would seem to indicate that this fraction of Co–A is inside structures (mitochondria?) in which the value of *P* is much smaller. The possibility of other explanations cannot be overlooked, but some variation in *P* from point to point in the cell is not unlikely.

For comparison, the values of *P* chosen by Zirkle and Tobias were in the range 200–300 Å.

Bearing in mind the reservations listed above, it would now seem possible to estimate the magnitude of the indirect effect on the molecular constituents of cells, using equation (3) and ionic yield values which will have to be measured in separate experiments. A suitable value for *P* from this work might be 30 Å. Calculating the direct effect from the size of the molecules involved (Pollard *et al.*, 1955), it is now possible to calculate the entire effects of irradiation on a given intracellular component. If the value of *P* of 30 Å is valid, it is easy to see that for globular proteins and larger spherical units, the effects of radiation will be 50% or more due to direct action, except in cases of enzymes, such as the sulphhydryl ones, which are particularly sensitive to water radicals. Smaller molecules will, in general, be more affected by the indirect effect.

REFERENCES

Barron, E. S. G., 1954. *Radiation Biology*, Vol. I, Part I, p. 299, ed. A. Hollaender, (New York, McGraw-Hill Book Co.).

Hayes, J. E., and Velick, S. F., 1954. *J. Biol. Chem.*, **207**, 225.

Kaplan, N., and Lipmann, F., 1948. *Ibid.*, **174**, 37.

Pollard, E. C., Guild, W. R., Hutchinson, F., and Setlow, R. B., 1955. *Progress in Biophysics*, **5**, 72.

Pollard, E. C., Powell, W. F., and Reaume, S. H., 1952. *Proc. Nat. Acad. Sci., Wash.*, **38**, 173.

Powell, W. F., and Pollard, E. C., 1955. *Radiation Res.*, **2**, 109.

Racker, E., 1950. *J. Biol. Chem.*, **184**, 313.

Sumner, J. B., 1921, 1924. *Ibid.*, **47**, 5; **62**, 287.

Wijsman, R. A., 1952. *Bull. Math. Biophys.*, **14**, 121.

Zirkle, R. E., and Tobias, C. A., 1953. *Arch. Biochem. Biophys.*, **47**, 282.

Discussion

Rajewsky. How long is the mean life time of the radicals?

Hutchinson. It is apparently the order of 10^{-8} seconds in the cell. But it clearly depends on the medium in which the radical is formed. John Ghormley at Oak Ridge has been able to get values as long as 10^{-3} seconds in very pure water.

Rajewsky. Have you some ideas about the nature of the radicals?

Hutchinson. No, but for a number of reasons we have a suspicion they are OH. But that's a long story.

Rajewsky. And what is P—is it measured from the centre of the spherical target or from the periphery?

Hutchinson. P is a parameter which is a measure of the mean distance which the radicals move in the solution.

Rajewsky. And r depends on the size of the molecule?

Hutchinson. Well, it depends in the following way. Experimentally what one measures is $4\pi\ (r^2P + P^2r)$Y. The radius r measures the size of the target attacked by the diffusing radicle. We have assumed that all the molecule surface is attacked and that r is the radius of a sphere of the same molecular weight as the molecule.

THE RELATIVE IMPORTANCE OF DIRECT AND INDIRECT RADIOCHEMICAL PROCESSES IN RADIOBIOLOGY

PETER ALEXANDER
Chester Beatty Research Institute, Institute of Cancer Research,
Royal Cancer Hospital, London, S.W. 3, England

The primary chemical changes which initiate biological lesions observed after irradiation, are brought about either by the absorption of energy within the molecules affected ('direct' action) or by the free radicals produced by energy deposited in the surrounding water ('indirect' action). The free radicals formed in water have to diffuse before they can react and their action can be modified by the presence of oxygen and be reduced by freezing or the addition of protective agents. The view is widely held that the chemical changes produced by 'direct' action cannot be influenced by the environment and that if a biological end effect can be altered by outside factors then the chemical reaction by which it was initiated must have been 'indirect'.

We have found a number of examples (Alexander and Charlesby, 1954 and 1955) during the irradiation of solids where energy taken up by one molecule is transferred to another. In this way added substances can protect by attracting energy from the test material to which they were added, or increase radiosensitivity by passing their energy on. Examples of both processes have been encountered (Setlow and Doyle, 1955, and Alexander *et al.*, 1954) and the concept that direct action cannot be influenced by external factors must therefore be rejected. We will show in this paper that changes in temperature and oxygen concentration also influence chemical changes produced in solids where the action must be 'direct' and that these factors cannot be used to distinguish between direct and indirect action.

Temperature effect

The efficiency of 'indirect' action falls off sharply below the freezing point of the system because the radicals are prevented from diffusing. We found (Alexander *et al.*, 1954) that the efficiency of 'direct' action in breaking the main chain of polymers decreases with decreasing temperature, but that there is no sudden discontinuity as in systems where the action is 'indirect'. The temperature effect when solids are irradiated is difficult to understand and may be due to the decrease in mobility of radical fragments. The writer prefers the explanation (Alexander *et al.*, 1955) that energy transfer processes are involved since the temperature dependence of a large number of different processes, all due to direct action, shown in Table I, is similar. But whatever the mechanism the important point is that a change in radiosensitivity with

TABLE I

System studied	Dose necessary to give effect at given temperatures / Dose necessary at − 190° C.				
	−190°	−89° C.	20° C.	37° C.	60° C.
Degradation of polyisobutylene	1·0	0·66	0·53	0·47	0·38
Degradation of polymethylmethacrylate	1·0	—	0·60	—	0·42
Crosslinking of polythene *	1·0	0·52	0·27	—	0·21
Inactivation of Dry T.I. phage † ...	1·0	0·70	—	0·48	—
Inactivation of Dry catalase ‡	1·0	0·61	0·61	0·61	0·38

* Black, R. M., 1956. *Nature, Lond.*, **175**, 365.

† Bachofer, C. S., Ehert, C. F., Mayer, S., and Powers, E. L., 1953. *Proc. nat. Acad. Sci. Wash.*, **39**, 744.

‡ Setlow, R. B., and Doyle, B., 1953. *Arch. Biochem. Biophys.*, **46**, 46.

temperature is not inconsistent with 'direct' action (e.g. difference in radiosensitivity of pollen and seeds at liquid air and room temperature (Rajewsky, 1952) so long as there is no sudden change at the freezing point.

Oxygen effect

One of the few generalisations of radiobiology is that oxygen increases the effectiveness of X-, gamma- and beta-rays. Gray (1954) has shown that the relationship between oxygen tension and relative radiosensitivity is quantitatively very similar for a large number of diverse systems and may be related to the formation of HO_2 radicals. We find that oxygen also effects the radiochemical changes produced in macromolecules by 'direct' action. In the first example studied (Alexander *et al.*, 1955; Alexander and Charlesby, 1955) polyisobutylene, the extent of main chain break-down was unaffected but the U.V. spectrum was quite different when the irradiation was carried out in vacuum or in air. We conclude that a different product is formed when oxygen is present, i.e. dissolved in the polymer, during the irradiation; exposure of the polymer immediately afterwards does not produce this effect. The U.V. absorption spectrum of irradiated polymethylmethacrylate, on the other hand, is altered by exposure to oxygen after the irradiation, suggesting that an unstable product is formed which is readily aerially oxidised. The most striking oxygen effect is found in those polymers which crosslink on irradiation. Figure 1 shows that in air the number of crosslinks introduced into polyethylene approaches a limiting value, while *in vacuo* there is no limit to the crosslinking. The behaviour of polythene in air is identical with that of a copolymer, one component of which crosslinks

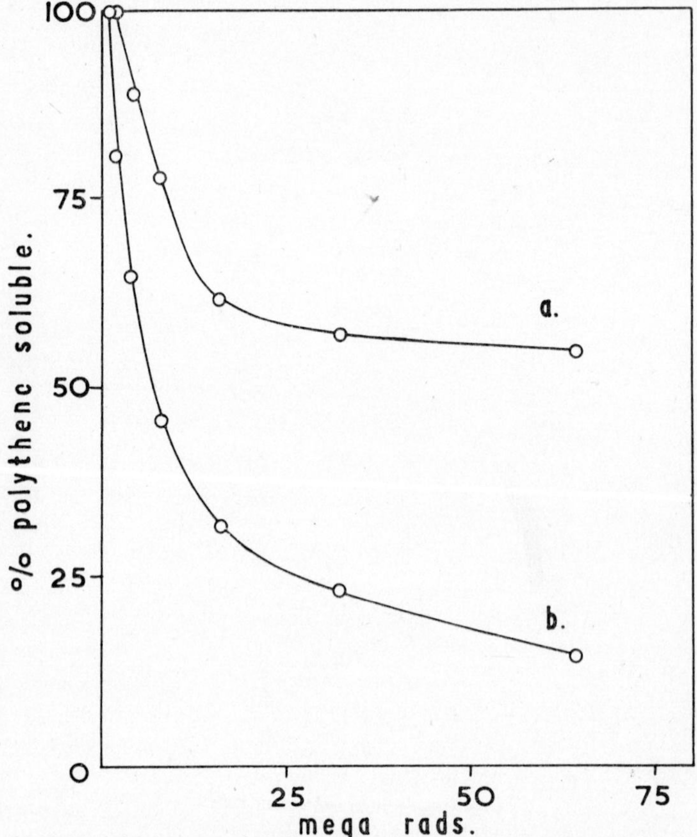

FIG. 1. Effect of oxygen on the crosslinking of thin polyethylene films by 2 MeV electrons. The crosslinking was measured by determining the fraction of the polymer rendered insoluble by extracting after irradiation with Xylene: (*a*) films irradiated in the presence of oxygen: (*b*) films irradiated *in vacuo*.

and the other degrades (Alexander and Charlesby, 1955). Analysis of the polythene data indicates that in air there is an additional process which brings about one break in a molecule for every two molecules involved in a crosslink. The actual number of crosslinks produced is the same in air as in vacuum (or nitrogen). In polystyrene sulphonate which crosslinks much less readily than polythene (i.e. requires a larger dose per crosslink) the number of main chain breaks produced in the presence of oxygen is greater than the number of crosslinks produced so that the net overall reaction is one of degradation (i.e. reduction in viscosity and no gel formation) in air and crosslinking *in vacuo* or under nitrogen (see Table II).

TABLE II

Oxygen Effect on Irradiation of Polystyrene Sulphonate with Gamma-rays

Dose (mega rads)	Viscosity
0	1·15
1·5 in air	0·53
5 in air	0·36
15 in air	0·14
15 *in vacuo*	1·1
46 *in vacuo*	1·4
230 *in vacuo*	79% crosslinked to an insoluble gel

The degradation may be due to combination of oxygen with the polymers to give an unstable product at a point in the molecule rendered reactive by irradiation; since the number of crosslinks is not decreased this centre cannot take part in crosslinking in the absence of oxygen. Another explanation is that oxygen combines with the low energy secondary electrons. The electron affinities of oxygen is much greater than that of hydrocarbons so that in the presence of air the very reactive negative ion O_2- is formed, while *in vacuo* a much less reactive negative ion will be formed (Burton and Magee, 1950). O_2- is closely related to the HO_2 radical which is known to degrade some polymers in solution (Alexander and Fox, 1952). Although no conclusive tests have yet been made to decide these two mechanisms the following experiment favours the O_2–hypothesis. Addition to polythene of substances such as allyl thiourea which are known to react readily with HO_2 radicals (Alexander *et al.*, 1952) reduces the degradation occurring in oxygen without influencing the crosslinking *in vacuo*. This protection can readily be understood on the basis of the competitive removal of O_2–radicals, but would not be expected if the other mechanism were operative.

A pronounced oxygen effect is also observed when solid proteins are irradiated but the complexity of the system makes it difficult to determine the molecular basis for the various changes. Solid serum albumin becomes insoluble in water after irradiation but the dose necessary to produce this effect is much smaller if the radiation is carried out in the presence of air than *in vacuo* or pure nitrogen. Also after irradiation in air the protein contains some new group which is capable of initiating polymerisation of vinyl compounds. For example, when 20 mg. of serum albumin which had received a dose of 2 mega rad of gamma-rays in air is added to a 10% solution of methacrylic acid in water, polymerisation sets in as soon as dissolved oxygen (a polymerisation inhibitor) is removed by bubbling nitrogen. The rate of polymerisation increases with the dose given to the protein until this becomes insoluble. The polymer formed contains protein which is firmly bound. Unirradiated protein or protein irradiated in the absence of oxygen does not initiate polymerisation at all. The most likely explanation is that in the presence of air a peroxide is formed which is sufficiently unstable to decompose within a day or so at room temperature in solution, but which is quite stable in the solid. Colour tests for peroxides were positive but a quantitative estimation has not yet proved possible.

Relative effectiveness of 'direct' and 'indirect' action

For specific chemical changes in large molecules 'direct' action is usually many times more effective, for example, the G value for main chain breaks in dry herring sperm DNA is about 7, while in solution where the process is 'indirect' it is about one fiftieth of this value (Alexander and Stacey, 1955). This means that indirect action only becomes significant for this reaction when DNA is irradiated at concentration of less than 2%. The local concentration of DNA in the chromosomes is of the order of 10%, so that for this reaction, which may be of biological importance, the contribution of free radicals formed in the surrounding water is negligible. With serum albumin the G value per molecule changes (as judged by disappearance of original ultracentrifuge peak) is 0·6 for 'indirect' action whereas for 'direct' action it is close to 5. Since many if not most of the critically important enzymes are present in sub-cellular structures where the protein concentration is high, it would again appear as if the free radicals produced in the surrounding water will not contribute greatly to their inactivation.

The reason why, in recent years, the role of free radicals from the water has received so much emphasis is that most biological systems showed temperature and oxygen effects and could be protected by added substances. The experiments with model substances described here show that these tests are no longer decisive and that considerations of relative efficiency must be used to determine the relative contribution due to energy absorbed by the damaged molecules themselves and by the free radicals from the surrounding water. The limited data available indicated that 'direct' action as defined here is the more important. We now know that energy transfer and related phenomena, such as the oxygen effect, make it necessary to revise earlier concepts. The difference between 'indirect' and 'direct' processes is not that the latter occurs only at the site of energy loss of the radiation and that the distribution of reactions is governed solely by the geometry of ionisation. In both cases the irradiated system itself has a part in determining where the chemical reactions occur; the significant difference is that in 'indirect' action the damage is brought about by free radicals formed in the surrounding water while 'direct' action implies that the energy was deposited in the molecules affected.

Acknowledgment

The work has been supported by grants to the Chester Beatty Research Institute (Institute of Cancer Research: Royal Cancer Hospital), from the British Empire Cancer Campaign, Jane Coffin Childs Memorial Fund for Medical Research, The Anna Fuller Fund and the National Cancer Institute of the National Institutes of Health, U.S. Public Health Services.

REFERENCES

ALEXANDER, P., BLACK, R. M., and CHARLESBY, A., 1955. *Proc. Roy. Soc.*, A. **232**, 31.

ALEXANDER, P., and CHARLESBY, A., 1954. *Nature, Lond.*, **173**, 578.

Idem, 1955. *Proc. Roy. Soc.*, A. **230**, 136.

Idem, 1955. *Radiobiol. Symp.* (1954), p. 49 (London, Butterworth).

ALEXANDER, P., CHARLESBY, A., and ROSS, M., 1954. *Proc. Roy. Soc.*, A. **223**, 392.

ALEXANDER, P., and FOX, M., 1952. *Nature, Lond.*, **169**, 572; *Trans. Faraday Soc.*, 1954, **50**, 605.

Idem, 1952. *Ibid.*, 170.

ALEXANDER, P., BACQ, Z. M., COUSENS, S. F., FOX, M., HERVE, A., and LAZAR, J., 1955, *Radiation Res.*, **2**, 392.

ALEXANDER, P., and STACEY, K. A., 1955. *Progress in Radiobiol.*, 1956, p. 105 (Edinburgh, Oliver and Boyd).

BLACK, R. M., 1956. *Nature, Lond.*, **178**, 305.

BURTON, M., and MAGEE, J. L., 1950. *J. Amer. Chem. Soc.*, **72**, 1965.

GRAY, L. H., 1954. *Radiation Res.*, **1**, 180.

LEA, D. E., SMITH, K. M., HOLMES, B. E., and MARKHAM, R., 1944. *Parasitology*, **36**, 110.

RAJEWSKY, B., 1952. *Brit. J. Radiol.*, **25**, 550.

SETLOW, R., and DOYLE, B., 1955. *Radiation Res.*, **2**, 15.

DISCUSSION

Zimmer. I fully agree with Dr. Alexander's remarks as to the importance of 'direct effects' in biological action of radiation. But the proposed definition of 'direct effects' as 'events within a molecule' does not seem to be easily applicable to biological problems nor can I see how to apply it unambiguously in cases such as (i) large molecules with bound water, or (ii) two kinds of chain molecules being twisted around each other. Under such circumstances it is not easy to discern within from without, especially when considering the smallness of molecular dimensions as compared to the cross-section of a traversing alpha track or to the size of a cluster of ions, that may well form the 'event'.

Alexander. I quite agree with Dr. Zimmer about the difficulty of defining direct and indirect action especially now that we know that energy transfer can occur within irradiated macromolecules. The problem of adsorbed water may not be serious since we have many indications that free radicles formed in the firmly bound water (water of hydration of proteins and nucleic acids) can only make a small contribution. In this case the definition for indirect effect can, I feel, be considered as the action of free radicles in surrounding water which have to travel several molecular diameters before they interact with the vital molecule.

Hutchinson. Some dry enzymes (invertase) are not markedly changed in radiosensitivity by traces of moisture. Other systems (alcohol dehydrogenase, Coenzyme A) are markedly increased in radiosensitivity by traces of water.

Swallow. I would like to ask Dr. Alexander whether he still believes the HO_2 radical can degrade polymethacrylic acid in solution, as implied in his paper. There seem to be three reasons why we should not believe that HO_2 can degrade polymers. Firstly, at the *p*H values used in Dr. Alexander's work the HO_2 radical would be in the form O_2^-. In this form the radical is a mild reducing agent, and it would be very unlikely to attack polymers. Secondly, Dr. Alexander's experiments have now been reinterpreted by a revised mechanism which does not involve attack by the HO_2 radical (Collinson and Swallow, 1956). The new mechanism fits the published experiments better than the HO_2 mechanism. Thirdly, Baxendale (1956) has obtained different results from Dr. Alexander on polymethacrylic acid so that the validity of the original experiments is now open to some doubt.

Alexander. Dr. Swallow has criticised the conclusions of earlier work of ours on the degradation of polymethacrylic acid in dilute solution. He challenged our deduction that this occurs by HO_2 radicles because, he maintains, these are dissociated at *p*H 7. The evidence for dissociation is very indirect and one certainly could not accept without evidence that HO_2 radical is dissociated at *p*H 7, but in any case this is quite unimportant since it would not matter if the attacking substances responsible for breaking C—C bonds is HO_2 or O_2^-.

He refers to an alternative explanation of our observed effects, which he claims is more consistent with the data. I would remind him that we have in our original publications fully considered a mechanism very similar to that proposed by him, and it was in fact the mechanism which we initially favoured. Detailed experiments to decide between the different mechanisms made it necessary to reject this possibility and forced us to adopt the hypothesis that the degradation is the result of attack by HO_2 (or O_2) radicles.

The whole of this discussion has been published in detail and little purpose can be served by discussing it more fully here.

Finally, Dr. Swallow refers to the paper by Baxendale who was unable to repeat our experiments concerning the necessity for oxygen to be present if polymethacrylic acid is to be degraded by X-rays in solution. Baxendale's polymer probably contained peroxide groups; when these are removed by heating, the polymer will only degrade in the presence of oxygen. This question has been discussed by Fox and myself (1955).

REFERENCES

Alexander, P., and Fox, M., 1955. *J. Chim. Physiol.*, **52**, 710.
Baxendale, J. H., 1956. *Chemistry and Industry.*
Collinson, E., and Swallow, A. J., 1956. *Chemical Review.*

THE BIOLOGICAL ACTION OF IONISING RADIATIONS FROM THE POINT OF VIEW OF RADIATION CHEMISTRY

GABRIEL STEIN AND A. J. SWALLOW

Department of Physical Chemistry, Hebrew University, Jerusalem, Israel, and Tube Investments Research Laboratories, Hinxton Hall, Cambridge, England

Relative biological efficiency

It is well known that for equal amounts of energy absorbed the biological efficiency of densely ionising radiations is nearly always greater than that of X-rays (Zirkle, 1954). On the other hand X-rays have the greater effect on aqueous solutions, and this is due to the fact that free radicals capable of producing chemical effects in solutes are produced in water in good yield from X-rays but in poor yield from α-particles and neutrons (Allen, 1955). It would be tempting to resolve the contradiction by ascribing the biological action of radiation to the densely ionising regions of the tracks only, where free radicals are unimportant, and to consider that free radicals are without biological action. However, oxygen is known to increase the biological effect of X-irradiation more than that of α-particle irradiation (Thoday and Read, 1949) so that it would be wrong to regard an X-ray simply as a diluted α-particle. Much more work is required before observations of this type can be regarded as completely general, but for the present it seems better to say that X-rays and α-particles act by quite different mechanisms. We consider that it is useful to follow Bonet-Maury's suggestion (1952) and to treat the trajectory of any given ionising particle as a composite of an X-ray part and an α-particle part, the chemical and biological action of the two being quite different.

In the case of aqueous solutions the 'ideal' X-ray would react with water only to give free radicals according to the Weiss processes (1944)

$$H_2O \xrightarrow{\times} \text{'H'} + OH$$

—molecular hydrogen peroxide and hydrogen would not be present. The closest experimental approach to the 'ideal' X-ray would be with very high energy X-rays at moderate or low dose rates (say$<$100,000r/min.) and using solutions concentrated enough to capture all the primary free radicals but not so concentrated that the direct effect becomes important.

The 'ideal' α-particle would produce only molecular products in the reaction—

$$H_2O \xrightarrow{a} \tfrac{1}{2}H_2 + \tfrac{1}{2}H_2O_2$$

—there would be no free radicals escaping from the track into the body of the solution. The closest experimental approach would be with low energy densely ionising particles such as α-particles, and using dilute solutions.

Chemical and biological actions depending on free radicals produced from water will therefore be more efficiently produced by radiations which increasingly approach 'ideal' X-rays in their properties. Conversely, if some biological actions of radiations are most efficiently produced by particles approximating to 'ideal' α-particles, then those particular actions cannot be considered to depend on the free radicals produced from water.

These considerations indicate a clear distinction between the action of different types of radiation, both on aqueous solutions and on biological systems.

Direct and indirect action. We may define direct action as that resulting from the formation of an ionisation or excitation within the affected molecule itself, whilst indirect action is the consequence of attack by active species formed from other molecules. This distinction is necessarily a crude one because it may happen that a positive ion is produced in one molecule, the ejected electron being captured by another molecule some distance away.

The equation relating the importance of the two types of reaction for an aqueous solution of a substance A is as follows:

$$\frac{\text{No. of A molecules destroyed by direct action}}{\text{No. of A molecules destroyed by indirect action}} = \frac{\text{G direct action}}{\text{G indirect action}} \times \frac{\text{Weight of A}}{\text{Weight of solvent}}$$

Indirect action, 'ideal' α-particle. Any indirect action of ideal α-particles would be entirely due to hydrogen, hydrogen peroxide and to a less extent other products formed from the organic constituents of cells. It might be thought that reactions which require two radicals to be present very close together might be produced effectively by α-particles. However, there is no chemical evidence that this type of reaction occurs. Moreover, it would be necessary for solute molecules to intersect the track, and in this case the effect produced would be a composite of direct and indirect action.

Actual slow α-particles and neutrons would approximate closely to the ideal, the δ-rays emanating from the tracks being the principle cause of non-ideality. The investigation of reactions of this type is a matter for conventional chemistry, not for radiation chemistry, except in so far as radiation is a useful method of producing products within a reaction vessel.

Indirect action, 'ideal' X-rays. The starting point of this discussion is the reaction

$$H_2O \rightsquigarrow H + OH$$

proposed by Weiss (1944). Although modification has since become necessary for real X-rays to allow for the molecular yield, the equation would still, of course, be valid for 'ideal' X-rays. There is good evidence that hydroxy radicals are present in irradiated water (Stein and Weiss, 1949) but valid evidence for hydrogen atoms does not exist, and in fact, Livingston *et al.* have recently shown that irradiated ice does not contain hydrogen atoms (Livingston *et al.*, 1955). Livingston's results support the idea that irradiated water contains hydrated electrons rather than hydrogen atoms (Jortner and Stein, 1955) and the chemical reactions in irradiated water can be quite well explained on this basis. We therefore consider that the primary act for 'ideal' X-rays may best be written

$$H_2O \rightsquigarrow H^+ + e + OH$$

Electrons would often react to give hydrogen atoms and would then be capable of dehydrogenating organic molecules of the general formula AH_2 (for example, ethanol):

$$e + H^+ + CH_3CH_2OH \rightarrow CH_3CHOH + H_2$$

OH radicals may also dehydrogenate organic material:

$$OH + CH_3CH_2OH \rightarrow CH_3CHOH + H_2O$$

—so that organic free radicals are produced. In many cases such radicals have been shown to be reducing agents. They have been shown to reduce diphosphopyridine nucleotide (Swallow, 1953 and 1955), ferrous ions (Baxendale and Smithies, 1955), robiflavin (Swallow, 1955) and methylene blue (Hayon *et al.*), and it seems most likely that they are at least partly responsible for reduction in other similar systems, especially in the cases of cytochrome C (Mee and Stein, 1956), vitamin B_{12} (Beaven *et al.*) and molecular oxygen itself (Johnson *et al.*, 1956). On the other hand, they appear to cause irreversible attack at the chromophores of thiamine and cocarboxylase (Ebert and Swallow).

The reactions of organic radicals of the general formula AH show how a component present in small amount can be attacked by radiation even in the presence of other acceptors. The work with riboflavin (Swallow, 1955) carries the argument still further. When AH radicals act as reducing agents the product must, in the first instance, itself be a free radical. In general these radicals would be unstable and would react still further, for example by dimerising or disproportionating. With riboflavin in acid solution, however, the radical is known on chemical grounds to be stable. This radical has now been prepared in quantitative yield by the irradiation of air-free acid aqueous solutions of riboflavin containing excess ethanol (Swallow, 1955).

As a result of this work we can picture how the highly active electrons and OH radicals are changed to the (often reducing) AH radicals, and how these in turn react to give organic free radicals which are either stable as in the case of riboflavin, or dimerise or disproportionate to give stable products. Such a picture is clearly relevant to radiobiology and shows how energy may be handed on from molecule to molecule until it finally reaches the substance whose destruction may be regarded as the biological primary act. Such reactions can also occur in semi-solid systems as illustrated by the case of methylene blue (10^{-5} M) incorporated in gels containing up to 40% of gelatin or agar-agar, where in the absence of free oxygen, methylene blue is selectively reduced by radiation (Day and Stein, 1951).

Direct action. It may well be that the radicals produced by direct action are those which would be produced by the action of hydrogen atoms or OH radicals on the same molecules. For example, in the first instance alcohols probably give the same AH radicals by direct irradiation as have been discussed above in connection with indirect action. Subsequent reactions of the radicals produced by direct action would depend on environment. In the presence of oxygen, for example, the usual reactions of radicals with oxygen would take place. In general one would expect the ultimate result to be chemical reaction with those groups possessing a high electron affinity (Stein, 1952). In an organised system of large molecules the conditions for all types of reactivity transfer (including ionisation, excitation and hydrogen atom transfer) would be especially favoured, and the effect of irradiation could be readily concentrated on potentially radiosensitive substances. These considerations are particularly applicable to the action of 'ideal' X-rays but are less certain for 'ideal' α-particles. We believe that understanding how α-particles work is one of the major unsolved problems in radiobiology.

The cell as an organised system

It is usual to assume that some component in the cell is especially radiosensitive. The above discussion attempts to explain how the primary effect of radiation could be selectively concentrated on such a component. There is, however, another non-selective type of action possible. Peters has recently emphasised that the cell is not a mere bag of enzymes, but a highly organised system, the individual enzymes being held in place in a 'cytoskeleton' (Peters, 1956). Some of the bonds holding the enzymes together may be ordinary chemical bonds; others may be hydrogen bonds. Now it has become apparent that hydrogen bonds can be broken by radiation. This has been discussed for proteins by Franck and Platzman (1954), and Cox, Overend, Peacocke and Wilson (1955) have recently shown that hydrogen bonds are broken when nucleic acids are irradiated. In fact, Cox has shown that hydrogen bond breakage is the main effect of irradiating aqueous nucleic acids, the yield being up to G = 65 (Cox). These results indicate the possibility of disrupting the organisation of the cell by breaking the bonds holding the individual enzymes together. If this were to occur, then the normal functioning of the cell would be upset, and damage would follow. No individual component, however, would have been inactivated.

Summary

It is considered useful in radiobiology and radiation chemistry to distinguish between 'ideal' X-rays and 'ideal' α-particles. In indirect action by 'ideal' X-rays, results can best be interpreted in terms of the primary formation of solvated electrons and hydroxyl radicals (not hydrogen atoms and hydroxyl radicals as previously assumed). In complex systems these primary products react with any substance present in excess to give organic free radicals, and the organic free radicals in turn react with minor constituents. Similar considerations apply to direct action. The effects of radiation are thus concentrated on minor components of complex mixtures. It is also suggested that in an organised system such as the cell, radiation may cause damage by producing disorganisation as well as by inactivating specific cell constituents.

Acknowledgment

The authors wish to thank the Chairman of Tube Investments Ltd., for permission to publish this paper.

REFERENCES

ALLEN, A. O., 1955. *Geneva Conf. on Atomic Energy*, No. A Conf. 8/P/738.

BAXENDALE, J. H., and SMITHIES, D., 1955. *Experientia*, **11**, 436.

BEAVEN, G. H., JOHNSON, E. A., and SWALLOW, A. J., unpublished.

BONET-MAURY, P., 1952. *Discuss. Faraday Soc.*, **12**, 72.

COX, R. A., personal communication.

COX R. A., OVEREND, W. G., PEACOCKE, A. R., and WILSON, S., 1955. *Nature, Lond.*, **176**, 919.

DAY, M. J., and STEIN, G., 1951. *Nucleonics*, **8**, No. 2, 34.

EBERT, M., and SWALLOW, A. J., unpublished.

FRANCK, J., and PLATZMAN, R., 1954. *Radiation Biol.*, **1**, 191 (New York, McGraw-Hill).

HAYON, E., SCHOLES, G., and WEISS, J., personal communication.

JOHNSON, G. R. A., SCHOLES, G., and WEISS, J., 1956. *Nature, Lond.*, **177**, 883.

JORTNER, J., and STEIN, G., 1955. *Nature, Lond.*, **175**, 893.

LIVINGSTON, R., ZELDES, H., and TAYLOR, E. H., 1955. *Discuss. Faraday Soc.*, **19**, 166.

MEE, L. K., and STEIN, G., 1956. *Biochem. J.*, **62**, 377.

PETERS, R. A., 1956. *Nature, Lond.*, **177**, 426.

STEIN, G., 1952. *Discuss. Faraday Soc.*, **12**, 227.

STEIN, G., and WEISS, J., 1949. *J. Chem. Soc.*, p. 3245.

SWALLOW, A. J., 1953. *Biochem. J.*, **54**, 253.

Idem, 1955. *Ibid.*, **61**, 197.

Idem, 1955. *Nature, Lond.*, **176**, 793.

THODAY, J. M., and READ, J., 1949. *Ibid.*, **163**, 133.

WEISS, J., 1944. *Ibid.*, **153**, 748.

ZIRKLE, R. E., 1954. *Radiation Biol.*, **1**, 315 (New York, McGraw-Hill).

Discussion

Bacq. I feel that I must call the attention of Dr. Swallow to the fact that the idea developed in the last part of his paper (release of enzymes by breaking bonds holding individual enzymes together within the cell) is exactly the hypothesis that Alexander and myself arrived at after careful examination of the available literature and our own work. You will find in our book (Bacq and Alexander, 1955), a long list of biochemical facts which are consistent with the view that liberation (which means activation) of enzymes in the cell, may be the primary biochemical lesion induced by ionising radiations. We are actively working along this line and so far experimental data agree with the hypothesis.

REFERENCE

Bacq, Z. M., and Alexander, P., 1955. *Principles of Radiobiology* (Butterworth, London).

SECTION II

BIOCHEMICAL IRRADIATION EFFECTS—ON ENZYMES AND OTHER CELLULAR CONSTITUENTS

X-RAY INACTIVATION OF THE COMPONENTS OF THE SUCCINIC ACID DEHYDROGENASE-CYTOCHROME-CYTOCHROME OXIDASE-SYSTEM

B. Rajewsky, G. Gerber and H. Pauly
Max Planck Institut für Biophysik, Frankfurt am Main, Germany

Most of the energy the cell gains by combustion of nutrients is liberated during the electron-transport via the cytochrome-cytochrome-oxidase-system. In the living cells, this enzyme-system is closely linked to the structure of the mitochondria. We investigated the sensitivity to radiations of these systems; as well as that of a particulate dehydrogenase in the mitochondria—the succinic acid dehydrogenase.

From the dose effect curves obtained, we calculated an 'inactivation volume' by means of the target theory (Lea, 1955, and Pollard, 1955) and tried to conclude the state of the single components in the cell and their change during the isolation of the mitochondria.

Figure 1 shows the single components of the system according to Slater's (1950) suggestion. The entire enzyme system—the so-called succinic acid

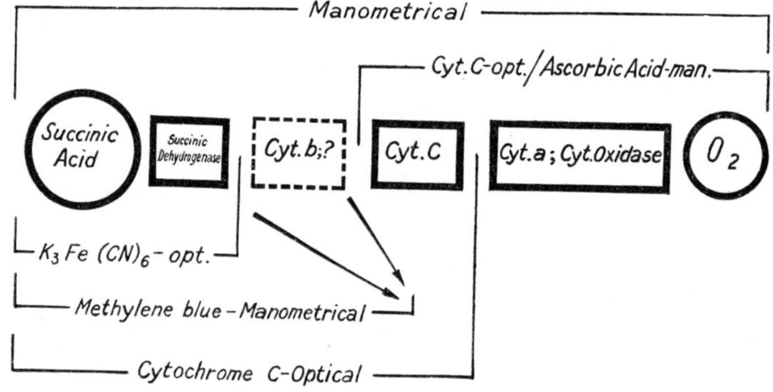

Fig. 1. Estimation of the components of the succinic acid oxidase system.

oxidase—found itself from succinic acid dehydrogenase, several less known links, Cytochrome *a*, *b*, *c*, and Cytochrome oxidase. The heavily edged components in Figure 1 were investigated.

Methods

1. *Tissue slices*: Freshly prepared tissue slices of mouse liver were irradiated, homogenised in the cold in 0·25 M sucrose solution by means of a Potter

and Elvehjem (1936) all-glass homogeniser, followed by determination of the activity of the single enzymes. The activity was calculated on the fresh-weight of the liver slices.

2. *The mitochondria*: fractions were isolated from a 20% homogenate (isotonic sucrose solution plus 0·01 M of versene) by fractionated centrifugation in a preparative Spinco-ultracentrifuge, Schneider and Hogeboom (1948).

3. *Method of Irradiation*: An X-ray tube with a beryllium window served as the X-ray equipment; the dose rate was 3×10^5 r/min. at 45 kV, 25 mA and a filter of 60 μAl. The half value layer in water is 1·5 mm. Tissue slices and the suspension of the mitochondria were 1 mm. high. The dose values plotted in the figures and the table refer to a 'medium dose' which includes 85% of the surface dose. Irradiation occurred in aluminium plates cooled in ice water.

4. *Assay of enzyme-activity*: The catalytic efficiency of the single components was determined either at 30° C. in a Warburg apparatus, measuring the speed of O_2-uptake, or optically at 20° C. by means of a Beckman spectrophotometer.

Succinic acid dehydrogenase (*SAD*): (*a*) Manometric measurement with addition of HCN and methylene-blue. (*b*) Optical measurement at 400 mμ with addition of HCN by reduction of $K_3Fe(CN)_6$ (Colowick and Kaplan, 1955), or optically at 550 mμ with the addition of HCN by reduction of oxidised Cytochrome *c* (Colowick and Kaplan, 1955).

Cytochrome a—Cytochrome oxidase—Complex (*CO*): (*a*) Manometric measurement with the addition of Cytochrome *c* and ascorbic acid as substrate (Colowick and Kaplan, 1955, and Schneider and Potter, 1943). (*b*) Optical measurement with reduced Cytochrome *c* as substrate.

Cytochrome c: Manometric assay of catalytic efficiency in the Cytochrome oxidase test.

Succinic acid oxidase (*SO*): Manometric measurements with succinic acid as substrate and the addition of different amounts of Cytochrome *c* (Colowick and Kaplan, 1955, and Schneider and Potter, 1943).

Results

1. *Irradiation of intact liver cells.* Figure 2 shows the almost exponential dose effect curve of the components. The same results were obtained by manometric and optical methods. Measurable inactivation is caused by X-ray doses from 10^6r onward. CO and SAD have the same dose dependancy and therefore the same sensitive region. During the manometric measurement, the dose effect curve of SO depends very characteristically on the Cytochrome *c* concentration. Measuring the dose relation of SO in the homogenate without addition of Cytochrome *c*, the speed of oxidation of

succinic acid then being limited by the endogeneous Cytochrome *c* of the mitochondria, one obtains a dose effect curve which can be interpreted as

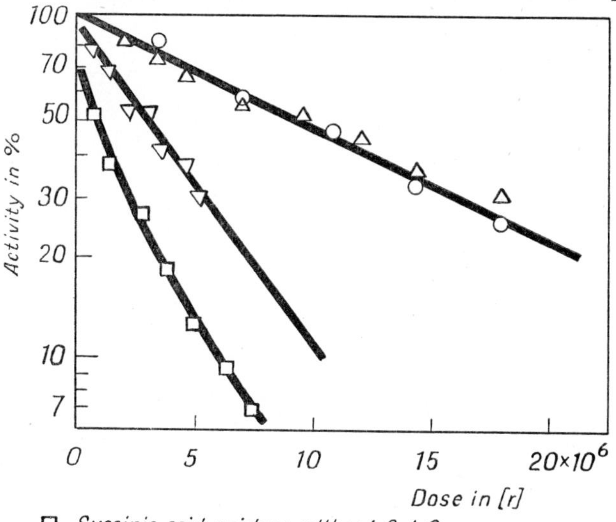

FIG. 2. Mouse liver slices.

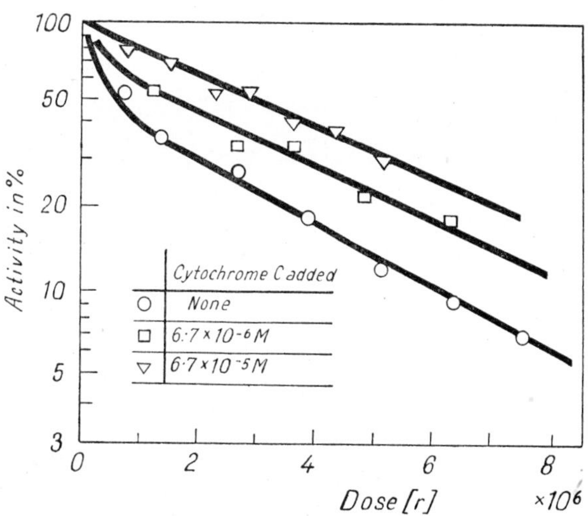

FIG. 3. Mouse liver slices with and without the addition of Cytochrome *c*.

a result of two or more exponential functions, Figure 3. Thus the SO-preparations seem to consist of fractions with differing radiation sensitivities.

Since Cytochrome *c* is the limiting factor in the chain, a rise of Cytochrome *c* concentration in the reaction vessel results in an increased SO activity. Simultaneously, the steep part at the origin of the dose effect curves disappears. Apart from the slope at the origin, the curves in Figure 3 run almost parallel. The slope at the origin of the curve which disappears after addition of Cytochrome *c* might also result from an increased diffusion of Cytochrome *c* out of the mitochondria. Further experiments are needed.

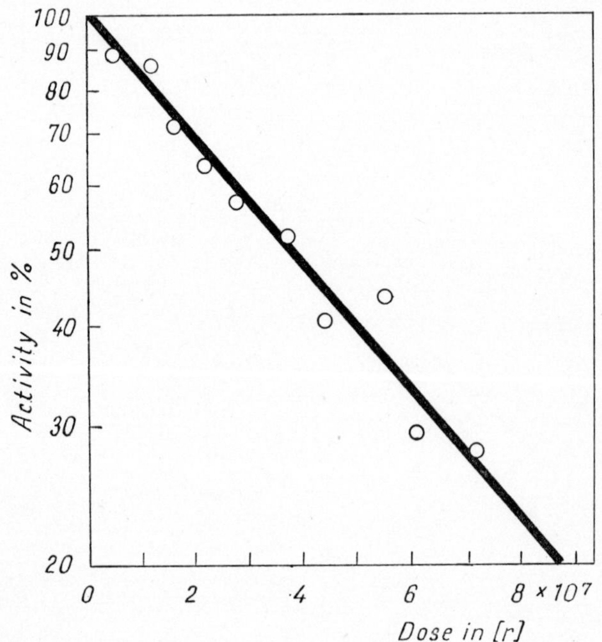

FIG. 4. Dry Cytochrome *c*.

We assume that in the sensitive part of the SO fraction, including 60–70% as to the curve, the sensitive part of the entire SO system is the limiting Cytochrome *c* link. The slope at the origin of the curve corresponds to a 37% dose of approximately 10^6r and an inactivation volume of $1{\cdot}7 \times 10^{-18}$ cm.^{3} (Lea, 1955).

This relatively large inactivation volume is in contrast to the volume of the small Cytochrome *c* molecule of only $1{\cdot}7 \times 10^{-20}$ cm.^{3} and a molecular weight of 14·000.

This contradiction can be interpreted by means of the indirect action of radiation.

The experimentally ascertained inactivation volume includes, besides the volume of the molecule itself, a 'diffusion volume', out of which energy-carriers formed by radiation absorption are able to diffuse to the molecule to inactivate it (Barron, 1952; Dale, 1952, and Rajewsky, 1952).

2. *Cytochrome c.* The following experiments show that actually the inactivation of the Cytochrome *c* molecule in aqueous solution is more probably due to indirect than to the direct action of radiation.

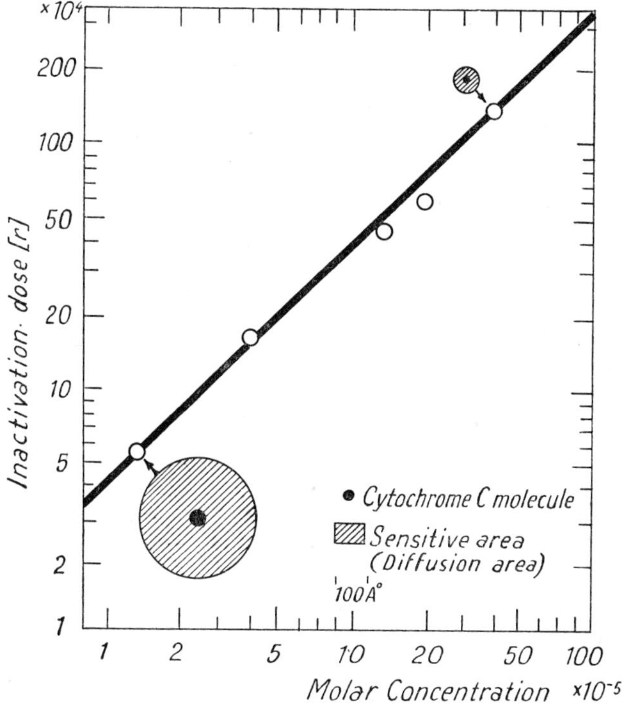

FIG. 5. Aqueous Cytochrome *c*.

Figure 4 shows the exponential dose effect curve for irradiated Cytochrome *c* in the dry state. A molecular weight of 16,000 corresponds to the inactivation dose of $5{\cdot}4 \times 10^7$r; a value agreeing with others obtained by different methods.

Figure 5 shows the inactivation dose of aqueous solution of Cytochrome *c*, dependent on the enzyme concentration. The inactivation is inversely proportional to the concentration of the Cytochrome *c*. The ionic yield amounts to 0·1.

3. *Irradiation of the mitochondria.* Isolation of mitochondria causes considerable structural changes, which can be observed by the electron microscope. Therefore, a change in the inactivation region of the particulate enzymes is to be expected.

Figure 6 shows the corresponding results. Surprisingly, the slope of the dose effect curves in the exponential part is slightly changed for SAD, almost unchanged for CO. The same result holds for SO.

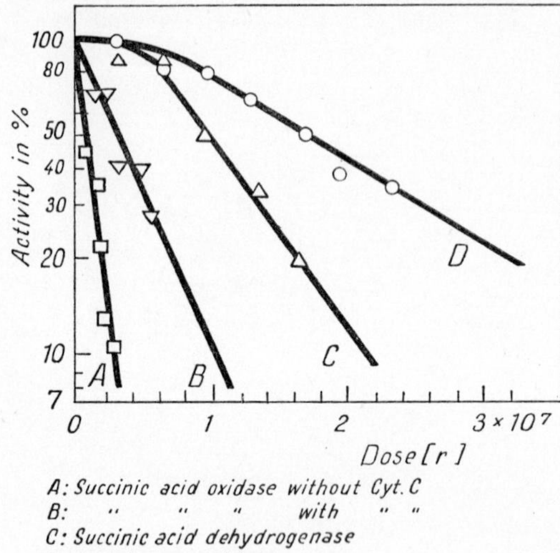

FIG. 6. Mouse liver—mitochondria.

Summary and Conclusions

The Table summarises the experimental results. The inactivation doses were calculated from the exponential part of the curve. This method, based

TABLE

Sensitive Area of the Components of the Succinic Acid-Oxidase System

	Tissue slices			Mitochondria		
	37%* dose 10^6r	Diam.† Å	Mol. wt. $\times 10^{-3}$	37%* dose 10^6r	Diam.† Å	Mol. wt. $\times 10^{-3}$
Succinic acid dehydrogenase ...	16	52	65	8	66	130
Cytochrome oxidase	14	54	70	17	50	60
Succinic acid oxidase + Cytochrome *c*	4·8	80	234	4·4	85	290
Succinic acid oxidase without Cytochrome *c*	1	150	1,500	1·5	120	790

* 37% dose deduced from exponential part of curve.

† Diameter calculated according to Lea.

on the formula $\frac{V}{V_o} = \frac{R+1}{e^{\alpha D} + R}$, permits us to describe the relative speed $\frac{V}{V_o}$ of a reaction running over intermediate, in dependance of the X-ray dose D. α designates the sensitive region of the most radiation sensitive enzyme of the chain, R means a constant (Pauly and Rajewsky, 1955).

The inactivation doses served to calculate the diameter of the inactivation volume, assumed to be spherical. The third column gives the molecular weight of a protein molecule (density 1·35 g./c.cm.) of the same size as the corresponding inactivation volume.

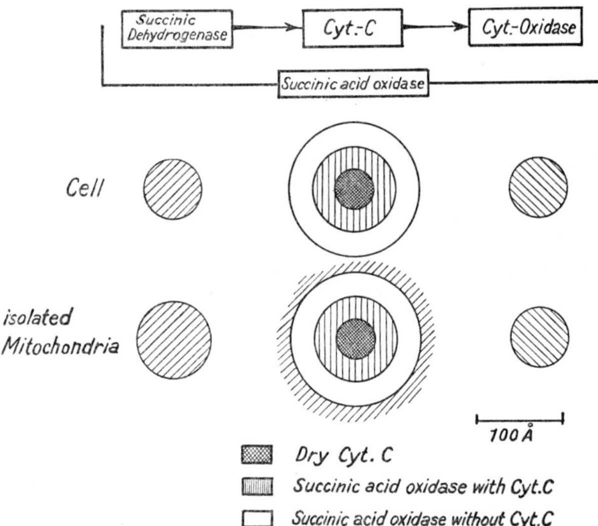

FIG. 7. Sensitive areas of components of succinic acid oxidase system of mouse liver.

Figure 7 shows besides a model of the enzyme chain, the inactivation volumes of the single components. The values of the inactivation volume are plotted in the correct proportions. The absolute values can be obtained by comparing with the scale drawn.

If the Cytochrome *c* molecule constitutes the most radio-sensitive link of this enzyme chain in the intact cell, one of the most important enzymes of the biological oxidation would be inactivated by means of the indirect effect of irradiation. We must postulate then that the Cytochrome *c* molecule in the mitochondria also exists in such a form as to be able to react with the energy-carriers (radicals). For irradiation effects we assume that the Cytochrome *c* molecule in mitochondria is present in a 'solute state', which agrees with biochemical experiences since Cytochrome *c* easily diffuses out of mitochondria.

REFERENCES

BARRON, E. S. G., 1952. *Radiation Biology*, Vol. I, p. 283 (New York).

COLOWICK, S. P., and KAPLAN, N. O., 1955. *Methods in Enzymology* (New York, Academic Press).

DALE, W. M., 1952. *Radiation Biology*, Vol. I, p. 225 (New York).

LEA, D. E., 1955. *Actions of Radiation on Living Cells* (Cambridge University Press).

PAULY, H., and RAJEWSKY, B., 1955. *Strahlentherapie*, **99**, 383.

Idem, 1956. *Progress in Radiobiology*, 1955, p. 32 (Edinburgh, Oliver and Boyd).

POLLARD, E., 1955. *Progress in Biophysics*, **5**, 72.

POTTER, V. R., and ELVEHJEM, C. A., 1936. *J. Biol. Chem.*, **114**, 495.

RAJEWSKY, B., 1952. *Brit. J. Radiol.*, **25**, 550.

SCHNEIDER, W. C., and POTTER, V. R., 1943. *J. Biol. Chem.*, **149**, 217.

SCHNEIDER, W. C., 1948. *Ibid.*, **176**, 259.

SLATER, E. C., 1950. *Biochem. J.*, **46**, 484.

DISCUSSION

Hutchinson. Applying the 'diffusion model' to the data on Cytochrome *c*, one finds that the 'action radius' of the active intermediates involved is of the order of 150 ÅE.

Hagen. I wonder whether the fact that the marked irradiation effects on Cytochrome *c* are mainly indirect is dependent on cytochrome being in a soluble state. Do we not have to consider the possibility of differing radiosensitivity when the enzyme is fixed in a natural state in the cell, as contrasted to the enzyme in watery solutions?

Rajewsky. It is true that the model experiments were performed on solutions, but in the corresponding investigations on cells and mitochondria we irradiated only the cytochrome in the cells; only thereafter have we added soluble cytochrome before the enzyme determinations. The very high radiosensitivity of cytochrome, which we mainly ascribe to indirect effects, is thus also a property of the cell-bound enzyme.

EFFECTS OF IONISING RADIATION ON THE *IN VIVO* INCORPORATION OF AMINO ACIDS INTO PROTEINS

J. A. V. Butler, P. Cohn and A. R. Crathorn
Chester Beatty Research Institute, Institute of Cancer Research, The Royal Cancer Hospital, Fulham Road, London, S.W. 3, England

Although numerous effects of ionising radiations on isolated cell constituents have been studied, in most cases a dose of radiation much larger than the lethal dose is required to produce observable reactions, and the nature of the initial processes which eventually lead to the death of the animal have not been identified. The possible sites of such changes are

(1) small molecules,
(2) soluble protein molecules, e.g. enzymes,
(3) cytoplasmic particles.

With regard to (1) no toxic substances formed by the action of small doses of radiation have been identified. In the case of (2) it is necessary to find a particularly sensitive protein which is essential to the life of the cell. As has been pointed out (Butler, 1956) there is little prospect that small soluble protein molecules such as enzymes are sensitive in this way. Cytoplasmic particles (3) such as mitochondria and microsomes, however, are of such a size that the chance of ionisation occurring within the particles with a dose of 1000r is reasonably large. It remains to be established if the biochemical effects of such a dose of radiation on these particles are sufficiently large to produce disturbances of the organisation or of metabolic processes which may result in death.

Experiments of this kind have been performed by van Bekkum (1954) who found that small doses of radiation have significant effects on the phosphorylating ability of mitochondria. As is well known, effects of similar doses on DNA metabolism have been observed (Euler and Hevesy, 1942; Hevesy, 1949; Abrams, 1951, and Hevesy and Forssberg, 1955).

Several workers (Hevesy, 1949; Abrams, 1951, and Holmes and Mee, 1952) have also attempted to look for effects by ionising radiations on the incorporation of amino acids into proteins of mammalian tissues. None of these results were, however, conclusive. Great interest attaches to the findings (Keller *et al.*, 1954) that labelled amino acids are very readily incorporated into proteins of the microsome fraction, which might thus be a main site of protein synthesis in the cytoplasm. Hence it was thought that more conclusive information could be obtained by an examination of the effect of X-rays on the incorporation of an amino acid ([3-^{14}C]-DL-phenylalanine) into proteins of cell particles subsequently isolated by fractional centrifugation.

Experimental procedure

August rats approximately 7 weeks old and about 120 gm. weight were taken and starved for 24 hours and then paired either to be irradiated or to act as controls. Within 3 minutes after irradiation (500r at 25r/min. of 250 kV X-rays) irradiated and control animals were each injected interperitoneally with 0·2 μc./g. body weight of [3-^{14}C]-DL-phenylalanine (2mc/mM). After a selected period of time the animals were killed by perfusion under anæsthesia and the required organs removed and frozen over-night to await analysis.

TABLE

Rats Irradiated 500r 250 kV X-rays injected 20 μc./100 g. [3-^{14}C]-DL-phenylalanine killed after 63 minutes

Radioactivity μc./g. Protein		
Liver fraction	Irradiated	Control
Nuclear	1·38	1·29
Mitochondria	2·16	1·84
Microsomes	3·25	3·20
Final supernatant	2·32	1·98

After some preliminary experiments it was decided to investigate amino acid incorporation in the proteins of the liver. The livers were each fractionated using the following standard procedure. To each gram of liver were added five millilitres 0·25 M sucrose solution and a suspension prepared in a homogeniser of the Potter-Elvehjem type. The nuclei were centrifuged down at 600×g for 10 minutes. The sediment was dispersed in sucrose (0·25 M), the suspension centrifuged at 600×g for 10 minutes and the supernatant liquid added to the previous one. The sediment containing the nuclei was suspended in cold water and citric acid (0·05 M) was added to lower the *p*H to 6·0–6·2. After centrifugation at 400–600×g for 8 minutes the supernatant liquid was discarded, the sediment resuspended in cold water and centrifuged again. The various fractions of the above separation were examined microscopically and appeared to be reasonably uncontaminated with other fractions. The supernatant liquid from this was centrifuged at 10,000 ×g for 15 minutes. After the supernatant liquid had been sucked off, the mitochondria were resuspended in sucrose (0·25 M) and centrifuged again at 10,000 × g for 10 minutes. The combined supernatant liquids were then centrifuged at 26,250 × g for a further 30 minutes to give a microsome fraction and a final supernatant fraction. During these operations the temperature was maintained as near as possible to 0° C.

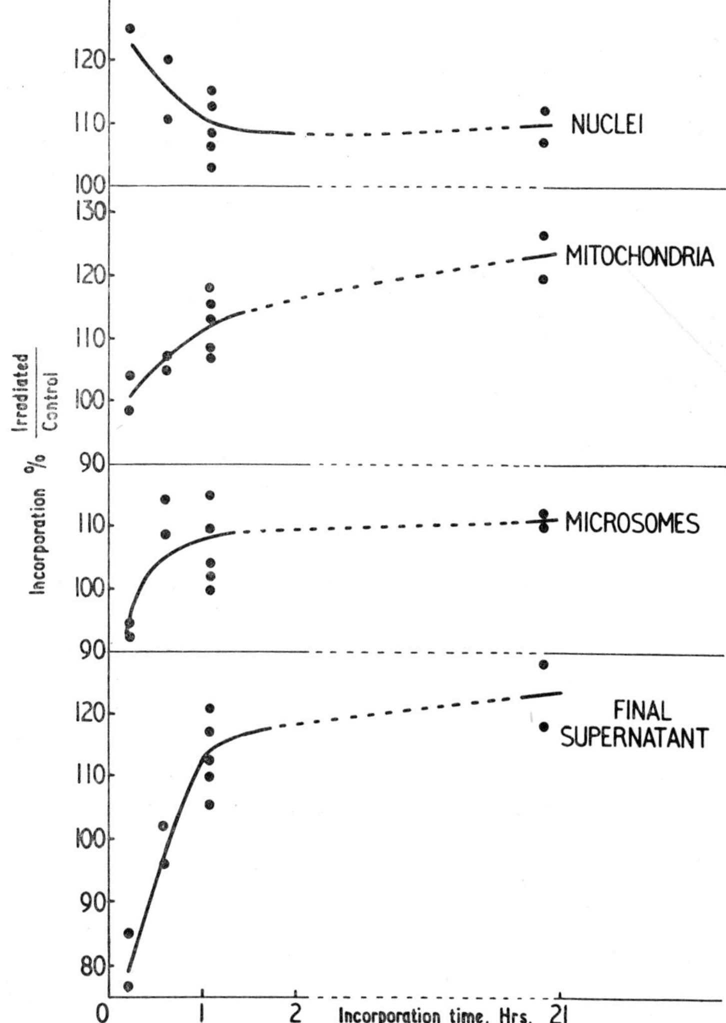

FIG. 1. Twenty-six rats irradiated 500r 250 kV X-rays injected 20 μc. per 100 g. [3-^{14}C]-DL-phenylalanine killed after 63 minutes.

Proteins were then precipitated from the four fractions with cold 10% trichloroacetic acid (TCA). Further washings with TCA, lipid extraction with chloroform-ethanol (1 : 1) and treatment with 5% TCA at 90° C.

yielded final protein precipitates which were washed and dried from acetone. They were then counted using a standard end-window counter assembly.

Results

Figures from typical experiments are shown in the Table. From several such experiments comprising 26 rats the results represented in Figure 1 are derived.

No attempt is made here to interpret these curves but from them it is possible to say, in confirmation of the observation already made (Hevesy and Forssberg, 1955) that irradiation generally produces an increase in incorporation into tissue protein and this effect has again been reported recently in an *in vitro* system (Kay and Entenman, 1956). It would also appear from this series of experiments that the increase first occurs in the nuclear fraction and is followed within an hour by increases in the incorporation in the other fractions.

During preliminary experiments it was observed: (i) For an incorporation period of not less than 40 minutes the effect was the same whether the amino acid was administered before or after the irradiation. (ii) In one series of experiments of 1 hour the effect was the same whether the animals received 500r or 1000r. (iii) No significant effect of radiation on the incorporation into proteins of similarly fractionated cell components of the kidney was found. (iv) In spleen, thymus and blood no effect on incorporation into total proteins was detected.

It has been shown that small doses of radiation have observable effects on the incorporation of amino acids into liver proteins. In view of the many biochemical entities that may be affected directly or indirectly by irradiation of living cells the causes of the effects reported here remain obscure.

Acknowledgments

We should like to thank Dr. L. F. Lamerton of the Royal Cancer Hospital for performing the irradiations, and Dr. V. C. E. Burnop of this laboratory for providing the [$3-{}^{14}C$]-DL-phenylalanine used.

This investigation has been supported by grants to the Royal Cancer Hospital and the Chester Beatty Research Institute from the British Empire Cancer Campaign, the Jane Coffin Childs Memorial Fund for Medical Research, the Anna Fuller Fund, and the National Cancer Institute of the National Institutes of Health, United States Public Health Service.

REFERENCES

ABRAMS, R., 1951. *Arch. Biochem.*, **30**, 90.
BUTLER, J. A. V., 1956. *Radiation Res.*, **4**, 20.
VAN BEKKUM, D. W., 1954. *Radiobiol. Symp.*, p. 201 (London, Butterworth).

EULER, H. I., and HEVESY, G., 1942. *Kgl. Danske Vidensk. Selskab. Biol. Medd.*, **17**, 8.

HEVESY, G., 1949. *Nature, Lond.*, **163**, 869.

HEVESY, G., and FORSSBERG, A., 1955. 3 *ème Congr. intern. Biochim. Brussels, Rapports*, p. 479.

HOLMES, B. E., and MEE, L. K., 1952. *Brit. J. Radiol.*, **25**, 273.

KAY, R. E., and ENTENMAN, C., 1956. *Arch. Biochem.*, **62**, 419.

KELLER, E. B., ZAMECNIK, P. C., and LOFTFIELD, R. B., 1954. *J. Histochem. Cytochem.*, **2**, 378.

Discussion

Forssberg. I noticed you spun down your microsomal fraction with 26,000×g. whereas Zamecnik gives values of about 100,000×g. in his papers. Are there any differences in the technique in the two cases?

Crathorn. We have gone up 100,000×g., too, but did not find any differences from 26,000×g.

THE COMBINED ACTION OF X-RAYS AND CHEMICAL INHIBITORS ON *IN VITRO* KIDNEY TRANSPORT AND RESPIRATION

M. L. MENDELSOHN *

Department of Biophysics, Walter Reed Army Institute of Research, Washington, 12, D.C., U.S.A.

The complex cytoplasmic process of para-aminohippuric acid (PAH) transport in kidney tubules can be assayed *in vitro* in terms of the ability of kidney cortex slices to concentrate PAH from the surrounding medium. This system represents a metabolic property of the intact cell, thus bridging the gap, from the radiobiological viewpoint, between cell survival on the one hand, and the reactions of isolated enzymes or cytoplasmic components on the other. As with most metabolic properties so far studied, it is considerably more resistant to radiation than either cell or organ survival. The feasibility of using this method quantitatively, the behaviour of it with several inhibitors and X-rays, and the proof that the X-ray effect observed is on active transport and not on increased leakage have already been described (Mendelsohn, 1955). The present study involves the effects of combinations of agents, and seeks to uncover possible analogies between the actions of X-rays and the chemical inhibitors. It was motivated to a large extent by the similarity of action of X-ray and dinitrophenol (DNP), in that both inhibit transport in the slice, and simultaneously increase respiration.

Methods

Concentration ratios are developed during a 4-hour incubation of rabbit kidney cortex slices at 25° C. in Warburg flasks in a buffered, PAH, acetate medium. The ratios, when converted to percents of control, and then to probits, yield a straight line when plotted against log dose. The 50% ED_{50} is defined as half of that dose which reduces the concentration ratio to 50% of control. Respiration is measured in conventional fashion during the incubation. The QO_2 is calculated from wet weights (which have not been found to be a function of any of the agents tested either before or after incubation), and is then expressed as percent of control. In most experiments, graded doses of an inhibitor are included in the medium, and half of the slices are exposed to 65,000r of 250 kVp X-rays (the 50% ED_{50}) before mounting. In the others, a fixed dose of inhibitor was used, again the 50% ED_{50}, and graded amounts of X-rays were given beforehand. One

* Present address: Department of Radiotherapeutics, University of Cambridge, England.

rabbit and 18 flasks were used per experiment. The dose response to the single agent used was defined from half the flasks. From this probit regression line the combined responses were then read off as dose equivalents. The difference between this equivalent and the given dose of inhibitor was taken to represent the contribution of the 50% ED_{50} of the second agent. This value was pooled as the geometric mean of all the doses tested in combination, and was finally expressed as the ratio between it, and the observed 50% ED_{50} of the single agent. A ratio of 1·00 would indicate exact addition of the two agents, while decreasing values imply less and less effective addition.

Results

NaFAc: Graded doses of sodium fluoroacetate (NaFAc) produce a roughly linear respiratory inhibition increasing with incubation time, and the characteristic sigmoid response of concentration ratio. The ED_{50} in 14 experiments averaged 8·7 uM per flask (SEM = 0·5). When combined with a fixed dose of X-ray, an equivalence ratio of 0·85 (SEM = 0·06) was obtained from 12 experiments. As shown in Figure 1, the respiratory increment of X-ray is inhibited by increasing doses of NaFAc. In one experiment

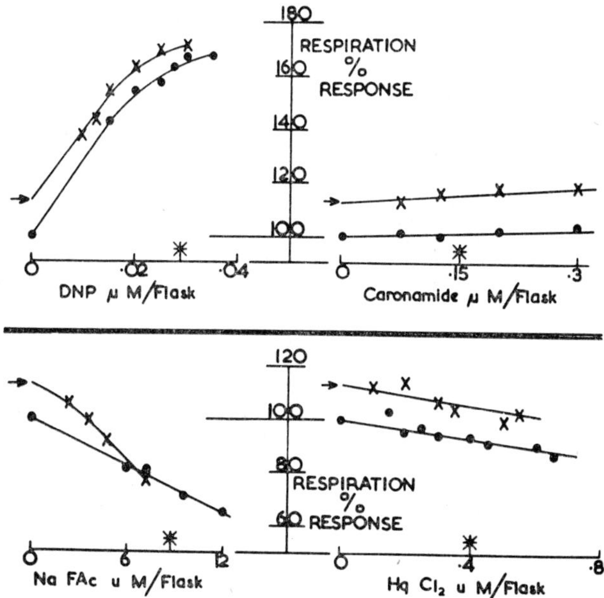

Fig. 1. Summary of respiratory results. The dots represent inhibitor alone; the crosses inhibitor plus 65,000r of X-rays. The arrows indicate the predicted respiratory effect of 65,000r alone. The data represent averages of several experiments: for DNP—7, for caronamide—4, for NaFAc—7, and for $HgCl_2$—12. For reference the asterisks are placed on the dosage scale at the mean transport ED_{50} for each inhibitor.

where NaFAc was combined with a fixed dose of DNP, the DNP induced respiratory increment was not influenced.

DNP: In 5 experiments with the X-ray dose fixed, DNP yielded a ratio of 0·54 (SEM = 0·08). With the DNP dose fixed, the ratio from 3 experiments was 0·52 (SEM = 0·14). Combining the two sets gives a value of 0·53 (SEM = 0·07). The respiratory increments of X-ray and DNP were additive (Fig. 1). Several experiments with larger doses of DNP suggested that X-rays did not produce a further respiratory increase when the DNP effect was already maximal. For any given effect on transport, the DNP has a greater effect on respiration than X-ray.

$HgCl_2$: The ratio obtained from 9 experiments with $HgCl_2$ and a fixed dose of X-ray was 0·70 (SEM = 0·05). No effect of $HgCl_2$ on the respiratory increment of X-ray was evident. The combination of $HgCl_2$ with a fixed dose of DNP gave the same type respiratory results as with X-ray, and a ratio of 0·49 (SEM = 0·05) for 4 experiments.

Caronamide: In 6 experiments, half with the X-ray dose fixed, and half with caronamide fixed, the ratio obtained was 0·32 (SEM = 0·05). The pooled respiratory results suggest a slight increment in respiration as dose increases with no effect on the increment due to X-ray.

Discussion

It is not to be expected that an indirect approach such as the study of mixtures of inhibitors would yield clear-cut biochemical information. At best, certain patterns of behaviour emerge that permit broad hypotheses about mechanisms of action. As a first approximation, the acceptance of certain biochemical concepts of action reduces considerably the vagueness, introducing at the same time some risk of bias.

The transport process will be considered as a complex mechanism being driven by the continued supply of high energy phosphorus. Caronamide, in that it competes with PAH, represents an agent acting on the transport mechanism *per se*, whereas the other inhibitors, with the possible exception of $HgCl_2$, interfere in various ways with energy supply. Caronamide and X-ray show little evidence of interaction; in fact, if one accepts a random distribution of sensitivity to each agent, the measured ratios would occur with no joint action.

NaFAc, on conversion to fluorocitrate, blocks aconitase in the Krebs Cycle, thus impairing both O_2 consumption and energy production. Since NaFAc abolishes the respiratory increment of X-ray, it is concluded that the Krebs Cycle is the pathway for this increased respiration. This is in contrast to the respiratory increment of DNP, which, so far as these experiments have gone, is not abolished by NaFAc, and hence must be at least partly independent of the Krebs Cycle. The DNP and X-ray effects on respiration are thus visualised as operating in parallel, a condition which is still consistent with them being additive. The action of $HgCl_2$ cannot be well defined, apart from the affinity of Hg^{++} for SH groups. The slight respiratory effects demonstrated by $HgCl_2$ do not interfere with either the X-ray or DNP effect, but $HgCl_2$ does show a relatively high degree of addition with X-ray in terms of transport inhibition. This addition is certainly compatible with an X-ray effect mediated via the SH group, but it should be equally emphasised that the respiratory data indicate the presence of X-ray effects quite distinct from a Hg^{++} inhibited moiety. One paradox inherent in this scheme of X-ray and chemical action is the NaFAc and X-ray addition with respiratory inhibition on the one hand, and the $HgCl_2$ and X-ray addition with no respiratory inhibition on the other. It can be avoided by arguing, as one is apt to anyway, that the X-ray effects are multiple; each inhibitor seeks out its own rate limiting potentialities best matched with those of X-ray. The $HgCl_2$ plus X-ray effect on transport would then logically fall (somewhere?) outside of the respiratory mechanism. One difficulty with this type of experiment, however, is the vulnerability to such paradoxes; a good example being the experiences of Davenport *et al.* with multiple inhibitors and gastric secretion (Davenport *et al.*, 1956).

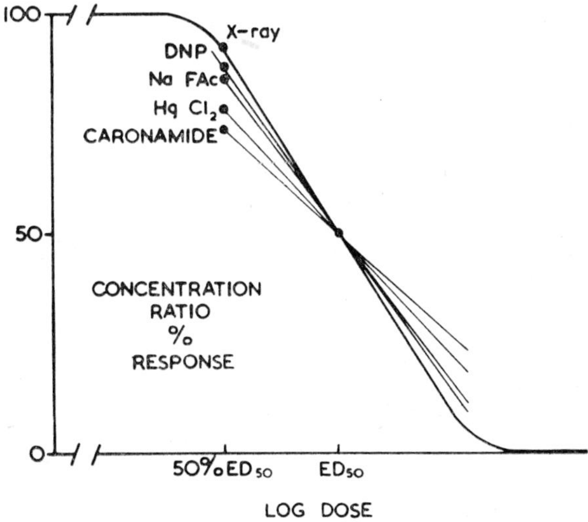

Fig. 2. The contrasting slopes of the various dose-responses tested. Each line was drawn through the 50% ED_{50} response for that agent, determined by averaging the values translated from the individual probit regressions.

The biometrics of addition experiments are almost prohibitive (Plackett and Hewlett, 1948; Finney, 1952). Along with the more usual problems, the transport data suffer from the additional drawback of a distinct lack of parallelism between the responses of the various agents (Fig. 2). X-rays demonstrate the steepest slope, much as they are known to do in whole animal survival studies (Mole, 1953). In the slice system, the simplest explanation of this relates to the random distribution of energy dissipation inherent in high energy radiations. Caronamide would be expected to have the widest variation in response, since along with its own variations in concentration within the slice, it is also subject to the competitive variations of PAH. By consistently choosing doses that straddle the 50% response, the effects of non-parallelism were largely minimised, and no evidence was found for a unique shape to the composite curves. In view of these limitations, and especially the lack of knowledge about the correlations of sensitivities within biochemical units, the equivalence ratios must be considered as having only relative significance; they do not represent finite degrees of addition.

Summary

The effect of combinations of X-rays and various chemical inhibitors have been tested on rabbit kidney slices by observing the responses of respiration and the ability to concentrate PAH. The inhibitors, listed in order of decreasing ability to add to the X-ray inhibition of transport, were NaFAc, $HgCl_2$, DNP, and caronamide. NaFAc was found to be the only agent of the group to inhibit the increment in respiration produced by X-rays. An attempt is made to translate the results into biochemical concepts of X-ray action.

REFERENCES

Davenport, H. W., Chavre, V. J., and Davenport, V. D., 1956. *Amer. J. Physiol.*, **184**, 1.

Finney, D. J., 1952. *Probit Analysis* (London, Cambridge Univ. Press).

Mendelsohn, M. L., 1955. *Amer. J. Physiol.*, **180**, 599.

Idem, 1955. *Ibid.*, **182**, 119.

Mole, R. H., 1953. *Brit. J. Radiol.*, **26**, 234.

Plackett, R. L., and Hewlett, P. S., 1948. *Ann. Applied Biol.*, **35**, 347.

Discussion

Hastings. Did you study the effect of different substrates in your incubating medium?

Mendelsohn. No, we have only had acetate present as substrate.

PRIMARY BIOCHEMICAL EFFECTS IN THE X-IRRADIATED *YOSHIDA* ASCITES SARCOMA

G. Höhne, H. A. Künkel, H. Maass and G. H. Rathgen
Universitats-Frauenklinik, Hamburg-Eppendorf, Germany

A great deal is known about the morphological and functional changes which occur in the nucleus and in the chromosomes as a result of ionising radiation; and the nucleus is known to be the most radiosensitive part of the cell. However, much less is known of the effects of irradiation of the plasma, and metabolic processes here are likely to affect the cell.

We have previously (1955 and 1956) observed a depression of carbohydrate metabolism in solid rat tumours after X-irradiation with 10–20,000r. Ascites cells are a more homogeneous biological material than solid tumours, and have other investigational advantages. This paper deals with some analogous investigations of the metabolic effects occurring in the *Yoshida* ascites rat sarcoma after irradiation.

The ascites cells were cultured by intraperitoneal implantation of the sarcoma tissue into Wistar rats. After sacrificing the animals, the cells were collected and washed with ice-cold Ringer's solution ($NaHCO_3$ added). The tissue metabolic rates were determined manometrically by Warburg's method. Triethanolamine-HCl-buffer (*p*H 7·4) Ringer's solution (Bücher, 1953) was used for the direct measurement of respiration. Aerobic and anaerobic glycolysis was determined with $NaHCO_3$ containing Ringer's solution. The irradiation with doses of 20,000r of 60 kV X-rays (dosage rate 4,000r/min.) was carried out at room temperature. Immediately after irradiation the reaction was started by adding glucose (final concentration 200 mg.%) and the metabolic rates were determined.

The results are summarised in Figure 1. An inhibition of the aerobic and anaerobic glycolysis of 30 resp. 50% was observed. In contrast to former determinations on solid tumours the respiration of the *Yoshida* ascites sarcoma decreases only for about 10% after 20,000r. The rate of glycolysis was determined at intervals of 5 minutes. Under anaerobic conditions the rate of glycolysis was generally constant (up to 180 minutes) if the suspension was not irradiated. In the case of irradiated cells, however, the production of lactic acid was first considerably depressed, then a gradual increase occurred. These findings seem to indicate that the X-ray-induced depression of glycolysis in the *Yoshida* ascites sarcoma may be a reversible process. Possibly an inhibition of one of the glycolytic enzymes is induced by the irradiation *in vitro*.

Therefore we tried to localise this effect. The best way to do this seemed to be the determination of steady-state concentrations of metabolites, as was recently carried out by Holzer (1956) to examine pathological metabolic

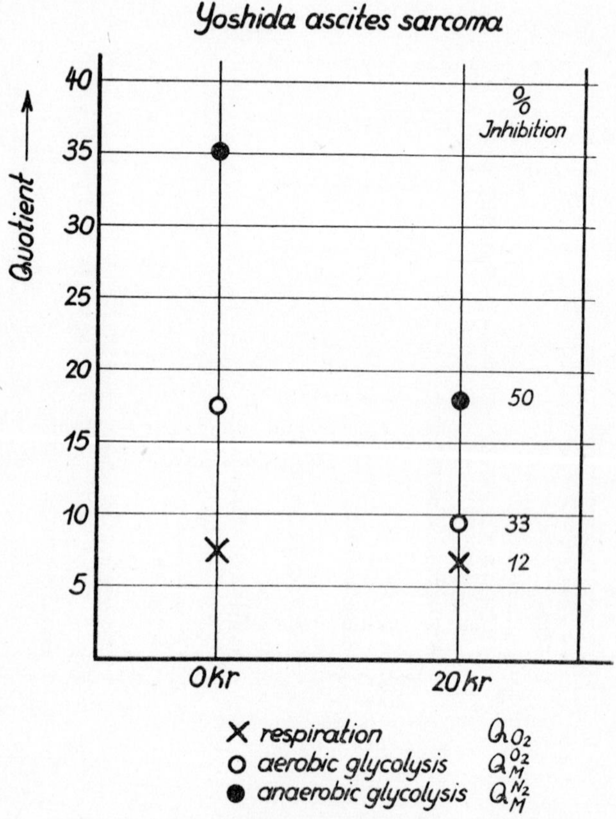

FIG. 1. Glycolysis and respiration of the *Yoshida* ascites sarcoma after X-irradiation with 20 kr.

states. In addition to the manometrical determination of the lactic acid produced the production of fructose-1,6-diphosphate (FDP), dihydroxyacetone phosphate (DAP) and pyruvic acid (PA) were measured in the glycolysing cells after irradiation with 20,000r. The ascites cells in the Warburg apparatus were disintegrated with $HClO_4$ (final concentration 5%) 60 minutes after the addition of glucose. The determination of FDP, DAP and PA in neutralised perchloric acid extract was performed photometrically (366 mμ) with DPNH by means of crystallised aldolase, α-glycerophosphate dehydrogenase and lactic acid dehydrogenase (Bücher *et al.*, 1953 and Holzer *et al.*, 1955).

As is shown in Figure 2 a considerable decrease of the steady-state concentration of pyruvic acid was observed 60 minutes after the addition of glucose, which corresponds to the reduced production of lactic acid. This

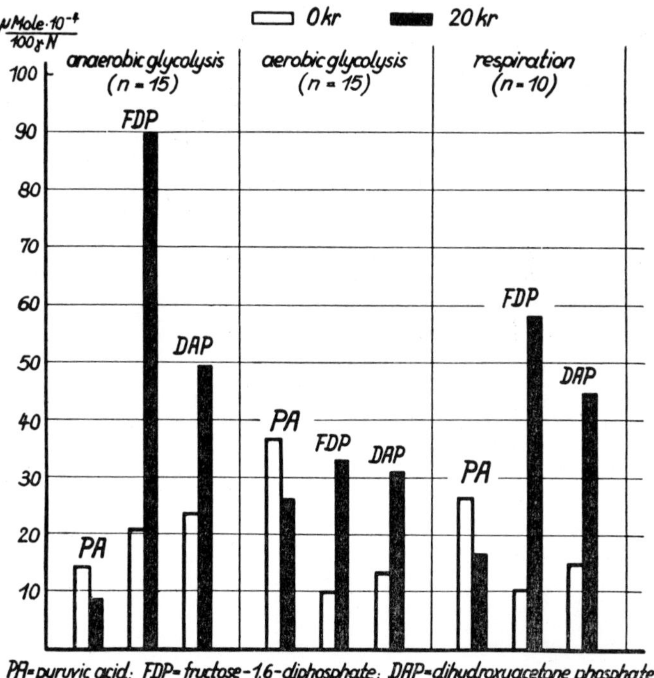

FIG. 2. Stationary concentrations of metabolites of the glycolysis in *Yoshida* ascites sarcoma after X-irradiation with 20 kr.

effect occurred in the *Yoshida* ascites tumour under aerobic conditions as well as under anaerobic ones. Simultaneously the concentrations of FDP and DAP markedly increased. The parallelism of the accumulation of FDP with DAP and the decrease in PA concentration, observed in all three groups of experiments, show significantly that the X-ray-induced inhibition of glycolysis is due to the blockage of a link in the reaction-chain which takes place beyond the triosephosphates. An inactivation of the aldolase can be excluded because of the accumulation of both FDP and DAP. An inhibition of the lactic acid dehydrogenase cannot be considered either, because the concentration of pyruvic acid declines proportionally to the production of lactic acid. From these findings it follows that the decomposition of triosephosphates is disturbed by the influence of irradiation.

Possibly one of the enzymes is inactivated between triosephosphates and pyruvic acid, e.g. the phosphoglyceraldehyde dehydrogenase. On the other hand it must be considered that it may not be an enzyme that limits the metabolic rate of the triosephosphate, but rather one of the coenzymes participating in the reactions. In this case, however, a decrease of the adenosine diphosphate level (ADP) would lead to a similar effect like an inactivation of the phosphoglyceraldehyde dehydrogenase. The investiga-

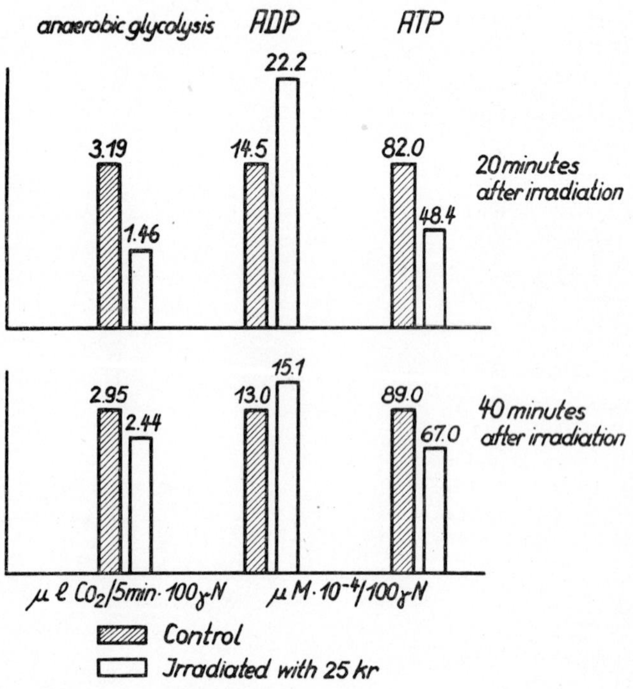

FIG. 3. Anaerobic glycolysis and adenosine diphosphate resp. adenosine triphosphate concentrations in *Yoshida* ascites sarcoma after X-irradiation with 25 kr.

tions of Ashwell and Hickman (1953) show that the X-ray-induced inhibition of the glycolysis of the spleen could be attributed to an effect as mentioned above. This inhibition, it is true, can be seen earliest a few days after irradiation. On the other hand, an inhibition of the phosphoglyceraldehyde dehydrogenase should produce just the opposite action with regard to the adenine nucleotides. In this case an increase of the steady state concentration of ADP and a decrease of ATP level should be expected.

In order to ascertain the point of attack by ionising radiations on glycolysis, determinations of ADP and ATP were carried out. These examinations were made by means of pyruvate kinase and lactic acid dehydrogenase respectively of phosphoglycerate kinase and phosphoglyceraldehyde dehydrogenase, as can be seen in Figure 3. These values show that there is really a slight increase in ADP corresponding to a decrease of ATP-concentration. This reaction can best be explained by the inactivation of an enzyme of the glycolysis. In all probability the phosphoglyceraldehyde dehydrogenase is inactivated by the action of the radiation. This can be deduced from the decrease of the concentration of ATP proportional to the inhibition of the glycolysis. If the glycolysis is depressed to a lesser extent, as for instance 40 minutes after the irradiation, the production of ATP accordingly increases. This may be demonstrated, furthermore, by the controls we carried out with the aid of iodine acetate which is evidently able to specifically inactivate the phosphoglyceraldehyde dehydrogenase, within a certain range of concentration.

Our experiments show that the inactivation of the phosphoglyceraldehyde dehydrogenase by the addition of iodine acetate is followed by analogous changes in the steady state concentrations of metabolites observed after an irradiation with 25,000r.

REFERENCES

ASHWELL, G., and HICKMAN, J., 1953. *J. biol. Chem.*, **201**, 651.

BÜCHER, TH. *et al.*, 1953. *Z. Naturforsch.*, **8**b, 555.

HÖHNE, G., KÜNKEL, H. A., RATHGEN, G. H., and UHLMANN, G., 1955. *Naturwissenschaften*, **42**, 630.

HOLZER, H., HAAN, J., and SCHNEIDER, S., 1955. *Biochem. Z.*, **326**, 451.

HOLZER, H., HOLZER, E., and SCHULTZ, G., 1955. *Ibid.*, **326**, 385.

HOLZER, H., 1956. *Ergebnisse der medizinischen Grundlagenforschung* (hrsg. von K. Fr. Bauer, Stuttgart, Thieme).

UHLMANN, G., and RATHGEN, G. H., 1956. In press, *Strahlentherapie*.

DISCUSSION

Forssberg. One reason why we are interested in biochemical effects of radiation is the possibility that such effects may form a link between the initial physical acts and microscopically or otherwise demonstrable structural lesions. If so, early biochemical changes, as being closer to the initial process, may possibly therefore tell us more about the primary chain of events than changes occurring later. It would also seem that supralethal doses should be avoided. Our experience is that already doses of 6,000–7,000r *in vitro* cause lethal damage—as such cell samples inoculated in mice do not give any growth of tumour cells.

I do not see how, quantitatively, one can correlate biochemical and biological data if the analyses are performed on a mixture of cells some of which are killed and others of which are in various stages of dying. Moreover, as time proceeds the percentage of virtually non-living material increases and this casts some doubt on time-effect relationship studies.

We have irradiated *Ehrlich* ascites cells *in vivo* (1,250r). This dose does not, as judged from supravital staining, protein synthesis and other data, by percentage kill any appreciable part of the cell culture for the first two post-irradiation days; although mitotic activity is inhibited for some 20 hours. Nevertheless with the first 15 minutes very marked changes in carbohydrate metabolism occur.

Höhne. Dr. Forssberg's remark as to the interpretation of biochemical changes after supralethal doses is, of course, of relevance when a certain percentage of cells are killed. We have found, however, that the changes in respiration and glycolysis which take place immediately after radiation are returned to normal value after some time. We believe this indicates that these changes are not simply due to a falling away of killed cells in the irradiated sample. In this connection it may be mentioned that Dr. Holzer, of the Institute of Physiological Chemistry in Hamburg, has found that monoiodoacetic acid, in non-lethal concentrations, decreases respiration and glycolysis of *Ehrlich* carcinoma cells—showing that decreased metabolism is not necessarily due to lethal cell damage.

Forssberg. Certainly biochemical effects do occur after non-lethal application of many chemicals as they do after irradiation at low dose levels, but this seems to be a good reason why supralethal doses should be avoided.

I wonder if the main interest lies not in the very heavy doses but in such low dose levels as give just traceable effects.

THE INCORPORATION OF ^{32}P INTO LIVER PHOSPHOLIPIDS AND RNA MONONUCLEOTIDES OF IRRADIATED AND NON-IRRADIATED MICE

F. G. SHERMAN AND A. B. ALMEIDA *
Brown University, Providence 12, R.I. and
Brookhaven National Laboratory, Upton, N.Y., U.S.A.

The incorporation of labelled precursors into various tissue fractions has been used in several laboratories to study the effect of X-radiation on metabolism (Euler and Hevesy, 1944; Holmes and Mee, 1952; Forssberg and Klein, 1954). Early studies by Hevesy and his collaborators showed that X-rays appear to reduce the incorporation of inorganic ^{32}P into RNA and DNA from *Jensen* sarcoma and non-tumour tissues. Examination of the total uptake of phosphorus of liver and other tissues by Forssberg and Hevesy (1952) revealed that irradiation with X-rays (2000r) resulted in marked increases in ^{32}P activity in the liver. About 40% more labelled phosphate was taken up by the liver during the first 15 minutes after irradiation than in the non-irradiated animals. A corresponding decrease in the ^{32}P incorporation was found in the muscle and bone from irradiated animals. Payne, Kelly and Entenman (1953) showed that total body irradiation of rats and mice resulted in an increase in the incorporation of ^{32}P into cytoplasmic RNA. A simultaneous reduction in the specific activities of ^{32}P in nuclear RNA and DNA was observed. Harrington and Lavik (1954) found that exposure of rats to X-rays resulted in either inhibition or enhancement of incorporation of the label into thymus DNA depending upon the label employed.

The experiments presented in this report are time course studies of the incorporation of inorganic $^{32}PO_4$ into various fractions of mouse liver from irradiated and non-irradiated mice.

Young male mice were irradiated with $508 \pm 5\%$ rads of filtered 200 kV X-rays (approximately 90% of the LD_{50}–30d. dose). Non-irradiated controls were given dummy irradiations for the same period as the irradiated animals. Immediately after irradiation all the animals were injected with about 2 μc. ^{32}P/gm. At intervals after injection the animals were sacrificed. Blood was taken from the heart for determination of plasma specific activity. The liver was removed at once and frozen in acetone-dry ice or liquid nitrogen. The frozen livers were extracted in the cold with ten volumes of cold 10%

* Present address: School of Medicine, University of Vermont, Burlington, Vermont, U.S.A.

TCA with the aid of a tissue homogeniser. Phospholipids were extracted from the residue with mixtures of ethanol-ether and methanol-chloroform. Solvents were removed under reduced pressure in a nitrogen atmosphere and the phospholipids taken up in petroleum ether. Further separation into lecithins, cephalins and sphingomyelin was carried out by adsorption on MgO.

Crude nucleates were extracted with 10% NaCl as described by Davidson and Smellie (1952). RNA was hydrolysed with dilute KOH under conditions which gave a quantitative recovery of the RNA mononucleotides. These were separated by paper ionophoresis by the method of Davidson and Smellie (1952).

Incorporation of inorganic $^{32}PO_4$ into inorganic phosphate, phospholipids and ribonucleic acid mononucleotides from the liver are given in the Table in terms of the percentage increases in specific activity in irradiated animals. The mean differences and standard error of the means were determined by paired comparison analysis according to Snedecor (1946). In each instance N was 4 except for the 360 minute series where N was 3.

Labelled inorganic phosphate was found to disappear from the plasma at approximately the same rate in irradiated and non-irradiated animals. Likewise there is no significant difference in the specific activity of inorganic phosphate in the liver of irradiated and non-irradiated animals. The values shown in the table have not been corrected for extracellular inorganic phosphate. The error introduced at 30 minutes after injection by the presence of plasma inorganic phosphate and other extracellular liver phosphate is probably considerable. At 60 minutes and thereafter, the specific activities of liver inorganic phosphate are nearly the same as those in plasma.

In contrast to similarities in the specific activities of liver inorganic phosphate and plasma inorganic phosphate, marked increases in the specific activity of phospholipids and RNA mononucleotides were observed. Among the phospholipids the percentage increase was highest in sphingomyelin, followed by lecithin. Although estimates of total amounts of phospholipids with precision sufficient to determine unequivocally if there was an increase in phospholipids synthesised was not possible, it was clear that there were no large-scale changes in total amounts.

The differences found between individual mononucleotides in the increases in incorporation of labelled phosphate are probably real. It is, however, not possible to interpret the significance of these differences because the sizes of the respective precursor pools are not known. The specific activities of RNA mononucleotides in non-irradiated animals are each different, and in our experience somewhat variable. This has been taken to indicate that the size of the various precursor pools is affected by the physiological state of the animal. It is clear, though, that exposure to X-irradiation has resulted in changes which have had marked effects on the incorporation of the label.

TABLE

Increase in Incorporation of $^{32}PO_4$ into Inorganic Phosphate, Phospholipids and RNA Mononucleotides in the Livers of Mice at Various Intervals after Exposure to 510 rads of X-radiation. Values shown are:

$$100 \times \frac{\text{Sp. Act. Irrad.–Sp. Act. Non-irrad.}}{\text{Sp. Act. Non-irrad.}} \pm \text{Standard error of mean.}$$

Time minutes	Liver inorganic P	Phospholipids		
		Cephalin	Lecithin	Sphingomyelin
30	2·7 ± 4·8	26·4 ± 5·3	43·8 ± 4·8	68·7 ± 6·2
60	3·0 ± 3·1	48·5 ± 12·2	21·0 ± 7·4	79·2 ± 20·8
180	5·1 ± 1·2	32·6 ± 10·1	33·0 ± 9·4	34·0 ± 7·4
380	3·1 ± 3·1	−9·9 ± 7·5	0·11 ± 0·05	− 43·3 ± 9·0

Time minutes	RNA mononucleotides			
	Cytidylic	Adenylic	Guanylic	Uridylic
30	38·2 ± 11·3	53·2 ± 15·1	31·6 ± 10·5	39·7 ± 12·4
60	86·0 ± 8·3	65·2 ± 14·1	37·5 ± 8·6	53·6 ± 20·1
180	50·0 ± 4·4	53·6 ± 5·9	56·0 ± 4·5	43·0 ± 3·0
360	117 ± 11·3	181 ± 15·2	82·1 ± 6·7	68·0 ± 8·4

The difference between the amount of ^{32}P incorporated into the RNA of irradiated and non-irradiated animals is about the same during the first three hours after irradiation. However the specific activity of RNA from non-irradiated animals remains at approximately the same level between 3 and 6 hours after application of the label. During this period the specific activity of liver RNA from irradiated animals continues to increase (Table). This suggests that high specific activity phosphorus continues to be available for incorporation into the RNA precursors. If in addition there was an increase in the total amount of RNA this could explain the continued increase in specific activity. Roth (1956) has reported data which indicates that X-radiation results in a decrease in rat liver ribonuclease activity.

If the increase in specific activity of the phospholipids and ribonucleic acid after irradiation were due to a net increase in phosphate transport into the cell, it would be expected that the specific activity of the liver inorganic phosphate would be significantly higher in the irradiated animals than in the controls (Table). Moreover, the specific activity of the plasma inorganic phosphate would be expected to be lower in the irradiated animals. Neither were found in our animals. Irradiation has been shown to result in an enhanced activity of ATP-ase and 5'nucleotidase in broken cell preparations of rat spleens and thymus glands as early as three hours after irradiation (Dubois and Peterson, 1954 and Ashwell and Hickman, 1953). This would have the effect of reducing the size of the ATP pool in the cell. In addition to a reduction of the ATP pool from enhanced phosphatase activity, precursor pool sizes may also be reduced by leakage from the cell. Billen *et al.* (1953) found that ATP and probably other substances absorbing at 260 mμ will diffuse out of X-irradiated *E. coli.* If the assumption of rapid equilibration of the various subcompartments which make up the acid soluble fraction of the cell is true, any reduction in the pool size of the precursors of phospholipids and RNA would be reflected in an increase in the specific activity of the label in the end product.

Finally, interference by X-rays in the role of the final product in the economy of the cell could be reflected by changes in the specific activity of the label. Entenman *et al.* (1955) observed increased concentrations of plasma phospholipids after irradiation in several species. If this led to a reduction in the concentration of phospholipids in the liver cell, it could explain the increased activity of phosphorus found in the phospholipid fraction in the experiments reported here. It would also explain the rapid decline in specific activity found at six hours after irradiation.

Summary

1. The removal of labelled inorganic phosphate from the plasma and the incorporation of the label into inorganic phosphate, phospholipids and mononucleotides of ribonucleic acid have been compared in irradiated and non-irradiated mice during the first 6 hours following exposure to X-rays.

2. The rate of disappearance of labelled inorganic phosphate from the plasma was similar in irradiated and non-irradiated mice.

3. The specific activity of liver inorganic phosphate was the same in irradiated and non-irradiated animals.

4. Marked increases in the specific activity of the phosphorus in phospholipids and RNA mononucleotides were observed starting at 30 minutes after irradiation.

5. The hypothesis is brought forward that the increases in the specific activity of liver phospholipids and RNA mononucleotides is due to a reduction of the amount of precursor substances available for synthesis.

Acknowledgment

This investigation was supported in part by a grant-in-aid to Brown University from the American Cancer Society upon recommendation of the Committee on Growth of the National Research Council. A portion of the research was carried out at Brookhaven National Laboratory under the auspices of the U.S. Atomic Energy Commission.

REFERENCES

ASHWELL, G., and HICKMAN, J., 1953. *J. biol. Chem.*, **201**, 651.

BILLEN, D., STREHLER, B. L., STAPLETON, G. E., and BRIGHAM, E., 1953. *Arch. Biochem. Biophys.*, **43**, 1.

DAVIDSON, J. N., and SMELLIE, R. M. S., 1952. *Biochem. J.*, **52**, 599.

DUBOIS, K. P., and PETERSEN, D. F., 1954. *Amer. J. Physiol.*, **176**, 282.

ENTENMAN, C., NEVE, R. A., SUPPLEE, H., and OLMSTED, C. A., 1955. *Arch. Biochem. Biophys.*, **59**, 97.

EULER, H., and HEVESY, G., 1944. *Arkiv. f Kemi. Mineral. o Geol.*, **17A**, 1.

FORSSBERG, A., and HEVESY, G., 1952. *Arkiv. f Kemi.*, Bd. **5**, 93.

FORSSBERG, A., and KLEIN, G., 1954. *Radioisotope Conf.*, **1**, 232 (London, Butterworth).

HARRINGTON, H., and LAVIK, P. S., 1954. NYO-4023.

HOLMES, B. E., and MEE, L. K., 1952. *Brit. J. Radiol.*, **25**, 273.

PAYNE, A. H., KELLY, L. S., and ENTENMAN, C., 1953. *Proc. Soc. exp. Biol. N.Y.*, **81**, 698.

ROTH, J. S., 1956. *Arch. Biochem. Biophys.*, **60**, 7.

SNEDECOR, G. W., 1946. *Statistical Methods* (Iowa, Ames).

DISCUSSION

Holmes. I should like to know whether Dr. Sherman has tried local irradiation to the liver instead of whole body irradiation. We have used local irradiation (450r) and obtained a very much smaller increase of ^{32}P uptake into RNA than does Dr. Sherman.

Sherman. We have not done that but it would be interesting.

Holmes. Mrs Kelly has found, after whole body irradiation, that the ^{32}P uptake into nuclear RNA is somewhat decreased and that into the cytoplasmic RNA somewhat increased. Has Dr. Sherman separated the nuclear and cytoplasmic fractions?

Sherman. These are total RNA determinations, but fractionated free from DNA. The mononucleotides were after hydrolysing, separated by einophoresis, and should be rather pure.

EFFECTS OF IRRADIATION ON DNA SYNTHESIS BY HUMAN BONE MARROW CELLS *IN VITRO*

L. G. Lajtha, R. Oliver and F. Ellis
Department of Radiotherapy, Churchill Hospital, Oxford, England

DNA synthesis has been studied by the incorporation of adenine ^{14}C and formate ^{14}C into DNA using a tissue culture technique and high resolution autoradiography. Formate ^{14}C is eminently suitable for DNA labelling since in this system it is preferentially utilised for thymine formation, Lajtha (1954). It thus has the advantage over adenine ^{14}C that no removal of RNA from the cells is necessary. Preliminary experiments have shown that adenine ^{14}C and formate ^{14}C are utilised in similar quantities for DNA synthesis.

Irradiation experiments with 1,000 rads given *in vitro* have shown that both adenine ^{14}C and formate ^{14}C uptake are depressed to about the same extent by radiation.

In human bone marrow cells the cell cycle consists of an approximately 12 hours DNA synthetic period (S period) which ends about 4 hours before mitosis, and which is preceded by a long resting period (G_1 period) of 12-24 hours' duration. In a short term culture experiment (5–7 hours), therefore, most of the cells which show an autoradiograph are those which during the time of incubation with the labelled precursor were actually in the S period. If, on the other hand, cultures are incubated without the addition of the labelled precursor for the first 15–17 hours after irradiation, and the isotope is only then added for a further 5–7 hours—in such experiments all the cells which at the time of irradiation (0 hour) were in the S period will have completed DNA synthesis before the isotope is added and the cells that are labelled between the 17–22 hours after irradiation represent cells which were in their G_1 period at the time of irradiation. This is based on the assumption that irradiation does not cause a significant hold-up of cells in the S period. It should be noted that for measuring S period effect the 'maximum grain count' (the average of the 10% highest counts) is a better indication than the average grain count since the former is a truer indication of whether cells, which normally would have completed DNA synthesis, have in fact achieved the same degree of incorporation (full diploid amount) or to what extent this has been able to proceed.

In previous experiments a dose of 5,000 rads irradiation was found to inhibit DNA synthesis in the S period immediately and to damage cells in the G_1 period so that they either did not enter the S period (decrease in % positivity) or they stopped synthesising DNA shortly after entering

S period (decrease of 'maximum' and average grain counts), Lajtha, Oliver and Ellis (1954).

These experiments were repeated using 1,000-1,600 rads and it was found that as with 5,000 rads both G_1 and S periods were affected, i.e. both the percentage of positivity and maximum grain counts have shown a decrease compared with the unirradiated controls. However, while 5,000 rads inhibited S period immediately, the effect of 1,000 rads was somewhat delayed.

Using smaller doses of radiation (400–750 rads) it was shown that even such doses affected DNA synthesis in the S period. At the dose level of 300 rads, however, the effect on the S period became variable; in some experiments a depression could be detected, in others not. Below 200 rads the S period was clearly not affected (Table I)

TABLE I

Effect of Irradiation on the S Period

Formate ^{14}C

Expt. no.	Timing and dose in rads	Proportion of positive cells	Average grain counts	Remarks
513	0—6	37% +	19	S period not affected
	X_{300}—6	40% +	18	
513	0—14	51% +	39	S period ? affected
	X_{300}—14	56% +	31	
514	0—14	36% +	27	S period affected
	X_{300}—14	23% +	21	
254	0—5	38% +	16	S period not affected
	X_{195}—5	33% +	15	
513	0—6	37% +	19	S period not affected
	X_{150}—6	43% +	21	
513	0—14	51% +	39	S period not affected
	X_{150}—14	57% +	36	

While 300 rads showed a variable effect on the S period it invariably affected the G_1 period in preventing or delaying cells from entering the S period (Table II). Experiments with 150 rads also indicated a damage to the G_1 period.

Finally, bone marrows were cultured before and after 400–500 rads given *in vivo* to the sternum and the results have shown a depression of the S period as well as G_1 period. It appears that *in vivo* irradiations of bone marrow cells have a similar effect on DNA synthesis in the cultures to comparable doses given *in vitro*.

In conclusion: small doses of radiation (150-300 rads) inhibit DNA synthesis if delivered to cells prior to the beginning of DNA synthesis, i.e. to cells in the G_1 period, but do not inhibit DNA synthesis once it has started, i.e. in the S period. These observations are in agreement with those of Howard and Pelc on bean root cells (1953) and suggest the presence

TABLE II

Effect of Irradiation on G_1 and S Periods

Adenine ^{14}C

Expt. no.	Timing and dose in rads	Proportion of positive cells	Maximum grain counts	Remarks
211	—2—20	53% +	30	G_1 affected
	X_{500}—2—20	38% +	30	S not affected
215	—3—20	60% +	35	G_1 affected
	X_{300}—3—20	46% +	30	S not affected
233	—2—24	67% +	50	G_1 affected
	X_{300}—2—24	51% +	40	S ? affected

Formate ^{14}C

Expt. no.	Timing and dose in rads	Proportion of positive cells	Maximum grain counts	Remarks
211	—2—20	71% +	50	Both G_1 and S
	X_{500}—2—20	49% +	30	affected
233	—2—24	63% +	45	Both G_1 and S
	X_{300}—2—24	18% +	20	affected
256	—17—22	50% +	17	S not investigated
	X_{300}—17—22	24% +	13	G_1 affected
256	—17—22	50% +	17	S not investigated
	X_{150}—17—22	34% +	17	G_1 affected
258*a*	—17—22	46% +	64	S not investigated
	X_{160}—17—22	24% +	57	G_1 affected
258*b*	—17—22	52% +	63	S not investigated
	X_{160}—17—22	28% +	51	G_1 affected

of a radiosensitive system in the cells connected with, but not identical with, the process of DNA synthesis. This system is apparently sensitive to smaller doses of radiation than DNA synthesis. The present experiments do not indicate whether this system is vulnerable only before DNA synthesis (G_1 period) or also during DNA synthesis (S period).

REFERENCES

Howard, A., and Pelc, S. R., 1953. *Heredity* (Suppl.), **6**, 216.

Lajtha, L. G., 1954. *Nature, Lond.*, **174**, 1013.

Lajtha, L. G., Oliver, R., and Ellis, F., 1954. *Brit. J. Cancer*, **8**, 367.

Discussion

Hevesy. Evidence has been brought forward suggesting that in different DNA fractions extracted from the same tissue ^{32}P is incorporated at a different rate and that incorporation in these fractions is also affected differentially by exposure to radiation. Would it be too exacting to try to fractionate DNA in Dr. Lajtha's experiment and to find out if the restricted (50%) effect of irradiation on ^{14}C incorporation into DNA, observed by him, is due to the presence of DNA fractions of different radiosensitivity?

Lajtha. The point raised by Professor Hevesy occurred to us, and it was a very tempting possibility. However, in our experiments, the amount of DNA synthesised by the individual cells was unaffected by 150r–300r, and the depression was produced by a proportion of cells (30–50%) not synthesising DNA at all. This suggested to us that it is not one type of DNA which is affected, unless one assumes that nearly half of the cell population contains a 'sensitive' DNA, while the other half contains a 'resistant' DNA.

Hevesy. Then a second question: Howard and Pelc observed a radiation effect already after an exposure to 35r. You did not observe an effect below 150r. Is it because you were not interested in reducing your dose further, or does the radiosensitivity of marrow cells differ from that of the meristemic cells of *Vicia Faba*?

Lajtha. We are interested in further reducing our dose of radiation, but the error range of the technique makes effects below 150r difficult to observe. The grain counting is accurate enough, but the 'per cent. positivity' counts have to be performed simultaneously with differential counts, since the marrow is a heterogenous population of cells. The error of differential counts is rather large. We are working on a method to overcome this difficulty, and we may then further decrease our radiation dose.

Holmes. We cannot altogether agree with one point in Dr. Lajtha's very interesting paper. In our own work with regenerating rat liver we have entirely confirmed the results obtained by Pelc and Howard and by Dr. Lajtha in that we find that DNA synthesis can be inhibited by small doses of X-rays given before the synthesis begins, but after synthesis has begun, large doses (say 2,000r) are required for inhibition. We have not, however, found any difference in sensitivity between different times in the pre-synthesis interphase; provided the X-rays were given before synthesis began it did not matter how long before. Inhibition of subsequent DNA synthesis could be caused by irradiation of the liver of a normal rat and afterwards stimulating synthesis and mitosis by partial hepatectomy.

Dr. Lajtha might be interested to hear of some Japanese work by Dr. Nagai and Dr. Matsuda of Osaka University. They found that 6 hours after irradiation of the liver with 600r some pyknotic or karyorrhectic nuclei could be detected.

If partial hepatectomy was performed either just before or just after the irradiation and the animal killed 6 hours later, a much higher proportion of such damaged nuclei was found.

Dr. Sibatani suggested to me that just those liver cells which have become so sensitive to radiation injury might later have been the first to synthesise DNA and to divide. Their destruction could account for the delay in DNA synthesis which has often been observed.

Lajtha. We have no real proof of a difference between sensitivity at different times in the pre-synthetic interphase (G_1 period). There are several possible explanations as to how the 50% inhibition may occur.

(1) One half of the G_1 population is sensitive, either the first or second half.

(2) One half of the G_1 population is sensitive in a random fashion, i.e. a cyclic metabolic process, one phase of which is sensitive.

(3) Or, as Dr. Holmes suggests, referring to the results of the Japanese workers, some cells are destroyed before entering the S period.

We think that in the case of (3) a significant drop in the total number of cells in the culture should be observed. For a 24 hours' culture over 50% of the cells are positive, and if half of the potentially positive cells are destroyed, then we should see at least a 25% drop in the total cell counts. Up to 300r we do not see such a drop.

There are several other theoretical explanations which we had to discard, and for the time being we favour possibility (1) or (2). Experiments to test them are in progress and we hope to have an answer in a few months' time.

REFERENCE

NAGAI, S., MATSUDA, H., AKITA, K., and KASUE, T., 1954. *Med. J. of Osaka Univ.*, vol. **5** and paper in press.

EFFECT OF VARIOUS DOSES OF X-RAYS—WHOLE BODY AND LOCAL IRRADIATION—ON THE NUCLEIC ACIDS OF THE BONE MARROW

P. MANDEL, C. M. GROS, J. RODESCH, C. JAUDEL AND P. CHAMBON

Institute of Biochemistry, Faculty of Medicine,
Strasbourg, France

The action of X-rays on the nucleic acids of the bone marrow has been studied by Lutwak-Mann (1951), Smellie *et al.* (1955) and by Mandel *et al.* (1951).

In all these cases the effect studied was of one or two doses of X-rays and whole body irradiation. The time interval between the observation and irradiation was mostly short and the results, except in our own experiments, expressed in units per fresh or dry weight of marrow. It is obvious that after irradiation other constituents of the marrow undergo changes as well as nucleic acids. In these conditions, by expressing the results in units per fresh or dry weight, we compare one variable, i.e. nucleic acids, to another, i.e. the substance of the marrow. This raises a difficulty in the interpretation of the results. We have therefore considered it of value to study variations in the absolute quantity of nucleic acids in a portion of a hæmopoietic system, such as the hind-limbs. In our homogeneous strain of rats the absolute quantities of RNA and DNA were very similar in this hæmopoietic system, for different animals of the same sex belonging to the same litter. There is evidence, too, from our laboratory that the absolute quantity of cells in the marrow of the hind-limbs is also very similar for animals of the same litter ($\pm 8\%$) (Mantz, to be published). Therefore animals of the same litter could be used as controls in the study of variations in the nucleic acids. The result is that these actual variations did not depend upon the variation of the other cellular constituents. In a series of experiments, we have studied the effect of various doses of X-rays from the time of irradiation until the marrow has fully recovered. In addition to work on the effect of a whole body radiation we have added work on local irradiation, so as to investigate direct and indirect effects of X-rays, as we did with skin (Rodesch and Mandel, 1955). The latter comparison has also been made for a dose of X-rays (700r), by following the incorporation of ^{32}P.*

Our nucleic acid analyses have been performed on a total of 385 animals.

* DNA: desoxyribonucleic acid. RNA: ribonucleic acid.

We first studied the action of five doses of X-rays; 325, 500, 700, 1,000, and 1,500r, and the subsequent variation in the absolute quantity of RNA and DNA in the bone marrow.

The radiation factors were 180 kV, 1 mm. of Cu, target distance 45 cm. and dose rate 25r per minute.

We then examined the effect of a local irradiation and of whole body irradiation of 700r and, finally, the incorporation of ^{32}P into RNA and DNA after a whole body or a local irradiation of one or both hind-limbs with the same dose (700r).

In all experiments the animals used as controls and those irradiated were of the same sex and litter.

There were at least 3 animals in each control group, and an equal number of animals for each type of irradiation. The methods of measurements and the chromatographic separation techniques have been described elsewhere (Mandel *et al.*, 1956). These have been slightly modified and adapted to the quantities of substances analysed. The results of our experiments are summarised in the figures below.

Results

Examination of Figure 1 shows that RNA undergoes a maximum decrease between the second and the fourth day after irradiation; and by 24 hours the reduction is close to this maximum; which is, for the five dose levels of 325, 500, 700, 1,000 and 1,500r; 42%, 55%, 60%, 65% and 72%.

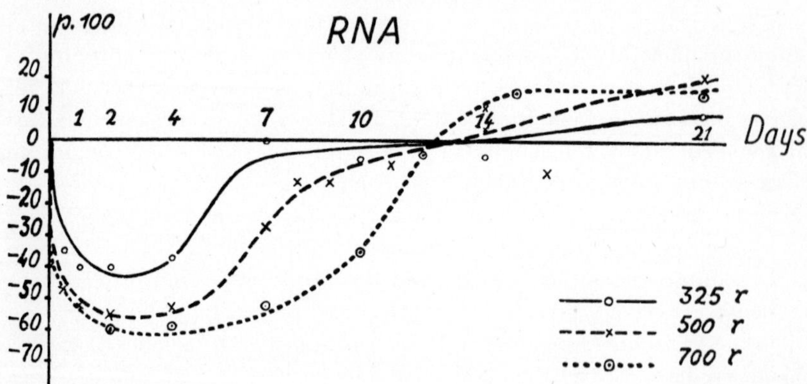

FIG. 1. Changes in the total amount of RNA, expressed as percentage of controls, in the bone marrow of rat hind-limbs after whole body X-irradiation (325r ———; 500r × -- × -- ×; 700r ○ -- ○ -- ○).

For measurements below 1,000r, the survival of the animals in good condition affords an opportunity to study the reconstitution of nucleic acids. From the 7th day onwards, there is a rise to normal values between the 10th and 14th day, and then to values higher than normal on the 21st day.

As far as DNA is concerned, the decrease after irradiation is almost as fast as that for RNA but is much greater.

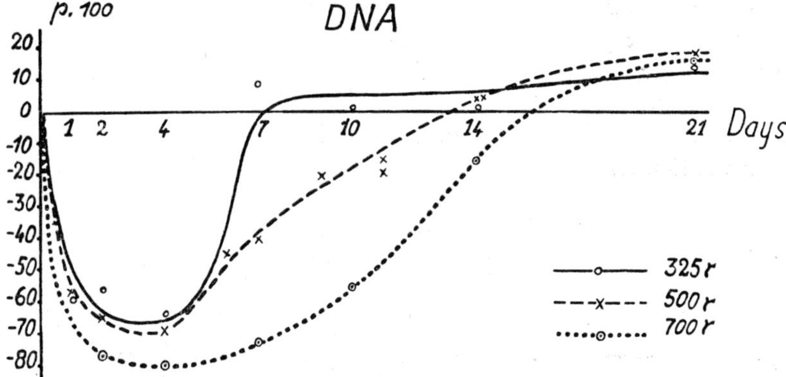

FIG. 2. DNA analyses on the same material as in Fig. 1.

The maximum decrease in DNA also occurs between the 2nd and 4th day. There is a reduction of 66% for 325r, 70% for 500r, 80% for 700r, 85% for 1,000r and 91% for 1,500r. For a given dose of X-rays the decrease of DNA is more important, and the differences between the effects of low and high doses are less marked than for RNA, Figure 2.

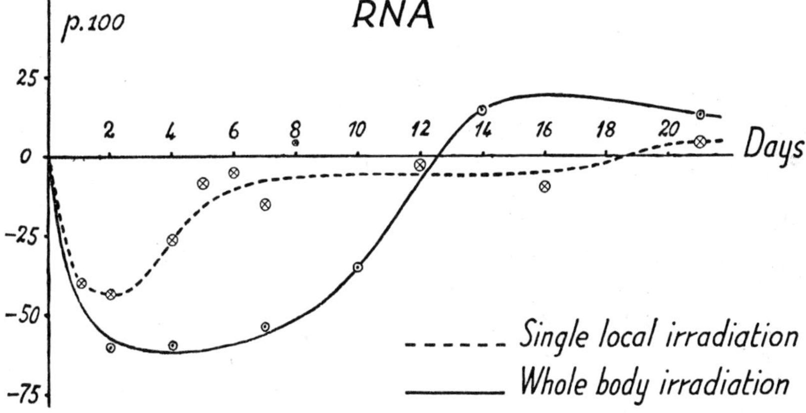

FIG. 3. Percentage changes in RNA concentration of the bone marrow of rat hind-limbs after: (*a*) single local irradiation (– – – –), and (*b*) whole body irradiation (———) with 700r X-rays.

The reconstitution of DNA starts on the 7th day, after doses ranging from 325r to 700r. After higher doses animals have not survived beyond 5 days. The rise of DNA to normal values is much slower for 500r and 700r than for 325r. But after 21 days the values are higher than those in the non-irradiated control animals. The reconstitution of DNA after 500r and 700r is slower than that of RNA. Comparative study of the decrease after total and local body irradiation shows for RNA, a maximal decrease of 45 and 60% respectively, and for DNA, 60 and 80%, Figures 3 and 4. The rise to normal values is faster after local irradiation and at the end of 21 days, but the values are higher in the case of total irradiation than in local irradiation. This data shows that the injury is greatest after whole body irradiation and is in agreement with the results obtained on the skin.

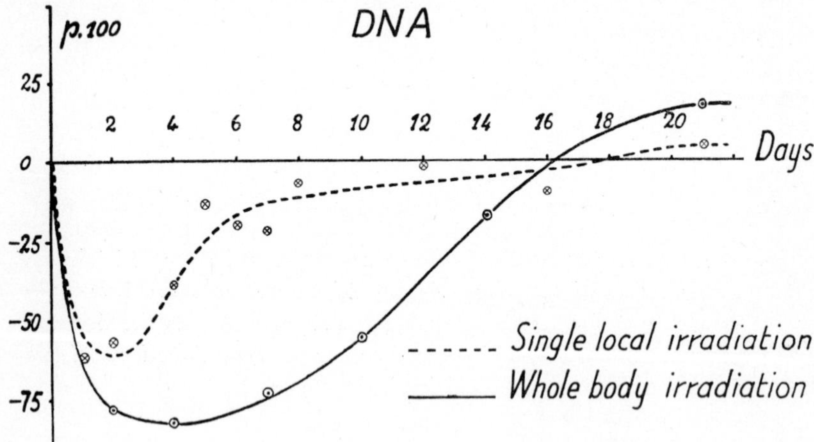

FIG. 4. DNA analyses on the same material as in Fig. 3.

The study of the incorporation of ^{32}P into nucleic acids provides evidence of reduction of the incorporation of RNA after local irradiation; (30% on the first day, 25% on the second day and 10% on the tenth day, and to an increase of 27% on the twenty-first day), see Table I.

After whole body irradiation reductions of 30, 50 and 30% appear on the first, fourth and tenth day, whilst on the twenty-first day an increase of 40% takes place.

The DNA decrease of incorporation is similar in both types of irradiation on the first day (60%). It continues to decrease after whole body irradiation on the fourth day but after local irradiation it rises on the fourth day to 40%, and finally, on the tenth day reaches normal values.

Study of the incorporation of ^{32}P into the nucleic acids corroborates the quantitative study above and proves that in the first 24 hours both local and whole body irradiation produce fairly similar effects. The differences occur only later, suggesting the action of intermediary elements.

In short, the study of the effects of various doses of X-rays as whole body irradiation, on the absolute quantity of the nucleic acids of the bone marrow shows a maximal decrease on the fourth day, which is of most significance for DNA, whose reconstitution is slower than that of RNA. Twenty-one days after irradiation, reconstitution leads to values higher than the normal ones for both RNA and DNA. Comparative study of the action of local irradiation of the hind-limbs and of whole body irradiation with the same dose of X-rays, reveals a similar effect in the first 24 hours, but the decrease is

TABLE

Incorporation of ^{32}P into the Bone Marrow RNA and DNA of the Hind-Limbs of Rat after Local and Whole Body X-irradiation (700r). Counts per minute per γP. The Percentage Change from Control Values are in Brackets ($\pm$)

Time (days)	RNA			DNA		
	Controls	Local irradiation	Whole body irradiation	Controls	Local irradiation	Whole body irradiation
1	156	109 (− 29)	112 (− 28)	150	60 (− 60)	64 (− 57)
4	220	187 (− 25)	116 (− 48)	265	164 (− 39)	95 (− 64)
10	156	143 (− 9)	111 (− 30)	158	185 (+ 7)	161 (+ 2)
21	124	157 (+ 28)	174 (+ 40)	—	—	—

more marked and the return to the normal values is slower after whole body irradiation. The study of the incorporation of ^{32}P supports the quantitative data revealing analogous variations.

Acknowledgment

To the National Institute of Hygiene.

REFERENCES

Lutwak-Mann, C., 1951. *Biochem. J.*, **49**, 300.
Mandel, P., and Rodesch, J., 1955. *J. Physiol.*, **47**, 234.
Mandel, P., Gros, C. M., Voegtlin, R., and Metais, P. 1951. *C.R. Acad. Sci., Paris, U.R.S.S.*, **233**, 1685.
Mandel, P., Weill, J. D., and Ledig, M., 1956. *Bull. Soc. Chim. biol., Paris*, **38**, 71.
Mantz, J. M. *Clinique Medicale A*, in press.
Rodesch, J., and Mandel, P., 1955. *C.R. Soc. Biol., Paris*, **144**, 415.
Smellie, R. M. S., Humphrey, G. F., Kay, E. R. M., and Davidson, J. N., 1955. *Biochem. J.*, **60**, 177.

DISCUSSION

Hevesy. May I ask Dr. Chambon if he found a difference between the rate of incorporation of ^{32}P into the DNA of the marrow of the protected and of the non-protected femur? My inquiry aims at the elucidation of the question of whether an indirect radiation effect, which has been observed by several workers a few hours after exposure, can be still observed under the conditions of his experiments; that is, after the lapse of a few days.

Chambon. Following irradiation of the femurs while the rest of the body was protected, we observed a 20% decrease in the ^{32}P incorporation into the nucleic acids 4 days after exposure. Thus an indirect effect could be observed even a few days after irradiation.

THE EFFECT OF X-RADIATION ON RAT THYMUS NUCLEIC ACIDS AT SHORT INTERVALS AFTER EXPOSURE *IN VIVO*

MARGERY G. ORD AND L. A. STOCKEN
Department of Biochemistry, University of Oxford,
Oxford, England

The inhibition of phosphorus incorporation into nucleic acids after irradiation *in vivo* has been known for a long time (Euler and Hevesy, 1942). A possible cause might be changes in the physical properties of DNA similar to those which occur *in vitro*. We therefore decided to look for any correlation in time between changes in physical and biochemical behaviour.

The physical properties of DNA are not well defined. We have used essentially the preparation III 2 (*d*) described by Chargaff (1955). Thymi from 20-30 rats (60–100 gm. body weight) were collected in semi-frozen 0·9% NaCl–0·05 M Na citrate, weighed, homogenised and centrifuged. After repeated washing, the DNA-nucleoprotein was deproteinised by 2·6 M NaCl and the DNA precipitated by alcohol (Crampton, Lipshitz and Chargaff, 1954). Two sodium dodecyl sulphate deproteinisations followed (Kay, Simmons and Dounce, 1952) after which the nucleic acid was dialysed overnight against CO_2-free water and dried from the frozen state. Control and irradiated DNA were always prepared simultaneously.

Table I shows the physical properties which were examined. The yields are expressed as dried DNA/100 gm. fresh thymus. We are indebted to Dr. Peacocke and Mr. Preston for measuring the molecular weights by the light scattering method. Titration data between *p*H 5·5 and 2·7 were obtained on DNA from control, 2 and 24 hours' irradiated rats, but no difference could be found. This is in agreement with the constancy of UV absorption but is in contrast with the results obtained by Cox, Overend, Peacocke and Wilson *in vitro* (1955). Heat sensitivity was determined by the increase in maximal UV absorption at 259 mμ measured at room temperature after 15 minutes at 100° (Laland, Lee, Overend and Peacocke, 1954). Neither the degree of sensitivity nor the rate at which the preparation became denatured was affected by exposure. In this connection, however, Butler (1956) has reported an increased sensitivity after 9,000r *in vitro*. Base ratios were obtained after hydrolysis with H.COOH and chromatographic separation (Wyatt, 1955). They confirm those reported by Wyatt but both here and with the N/P ratio it is unlikely that the methods used are adequate to establish small changes in the molecule. Limperos and Mosher (1950) reported a decreased N/P ratio in rat thymus DNA isolated immediately after exposure to 1,000r; a decrease in viscosity was not detected but was marked in DNA isolated 24 hours after radiation. Our 24 hour samples also have a decreased viscosity and a smaller molecular weight while UV absorption and titration data do not show any reduction in H bonding.

We are extremely grateful to Drs. V. Luzzati and M. Bertinotti for examining the control, 2- and 24-hour samples of DNA. They concluded that the X-ray diffraction patterns obtained under various conditions were essentially the same.

TABLE I

Physical Properties of Rat Thymus DNA at Varying Times after Exposure to 1,000r

	Control	½	1	2	24 hr.
Yield	1·09%	1·21%	0·88%	1·14%	0·53%
Molecular weight ...	9·1 and $7{\cdot}4 \times 10^6$	—	—	$8{\cdot}2 \times 10^6$	$3{\cdot}1 \times 10^6$
E(P) Max.	6,600	6,390	7,350	6,755	6,960
E(P) Min.	2,960	2,780	3,190	2,935	2,830
Heat sensitivity ...	29%	31·5%	31%	29%	28%
N/P ratio	3·68	3·58	3·8	3·72	—
Base ratios ... ... Ad : Thy : Gua : Cyt :	10 : 10·2 : 7·2 : 7·2			10 : 10·2 : 7·5 : 7·2	

The data do not indicate profound alterations in DNA until sometime later than 2 hours after exposure. It is of course possible that damaged DNA may be excluded in the method of isolation (Butler, 1956) which is known to affect the activity of transforming principle (Zamenhof, 1956, but *cf.* also Shooter and Butler, 1956). Cole and Ellis (1956) have reported an increase in the solubility of mouse spleen DNA in phosphate buffer 2 hours after irradiation, but further details of the experiments are necessary before this effect can be correlated with our results.

One further property relating to DNA was examined. Thymus nucleoprotein has been shown by Stern, Goldstein and Albaum (1951) to have ATP-ase activity. The enzyme is not an SH enzyme and is activated by 4×10^{-3} M Ca^{++}. The nucleoprotein was obtained from one or two rats as in the initial stages of DNA preparation. No increase in ATP-ase activity was detected at 2 hours but by 24 hours there was a 42% increase (*cf.* Dubois, Derion and Petersen, 1953).

When short-time incorporations are considered the time of administration of the isotope prior to irradiation and the duration of exposure may both alter the response obtained. In the experiments in Table II at 30 minutes and at 2 hours the animals were injected 5 minutes before irradiation, exposed for 14 minutes 38 seconds and killed at the stated time from the start of exposure. For the 24-hour experiment the animals were given 50 μc. ^{32}P/100 gm. body weight intramuscularly 2 hours before death. In the 3-minute experiments the rats were exposed at half the distance so that 1,000r was received in 3 minutes 40 seconds. They were then immediately injected with 100 μc. ^{32}P and killed 3 minutes later. Nuclei were prepared in a modified tissue culture medium as developed by Mr. M. P. Esnouf and the supernatant taken for cytoplasmic RNA. Acid-soluble phosphates

TABLE II

Effect of 1,000r X-rays on ^{32}P Incorporation into Rat Thymus Nucleic Acids

Time after exposure and number animals	DNA	Nuclear RNA	Cytoplasmic RNA
3 min. (4 × 1) ...	49	—	—
30 min. (4 × 2) ...	61	72	76
2 hr. (4 × 2) ...	46	57	65
24 hr. (4 × 4) ...	2	5	6

Data expressed as % of control values.

and lipids were removed, the nucleic acids obtained by the Hammarsten method and separated by Schmidt-Thannhauser alkaline hydrolysis (Deluca, Rossiter and Strickland, 1953). The activities are given relative to the specific activity of the plasma inorganic phosphate.

Inhibition of ^{32}P incorporation into rat thymus nucleic acids can be detected very soon after exposure and there is little progressive effect in the next 2 hours. By 24 hours, when profound histological changes are evident, incorporation has practically ceased. Nuclear and cytoplasmic RNAs are less affected than DNA although no causal relation need be inferred. It is not yet possible to say whether the inhibition is due to an effect on enzyme(s), to a failure to obtain energy or to an undefined disorganisation in the state of the system. We have been unable to find any change in thymus DNA but it must be remembered that it was isolated by a procedure which might effect fractionation or partial denaturation of the material originally present. However, if the isolated DNA is representative of the nuclear material, it seems that within the limits of sensitivity of the physical measurements it is initially unaffected. This suggests that the failure to incorporate precursors into nucleic acid is not due to changes in

DNA structure. Since ^{32}P incorporation into n-RNA and glycine incorporation into nucleohistone (Dr J. E. Richmond, unpublished observation) are also affected, inhibition of utilisation of energy or its reduced generation are possible explanations.

REFERENCES

Butler, J. A. V., 1956. *Radiation Res.*, **4**, 20.

Chargaff, E., 1955. *Nucleic Acids, New York Acad. Press*, **1**, 307.

Cole, L. S., and Ellis, M. E., 1956. *Fed. Proc.*, **15**, 1340.

Cox, R. A., Overend, W. G., Peacocke, A. R., and Wilson, S., 1955. *Nature, Lond.*, **176**, 919.

Crampton, C. F., Lipshitz, R., and Chargaff, E., 1954. *J. biol. Chem.*, **206**, 499.

DeLuca, H. A., Rossiter, R. J., and Strickland, K. P., 1953. *Biochem. J.*, **55**, 193.

Dubois, K. P., Derion, J., and Petersen, D. F., 1953. *U.S.A.F. Quart. Prog. Rep.*, vi, 1.

Euler, H., and Hevesy, G., 1942. *Biol. Medd., Kbh.*, **17**, 8.

Kay, E. R. M., Simmons, N. S., and Dounce, A. L., 1952. *J. Amer. chem. Soc.*, **74**, 1724.

Laland, S. G., Lee, W. A., Overend, W. G., and Peacocke, A. R., 1954. *Biochim. biophys. Acta*, **14**, 356.

Richmond, J. E., unpublished observations.

Shooter, K. V., and Butler, J. A. V., 1956. *Nature, Lond.*, **177**, 1033.

Stern, K. G., Goldstein, G., and Albaum, H. G., 1951. *J. biol. Chem.*, **188**, 273.

Wyatt, G. R., 1955. *Nucleic Acids, New York Acad. Press*, **1**, 243.

Zamenhof, S., 1956. *Progress in Biophys.*, **6**, 86.

Discussion

Hevesy. May I ask Dr. Stocken how he interprets Anderson's result, obtained in Hollaender's laboratory, who found a marked decrease in the viscosity of diluted thymus homogenates after an exposure to a 25r dose only?

Stocken. Anderson's results were obtained after irradiation *in vitro*. The reduction in viscosity with small doses was not an immediate effect and the concentration was very low in comparison with that in the nucleus. This reminds us of the effects found with enzymes *in vitro*, where the indirect action predominates.

Hevesy. What is the ratio between the nuclear and cytoplasmic volume of the very sensitive marrow cells? A question which presumably Dr. Lajtha can answer.

Lajtha. There is a great variation between different cell types, but an approximate nuclear/cytoplasm volume ratio of 1 : 5 seems to me reasonable.

Hevesy. A further question of mine relates to an earlier, very interesting remark made by Dr. Stocken. You interpreted the great radiosensitivity of thymus as being due to the extremely large ratio between the size of the nucleus and the cytoplasm of the cells of that organ. Had you in view the assumption that a large nucleus has a greater chance of being reached by damaging radicals than a small one?

Stocken. The idea we originally put forward was that the volume and metabolic activity of the cytoplasm would perhaps control the recuperative power of the cell, as well as protecting against injury.

Bacq. We have reason to believe that the liberation in the cytoplasm of DNA-ase, by X-irradiation, from the mitochondria and microsomes in which they are concentrated and inactive, may be one of the factors contributing to the damage of the nucleus after irradiation. Altmann *et al.* have recently shown a marked increase of DNA-ase activity both in plasma and urine 24 hours after irradiation of rats. Danielli and Ord have demonstrated beyond doubt the deleterious actions of irradiated cytoplasm on the non-irradiated nucleus of the amœba.

As shown by Mrs. Goutier in my laboratory, the optimum *p*H of spleen mitochondrial DNA-ase is significantly changed 30 minutes after a total-body dose of 800r in the rat. If rat's spleen homogenate is prepared according to Schneider-Hogeboom, in water (not in iso- or hypertonic sucrose), much more DNA-ase activity is found in the supernatant if the spleen is examined 30 minutes after a total-body dose of 800r (250 kV X-rays) than in a control spleen.

The release of enzymes from the mitochondria or microsomes in which they are concentrated is only a matter of minutes. For instance, the DNA-ase activity of the plasma is much higher (sometimes 10 times) in rats killed by a blow on the head, than in decapitated animals. Some of these findings are published in the *Acta Biochim. et Biophys.*

SECTION III

PHYSIOLOGICAL AND MORPHOLOGICAL IRRADIATION CHANGES

PROLONGATION OF THE CLOTTING TIME OF PERITONEAL FLUID AFTER X-IRRADIATION

DOUGLAS E. SMITH AND YEVETTE S. LEWIS
Argonne National Laboratory, P.O. Box 299
Lemont, Illinois, U.S.A.

The present observations stem from experiments on hamsters in which the free cells of the peritoneal cavity were collected by injecting small volumes of Tyrode's solution into the cavity and subsequently withdrawing the fluid. Within fluid so collected from otherwise untreated animals fibrin clots readily formed. In fluid similarly collected from X-irradiated animals the formation of such clots was impaired. The characterisation of this impairment is the subject of the present paper.

Materials and methods

Adult male Syrian hamsters (100 gm.) and Sprague-Dawley rats (200 gm.) were employed. To obtain fluid for study, 5 ml. of Tyrode's solution was injected intraperitoneally and withdrawn about five minutes later immediately after sacrificing the animal by ether anæsthesia. The fluid was pipetted from the opened abdominal cavity into glass tubes which were placed in a water bath at 37° C. Clotting time was taken as the time of appearance of the first visible strands of fibrin, observations being made at 30-second intervals. Clotting times were determined on fluid from non-irradiated rats and hamsters and from animals subjected to single total-body exposures 600r or 1200r of X-radiation (200 kV, 15 mA; 0·5 mm. Cu and 3·0 mm. Bakelite filters; 72·5 cm. target distance; 0·9 mm. Cu half-value layer; 15–20r/min.).

In about twenty instances, one or two drops of protamine sulphate (1% or 0·1% in normal saline) was added to fluid which had failed to clot within 15 to 30 minutes after collection from irradiated animals.

Results and discussion

The existence of a clotting defect in the peritoneal fluid of a large percentage of X-irradiated rats and hamsters is apparent from examination of Tables I and II. The impairment in clotting is detectable as early as one day and persists as long as 11 days (the limit of the study). It does not seem to be related to dose over the range of 600 to 1200r.

The addition of small amounts of protamine sulphate to fluid (from X-irradiated animals) which remained unclotted for 15 to 30 minutes after withdrawal resulted in the immediate appearance of strands of fibrin. This suggests that the clotting impairment may be attributed to increased amounts

TABLE I

Influence of Total-body X-irradiation upon Clotting Time of the Peritoneal Fluid of the Albino Rat

Time after irradiation (day)	Clotting time (minutes) *	
	600r	1200r
1	—	1
2	120, 2	120, 6
3	—	>180, 50, 39, 30, 20
4	10, 5, 2	45, >180
5	10, 4	—
6	>180, 6	—
7	11, 7	—
10	>180, >180	—
11	7, 5, 3	—

Control values—5, 4, 3, 2·5, 2·5, 1·5, 0·5

* Each value is from an individual animal.

TABLE II

Influence of Total-body X-irradiation upon Clotting Time of the Peritoneal Fluid of the Syrian Hamster

Time after irradiation (day)	Clotting time (minutes) *	
	600r	1200r
1	—	120
2	>180, >180, >180	1, 1, 3
3	>180, >180	>120, 7, 5, 4
4	30, 3	>180, >180
5	>180, 4	—
6	>180, 35	—
7	15, 6	—
10	10, 15	—
11	25, 5	—

Non-irradiated control values—2, 2, 1, 1, 0·5, 0·5

* Each value is from an individual animal.

of material having the properties of heparin. Attempts to measure the heparin content of the fluid, using paper electrophoresis and paper chromatography techniques, were unsuccessful. Apparently the amounts of heparin in the fluid from both control and irradiated animals were too small to be detected by the technique employed.

The time of appearance of the impairment in clotting time of the peritoneal fluid coincides with the time during which widespread disruption of mast cells is evident in the serous membranes of the peritoneal cavity, Smith and Lewis (1953). It is possible that the heparin, Jorpes (1946), contained in the mast cells is freed upon their disruption and this accounts for the coagulation defect observed.

Acknowledgment

This work was performed under the auspices of the U.S. Atomic Energy Commission.

REFERENCES

SMITH, D. E., and LEWIS, Y. S., 1953. *Proc. Soc. exp. Biol. N.Y.*, **82**, 208.
JORPES, J. E., 1946. *Heparin*, 2nd ed. (Oxford University Press).

DISCUSSION

Hutchinson. Have you studied the relationship between changes in clotting time and changes in the number of mast cells in individual animals?

Smith. We have not as yet attempted to do this. Our previous experience indicates that it would be difficult to find real differences in counts of total mast cells, because of the high variation in mast cell counts from animal to animal and even in various parts of the same tissue. The frequency of abnormal or degenerate mast cells, however, is low and constant (1–2%) in untreated animals and increases markedly after irradiation. It is possible that there may be a correlation in individual animals between the percentage of degenerate mast cells and the clotting change and we plan to explore this possibility.

THE PARTICIPATION OF THE PERIPHERAL NERVOUS SYSTEM IN THE REACTION TO IRRADIATION

VLASTIMIL SLOUKA
Biophysical Institute of the Czechoslovak Academy of Science,
Brno, Czechoslovakia

The part played by the peripheral nervous system in the post-irradiation reaction has been little studied up to the present time. Bade (1939) reported work on rats, in which following local skin irradiation he determined the latent period prior to the development of moist desquamation, and how this was changed by operations on the spinal cord. The latent period was doubled after division of the cord in the thoraco-lumbar region, and after division of the anterior and posterior spinal nerve roots the latent period was prolonged from 4 days to 6–8 days. These results are not entirely convincing in view of the fact that the rather rough operative technique involved produces a serious disturbance in the animal with paralysis and trophic changes in the hind legs and circulatory disturbances in these limbs.

Capua (1940) studied neural control of hæmatological changes in dogs, after cutting the femoral, obturator and ischial nerves, after periarterial femoral sympathectomy and after X-irradiation. No changes were demonstrated and the author doubted whether the nervous system was involved.

Meissel (1930) proved the active participation of the peripheral nervous system in the skin reaction by morphological studies in albino mice. After doses of 1,000r to 3,000r on the skin of the back he observed nerve growth and reconstruction and the development of atypical nervous structures even in the sites of serious necrosis.

We have studied the part played by the peripheral nervous system in the local skin reaction to irradiation.

Material, methods and results

Male Wistar rats 2–3 months old were used in most of our experiments. In the others we used rabbits.

Ether anæsthetics were administered to the rats and after laminectomy and identification of the spinal roots, unilateral anterior or posterior root divisions were performed in the lumbar region beginning with Th 11. Two corresponding gluteal areas 15 × 15 mm. were irradiated with single doses of 2,000r–3,000r; the experimental and control areas were thus in the same animal. A control group of animals was used with each tested group. Irradiation conditions: 60 kV, 4·5 mA, filtration inherent in the tube wall, F.S.D. 8·5 cm., dose rate 300r/minute (Victoreen).

The reaction is intense and its occurrence is regular. After early erythema about the 4th or 5th day a slight serous exudate occurs on the 6th or 7th day, and the reaction proceeds by the formation of crusts, cracks, hæmorrhages, ulcers and exudation, reaching a peak on the 8th–10th day. After this time the reaction begins to subside.

Healing begins from the edges of the defect about the 13th–15th day and by the 30th–35th day the defect has healed, the site being marked by a small shallow scar with permanent alopecia.

Four hundred rats were used, we have studied the time-course of the reaction, and its intensity, and we have carried out histological examinations.

Results

1. Unilateral anterior and posterior lumbar root section before irradiation.

The rats were irradiated on the 5th–8th day after denervation, and the operation is followed by a considerable prolongation of the latent period before the skin reaction occurs, together with a retardation of its course and delay in healing. In 24% of the irradiated animals no exudative dermatitis developed on the side of the experiment; this phenomenon has never been observed in controls.

2. Unilateral, anterior and posterior root section 10–20 minutes and 24 hours after irradiation has no effect on the radiation reaction.

3. Detailed analysis of the nervous control showed that after efferent denervation the onset of the principal reaction is delayed, the course of the reaction is prolonged and more marked; while after peripheral denervation these changes are still more marked. On the contrary, after afferent denervation there is a considerable weakening of the skin reaction.

4. We then studied the effect of cervical sympathectomy on the reaction occurring in the rabbit's ear after X-irradiation in 26 rabbits. The animals were silver grey and black haired males, 2,300–2,800 gm. in weight. The superior cervical ganglia were extirpated under pentothal anæsthesia and areas 25 × 25 mm. were irradiated on the 4th post-operative day. Conditions of irradiation were 80 kV, 4·5 mA, filtration inherent in the tube wall, F.S.D. 8·0 cm., dose rate 220r/minute (Victoreen). Radiation doses of 2,500r–3,000r were given in a single exposure.

Epilation occurred in controls between the 14th and 28th day, the maximal reaction being present about the 10th–12th week; exudation with crust formation occurred frequently. On the operated side of the animal the reaction is more marked, the maximum occurs earlier and is prolonged in duration.

5. The use of circular novocain blocks of the irradiated area considerably reduces the local reaction to ultra-violet irradiation. This is considered by some authors (Braun, 1954) to be explained by a simple absorption of U.V. by novocain. However, we have achieved similar results after X-irradiation using paravertebral blocks with 2% novocain. The novocain block experiments show the feasibility of their application to X-ray therapy with massive doses of irradiation.

6. Local application of 0·05% epinephrine to the isolated nervous auricularis magnus of the rabbit reduces the reaction on the operated side. The reaction is also considerably weakened in rats following the application of 0·25 mg./100 g. s.c. epinephrine and after ergotamine.

Two factors are important as far as the drugs of the vegetative system are concerned, namely, the vasomotor effect and the changes of local oxidation in tissues. It follows, both from our own and other authors' work, that the peripheral nervous system plays, in the first instance, an important part in regulating vessel dynamics in these circumstances.

In summary we conclude that various metabolic changes occur directly after irradiation in the irradiated area; in particular, there are changes of permeability and the formation of atypical metabolic products resulting in an early reaction. In addition, histamine and H-substances are formed. These changes mainly occur in the irradiated area, but subsequently spread to the whole organism. The nervous system takes part in this reaction in such a way that the receptors signalise peripheral irritation and activate central reflex mechanisms. From the central nervous system impulses regulating the degree of local reaction are emitted through the efferent tract. These impulses are mainly of an inhibitory character.

Together with the other tissue elements the components of the nervous periphery are affected, especially the nerve endings in the skin and in the walls of the arterioles and the capillaries. This is the second form of the participation of the peripheral nervous system in the reaction to irradiation.

Conclusions

The participation of the peripheral nervous system in the local post-irradiation reaction has been demonstrated in the experiments involving surgical denervation, novocain blocks and with drugs of the vegetative nervous system.

REFERENCES

Bade, H., 1939. *Strahlentherapie*, **66**, 50.

Braun, A. A., and Prizivojt, I. F., 1954. *Bjull. eksp. Biol. i Med.*, **39**, 73.

Capua, A., 1940. *Scritti Ital. radiobiol.*, **7**, 346.

Meissel, M. N., 1930. *Virchows Arch.*, **276**, 77.

Discussion

Bacq. It seems to me that what you have investigated is not necessarily a specific reaction to radiation but a general scheme of reactions to noxious agents. If you had taken, for example, mustard gas instead of X-radiation would you not have had the same results?

Smith. Yes; is it not true that a great many physiological changes following irradiation are of a non-specific nature?

Slouka. I think that the radiophysiological response is of a non-specific nature. In partial manifestations, it is possible to provoke this reaction by different agents, including radiomimetics. I should like to emphasise that a local reaction is controlled by complicated nervous and neurohumoral relations.

Marcovich. Your results could be explained by a modification in the oxygen tension in the tissues following the nervous alteration.

Slouka. I quite agree with you. In my opinion, the peripheral nervous system plays a very important part especially in regulation of vessel dynamics and in changes of local oxidative processes in tissues.

THE EFFECTS OF LETHALLY DAMAGED CELLS UPON SURVIVORS IN X-IRRADIATED EXPERIMENTAL TUMOURS

L. Révész
Institute for Cell Research and Genetics,
Karolinska Institutet, Stockholm, Sweden

Tumour cell populations that have been irradiated by X-ray doses falling below the dose necessary for complete inactivation, can be schematically considered as mixtures of two kinds of cells, differing in prospective reproductive integrity. One fraction is composed of cells that have been more or less lethally damaged and have lost their capacity for continued reproduction, either immediately or after a number of cell divisions. Another part of the cell population is composed of cells that have only been damaged in a reversible way and are still able to undergo serial proliferation without restraint. The relative proportions of these two fractions must be related in some way to the magnitude of the X-ray dose as well as to various environmental conditions during irradiation. Very little is known about the influence that the lethally damaged part of the population may exert on the continued proliferation of the surviving fraction, either directly or indirectly via the host organism. It would be conceivable that the dying cells stimulate the growth of the survivors in a variety of ways, but it would be also possible that they inhibit them, possibly by way of the intense inflammatory reaction they provoke on the side of the host.

Experimental studies were carried out to determine the effect of lethally X-ray-damaged cells on admixed living cells of the same neoplasm in various experimental tumours.

The tumours used can be divided into two groups for the present purposes. The first consists of neoplasms that arose spontaneously or have been induced by carcinogens in mice of inbred strains, or F_1 hybrids produced by crossing two inbred strains. These tumours were used in their first transfer generation, being always propagated in their genotype of origin. A separate category is represented by the *Ehrlich* ascites tumour that arose in a mouse of unknown genetic composition and is being carried in heterozygous animals.

Suspensions were made from the various tumours according to the Kaltenbach method (1954) and the number of cells counted in a hæmocytometer after dilution with an eosin solution as described by Schrek (1936). One aliquot of each suspension was irradiated *in vitro* with 12,000r in a nitrogen atmosphere, a dose much higher than is needed to inactivate the tumour completely. This irradiated population was mixed in various proportions with another, untreated aliquot of the same suspension. Three groups of mice were inoculated. Group A received lethally damaged cells alone. Group B was inoculated with different dilutions of the untreated cells alone. Group C received mixtures of the two cell types in various

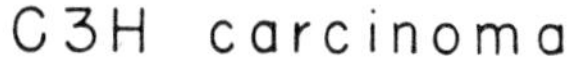

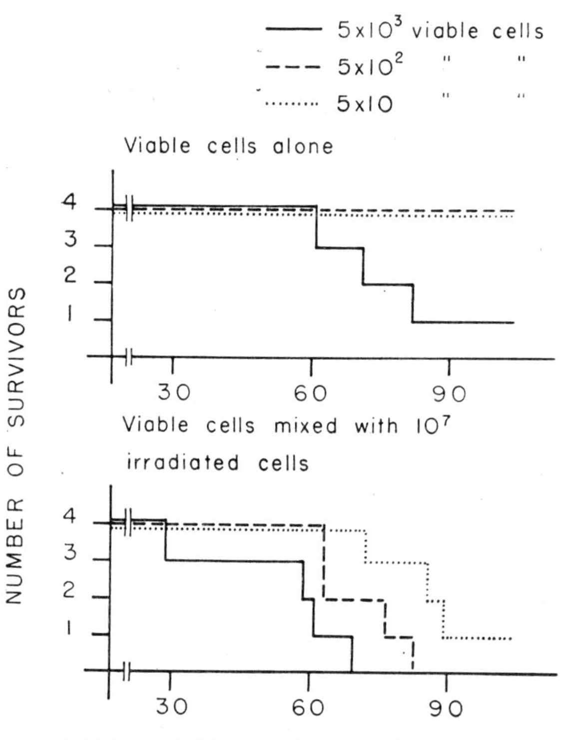

Fig. 1. Survival of mice after inoculation with viable cells alone and with viable cells mixed with irradiated cells.

proportions; the number of lethally damaged cells was kept constant and the untreated cells were admixed in the same proportions as used for group B. The latency period elapsing before palpable tumours appeared and the survival time of the tumour-bearing animals was determined. Surviving animals were followed for 3 to 6 months. In the case of the *Ehrlich* ascites tumour the total number of free tumour cells in the peritoneal cavity was followed by quantitative growth curves (Révész and Klein, 1954).

The results can be summarised as follows:

(1) With the tumours that have originated in the inbred C3H strain or in F_1 hybrids of the inbred lines A and ASW, the presence of lethally damaged cells had an intense stimulating effect on the growth of admixed cells in all cases when inoculated into the strain of origin (Figs. 1 and 2). This stimulation was most apparent when the number of viable cells was comparatively small (10 to 10^3 viable cells mixed with about 10^7 lethally damaged cells). In such cases there was a considerable difference between the median survival times of the two groups. With a higher proportion of living cells the stimulation was less intense, although daily caliper measurements of tumour size also showed enhancement.

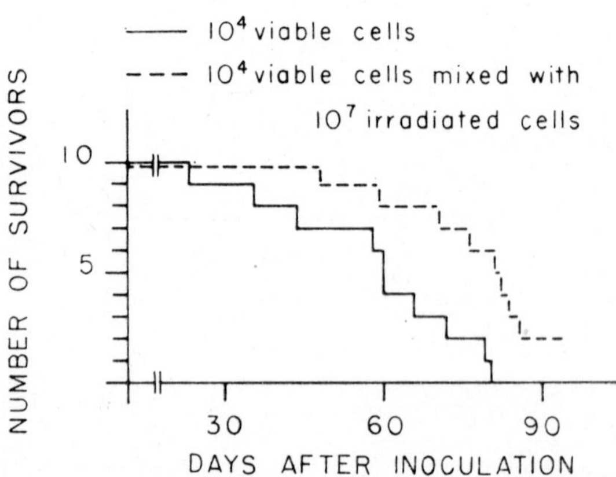

FIG. 2. Survival of mice after inoculation with different doses of viable cells alone and with the same doses cf viable cells mixed with irradiated cells.

(2) Different results were obtained with the *Ehrlich* ascites tumour (Révész, 1955). Here stimulation was apparent only if the proportion of living cells was rather high (e.g. 20×10^6 killed cells admixed to 2×10^6 living ones). Decreasing the proportion of living cells to 100 resulted in immunisation of the genetically foreign animals by the dead cells which thus prevented the growth of the small viable inoculum, capable of giving rise to 100% tumours if inoculated alone (Fig. 3).

These results seem to have two main implications. The first is a purely technical point, of interest for experimental investigations on tumour therapy. It shows how fallacious results may be that were obtained by using tumours growing in genetically foreign hosts. It seems that a treatment that damages only part of the cell population may bring about tumour regression after receiving help from the histoincompatibility-reaction of the mouse. Of greater interest is the fact that damaged cells always stimulate the growth of a very small surviving fraction in a tumour where host and neoplastic cells are genetically identical. This does not only apply to the case of X-ray-killed cells; similar results have been obtained by Eva Klein when mixing a small fraction of genetically compatible cells with a large fraction of genetically incompatible cells. In this case the incompatible cells were completely destroyed by the host reaction but this not only did not prevent the growth of a small viable inoculum but even stimulated it. Thus it would seem that tumour therapy has to aim at the complete eradication of all tumour cells and not only a partial damage even of the majority of the population.

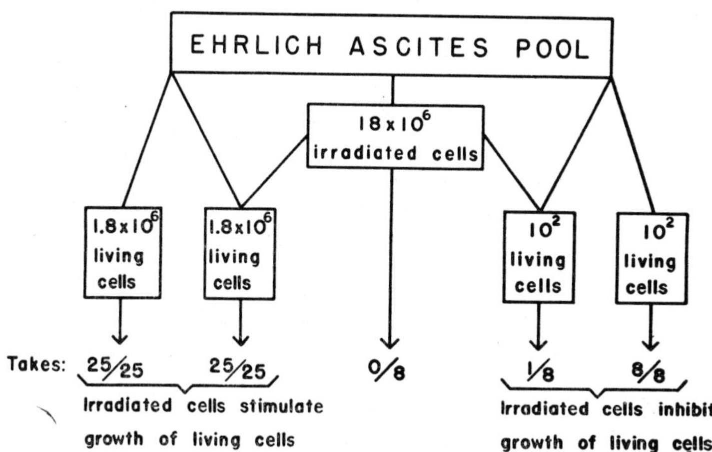

FIG. 3. Inhibiting or enhancing effect of irradiated *Ehrlich* tumour cells on viable cells depending on their quantitative proportions.

The question arises as to the nature of the stimulating effect exerted by dead cells. In our continued experiments we are trying to distinguish between the following three possibilities: (1) specific stimulation by homologous cell products (Weiss, 1955), (2) a stimulation by providing essential nutrients, and (3) stimulation through provoking an inflammatory reaction from the host. These experiments are in progress and have not yet yielded any definite results.

One last point has to be made in this connection. It is remarkable how little the intense and histologically very impressive host reaction can do about the destruction of genetically compatible tumour cells. It would seem that this reaction is simply an answer to the products of radiation-damaged or killed cells. This is also supported by the previous finding (Klein and Forssberg, 1954) that inflammatory reaction in different ascites tumours is proportional not to the X-ray dose but to the amount of cell damage inflicted by a given dose in different tumours.

REFERENCES

Kaltenbach, J. P., 1954. *Exp. Cell. Res.*, **7**, 568.
Klein, G., and Forssberg, A., 1954. *Ibid.*, **6**, 211.
Révész, L., 1955. *J. nat. Cancer Inst.*, **15**, 1691.
Révész, L., and Klein, G., 1954. *Ibid.*, **15**, 253.
Schrek, R., 1936. *Amer. J. Cancer*, **28**, 389.
Weiss, P., 1955. *Biological Specificity and Growth*, p. 195 (Princeton University Press).

Discussion

Ebert. Why did you irradiate your cells in nitrogen and not in air?

Révész. Our purpose was to have easily reproducible and controlled experimental conditions.

Hutchinson. Is the stimulating effect specific for the X-ray-killed cells or could cells killed in different ways also exert stimulation?

Révész. In this respect some experiments are going on which have not yet yielded any definite results. There are some indications, however, that the stimulating effect described is not X-ray specific, since similar stimulation could be obtained with heat-killed cells.

Howard-Flanders. I wonder whether the time at which you admix dead cells adds to the different effects on growth?

Révész. No special study of time-factor was made. In all cases, the untreated cells were stored in an ice-box for a period of about two hours during which irradiation, cell counts, etc. were performed. Afterwards, the mixture of untreated and X-ray-killed cells was prepared and inoculated into mice simultaneously with the inoculation of another group of animals that received untreated cells alone.

Marcovich. Could a decreased host defence against transplanted tumour effected by the damaged cells be responsible for obtaining a response which simulates an enhancing effect?

Révész. A host defence which inhibits the growth of the transplant occurs in the case of the *Ehrlich* tumour that arose in a genetically known mouse and has since for about 60 years been propagated in heterozygous animals; a successful immunisation of the host against this genetically foreign tumour tissue was observed and can be responsible for the inhibiting effect. No similar immunological phenomena can be expected against isologous transplants, however. In fact, using inbred strains of mice and the first transplant generation of a tumour of the same inbred strain we have never observed inhibition and dead cells.

Lajtha. There is a similar stimulating mechanism known in tissue culture. Some tissues which are difficult to grow in fresh medium grow well in 'conditioned' medium. Such conditioned media consist of a mixture of fresh medium and old medium (in which cells were grown previously). I wonder whether this conditioning may be correlated with your observations on stimulation of tumour growth?

Révész. One of the possibilities we consider in our continued studies about the nature of the stimulation is the 'feeder effect' described by Puck who used X-irradiated cells to supply conditioning factors in tissue culture of *HeLa* cells. These observations would be in line with the concept of growth stimulation by homologous cell products studied by P. Weiss and cited earlier.

Scott. I have performed experiments with the *Ehrlich* ascites tumour in which the effect of adding heavily irradiated cells to undamaged cells was examined. However, I used washed tumour cells suspended in Ringer-phosphate. I observed the same stimulating effect observed by Dr. Révész and the absence of ascitic fluid in this work suggests that the stimulating effect is due to the presence of damaged cells.

REFERENCE

Puck, T. T., and Marcus P. I., 1955. *Proc. nat. Acad. Sci., Wash.*, **41**, 432.

INDUCTION OF CHROMOSOME BREAKS IN THE *YOSHIDA* SARCOMA ASCITES CELLS OF THE RAT BY X-RAYS AND FAST ELECTRONS AT DIFFERENT OXYGEN PARTIAL PRESSURES

W. Dittrich

Research Department, University Frauenklinik, Hamburg-Eppendorf, Germany

Experiments dealing with the oxygen effects may contribute to a better understanding of the basic mechanism by which ionising radiations produce chromosome aberrations. The effects of oxygen are usually explained by an increased production of free radicals around the irradiated radio-sensitive structures. In general they are largely dependent upon the concentration of oxygen during the irradiation. Some authors discuss more indirect effects, e.g. the possibility that oxygen produces a change in the radio-sensitivity of chromosomes.

There is a close relationship between the frequency of radio-induced chromosome aberrations on the one hand, and the oxygen concentration on the other. In different radiobiological experiments, the question arises whether the frequency of primary induced breaks increases, or whether the number of restitutions of these breaks decreases, with the concentration of oxygen in the cell.

In *Tradescantia*, a variation of the oxygen concentration following irradiation produces no effect, not even during the restitution period.

Usually oxygen effects decrease with the linear energy transfer and are not detectable at all in the case of alpha-rays. The restitutional processes in the induction of chromosome aberrations are important, no matter whether X-rays or alpha-rays are used for the induction; the restitution hypothesis is hardly compatible with this fact.

The most convenient material for the *in vivo* investigation of the effects of oxygen on different types of radio-induced chromosome aberrations in animal cells, are the transplantable ascites tumours; such as the *Ehrlich* ascites carcinoma of the mouse or the *Yoshida* sarcoma of the rat. The oxygen concentration in tumour ascites can easily be varied during the irradiation by giving the animals oxygen at different pressures. Twenty-four hours after total-body irradiation in the cells of the ascites, two types of pathological mitoses can be differentiated: chromosomal bridges and chromosomal fragments, the chromosomal bridges being formed by recombination of break ends, and chromosomal fragments by lack of union of acentric fragments. Any accumulation of these two types of pathological mitoses, due to increased oxygen concentration during irradiation, could easily be explained as an increase in the number of primary breaks; but this would hardly be in accordance with the restitutional hypothesis.

The use of radiations with different specific ionisations makes it possible to obtain information regarding the oxygen effects and their relation to varied linear energy transfer.

Experimental

Four or five days after intraperitoneal inoculation of 0·15 ml. of *Yoshida* sarcoma ascites into rats of an inbred strain, the animals were totally irradiated with X-rays (200 kV, 0·5 mm. Cu, 20 mA, 200r/min.) or with the same dose of 186±20 rads of fast betatron electrons (15 MeV, 186r/min.). The dose was measured with a Siemens dosimeter and a thimble chamber.

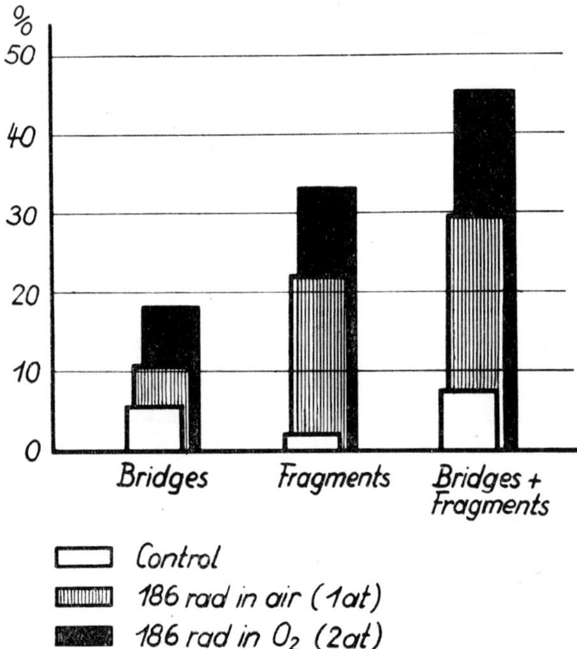

FIG. 1. Relative frequency of chromosomal aberrations in ascites cells of *Yoshida* sarcoma in mitosis (late anaphase and early telophase) 24 hours after X-irradiation (186 rad) in air and oxygen (2 at.).

Figure 1 gives the frequency of pathological mitoses 24 hours after X-irradiation. The percentages refer respectively to 1,330 bipolar mitotic cells in late anaphase and early telophase, the total number of cells was about 4,000.

As in *Ehrlich* carcinoma, the frequency of tumour cells with at least one chromosome fragment depends on the oxygen pressure during irradiation: fragment cells are much more frequently induced under two oxygen atmospheres than in air at normal atmospheric pressure. In *Yoshida* sarcoma cells respiring oxygen during the irradiation, both types of chromosomal aberrations, i.e. fragments and bridges, were induced more frequently and the differences are significant.

Under irradiation with fast electrons, as with X-irradiation, the same close relationship to oxygen pressure in regard to the induction of chromosome bridges and fragments is observed. Figure 2 shows the results of betatron experiments. Percentages refer respectively to 750-775 bipolar mitotic cells in late anaphase and early telophase. Figure 2 represents a total of more than 2,200 cells.

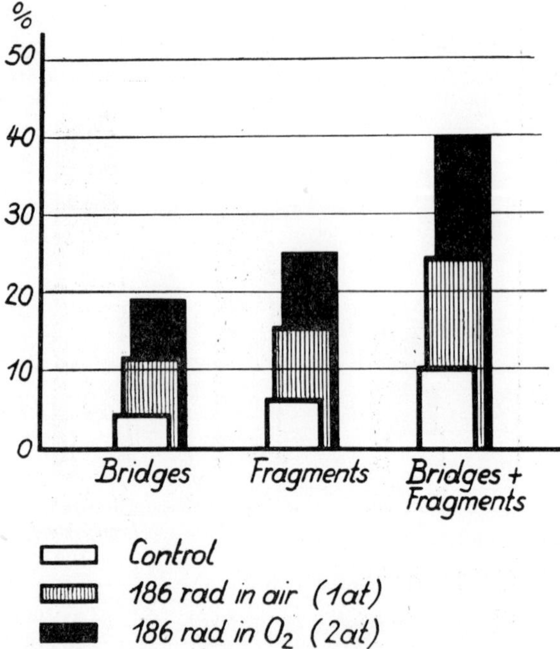

FIG. 2. Relative frequency of chromosomal aberrations in ascites cells of *Yoshida* sarcoma in mitosis (late anaphase and early telophase) 24 hours after irradiation with fast electrons (15 MeV, 186 rad) in air and oxygen (2 at.).

Mitotic cells with at least one chromosome bridge or fragment were observed more frequently after irradiation in oxygen at two atmospheres, than after irradiation in air.

Comparative study of Figures 1 and 2 shows only small differences in the respective frequencies of cells with chromosome bridges. Different results are obtained with regard to fragment cells, because the latter are observed less frequently after betatron irradiation than after X-irradiation with the same rad-dose. This is in keeping with the fact that fast electrons are biologically less effective than X-rays in the majority of radiobiological reactions. The percentage of additional fragment cells which appears after irradiation in oxygen is almost the same in the case of electrons as it is for X-rays. This indicates that fast electrons have a more distinct oxygen effect than X-rays, and is in accord with the experience that oxygen effects increase with decreasing linear energy transfer.

The above experiments are important for a better understanding of the basic mechanism relating to the mode of action of ionising radiations. They suggest that an increase of oxygen concentration in the cell accumulates free radicals, which further leads to primary chromosome breaks by chemical 'light' hits and to structural chromosome changes. As regards the other possibility, that the radiosensitivity of chromosomes is changed by oxygen, these results are hardly compatible with the restitution hypothesis, but tend rather to support the break hypothesis of oxygen effect.

Acknowledgment

The experimental work has been supported by the 'Deutsche Forschungs-gemeinschaft'.

REFERENCES

DITTRICH, W., 1955. *Strahlentherapie*, Bd. **35**, 177.

DITTRICH, W., and HEINRICH, H., 1956. *Z. Krebsfschg.*, **61**, 38.

DITTRICH, W., and STUHLMANN, H., 1954. *Naturwissenschaften*, **41**, 122.

GILES, N. H., 1955. *J. cell. comp. Physiol.*, Vol. 45, suppl. 2.

GILES, N. H., Jr., and RILEY, H. P., 1950. *Proc. nat. Acad. Sci. Wash.*, **36**, 337.

GRAY, L. H., and CONGER, A. D. EBERT, M. HORNSEY, S. and SCOTT, O. C. A., 1953. *Brit. J. Radiol.*, **26**, 312, 638.

LEA, D. E., 1955. *Actions of Radiations on Living Cells* (Cambridge, Cambridge University Press).

READ, J., 1952. *Brit. J. Radiol.*, **25**, 89.

TIMOFÉEFF-RESSOVSKY, N. W., and ZIMMER, K. G., 1947. *Zimmer Biophysik, Leipzig.*

DISCUSSION

Rajewsky. How long before the irradiation did you change the O_2-tension and what is your opinion of the O_2 effect?

Dittrich. We switched on the oxygen 5–10 minutes before irradiation. We are inclined to believe the effect is an indirect one, possibly mediated through HO_2. The possibility that oxygen may increase the radiosensitivity of the chromosomes and consequently the number of direct hits must also be considered, however.

OBSERVATIONS ON BACTERIAL GROWTH AND MORPHOLOGY SHORTLY AFTER IRRADIATION AND SOME REMARKS ON THE OXYGEN EFFECT

TIKVAH ALPER

Medical Research Council
Experimental Radiopathology Research Unit,
Hammersmith Hospital, London, W.12, England

For assessing radiation effects on micro-organisms, a fruitful and widely used technique is to count numbers of visible colonies appearing after, say 24 hours' incubation on the surface of nutrient agar. The number of colonies is equated to the number of 'survivors' of any particular treatment, so that a survivor is defined as a micro-organism which has ultimately given rise to millions of progeny. Very few comparable experiments have been done with higher cells, as it is only in exceptional cases that techniques can be used which give definite information on the numbers of cells which have succeeded in multiplying for many generations. Two types of experiment will be described in which radiation effects could be observed on bacteria which would be counted as survivors, in terms of colony counts. These effects may be analogous to 'non-lethal' effects in higher cells.

Growth curves in liquid culture

After subjecting washed suspensions of *E. coli* to X-radiation, small samples were inoculated into tubes of broth which were incubated at 37° C. At inoculation, and at short intervals thereafter, samples were plated on nutrient agar, so that multiplication could be followed. Figure 1 illustrates growth curves after various doses of radiation. There are two main points of interest:

(*a*) The length of the lag phase was sensitive to quite small doses of radiation: thus, with only 300 rads, the lag phase was increased from a normal 40 minutes to 63 minutes. In terms of colony counts, this dose gave over 90% survivors. The prolongation of the lag phase is presented in Figure 2 as a function of dose.

(*b*) When multiplication of the irradiated bacteria began the initial average generation time was less than in the controls. The period of decreased generation time was dose-dependent, being longer for larger doses. The extent to which the generation time was reduced, however, was greater for smaller doses. The decrease in generation time is easily seen in Figure 3, from an experiment in which *E. coli* B were irradiated at the end of the normal lag phase, when, as shown by Stapleton (1955), the sensitivity of the cells to ionising radiation is considerably less than in the resting stage.

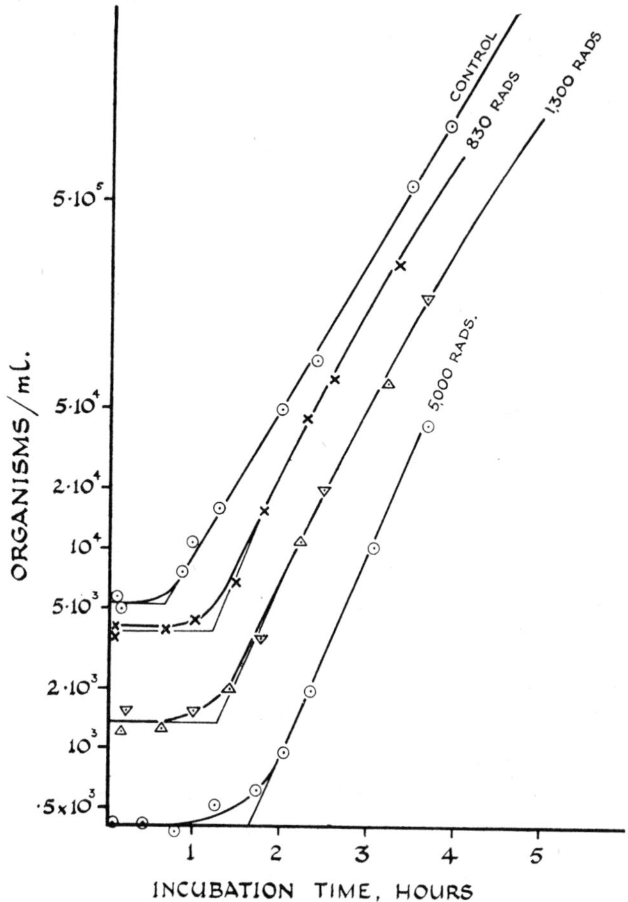

FIG. 1. Growth curves, *E. coli* B: survivors of 830, 1,300 and 5,000 rads, delivered in stationary stage.

After the additional lag period induced by 3,000 rads, the generation time of the 80% survivors was about half that of the normal bacteria; this continued for about four generations, after which the slopes of the growth curve became equal. The prolonged lag phase, followed by a much reduced generation time, is reminiscent of the growth curves obtained after cold shock, as used for synchronisation of bacterial division (Lark and Maaløe, 1954).

Quantitative morphological observations

A technique has been evolved which has made it possible to examine micro-colonies within any desired period after seeding on to nutrient agar.

Small measured droplets (0·001 ml.) of irradiated and control *E. coli* suspensions are put down onto pieces of cellophane which have been laid on the surface of well dried nutrient agar plates. After the moisture has been absorbed, the plates are incubated, and fixed at any desired time. The pieces of cellophane may then be lifted from the agar and stuck down on microscope slides with small drops of glycerin-albumen. Further fixing

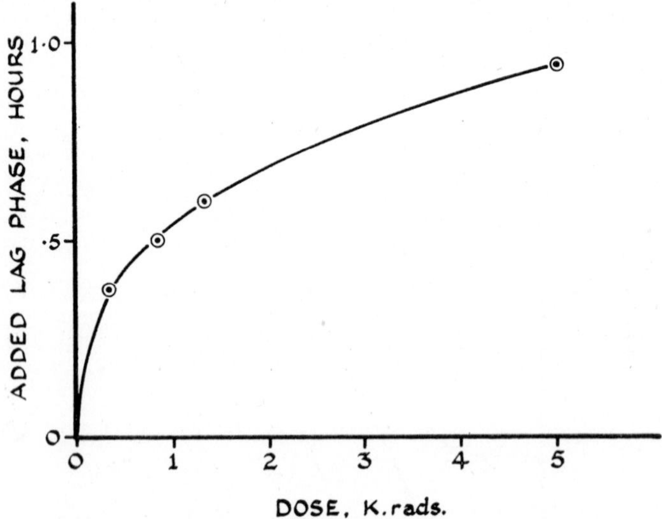

FIG. 2. Prolongation of lag phase as a function of dose.

of the organisms to the cellophane is attained by gentle heating (at about 80° C.) for a few minutes. After this treatment the bacteria can be stained as desired without disturbing the micro-colonies, an important requisite of this method (see Fig. 3*a*). Variables such as numbers of organisms per colony, production of filamentous forms, and so on, can then be investigated as functions of radiation dose and of incubation time (Figs. 3*b* and *c*). Preliminary work has shown that five criteria of radiation damage may be established. In order of sensitivity to radiation dose, these are:

(*a*) Increase in length of lag period (perhaps analogous to inhibition of mitosis in higher cells).

(*b*) Fraction of bacteria growing into normal colonies, containing no filamentous forms (perhaps analogous to fraction of higher cells with no nuclear abnormalities).

(*c*) Total fraction growing into viable colonies, some of which have

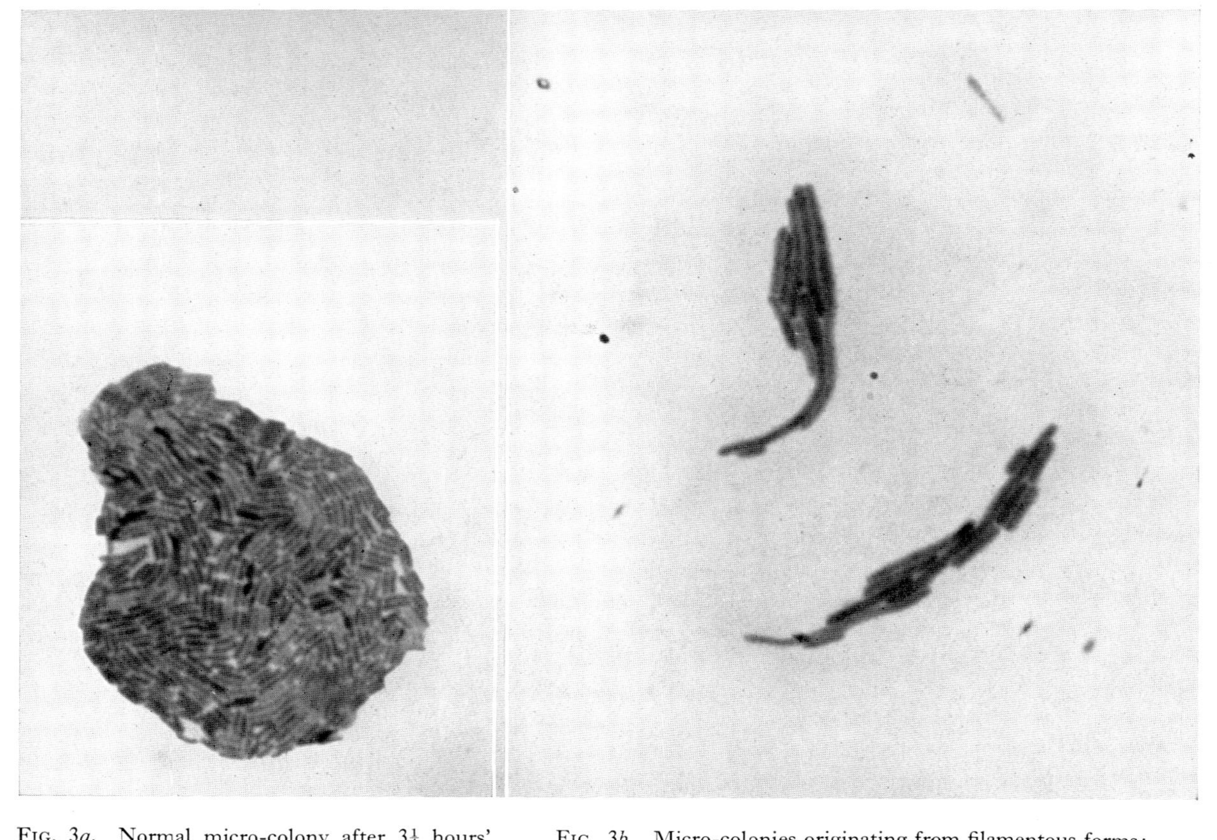

FIG. 3*a*. Normal micro-colony after $3\frac{1}{4}$ hours' incubation.

FIG. 3*b*. Micro-colonies originating from filamentous forms: $2\frac{3}{4}$ hours' incubation—2,500 rads.

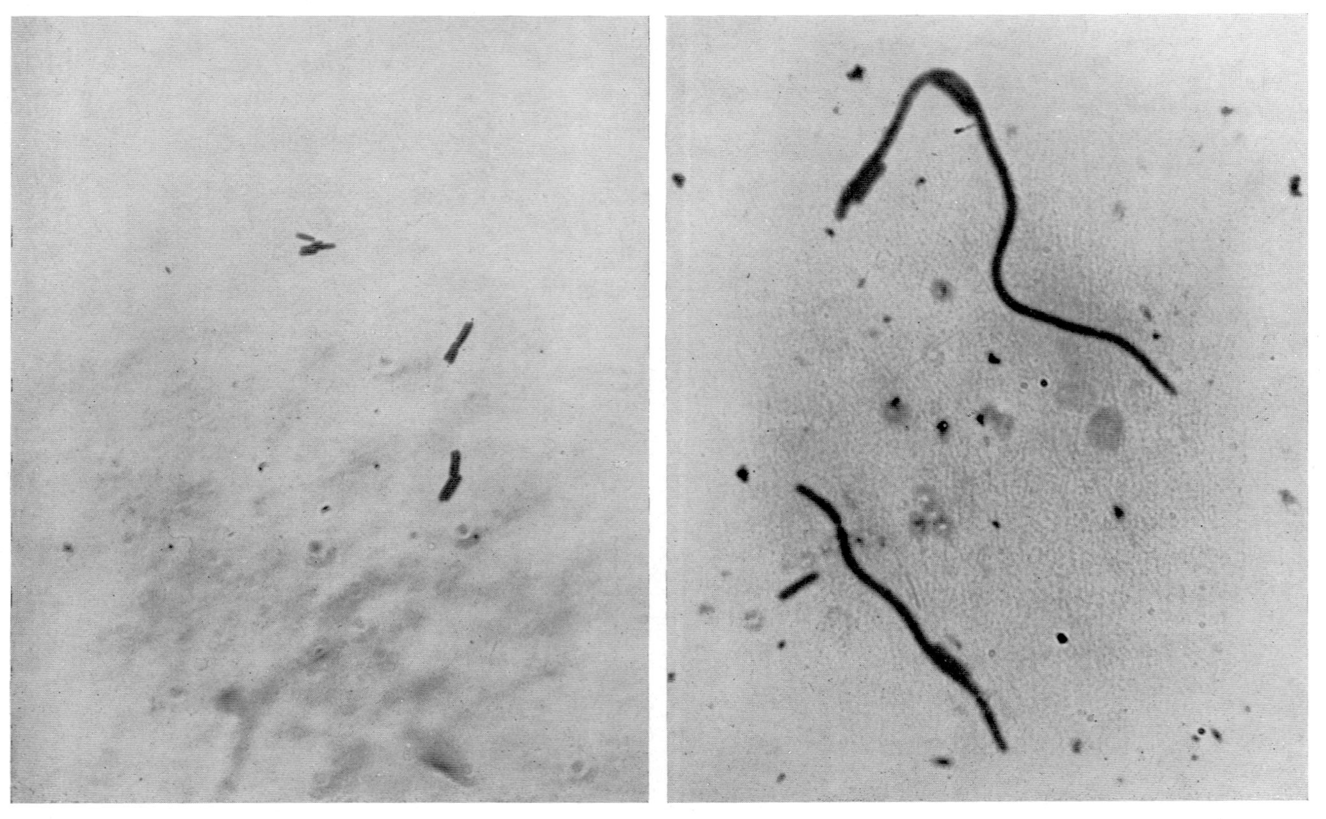

Fig. 3*c*. Filamentous forms which have failed to initiate colony formation in $2\frac{3}{4}$ hours' incubation: 2,500 rads.

Fig. 3*d*. Unirradiated bacteria—field showing several four-celled colonies after $1\frac{3}{4}$ hours' incubation.

started from filamentous forms. (Corresponds to ordinary 'viable' or 'plate' count.)

(*d*) Fraction growing into filamentous forms which fail to divide.

(*e*) Fraction showing no cytoplasmic increase.

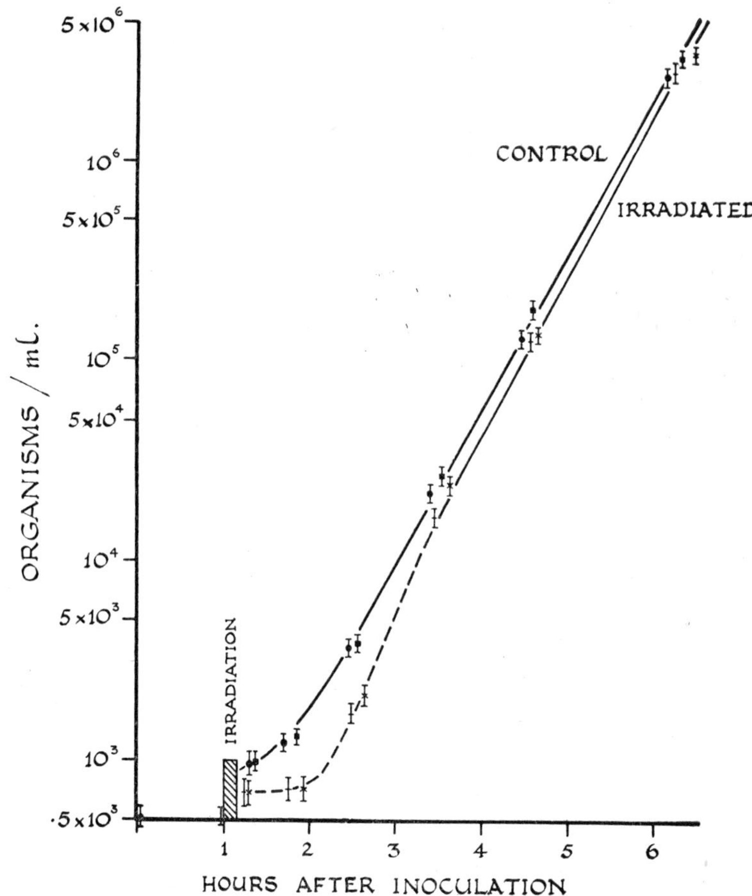

FIG. 4. Growth curves, *E. coli* B: survivors of 3,000 rads delivered at end of lag phase.

The method of assessing the increase in the lag period was, of course, different from that for growth in liquid culture. It was found that the non-irradiated cells were fairly synchronous for the first few divisions—at $1\frac{3}{4}$ hours after commencement of incubation, for example, over 60% of the micro-colonies contained four cells each (Fig. 3*d*). The alteration in lag phase with dose was therefore measured in terms of the ratio of irradiated to non-irradiated bacteria which had formed micro-colonies of n cells or more at that time.

The counting of a series of slides gave data which are presented graphically in Figure 5. If the slopes of the curves for 'fraction of bacteria forming

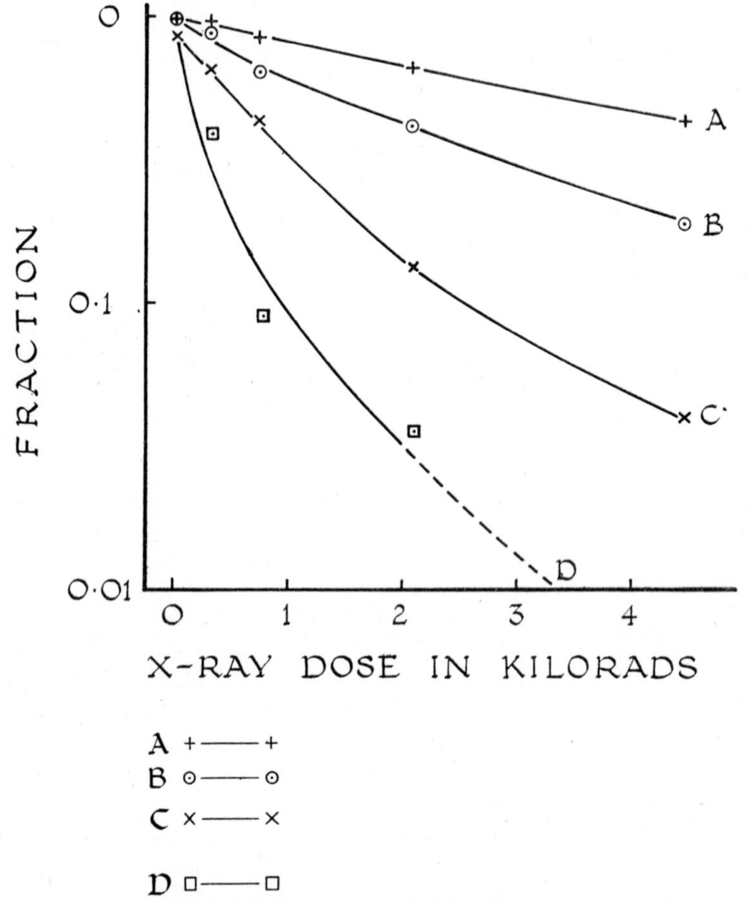

FIG. 5. Radiation damage to *E. coli* B: assessed by various criteria.

A + - - - +: Total fraction showing cytoplasmic increase.
B ⊙ - - - ⊙: Fraction forming colonies (ordinary 'viable count').
C × - - - ×: Fraction forming colonies consisting only of normal-sized bacteria (i.e. no filamentous forms).
D ⊡ - - - ⊡: Proportion of colonies with 8 cells or more each, after $2\frac{1}{4}$ hours' incubation, as fraction of controls.

colonies' and 'fraction of bacteria forming colonies with no filamentous forms' are compared, it is seen that the ratio is about 1 : 2, i.e. about half the colonies which would finally appear as visible colonies originate from bacteria which have grown into filaments before dividing. Thus, the preliminary studies here described deal in part with two types of 'restorable damage' to bacterial cells, namely, prolongation of the lag phase of survivors, and formation of abnormal forms which ultimately divide and give rise to colonies of normal macroscopic appearance.

That some types of damage to vital targets in living cells are restorable is an important aspect of a hypothesis which P. Howard-Flanders and I (1956) have recently formulated to account for the oxygen effect: we have postulated that, in living cells, oxygen acts not through the medium of free radicals

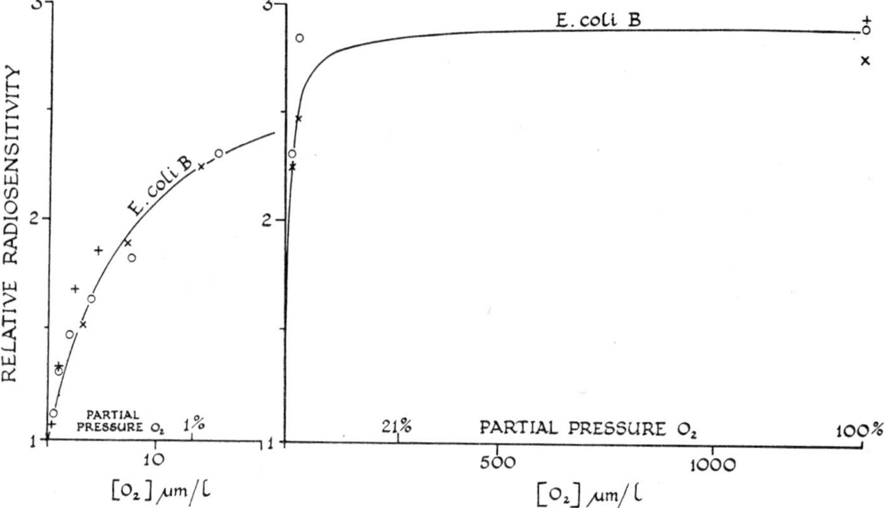

FIG. 6. Variation with oxygen tension in radiosensitivity of *E. coli* B.

derived from water decomposition, but by reacting with a directly ionised target molecule, to give a peroxide which leads to non-restorable damage. The difference in radiosensitivity of cells irradiated in the presence or in the absence of oxygen is, by this hypothesis, accounted for by the probability that many molecules which suffer ionisation can be subsequently restored to normal function, if oxygen is absent at the time of irradiation. This hypothesis was suggested in part by investigations on the variation with oxygen concentration of the radiosensitivity of *E. coli* B (Fig. 6). The initial slope of the curve is much steeper than has been indicated by investigations on other materials: the radiosensitivity of the bacteria is doubled at about 0·5% partial pressure of oxygen, and 90% of the 'full oxygen radiosensitivity' is reached at a partial pressure of only 3% oxygen. Thus the bacterial cell suffers a large increase in radiation damage at oxygen tensions which are low, compared with those which are effective in some radiation chemical experiments. It would seem improbable that HO_2 radicals can be the operative mechanism for the oxygen effect in living cells, in view of these results, and of the facts that (*a*) in conditions where 'indirect action' operates, the immediate effects of radiation on *in vitro* preparations of DNA (biologically inactive as well as active) are not enhanced by the presence of oxygen; in several instances, in fact, oxygen is protective (Alper, 1952; Daniels, Scholes and Weiss, 1953; Kimball, 1955): (*b*) it is not now considered likely that, at neutral *p*H, HO_2 radicals would remain undissociated or would be able to act as oxidants of organic molecules (Dainton, 1955).

It may be that the sharp increase in radiosensitivity at low oxygen concentrations is a special feature of bacterial cells, or of the test of damage used, which was failure to form macro-colonies. Other investigations of change in radiosensitivity with oxygen concentration have been made with organised tissues, or with different tests of radiation damage, such as the production of chromosomal aberrations. It is possible, however, that with dilute bacterial suspensions, through which gas mixtures can be vigorously bubbled, the free oxygen concentration in the cells is much more nearly that of the suspending fluid than has been the case in other experiments. Where organised tissues have been used, such as bean roots, Read (1952) or plant inflorescences, Giles and Beatty (1950) the oxygen concentration in the sensitive cells is certainly considerably lower than that outside the tissue as a whole, since the oxygen must diffuse through respiring cells. If corrections were made for this and for other circumstances leading to oxygen tensions in cells being very much lower than those actually measured, in the experiments concerned, it might be found that the constants of the oxygen effect curve for other materials are not substantially different from those found with bacteria.

REFERENCES

ALPER, T., 1952. *Discussions Faraday Soc.*, No. 12, p. 234.

ALPER, T., and HOWARD-FLANDERS, P., 1956. *Nature, Lond.*, **178**, 918.

DAINTON, F. S., 1955. *Progress in Radiobiol*, 1956, p. xix (Edinburgh, Oliver and Boyd).

DANIELS, M., SCHOLES, G., and WEISS, J., 1953. *Nature, Lond.*, **171**, 1153.

GILES, N. H., and BEATTY, A. V., 1950. *Science*, **112**, 643.

KIMBALL, R. F., 1955. *Ann. N.Y. Acad. Sci.*, **59**, 638.

LARK, K. G., and MAALØE, O., 1954. *Biochim. biophys. Acta*, **15**, 345.

READ, J., 1952. *Brit. J. Radiol.*, **25**, 89 and 154.

STAPLETON, G. E., 1955. *Bact. Rev.*, **19**, 26.

Discussion

Howard-Flanders. Miss Alper has shown a graph indicating the change of radiosensitivity of *E. coli* B. with change in the concentration of dissolved oxygen in the cell suspension. The experimental points are fitted closely by a curve representing the equation

$$\frac{S}{S_N} \frac{k \times m[O_2]}{k + [O_2]}$$

where $[O_2]$ = the concentration of dissolved oxygen;
S_N = the radiosensitivity in the absence of oxygen;
S = the radiosensitivity at the oxygen concentration $[O_2]$;
k = a constant, 7 μ Moles per litre;
m = a constant 2·9.

A relation of this kind may be derived from either of a number of hypotheses as to the manner in which oxygen affects the radiosensitivity, and it may be applicable to other bacteria and to cells in higher organisms. Recently, Dr. E. A. Wright and I have obtained some evidence that it may be applicable to a mammalian tissue *in vivo*; growing bone in the tail of the young mouse.

We have already reported (1955) values for the radiosensitivity of growing bone in the mouse tail, while the animal was breathing various partial pressures of oxygen up to 3 atmospheres, or with the tail made anoxic. Recently we have been able to make a crude estimate of the oxygen pressure in the sensitive tissue while the animals were breathing air. This estimate depends upon timing the fall in the radiosensitivity of the tissue following the occlusion of the blood by pressure. In experiments carried out with 1·6 MeV electrons, doses of 2,000 to 4,000 rads were given to the tails of 7-day-old mice in about 1 second, at various times a few seconds before or after the application of pressure to occlude the blood from the tail. The results shown in Figure 1 indicate that the radiosensitivity falls by a factor of 1·8 in 4 seconds and then remains constant until the blood is released. This we interpret as indicating that the dissolved oxygen in the sensitive tissue is metabolised in less than 4 seconds. Only a certain range of values for the oxygen consumption of the tissue are consistent with this observation, and the need for oxygen to penetrate through the whole vertebral epiphysis and inter-vertebral disc. Using these values for the oxygen consumption, and assuming that the solubility of oxygen in a cell is not very different from that in water, we conclude that the partial pressure of oxygen at the radiosensitive layer of cells within the epiphysis probably lies between 5 and 20 mm. Hg. when the animal is breathing air. From this we deduce that k lies between 4 and 15 mm. Hg. (6 to 30 μM.). The curve is shown in Figure 2, where the sensitivity is plotted against the estimated partial

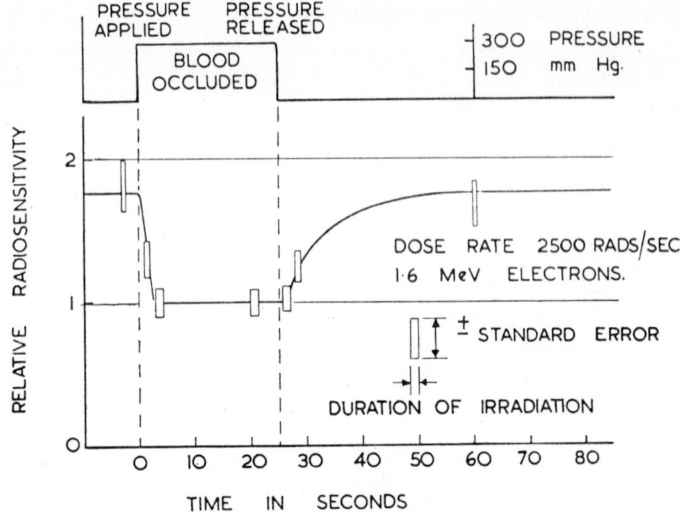

FIG. 1. The change in the radiosensitivity of growing bone in the mouse tail following the occlusion and release of the blood.

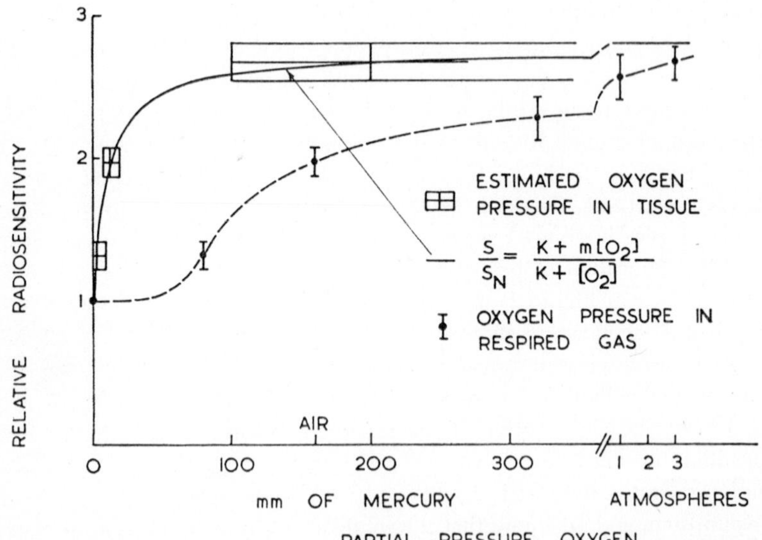

FIG. 2. The effect of oxygen on the sensitivity of growing bone in mouse tail to X-rays. The lower curve shows the relative radiosensitivity of growing bone in the mouse while the animal breathes various partial pressures of oxygen. The anoxic point was obtained with the animal breathing air but the blood supply to the tail occluded.

pressure of oxygen both in the tissue and in the respired gas. The upper curve is calculated using the values $k = 12$ mm. Hg. and $m = 2{\cdot}8$. Evidently the radiosensitivity of growing bone in the mouse tail is increased by remarkably small partial pressures of oxygen in the tissue.

REFERENCE

HOWARD-FLANDERS, P., and WRIGHT, E. A., 1955. *Nature, Lond.*, **175**, 428

Kanazir. A striking effect of radiation on micro-organisms is the immediate alteration of DNA-synthesis causing inhibition of cytoplasmic division and leading to important changes in the metabolism as well as induction of mutations. We have used *Salmonella typhimurium* and observed, jointly with Kacauski and Krajincanic, a continued death of UV-irradiated cells in spite of an almost normal DNA-synthesis. The experiments suggested that DNA-synthesis may be altered after irradiation in a manner which involves a modification of the heredity too. Doses which inhibit DNA-synthesis also cause structural abnormalities of DNA-molecules which may cause a blocking of cell division. We observed that highly polymerised homologous DNA (extracted from normal cells) is able to restore the radiation damage in *Salmonella* so that the cell-division starts at a normal rate again.

It is suggested that UV-irradiation produces structural alterations which lead to the formation of incomplete DNA-molecules, being unable, in spite of the reduplication, to transfer the genetic information to the progeny. Normal DNA is able by an unknown mechanism, perhaps similar to that of the transforming principle, to induce its own reduplication in irradiated cells. This leads to the formation of DNA-molecules with the potency to transfer genetic information to the offspring of irradiated bacteria.

Swallow. In our paper (Stein and Swallow, this conference) we stress that the direct action (of ideal X-rays especially) on an organic substance may well give the same free radical as would be produced by indirect action. Thus on our nomenclature, Miss Alper's 'ionised target molecule in a reactive state' would be a free radical, and her 'TO_2' would be our 'AHO_2'. We are therefore in fundamental agreement. On the case of irradiation with ideal alpha-particles, if the action of alpha-particles on water is anything to go by, organic free radicals capable of reacting with oxygen would not be formed, and so there would be no oxygen effect. I think this basically resembles Miss Alper's explanation of the absence of an alpha-particle oxygen effect.

Alexander. Swallow said in discussion that direct and indirect action produces the same chemical changes in single substances. This is not true, there is abundant evidence which shows the differences.

Swallow. Some surprise has been expressed that we should assume that direct and indirect action on an organic molecule may give the same free radical. However, this does in fact seem to happen. Possibly the best example is in the irradiation of ethanol. Aqueous ethanol gives the CH_2 CHOH radical by indirect action, and this is a reducing agent capable of reducing many substances, for example ferric ions or methylene blue. When pure ethanol is irradiated, the same free radical appears to be produced, and once again it can reduce ferric ions or methylene blue. There are many examples of this type of behaviour, but there may also be exceptions. The final products of direct and indirect action may be different, but this is because the conditions are different not because the organic free radicals themselves are different.

Sobels. Could you give us an estimate of the lifetime of such peroxides? In a recent paper Clark presented evidence for the genetic effects in *Drosophila* of active chemical substances produced by high-intensity irradiation. According to Clark's estimates the duration of life of such substances is at least five minutes. In an attempt to verify whether these chemical radiation products are of a peroxide-like nature we subjected flies to cyanide after high-intensity irradiation (2,200r/min.). It was observed then that the frequency of induced lethals is higher in the post-treated group than in flies which received irradiation only. Exposure to cyanide following lower intensity of irradiation (460r/min.) does not, however, raise the frequency of induced lethals.

REFERENCE

CLARK, J, 1956. *Nature, Lond.*, **177**, 787.

Alper. I do not know of any firmly based estimates of the lifetime of organic peroxides in the complicated environment which the living cell provides. It would seem that Dr. Sobels' own elegant experiments may provide a very good basis for such an estimate, particularly as the frequency of radiation-induced lethals may be influenced by treating flies after irradiation. It should perhaps be pointed out that the interpretation of experimental data obtained by using enzyme inhibitors which affect respiration is not always clear-cut. When cells receive their oxygen by diffusion through neighbouring respiring cells, less will be used up in the presence of respiration inhibitors, so that the free oxygen in the cells at the time of irradiation, and therefore their radiosensitivity, may be increased. This would of course not apply to Dr. Sobels' experiments in which cyanide is applied after irradiation.

Alexander. Once a macro-molecule is peroxidised it is in our experience not possible to restore the molecule to its original state. But when these molecules are irradiated in the absence of oxygen then an added molecule can repair the macro-molecules in many cases.

Tobias. Beside the factors discussed here, that is primarily the modifying effect of oxygen, one should mention other considerations that enter into determination of the nature of the site of radiation action. I mean here that in some instances information is available of the number of ionising particles crossing the biological test objects per unit area. In some such instances it is possible to determine that the ionising events, at least in part, must occur outside the molecules affected. An example for such consideration was given in Dr. Hutchinson's paper to-day. After all, the fact that certain physical and chemical factors can modify radiation sensitivity in the post-irradiation period, should not be taken as direct evidence for the nature of the primary events. Better evidence would be obtained from demonstration of dose-rate effect and post-irradiation oxygen effect. At the present time many reasons still exist for admitting indirect as well as direct immediate action, and in addition direct as well as indirect post-irradiation recovery or enhancement effects. More experimental information is needed to allow one to decide how much of each of the above alternatives participates in radiation effects on a given living object.

Howard-Flanders. I agree with Dr. Tobias that there is ample evidence of the indirect action of radiation on enzymes in aqueous solution, and that Dr. Hutchinson has given evidence that migration of chemical intermediates can play a part in enzyme inactivation within whole yeast cells.

It should be borne in mind, however, that when enzymes are irradiated in solution, the relative efficiency of heavy particle radiation (i.e. having a high linear energy transfer, L.E.T.) is very much lower, in comparison with that of X-rays, than has ever been found with the lethal effect on cells. Indeed, the relative biological efficiency for lethal effects has frequently been found to increase with a rise in L.E.T. This difference argues against there being a mechanism in common.

Miss Alper has already mentioned the hypothesis of direct action and restoration, which we believe gives an explanation both of the high efficiency of the densely ionising radiations, and of the modifying effect of dissolved oxygen. It is supposed that the primary action of the radiation in preventing cell multiplication is to ionise a vital target molecule or structure, which is then left in a highly reactive state, so that it will be involved in a chemical reaction within a very short time. If the L.E.T. of the ionising particle is low, there may be a high probability that the target molecule is restored to a functional state. For radiation of high L.E.T. the likelihood of restoration is much smaller, so that, following the passage of atomic fragment such as C^{6+}, there may be hardly any chance of restoration occurring. It is supposed that the nature of the chemical change which occurs following ionisation is influenced by the presence of water, oxygen and other agents. Since the chance of restoration is relatively small for high L.E.T. radiations, the modification by dissolved substances is relatively slight, which accords with a number of observations.

While this hypothesis was formulated in relation to micro-organisms, it may well be that the direct action and restoration hypothesis provides an explanation of effects observed in the cells of higher organisms.

SECTION IV

ALPHA-PARTICLE IRRADIATION OF SINGLE CELLS

THE 'SHOOTING' OF BACTERIA ONE AT A TIME BY SINGLE ALPHA-PARTICLES

R. J. Munson

Medical Research Council, Radiobiological Research Unit,
Atomic Energy Research Establishment,
Harwell, Berkshire, England

Zirkle and Bloom (1953) have shown that a microbeam of fast protons can be used as a probe to determine the relative radiosensitivities of different parts of cells. Within a cell the radiosensitivity varies from point to point and the details of its pattern will only be evident if the resolving power of the microbeam is adequate. Since only one point on such a pattern is provided by the irradiation of one cell, the complete pattern can only be built up by irradiating a large number of cells of the same age and in the same stage of mitosis. The factors which determine the detail which such a microbeam can be expected to show include:

(1) the diameter of the aperture which collimates the beam of ionising particles,

(2) the extent to which the particles are scattered by the walls of the collimator; this amounts in practice to at least a few degrees,

(3) the distance between the end of the collimator and the point of interest in the cell,

(4) the length of the particle track within the cell, since this determines its average deviation by multiple scattering ($\sim$1 micron in 20 microns of track for a 5 MeV-alpha particle),

(5) the lateral diffusion of active radicals and other radiation products from the track of the particle. With a densely ionising particle, most of these products will recombine within a few millimicrons of the track, Lea (1947), but a small proportion may remain uncombined in pure water at much larger distances.

It is clear, therefore, (see Fig. 1) that the highest resolving power can only be attained by using the smallest possible collimating aperture in contact with the thinnest cell preparation.

Collimators of about 1 micron diameter in microscope coverglasses have been made, Munson (1955), and others of 0·5 micron or less seem quite feasible. With a small cell, such as a bacterium, arranged close to an aperture of this size, it should be possible, at least in principle, to achieve resolution of details in the pattern of radiosensitivity which are almost as small as the details shown by an optical microscope. It would not be worthwhile to attempt to improve resolution beyond this, since the pattern of radiosensitivity can only be fixed in relation to visible structures of the cell. However, with a large bacterium such as *B. megatherium*, a resolving power of 1 micron might yield some results of interest. It should be possible, for example, to determine whether the supposed nuclear bodies, which can be demonstrated by staining, are also regions of high radiosensitivity.

Since the energy lost by a charged particle when it passes through a bacterium is relatively small, a suitable counter can be used to record particles after their passage. This means that it should be possible to 'shoot' a bacterium by one or more particles as desired.

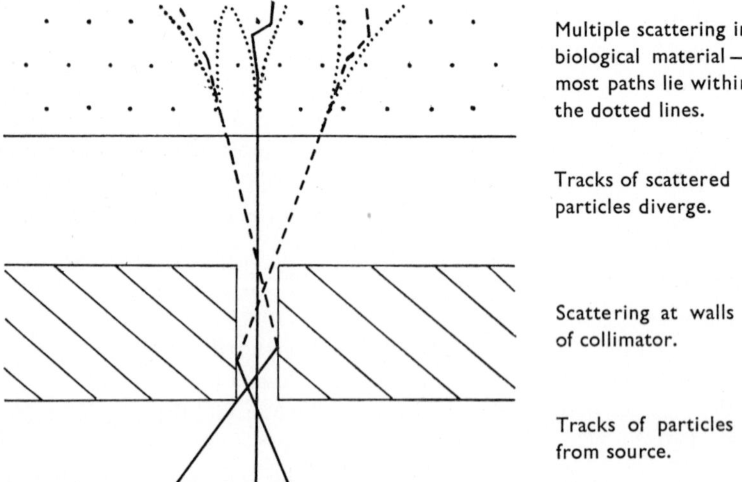

FIG. 1. Schematic diagram to illustrate the broadening of a microbeam by scattering.

The apparatus for experiments based on these ideas has been constructed but results are not yet available.

Before and during irradiation the bacteria in liquid suspension ($\sim 10^9$ cells per ml.) are contained in a thin-walled glass tube which, over a length of about 100 microns has a nearly uniform bore just large enough to accommodate a bacterium. The smaller end of this irradiation tube enters another tube of about 100 microns' bore through a hole in its wall (Fig. 2). During the irradiations a steady flow of sterile water will be maintained in this larger or flow tube.

The positions of both tubes can be adjusted independently and their watery contents can be driven by hydrostatic pressure in either direction as required. Surface tension seals the joint between them very efficiently.

At the start of each experiment a small quantity of a suspension of bacteria in the logarithmic growth phase will be drawn through the flow tube into the irradiation tube. When enough bacteria are in the irradiation tube they will be held there whilst the flow tube is thoroughly flushed with sterile water. After each bacterium has been 'shot' it will be forced along the irradiation tube and into the flow tube where it will be carried along in the steady stream and finally collected at the far end and incubated. The average spacing of bacteria in the irradiation tube will be so large that each irradiated bacterium will usually have left the irradiation tube before the next one comes into position.

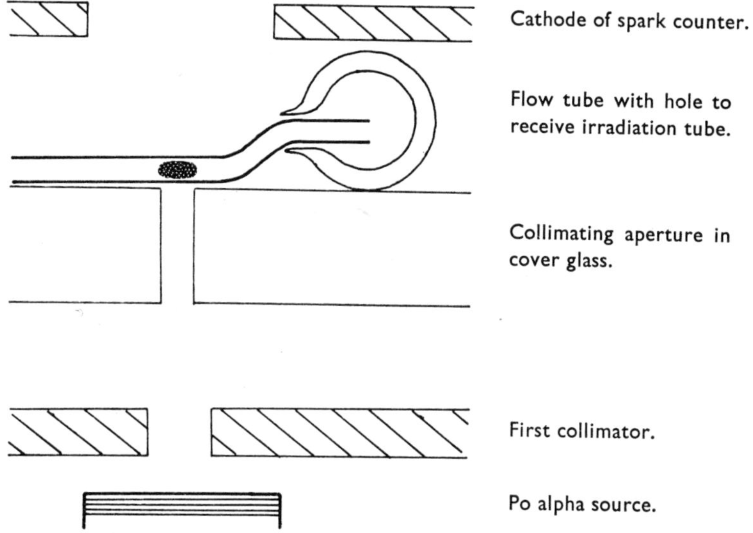

FIG. 2. Diagram of apparatus, not to scale.

The irradiation tube lies across and in contact with the end of the aperture which collimates the alpha-particles. These originate from a ^{210}Po source below, and pass through the collimator at the rate of about 20 per hour. After crossing the irradiation tube diametrically each particle is recorded by a spark counter, Munson and Garrett (unpublished), supported just above the tube. A microscope fitted with a $\times$ 52 reflecting objective (N.A., 0·65) having a working distance of 4 mm., provides sufficient space for this counter and permits continuous observation of the selected bacterium through it. A cell can be maintained in position over the collimator by fine adjustment of the pressure difference across the irradiation tube, or by moving the tube slightly backwards or forwards parallel to its axis.

REFERENCES

LEA, D. E., 1947. *Brit. J. Radiol.*, Supplement No. 1.
MUNSON, R. J., 1955. *Rev. Sci. Inst.*, **26**, 236.
MUNSON and GARRETT. Unpublished.
ZIRKLE, R. E., and BLOOM, W., 1953. *Science*, **117**, 487.

ALPHA-IRRADIATION OF PARTS OF SINGLE METAPHASE CELLS IN CHICK TISSUE CULTURES

ROSS MUNRO

British Empire Cancer Campaign Research Fellow,
Tissue Culture Laboratory, Christie Hospital and Holt Radium Institute,
Manchester, 20, England

Bloom, Zirkle and Uretz (1955) have irradiated parts of amphibian cells in tissue culture with protons, and Simons (1955) has described a technique for generating a micro-beam of alpha-particles. These workers produced collimated beams of particles by means of very small apertures.

Here, chick cells in tissue culture have been irradiated by placing small polonium sources near them. Polonium alpha-particles have a well defined range of 39·9 microns in water (Aniansson, 1955), and so by placing a source at an appropriate distance from a cell, a part of it can be irradiated (Fig. 1).

Techniques

(1) *Preparation of sources.* The sources are made by depositing polonium on tungsten micro-needles. It is deposited along a line about 10 microns long on one side of the tip of a needle, and is sealed in position by an electroplated layer of chromium. The alpha-particles lose some of their energy in passing through the chromium, and emerge with a range which is usually between 20 and 35 microns, according to the thickness of the plating. The variation in dose rate with distance from a needle is measured by a scintillation counting technique.

(2) *Dosimetry.* An approximately cubical zinc sulphide crystal, with sides 8–10 microns long, is stuck to a plain tungsten micro-needle with 'Araldite D'. This crystal is placed on the stage of a microscope, under a cover-slip, and dipping into a hanging drop of water. The needle which is to be measured is placed near the crystal (Figs. 2 and 5). It is manœuvred into position with a micromanipulator. The flashes of light which the crystal emits when alpha-particles strike it are detected by a photo-multiplier mounted over the microscope, and counted by a scaler. The microscope can be used either to measure the distance from the micro-needle to the crystal, or to guide the light emitted by the crystal to the photo-multiplier.

From the measured counting rates and a knowledge of the area of the face of the crystal which the particles strike, the mean number of particles crossing a square micron of this face in a minute is calculated. Dose rates are normally expressed as particles per square micron per minute. They can be converted to rads per minute by multiplying by a factor which varies

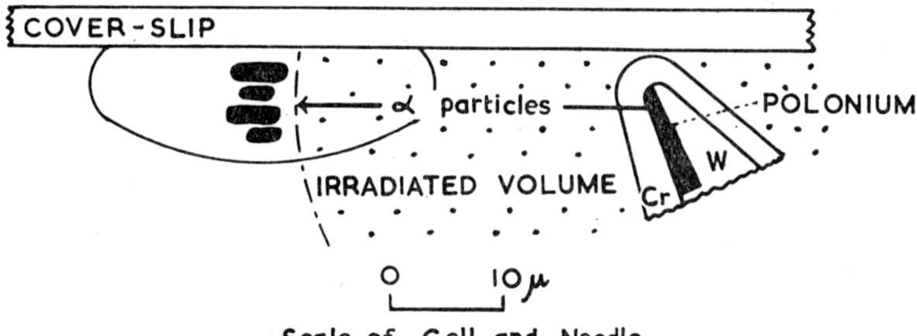

FIG. 1. Irradiation of part of the cytoplasm and spindle of a cell at metaphase. Side view.

between 1,700 and 4,100, according to the energy of the particle at the point at which the dose rate is measured. Figure 3 shows how the dose rate varies with the distance from a typical needle.

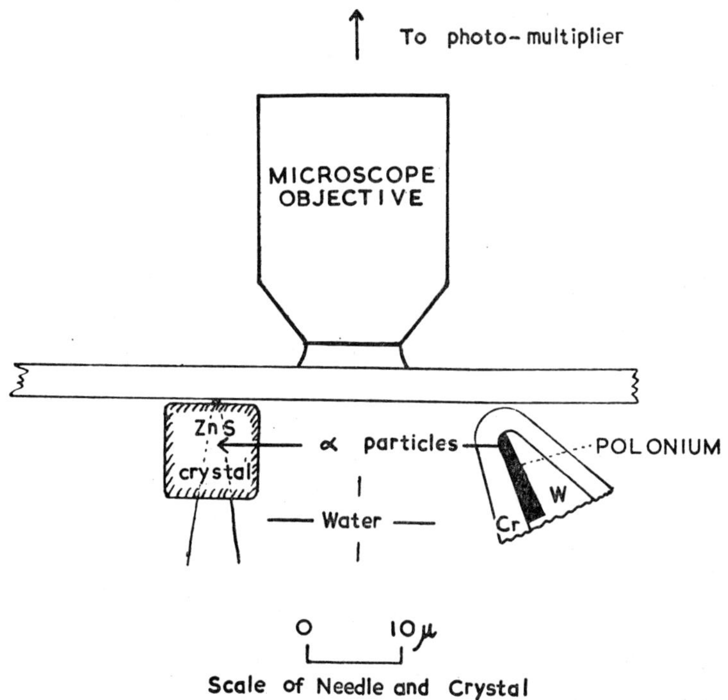

FIG. 2. Technique for dosimetry.

A particle which strikes a crystal has to dissipate a certain minimum amount of energy in it (about 0·5 MeV with the present apparatus) if a count is to be registered. A correction which allows for this is applied to the graph of dose rate against distance. The needles are tested frequently by a technique which will be described elsewhere, to ensure that polonium is not escaping through the chromium coating. Control experiments are performed with needles which do not contain polonium, to ensure that, for instance, chromium ions liberated from the needles do not disturb the cells.

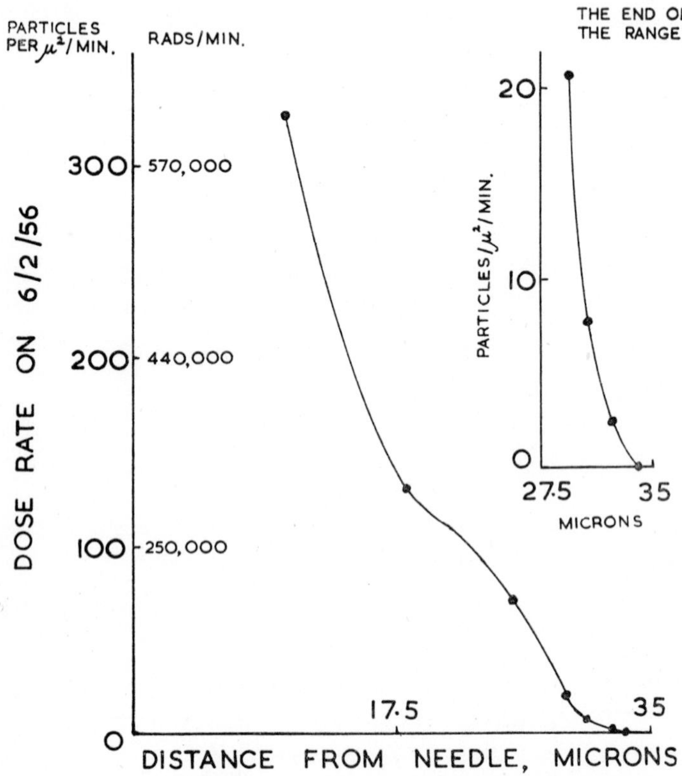

FIG. 3. Variation of dose rate with distance from a needle.

(3) *Tissue culture.* Cultures of fragments from the hearts of 7-day chicken embryos are used. The cells studied grow in hanging drops of a mixture of chicken serum and embryo extract. For micromanipulation they are grown at 37° C. over medicinal paraffin, on chambers of the kind devised by Commandon and de Fonbrune (1949).

In a typical experiment, a cell at the stage of division which is to be studied is found, and photographed on 16 mm. film by phase contrast. A micro-needle is placed near it, in a position such that the required region is irradiated. In the experiments described here, the irradiation time was usually one minute, and always between one and one and a half minutes. After irradiation, cells were followed for a period of up to an hour, and then fixed and stained.

Results

(1) *General.* At metaphase, the cells appear approximately elliptical, with the chromosomes in a row along the minor axis of the ellipse. At this stage, one can irradiate part of the cytoplasm and spindle without irradiating the chromosomes, by placing a needle in the position shown in Figures 1 and 6 (*a*). If the needle is placed nearer the cell, or at the side of the row of chromosomes, the chromosomes are irradiated as well (Fig. 7 (*a*)).

The cells considered here were irradiated at metaphase. The effects upon anaphase and upon nuclear reconstruction at telophase were studied.

Anaphase usually began a few minutes after irradiation had finished. If this stage was under study, cells were fixed 2–4 minutes after it had begun; if telophase, they were fixed 20–50 minutes after the beginning of anaphase, by which time normal cells would have passed through telophase and reconstructed pairs of new daughter nuclei.

If the chromosomes were irradiated, the dose always varied somewhat over the region which they occupied. In the description which follows, the figure given for the dose is the maximum received by any of the chromosomes.

(2) *Irradiation of the region containing the chromosomes with small doses.* When the chromosomes of cells at metaphase were irradiated, in such a way that the maximum dose was from 0·25 particles/μ^2 upwards, sticky bridges between the two separating groups of chromosomes were produced at anaphase. The bridges could be seen in stained cells or, with careful focussing, in living cells by phase contrast.

In order to study the effect on telophase, doses of 0·2–8 particles/μ^2 were delivered to the chromosomes of some further cells at metaphase. They were followed for 20–36 minutes after anaphase began, and then fixed and stained. All passed successfully through telophase; the daughter nuclei were surrounded by nuclear membranes and contained nucleoli. They were not always entirely normal; sometimes they had the shape shown in Figure 4, where the pointed ends correspond to the remains of sticky anaphase bridges.

(3) *Irradiation of the chromosomes with larger doses.* A third batch of cells was irradiated at metaphase with larger doses. The maximum dose delivered to the chromosomes ranged from 11 to 26 particles/μ^2. Again, the chromosomes were sticky at anaphase. Some of the chromosomal material usually failed to move, and remained in its metaphase position, midway between the two groups of anaphase chromosomes and joined to them by sticky bridges (Fig. 7 (*b*)).

The cells were fixed 41–49 minutes after anaphase had begun. Whereas the nuclear membrane becomes visible in normal cells 8–9 minutes after the beginning of anaphase, none of these cells had passed through telophase. Figure 7 shows one of them. At anaphase, less than half of the chromosomes separated. The remainder, which were in two clumps, eventually moved into the right-hand half cell and joined up with the group that moved normally, forming the clover leaf like structure on the right hand side of Figure 7 (*c*). The fluffy structure on the left hand side of the picture is the group of chromosomes which moved in the opposite direction. There is no sign of nuclear reconstruction.

In other cases, the chromosomes have been spread out over the cells, or arranged in a line resembling a long, dense sticky bridge.

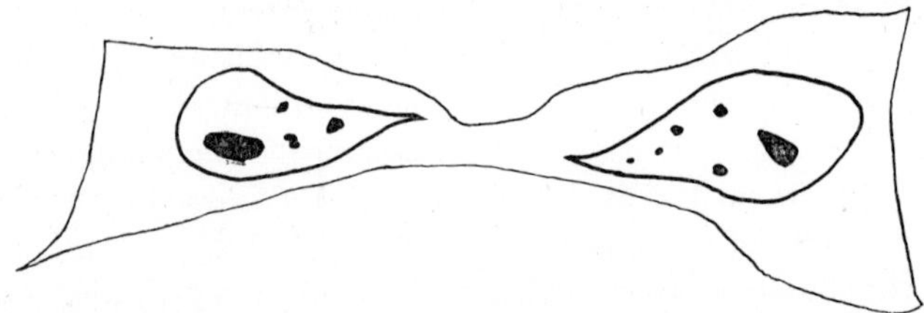

FIG. 4. Binucleate cell at early interphase, with pointed nuclei, following irradiation at metaphase and sticky bridges at anaphase.

(4) *Irradiation of part of the cytoplasm and spindle.* It has been shown that irradiation of metaphase chromosomes can produce sticky bridges at anaphase, and that larger doses will inhibit telophase. The next step was to determine whether irradiation of part of the cytoplasm and spindle, in the manner shown in Figure 1, could produce either of these effects.

Four cells were irradiated in this way and fixed during anaphase. The dose to the chromosomes was zero; at a distance of 5 microns from the nearest chromosomes it was between 4·3 and 8·5 particles/μ^2. No sticky bridges were seen in the stained cells.

Five cells were irradiated similarly and fixed 24–36 minutes after anaphase began, to determine whether nuclear reconstruction had taken place. Since it was already known that small doses to the chromosomes did not stop nuclear reconstruction, and since the maximum possible dose was to be delivered to the cytoplasm, the needle was placed slightly nearer the chromosomes, so that they did, in some cases, receive a small dose. The dose delivered to the cytoplasm at 5 microns from the chromosomes was 11–36 particles/μ^2.

All these cells passed through telophase and reconstructed new resting nuclei. Only one was abnormal; in this, irradiation of the chromosomes produced sticky bridges at anaphase, and the daughter nuclei had pointed ends like those shown in Figure 4. One of the cells which behaved normally is shown in Figure 6.

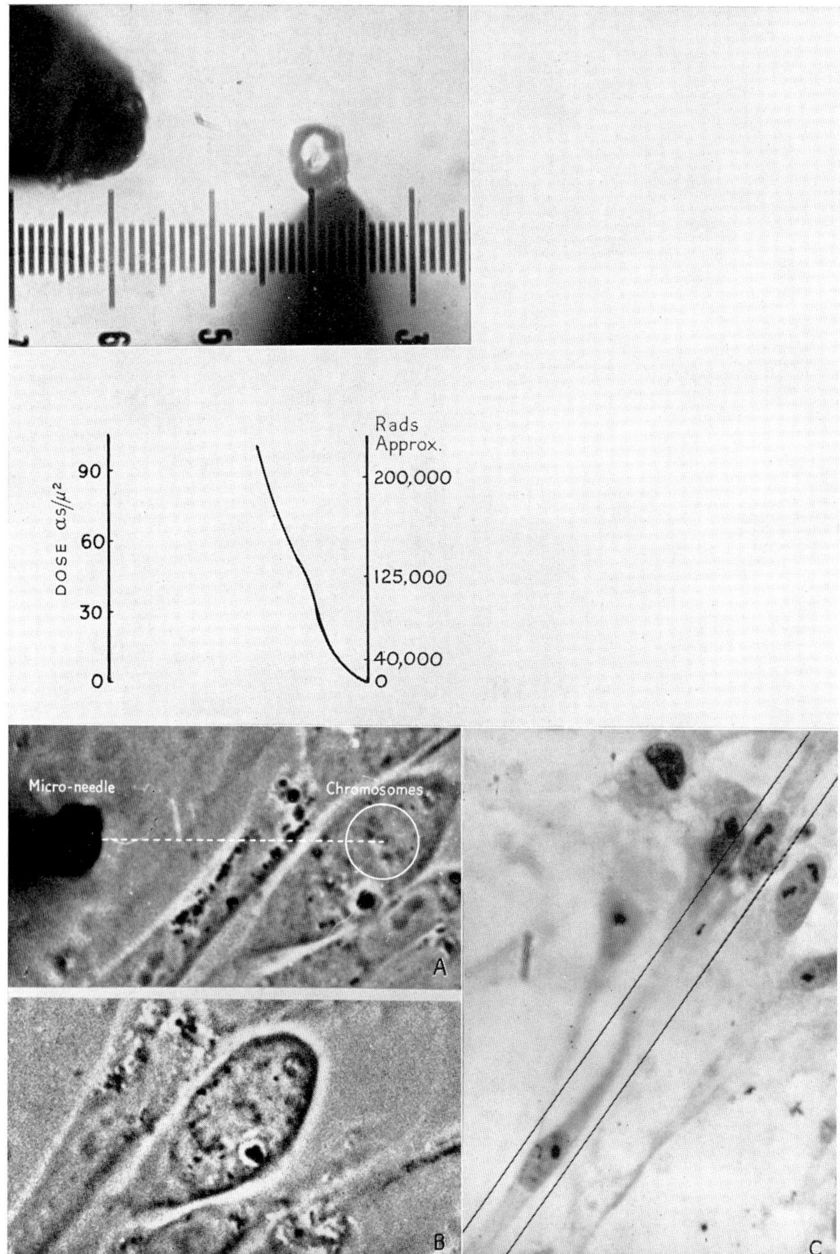

FIG. 5. Micro-needle, zinc sulphide crystal and eyepiece graticule, as seen with dosimetry microscope. Scale divisions 1·8 microns.

FIG. 6. (*a*) Irradiation of part of the cytoplasm and spindle of a cell at metaphase, 8 minutes before anaphase began. The graph shows the variation in dose along the dotted line.

(*b*) Mid-anaphase, 9 minutes after (*a*). Normal. (*a*) and (*b*) are phase contrast pictures of the living cell, ×1,350.

(*c*) Fixed and stained, 36 minutes after anaphase began. Nuclei appear normal. ×980.

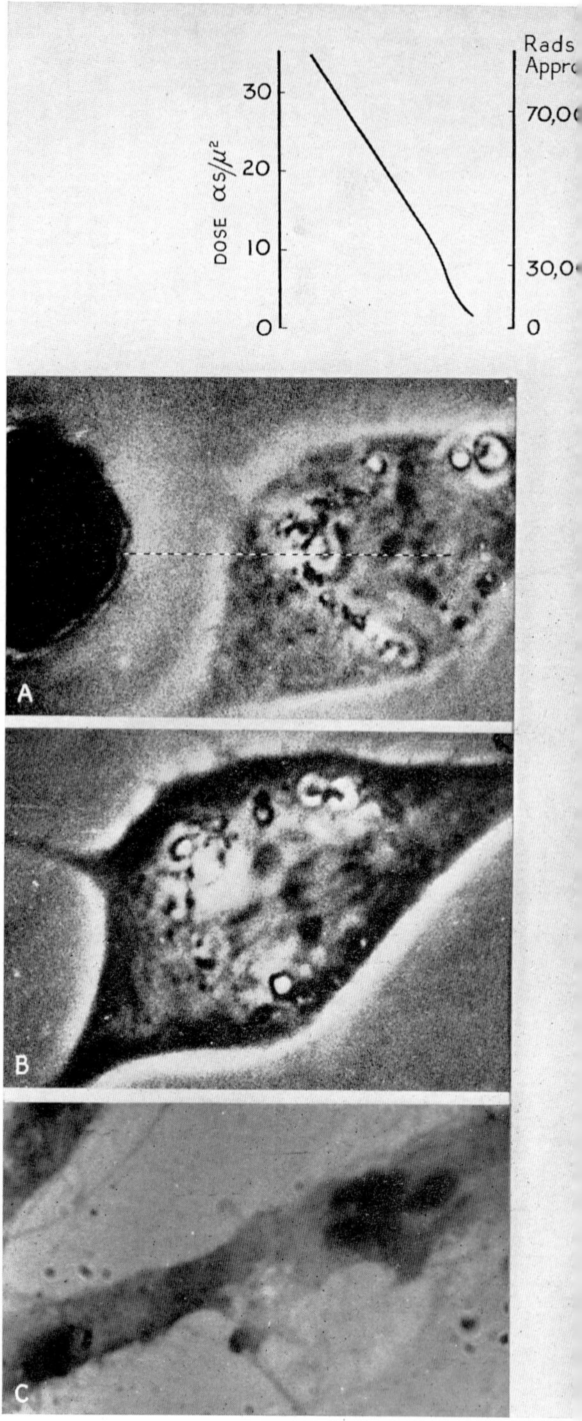

Fig. 7

(*a*) Irradiation at metaphase, including the chromosomes, 1 minute before anaphase began.

(*b*) Mid-anaphase, 2 minutes after (*a*), showing some chromosomes left in the position of the metaphase plate, and sticky bridges joining them to the separating anaphase groups. (*a*) and (*b*) are phase contrast pictures of the living cell. ×2,700.

(*c*) Fixed and stained, 41 minutes after anaphase began. New nuclei have not been reconstructed. ×2,700.

Summary

Irradiation of the chromosomal region of chick cells in tissue culture at metaphase with small doses of alpha radiation produced sticky bridges at anaphase. When the radiation was confined to part of the cytoplasm and spindle no abnormalities were seen, although the dose at a point distant 5 μ from the chromosomes was up to thirty times that needed to produce sticky bridges when the chromosomes were irradiated.

After small doses, cells which showed sticky bridges at anaphase passed through telophase and reconstructed new resting nuclei. These were sometimes abnormally shaped.

When larger doses were delivered to the chromosomes at metaphase, nuclear reconstruction did not take place, at least up to 40 minutes after the beginning of anaphase. It was not possible to inhibit reconstruction by irradiation of part of the cytoplasm at metaphase, when the dose at 5 μ from the chromosome was up to three times that which had stopped reconstruction when applied to the chromosomes.

Acknowledgment

I should like to thank the Prophit Trust for a Studentship in Cancer Research, under which this work was begun.

REFERENCES

ANIANSSON, G., 1955. *Physical. Rev.*, **98**, 300.
BLOOM, W., ZIRKLE, R. E., and URETZ, R. B., 1955. *Ann. N.Y. Acad. Sci.*, **59**, 503.
DE FONBRUNE, P., 1949. *Technique de Micromanipulation* (Paris, Masson).
SIMONS, H. A. B., 1955. *J. Sci. Instrum.*, **32**, 21.

THE IRRADIATION OF SINGLE CELLS AND PARTS OF SINGLE CELLS IN TISSUE CULTURE WITH MICROBEAMS OF ALPHA-PARTICLES

MARGUERITE I. DAVIS, (MRS) I. SIMON-REUSS AND C. L. SMITH

Department of Radiotherapeutics, Cambridge University, England

An uncommon but interesting use of radiation as a tool in biological research is to employ it as a probe into the structure and organisation of living organisms. By the use of a very fine beam of radiation it is possible not only to observe its action on a particular cell component, but to determine the relationships of the cell components to normal cell function. By damaging or totally destroying a definite part of the cell, the resulting behaviour may be ascribed to the impaired activity of this particular cell component, and information can thus be gained about its function in the normal pattern of cell development. As early as 1912, Tschakotin used a beam of ultra-violet light focussed down to a spot 1 μ in diameter in order to irradiate the nucleus or cytoplasm of large unicellular organisms. It was not until the phase contrast microscope came into use that microbeam irradiation of living cells in tissue culture became feasible. Zirkle and Bloom (1953) irradiated parts of cells of newt hearts grown in tissue culture, using a beam of protons from a Van de Graffe generator, collimated down to 2·5 μ in diameter. This work was correlated with that of Uretz *et al.* (1954) who, working with the same material, irradiated the cells with a beam of ultra-violet light focussed to a spot 5 μ in diameter. Their results showed that cell abnormalities were produced when the beam of either ultra-violet light or protons was incident upon the chromosomes. While ultra-violet light incident on the extra-chromosomal area caused delayed and abnormal divisions, large doses of protons to this region were without effect.

In the present work, α-particles emitted by ^{210}Polonium have been collimated into a beam 1 μ–1½ μ in diameter and used to irradiate discrete portions of chick fibroblast cells in tissue culture. These α-particles are almost wholly of a single energy, practically uncontaminated with either β- or γ-radiation and are produced by a source with a reasonably long half life (138 days). The range of the α-particles emitted is adequate for single cell irradiations if care is taken to keep to a minimum the total absorber between the source and the target. By using a radioactive element as a source of

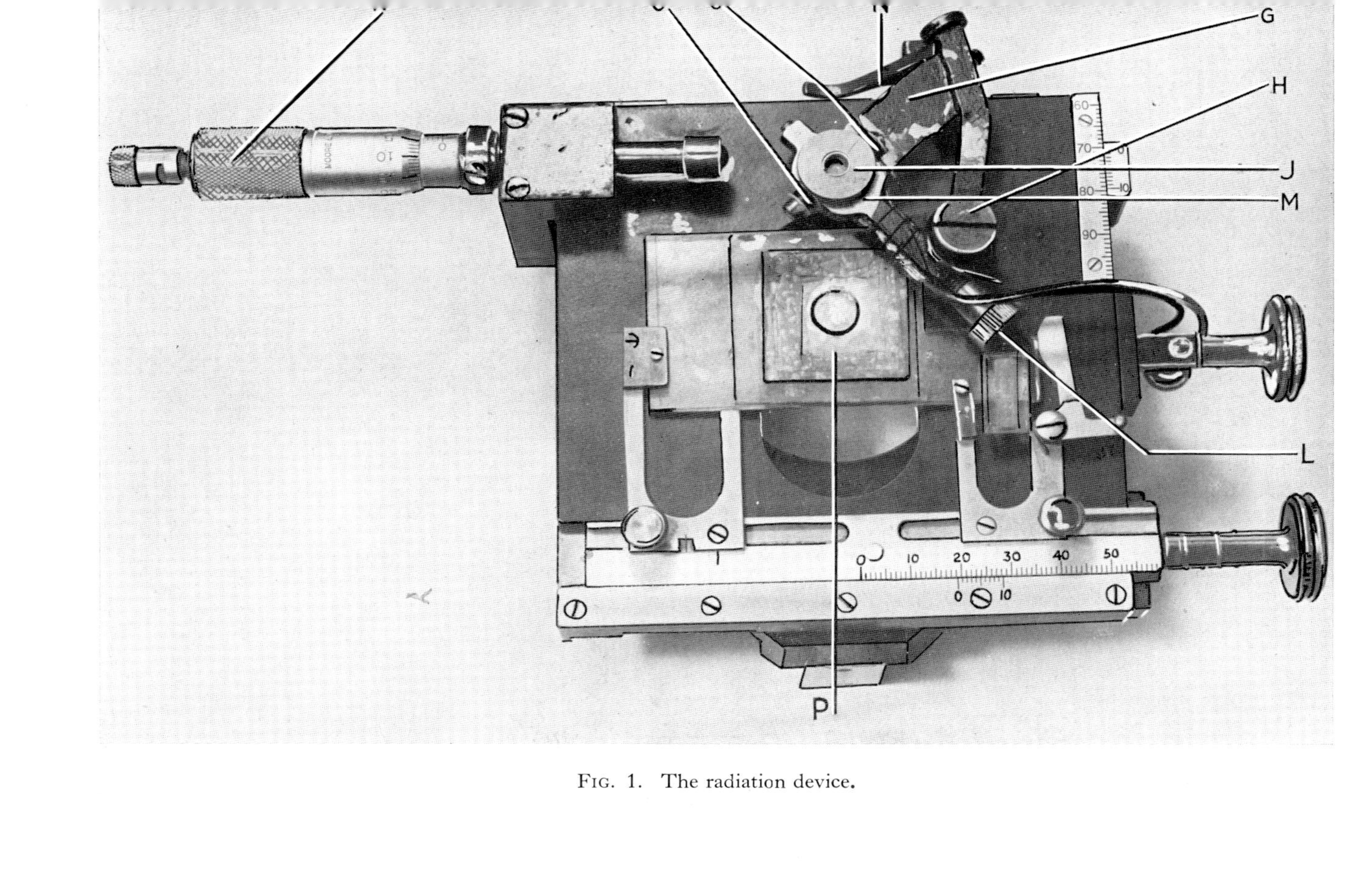

FIG. 1. The radiation device.

ionising particles the apparatus may be made small enough for the irradiations to be carried out on the microscope stage.

The cultures of chick fibroblasts were grown by the usual technique of hanging drop cultures in embryo extract and plasma clot; all cultures were at least in their fourth passage and were used 24 hours after sub-culturing. The cultures were grown on mica coverslips of 1 cm. air equivalent stopping power (approximately 5 μ thick) in specially designed culture chambers. The culture was searched prior to irradiation by using as a microscope objective the Newton Long Working Distance attachment fitted with a Cooke, Troughton and Simms 4 mm. phase contrast air objective. This attachment increases the working distance between the objective and the specimen to 12·8 mm. allowing the collimated α-particle source to be inserted between the specimen and objective without moving any microscope part in the process.

The irradiation device consists of a swinging arm G (Fig. 1) mounted on a bearing H attached to the fixed base plate of the microscope stage. The movement of the specimen is achieved independently of this plate. The arm may be held back, when not in use, by the hook N out of the field of view of the microscope. For an irradiation it is moved into position across the stage to rest against the micrometer screw K. Adjustment of this screw causes the swinging arm to describe an arc across the stage; movement of less than 1 μ can be achieved with this screw. The screw L pushes a brass collar, into which the collimated source J is inserted, against a stiff wire spring thus producing movement of the source in a direction approximately at right angles to that produced by the micrometer K. The combination of these two movements permits adjustment of the collimating hole with an accuracy of about 1 μ.

The polonium source, prepared by the Radiochemical Centre, Amersham, England, was deposited on 1 mm.^{2} of platinum foil and was initially of 8 mc. strength. It is mounted on a glass cover slip in the source holder as shown in Fig. 2. A small shutter C, made of steel, has a mica window over one hole sufficiently thick to stop the α-particles. By passing a current through one or other of the two coils O (Fig. 1) the shutter may be moved so that the α-particle beam either is, or is not, incident on the collimating hole. The collimating hole is mounted on the bottom of the source holder. The holes, 1 μ in diameter, are produced by evaporating tin in a vacuum on to copper discs 3 mm. in diameter and 20 μ thick with apertures at the centre 5 μ–8 μ diameter. By this means the central holes are reduced in diameter to 1μ. The collimating hole may be seen under a microscope through the glass cover slip and shutter.

In order to perform an irradiation, the target area of the cell is aligned with cross hairs in an ocular graticule, using the Newton Long Working

Distance attachment on the objective. A brass wedge is inserted between the source holder and the swinging arm in order to raise the former by a few millimetres. The swinging arm is then moved into position in the centre of the stage and when above the culture, the wedge is removed and the source slides down, under its own weight, to rest on the mica cover slip of the tissue culture. With the source in such a position it is then possible to see the collimating hole as a pinpoint of light in a dark field. The image of the hole is then aligned with the same ocular cross hairs as was the cell target. The irradiation is commenced by activating the proper electromagnet

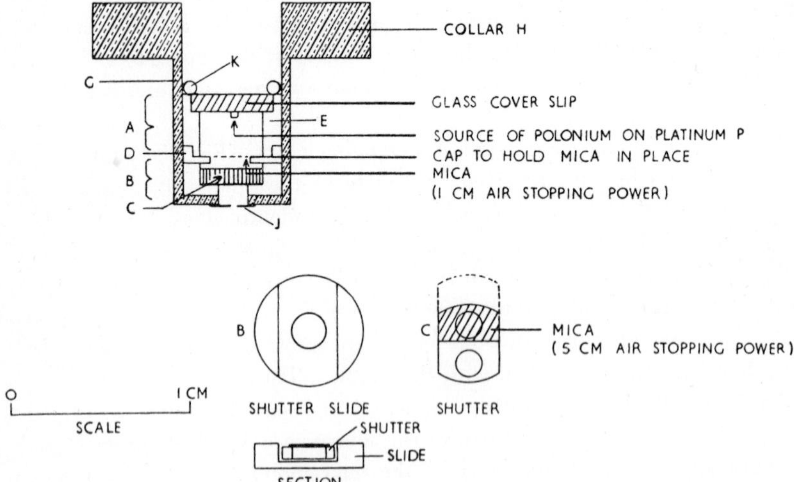

FIG. 2. Mounting of polonium source.

to open the shutter. After the desired irradiation time, which is normally a matter of seconds, the shutter is closed, the source raised by the brass wedge and swung away.

The alignment of the target is rechecked. The Long Working Distance objective is replaced by an oil immersion objective and filming of the cell may be commenced within two minutes after irradiation. This is done on 16 mm. film with exposures generally made at 4-second intervals.

The irradiations can be divided into three groups: prophase irradiations —in these the chromosomes were in varying stages of condensation; metaphase plate irradiations and extra-chromosomal area irradiation of metaphase cells.

The results can be summarised as follows:

1. The sensitivity of a cell increases by a factor of ten from prophase to metaphase. The sensitivity seems to increase with the state of condensation of the chromosomes.

2. A change in the character of the cytoplasm and abnormalities in cleavage are observed in the chromosome irradiations in metaphase, phenomena which are never observed after prophase irradiation.

3. Irradiation of the spindle produces changes in the cytoplasm and abnormal cleavage with no effect on the chromosomes which do not separate.

4. Irradiation of the extra-chromosomal region away from the spindle area indicates the presence of a radiation sensitive structure responsible for the active pulling apart of the two daughter cells and characteristic form of the resting fibroblast cell.

REFERENCES

Tschakotin, S., 1912. *Biol. Zbl.*, **32**, 623.

Uretz, R. B., Bloom, W., and Zirkle, R. E., 1954. *Science*, **120**, No. 3110, 197.

Zirkle, R. E., and Bloom, W., 1953. *Science*, **117**, No. 3045, 487.

Discussions

Howard. May I ask Dr. Munson what his irradiation tube is made of? Is the apparatus also suitable for irradiation with radiations other than polonium alpha's?

Munson. The material is glass, very thin. The wall thickness is of the order of 1 micron or rather less and the total thickness of material the particles go through will be, say, 3 microns, and the range in unit density material is about 35 microns so there will be no trouble with penetration. Other particles, yes, protons for example, though they do have to be heavy particles.

Burns. Have you considered and calculated the amount of scattering from the walls of the holder?

Munson. Provided the bacterium is really sitting close to the end of the defining aperture the amount of scattering is small. You still have scattering from walls of the defining hole but if the thickness of the bacterium is only of the order of twice the diameter of the defining hole then one can of course ignore scattering effects, since we are dealing only with small angle scattering. We are also of course dealing with a single particle and not with a diffuse beam. We are just uncertain where the particle goes in the bacterium to the extent of the diameter of the hole.

Sherman. Have you one bacterium in the tube at a time?

Munson. No, we try to have a whole lot arranged like peas in a pod and push them along one at a time. The hole of the tube is about 100 microns and one has no difficulty in seeing the bacteria and following them along. The only risk is that they may stick to the walls, but provided the bacteria are in the middle of the stream and the flow is not too turbulent the chances of this are not great.

Crathorn. Is it your intention to examine changes in morphology, mutations, and so on?

Munson. The main idea is to determine whether or not the bacterium is viable, the test being the capacity to form colonies on the plate.

M. Davis. What sort of detail do you hope to be able to resolve in your observations of the cell in the microscope?

Munson. The numerical aperture of the microscope we use is about 0·65. We can probably see sufficient detail to locate to within perhaps 0·25 micron.

Hutchinson. Can Dr. Munro say how many alpha-particles must pass the chromosomes to produce observable aberrations?

Munro. That is difficult to answer because experiments of that kind have to be statistical. All I can say is that dose rates as low as one alpha-particle per 4 square microns do produce sticky bridges. It presumably depends whether the particle happens to pass through the right part of the chromosome.

Hutchinson. So one alpha-particle per 4 square microns is roughly the limit for having an effect?

Munro. It is certainly about the right order of magnitude. 0·1 does not produce any effect and 10 produces a considerable effect.

Lajtha. One assumes that chromosome stickiness impairs cell-viability, I wonder whether you have any observations on following up cells with sticky chromosomes and whether they can restitute normally, or only after a delay, or whether they are doomed in the long run?

Munro. That is a point which I hope to study one day but so far I have not followed them for more than, I think, 50 minutes after the beginning of the anaphase, at which time they are still alive though perhaps oddly shaped.

Hevesy. Applying 100 alpha-particles some 2×10^7 ion pairs are produced. What percentage of these reaches a chromosome and what is the minimum number of ions which suffice to produce observable changes in the chromosomes?

Davis. The chromosomes receive a tremendous amount of radiation. A cylinder of about 2 microns in diameter and 10 microns in length receives about 1,000r per alpha-particle passing along its axis. Since between 10 and 100 alpha-particles were used in the irradiations between 10^4 and 10^5r was given to this small volume of the cell in a few seconds.

Hevesy. Well, then 100 alpha-particles must produce some 10^7 ion pairs but only a fraction reaches the chromosome.

Davis. When the dose is said to be 100 alpha-particles that is the number of alpha-particles which have sufficient energy to penetrate the mica cover-slip and to enter the cell. If the irradiation occurs when the cell is in prophase, the entire cell volume is occupied by the condensing chromosome material, and there is a chance that any one of the incident alpha-particles may do damage. If the irradiation takes place in metaphase, the chromosomes are arranged on the metaphase plate with the chromatids arranged radially like the spokes of a wheel in plane perpendicular to the cover-slip. The ends of the chromosomes will be only a few microns below the cell membrane, and therefore will receive a large dose of alpha-particles. The dose will, of course, be decreased along the length of the chromatid, and the portion of the chromosomes near the centre of the cell will receive a much smaller dose than the ends.

Smith. If a chromosome may be considered to be about 2 μ across and the bombarding area 1 μ^2 to 2 μ^2, then each alpha-particle in passing through a chromosome will produce about 8,000 ionisations at the end of its range.

Rajewsky. But what about the fact that there is heavier ionisation at the end of an alpha-particle beam?

Smith. These mental calculations are approximate. In the case of alpha-particles there is only a factor of about 2 between the rate of ionisation at the beginning and end of the range.

Rajewsky. How were you able to measure the diameter of the alpha-particle beam?

Smith. We measured the diameter of the beam by the following method: A perspex slide 3 in. by 1 in. with a hole in it had a piece of autoradiograph stripping film mounted over the hole and also a piece of mica of 1 cm. air stopping power—the same thickness as that on which the cultures were placed. This slide was placed on the microscope stage and the emulsion was irradiated through the mica. After development it was very difficult to find the alpha-spots unless a series of irradiations along a straight line had been made.

Rajewsky. And now to the biological implications. Have you used the single cells in your tissue cultures or groups of cells?

Davis. We pick out a cell which is in the stage of mitosis in which we are interested, from a number of cells in the hanging drop which is growing on the very edge of the culture, against the mica cover-slip.

Rajewsky. A cell being in connection with the bulk of tissue culture?

Davis. Yes. I have irradiated the plasma clot with a very large dose of alpha-particles in the region of mitotic cells, and then have observed the cells for some hours, and found no aberrations. I have also observed cells adjacent to the irradiated cells and very frequently these neighbouring cells are seen to go through mitosis normally: if the cells do not happen to enter mitosis, but remain as resting cells, they display no abnormal appearance.

Rajewsky. We have given isolated cells of a tissue culture very high X-ray doses, but they did not show any morphological damage.

Munro. How long did you observe them?

Rajewsky. Perhaps for 20 passages. But this is in contrast to cells which are connected to the whole piece of tissue culture irradiated; in this case we observed very marked lesions. It seems as though the radiation reaction of the cell is not only a function of the energy the cell itself absorbs.

Hutchinson. Zirkle and Bloom, using protons, report factors the order of thousands between the dose to cytoplasm plus chromosomes for observable effects. Your results are quite different with the factor of only 10 or less.

Davis. My results do differ from those of Zirkle and Bloom but not only has a different type of irradiation been used but also the material irradiated is different. I have worked entirely with chick fibroblasts while Zirkle and Bloom have worked with cultures of newt hearts. I hope to irradiate the newt heart material with the alpha-particle beam and then I feel that the results can be compared and contrasted. It is quite possible that the chick is more sensitive and shows more marked irradiation effects.

Gopal-Ayengar. You have irradiated cells at different stages of the cycle. Have you followed the results for longer times?

Davis. I have irradiated resting cells as well as mitotic cells and have observed effects on the resting cell such as bubbling and loss of cytoplasm when the cytoplasm alone is irradiated and cell œdema followed by pyknosis if the nucleus is irradiated. These cells have not been followed for long periods of time. It must be borne in mind that these results from all of the present microbeam work demonstrate only the immediate effects of the irradiation. No attempt has been made thus far to follow the irradiated cells through a second cell division. The cells which appear normal after the low doses of irradiation in some stages of cell division may possibly be damaged in such a way that they are unable to divide again or will show abnormalities in some subsequent division.

SECTION V

MODIFICATION OF SYSTEMIC IRRADIATION EFFECTS

I. GENERAL PRINCIPLES

THE EFFECTS OF PRE- AND POST-TREATMENT ON THE RADIATION SENSITIVITY OF MICRO-ORGANISMS

ALEXANDER HOLLAENDER
Biology Division, Oak Ridge National Laboratory,
Oak Ridge, Tennessee, U.S.A.

One of the crucial problems in radiation biology concerns the first damage that occurs immediately after radiation is absorbed. If these immediate steps could be understood, experiments could be directed towards the prevention of this damage and possibly counteracting it after radiation has stopped. This is, of course, the problem radiation chemists have been most interested in. Naturally, the radiation chemists have selected the process which is most easily studied; that is, the effect of radiation on water or on enzymes in very dilute solutions. This approach is of basic importance. However, we have in the living cell, not pure water, but conditions which are much more complicated. There is the cell wall, the cytoplasm, the nuclear material surrounded by a nuclear membrane, and a wide variety of particles like mitochondria, plastids, and many more which are still unknown. The diffusion of biologically important compounds, especially water and solids, across membranes or the establishment of diffusion gradients inside the living cell are just as important as the changes which take place in the water content of the cell. There are conditions in the cell which are often difficult to define but it is important that we pay more attention to them if we want to understand the many effects of radiation. This is especially true if we are concerned with the effects of radiation on the mitotic apparatus. In this case, we have a great variety of biocolloids, some of which are free, others which are fitted in a loose pattern; and to further complicate the problem, the conditions of these biocolloids change with the different steps of mitosis. The pattern of organisation which leads to the successive steps of mitosis in these loosely defined structures and which is highly controlled in the living cell is most easily upset by the absorption of radiant energy. That these steps in the development of the cell are highly sensitive was established many years ago by Dr. Carlson (for review, see Carlson, 1954). As a matter of fact, the effect of as little as 1r on mitotic rate can be recognised (Gaulden, 1956). It has also been established that if the damage is not too severe, the repair will take care of the interference which has been produced by radiation and the cell will finally return to its normal rate of mitosis. I am giving this discussion not to discourage chemists from continuing the study of effects of radiation on water, since it is essential for understanding radiation phenomena, but I believe that in addition to this another approach should be used which will permit us to recognise changes in the irradiated material for which our chemical tests are not yet sensitive enough. It is here that several members of our laboratory have concentrated some of their efforts. It now appears to us that the initial steps of radiation damage probably are rather simple and that what we usually recognise as radiation damage, through observation at a considerable time after exposure, are late steps in the process of radiation damage. I will return to this point later on.

It has been known for years that if you want to grow the grasshopper neuroblast in tissue culture you have to use a certain type of isotonic salt solution. Recent studies have shown that if the salt solution is slightly hypertonic, the successive steps of mitosis which are exactly timed can be speeded up to a slight degree. This effect of the hypertonic salt solution can be determined only if you have an exact method for timing the different steps of mitosis. It has also been observed that the most striking effect of radiation on the rate of mitosis is a slowing down—as a matter of fact, the cells will pile up in late prophase and go ahead again as soon as recovery has started. Mitosis in the grasshopper neuroblast is very radiation sensitive and, as I mentioned before, will react to as little as 1r. When Dr. Gaulden (1956) in our laboratory used hypertonic salt solution (which by itself slightly speeds up the rate of mitosis) immediately after radiation was stopped, it was possible for the cell to overcome the radiation damage so that no slowing down of the rate of mitosis could be observed. In other words, hypertonic salt solution would, in effect, stimulate recovery from radiation damage. This was successful to some extent even after exposure of up to about 50 or 60r. However, the hypertonic salt solution is successful in producing recovery only if it is used immediately after radiation; that is, less than 60 seconds after exposure (Fig. 1). If the hypertonic solution is added later than 60 seconds it will have no direct effect, because by then, apparently, the initial steps of radiation damage have been carried through so that other processes in the cell have been interfered with and nothing can be done by simple means to repair the damage. These experiments assume even more significance in view of the fact that most cells take considerably more time to go through mitosis than does the grasshopper neuroblast. Consequently, there is the possibility that radiation damage induced in cells with a considerably longer mitotic cycle might be even more amenable to recovery agents. Many cells in the animal body take hours before they complete one cycle and it might be possible if one enters immediately (and 'immediately' may be only relative) after radiation into the animal body with certain types of solution, one could counteract at least part of the radiation damage. I bring this example from our laboratory to point out that the initial steps of radiation damage are possibly relatively simple; perhaps they affect the permeability of the cell walls or change in some way the concentration gradient of different solutions in the cell. If we could go immediately after radiation into the cell and study these concentration gradients or the change in permeability, we might be able to get closer to the immediate effects of radiation damage. This new development in the study of radiation damage

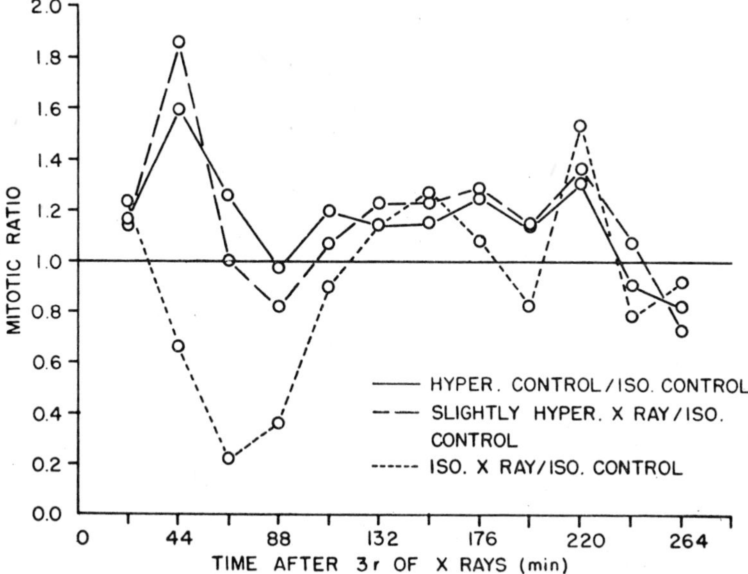

FIG. 1. Mitotic ratio for grasshopper neuroblast versus time after exposure to 3r of X-rays (Gaulden, 1957).

is not as incredible as it may appear at first sight, because it has been known for many years that certain salt solutions will have a slightly stimulating effect on bacteria and bring about recovery in bacteria not only from UV radiation damage but also from certain toxic conditions. I refer especially to the work which was done by Dr. C. E. Winslow (1918) more than 35 years ago and which is buried in the literature. It may be that changes in the synthetic activity of the cell which are affected by radiation, such as: changes in nucleic acid synthesis or breaking down of certain structures, or different steps in the mitotic cycle, or leakage of material out of the cell, are possibly secondary events—events which take considerable time to occur and which could be stopped if we were able to enter the cell soon enough and counteract the damage.

I would like now to discuss some of the newer developments in radiation protection and recovery. Cysteamine continues to be our best protective compound as far as *Escherichia coli* is concerned—less so, in regard to mice. With broth-grown *E. coli*, we have found a dose-reduction factor (DRF) of 12 with cysteamine. There are still some complications in regard to its protective ability since we do not always get a straight-line relationship in regard to protection and the amount of energy used. Dr. Stapleton in our laboratory is now in the process of investigating this in greater detail, especially with regard to dose dependence and rate of diffusion.

The protective compound which has turned out to be most successful in mammals is S, β-aminoethylisothiuronium·Br·HBr (AET), but unfortunately it does not protect as well as cysteamine in *E. coli*, probably because it becomes toxic (Doherty and Burnett, 1955). I will not go into a detailed discussion on this point but I hope it will come up again during discussion of some of the papers which will be given in other sections, especially since very different ideas have developed in regard to protective ability of AET and the mechanisms which are involved.

I would prefer to concentrate a good part of my discussion on the mechanism of bacterial recovery from radiation damage. As we reported earlier, Dr. Stapleton and his co-workers found that if *E. coli* are grown in broth and plated out after exposure on a basal medium consisting of inorganic salts and glucose, a much lower rate of survival is obtained than if the plating out is done on a complete medium (Stapleton *et al.*, 1953). They also found that the radiation incubation temperature is a significant factor: that is, for certain strains of *E. coli*, they had a very much higher survival ratio at 18° C. than at the optimal growth temperature (Stapleton *et al.*, 1955). Further studies showed that if before irradiation bacteria are grown on a basal medium—that is, inorganic salts and glucose, in which normal *E. coli* grow quite rapidly—these cells, after exposure to radiation, do not require the complex medium required by cells which had been grown before exposure in a complete medium. It has also been found that the temperature and the medium effect are additive and that certain natural materials like beef or yeast extracts can be substituted by three simple compounds; i.e. guanine, uracil, and glutamic or aspartic acids, all precursors of ribonucleic acid or protein synthesis. If these amino acids or purines and pyrimidines are added to a basal medium before irradiation, the cells harvested from this medium require these same compounds for increased survival. In these cases amino acids were most effective. It appears from the data on growth conditions prior to irradiation, and the subsequent response to certain types of media, that a good part of the so-called recovery process is related to adaptation, perhaps even to synthesis of enzymes. The cells which have been grown on a basal medium, and which do not require the supplementary materials after irradiation, apparently have all the enzymes necessary to construct these compounds on their own. In contrast to this, the cells which have been grown on a complete medium have lost the ability to synthesise these complicated compounds which are necessary for the growth of the cell. Further evidence for this can be found in the fact that if irradiated cells are incubated for a short time in extracts of natural materials and then removed by centrifugal washing and plated out on a basal medium, they will respond just the same as if they had been plated out on a complete medium. The contact with the natural materials apparently is sufficient to initiate the process or leave behind some substances which enable the cells to recover from the radiation damage. A further point was carefully investigated: the synthesis of ribonucleic and deoxyribonucleic acids in irradiated and non-irradiated cells under a variety of conditions; that is, incubation at 37° C. and 18° C. in both a complete medium and a basal medium. These cells were harvested after various incubation periods, washed and extracted with trichloracetic acid and analysed for nucleic acid by the Schneider technique. Non-irradiated cells show the same synthesis of these compounds in both media at 37° C. At 18° C. there is, as one would expect, a reduction in the rate of synthesis of all compounds. The rate of reduction is not affected by the medium—it is the same for both. In irradiated cells, incubated at 37° C. in a basal medium, no net synthesis of DNA and a substantial reduction of synthesis of the other compounds was observed. However, irradiated cells incubated at 37° C. in a complete medium show a synthesis of all these compounds but a lower rate than takes place in non-irradiated cells; while at 18° C. in a complete medium the cells showed synthesis of all compounds at a rate approaching that in non-irradiated cells. We are investigating this further to determine whether irradiating cells at different phases of the growth cycle will affect the ability of the cell to recover. This can be accomplished only by having synchronised cell division and techniques are now available for doing this. (For a more detailed discussion, see Hollaender and Stapleton, 1956.)

It is probable that some of the most important aspects of radiation damage to be considered are the genetic effects, which may be responsible not only for modification of the genetic make-up of the cell but possibly also for the death of the organism.

As far as chromosome aberrations are concerned, it will be shown at one of the later meetings that, in bean roots and *Drosophila*, modification of this damage by certain treatment after irradiation is quite possible. Although this fits very well into my discussion, I would rather not go into it here since it will be discussed by some of the other speakers.

I would like to emphasise at this time the results which have been obtained with bacteria. The two points I will discuss here are:

1. Are the genetic changes produced by radiation modified by treatment before irradiation as well as by post-treatment?
2. If micro-organisms are protected against radiation damage by reduced oxygen tension or by chemical protection during exposure, will the mutation rate of such organisms be determined by the amount of radiation they have been exposed to, or will the protection for survival extend also to mutational changes?

The use of certain chemicals for radiation protection will be discussed in some of the other papers and I will not go into detail about it here; however, as I have mentioned before, it has been quite well established that one can modify the survival ratio of living cells by using certain chemicals.

The relationship of the mutation rate to the survival ratio can be checked quite readily, especially in regard to chemical protection, since we can have a DRF as high as 12; e.g. using cysteamine on *E. coli*. With a DRF this high, it should be easily possible to determine whether the protective devices will affect the mutation rate as well as the survival ratio. Tests showed that in bacteria which were protected there is a highly significant reduction in the mutation rate; that is, the mutation rate is reduced to about the same extent that the survival ratio is increased. These are experiments which have been done with nutritional reversions and which lend themselves rather readily to an experimental check, going from the requiring state to the non-requiring (Girolami and Hollaender, 1957) (Fig. 2).

Bacteria which are protected against radiation damage by chemical means show a considerably lower mutation rate than one would expect on the basis of the amount of radiation they have received. The mutation rate is somewhat higher than in the case of unprotected bacteria which were given radiation sufficient to yield a survival ratio equivalent to the ratio for the protected bacteria. But the significant factor is that the mutation rate in the protected bacteria is still much lower than it would have been if they had been given the same amount of radiation without protection. Thus we have protected significantly against mutation production in bacteria when we protected them against the killing effect of X-rays.

Now what happens if we permit bacteria to recover from radiation damage by supplying certain nutritional factors? This is the process which I have described before whereby we can get a significant increase in survival ratio by giving guanine, uracil, and glutamic acid. The data are not as plentiful as we have obtained in regard to chemical protection but the indications are that if we add these nutritional factors to broth-grown bacteria after X-irradiation, the result is a lower mutation rate than one would expect from the amount of radiation the cells have received. Our most striking results were obtained when we combined chemical protection with certain nutritional support after irradiation. Unfortunately, on account of certain technical difficulties, the supplying of these nutritional factors is actually a more complicated process than one would visualise, since this interferes with the nutritional deficiencies which must be present for recognising certain types of mutations. However, the indications are that as a result of treatment after exposure, the mutation rate in the recovered cells is considerably lower than one would expect from theoretical reasons.

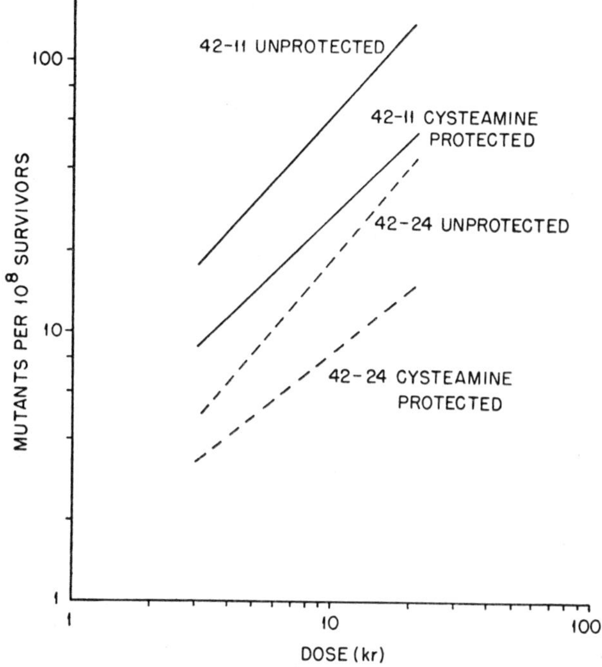

FIG. 2. Mutants per 10^8 survivors versus dose, in kiloroentgens, for two *E. coli* auxotrophs pretreated with cysteamine and unprotected (Girolami and Hollaender, 1957).

What is the mechanism which would account for this peculiar behaviour of the mutations which does not fit exactly into the type of pattern which it was formerly thought existed when cells absorbed ionising radiation? I believe these results indicate that the mutational process, at least in bacteria and probably in other cells, is not a direct one, but that after irradiation is absorbed some physical-chemical change takes place in the cells which requires a certain time for completion. During this time, one can enter the cell and help it to recover from radiation damage. This explanation could serve very well for the effectiveness of certain chemicals on the modification of mutations. It is possible that the steps from the absorption of the radiation and ionisations, which takes place in a fraction of a second, to the chemical change which finally leads to a mutation, require time and are reversible for at least a certain period of time after exposure.

I have emphasised the reversibility of radiation damage in every part of my discussion; the reversion of radiation damage in the grasshopper neuroblast is a very good example of what could be done to modify radiation damage by treatment immediately after exposure. I have emphasised this in regard to chemical protection on bacteria and shall again emphasise here the point that one can reverse certain of the activities which are initiated by radiation if one enters the cell soon after irradiation is completed. To clear this further, we have investigated the time limits governing modification of radiation damage in *E. coli*, and we find that it is not possible to modify radiation damage after cell division has taken place. In general, when this process is initiated, things become irreversible since so many steps are necessary for the growth of the cell, and it would be difficult to reverse this. I believe the cell itself contains in its make-up enough potential materials to counteract a certain amount of radiation damage and that the activities which we initiate by proper treatment are only an encouragement of the protection and recovery potential within the cell (Hollaender, 1956).

Now, what happens in the case where one can show that the effects of onising radiation on mutation production are cumulative? Does this in any way affect the interpretations set forth here? I do not believe this is so. I can visualise that the processes which lead to mutations are statistical processes and that these could be built up step by step and, after they have been completed and established on the chromosome reversion, may no longer be possible. By this I mean that each microscopic step that finally leads to mutation as such may have reversibility for only a limited time and that the final step which we recognise as a mutation may be an accumulation of processes which are irreversible only if nothing is done about them immediately after irradiation. In other words, the philosophy on which we are working now is that it might be possible to reverse individual steps which lead finally to the production of mutations which we recognise, but that treatment must take place at the stage or step when the process is reversible. I do not know how general the applicability of such an idea might be and it may apply only to a certain limited number of cases and special techniques may have to be designed to recognise these (Hollaender and Kimball, 1956).

I would like now to return to the initial theme which was emphasised in the introduction; that if we could understand the immediate steps and initial processes which take place when radiation is absorbed, it might be much easier for us to design means of counteracting radiation damage.

I would like to add a word of caution in regard to the general applicability of this idea. Our data are quite limited and we do not know to how many organisms this can apply; however, we do have at least a beginning of a type of investigation which may lead us to a better understanding of radiation mechanisms, and eventually to a means of modifying radiation damage to living cells.

Acknowledgment

Work performed under Contract No. W-7405-eng-26 for the Atomic Energy Commission.

REFERENCES

CARLSON, J. G., 1954. *Radiation Biology*, Vol. I, ed. A. Hollaender (New York, McGraw-Hill Book Co.).

DOHERTY, D. G., and BURNETT, W. T., Jr., 1955. *Proc. Soc. exp. Biol. N.Y.*, **89**, 312.

GAULDEN, M. E., 1956. Manuscript.

GIROLAMI, R. L., and HOLLAENDER, A., 1957. Manuscript.

HOLLAENDER, A., 1956. *Proc. Int. Conf. Peaceful Uses of Atomic Energy*, Vol. 16 (United Nations, New York).

HOLLAENDER, A., and KIMBALL, R. F., 1956. *Nature, Lond.*, **177**, 726.

HOLLAENDER, A., and STAPLETON, G. E., 1956. *Ciba Foundation Symposium—Influence of Ionising Radiation on Cell Metabolism* (London, Churchill).

STAPLETON, G. E., BILLEN, D., and HOLLAENDER, A., 1953. *J. Cell. Comp. Physiol.*, **41**, 345.

STAPLETON, G. E., SBARRA, A. J., and HOLLAENDER, A., 1955. *J. Bact.*, **70**, 7.

WINSLOW, C. E. A., and FALK, I. S., 1918. *Proc. Soc. exp. Biol. N.Y.*, **15**, 67.

DISCUSSION

Alper. Conditions which can ordinarily be regarded as optimal for the growth of micro-organisms are apparently not necessarily those which will enhance recovery from lethal or genetic effects—in fact several reports indicate that the reverse is true. From the work of Dr. Hollaender and his collaborators we know that holding bacteria at sub-optimal temperatures after irradiation increases the number of viable colonies formed. Dr. Kimball has shown that in conditions which slow down the synthetic processes in *Paramecium*, after irradiation fewer dominant lethals are produced; and recently Dr. Witkin has made a careful investigation of mutation production by UV in *S. typhimurium*. She found that expression of mutants could be very greatly suppressed by conditions which prolonged the log phase of the organisms.

Hollaender. I agree with Dr. Alper that conditions which are optimal for growth are not necessarily the best for recovery. Our interpretation of this is that when cells divide at the maximum rate a great strain is put on all of the cell's resources. Radiation has interfered with several steps of the synthetic function of the cell and has made it difficult to supply the necessary building stones for growth. If we incubate our cells at sub-optimal temperatures—for instance, *E. coli* at 18°—the enzymatic functions of the cell continue but growth is slowed down so much that a sufficient supply of enzymes can be built up thereby catching up with cell division.

Alexander. I wonder whether Dr. Hollaender would consider the following as a possible explanation for the remarkable restoration of radiation damage by changing salt concentration, which he has just described.

Bacq and I have suggested that an important factor in the cellular injury of radiation may be the release of enzymes from their special sites. These released enzymes can then come into contact with sites from which the cell normally takes great care to keep them away, and in this way produce serious damage. By making the external solution which surounds the cell more hypotonic immediately after irradiation it may be that these released enzymes are washed out in the exchange of liquid which must occur in changing the osmotic pressure. In this way the released enzymes are prevented from producing the damage.

Hollaender. The scheme Dr. Alexander suggests, I believe, is reasonable. However, I think the conditions are even simpler: the concentration of toxic substances might be high in a very small localised area, but might be diluted when spread out over all the cell and thus become less toxic. It should be kept in mind that these are all pure speculations.

Dale. I should like to ask Dr. Hollaender whether the compensating effect of hypertonicity can be substituted by dehydrating procedures?

Hollaender. This has not been tried because of the high sensitivity of neuroblasts to water removal. Also bacteria become very sensitive to water removal after irradiation. I think in other cells this can be tried, on *Neurospora* spores or other organisms.

Auerbach. Are there indications that different mutational steps require different growth factors for recovery? This might have a bearing on the preferential production of certain mutations by given mutagenes.

Hollaender. We have no real information on this point at the present time, but we have some indication that this can take place.

Marcovich. Did you try the cysteamine in the case of a streptomycin-dependent strain, which has been shown by Anderson to be unaffected by oxygen tension?

Hollaender. We have tried using cysteamine on the streptomycin-dependent strain, but have got some contradictory results. We have found that the streptomycin-dependent strain is the most difficult one to handle.

van Bekkum. With regard to the experiments with neuroblasts, I should like to ask you what happens when the neuroblasts are irradiated in hypertonic solutions?

Hollaender. Dr. Gaulden uses a slightly hypertonic solution during irradiation. This solution, if it is not too high, will speed up mitosis to a slight degree. As a matter of fact, the safest way to get good effect of hypertonic salt solution after irradiation, which in this case is about 1·2, is to use hypertonic salt solutions—for instance, 1·02 during irradiation.

Swallow. Dr. Hollaender has emphasised the importance of the complexity of the cell. One of the reasons why radiation chemists have so far neglected this aspect has been the difficulty of knowing how to tackle a problem like organisation in chemical terms. It now seems that the answer has been found in the study of hydrogen bonding. Hydrogen bonds may play an important part in maintaining the loose structure in the cell. Now Cox and Peacocke have shown by an elegant new titration method that hydrogen bonds are exceptionally sensitive to the action of radiation (paper in *Nature*, *Journal of the Chemical Society*, *Proceedings of the Royal Society*). Radiation chemists now have available a purely chemical technique for studying what seems to be a vital aspect of the cell. Work of this type may bridge the gap between the sort of chemical work that has been done so far and some of the biological work.

Hollaender. I like the suggestions of Dr. Swallow very much and the technique he suggests may be of great use in radiation biology.

THE PRESENT STATUS OF RADIATION PROTECTION BY CHEMICAL AND BIOLOGICAL AGENTS IN MAMMALS

J. A. Cohen, O. Vos and D. W. van Bekkum
Medical Biological Laboratory of the National Defence Research Council T.N.O., Rijswijk-Z.H., Netherlands

The injury caused by total-body irradiation in mammals can be significantly reduced by two forms of treatment which are commonly known as chemical protection and bone marrow therapy. Both methods of treatment may be considered of value in counteracting the lethal effects of ionising radiation. Moreover the study of the mechanisms involved may be expected to yield additional information on the nature of the radiation injury. In this paper the two methods of treatment will be discussed in the light of results obtained recently in our laboratory.

Chemical protection

The so-called chemical protection consists in the administration of certain chemical substances prior to irradiation and results in a decrease of the damage to a number of radiosensitive tissues. So far the exact degree of protection has only been estimated satisfactorily with regard to mortality. These data probably reflect the protection afforded to the critical tissue, which under these conditions would be the hæmopoietic system. The protection afforded to other tissues has not been investigated systematically but the available data indicate an unequal degree of protection. Whether these differences are adequately explained by a selective concentration of the protective substance, Eldjarn (1956), Verly, Koch and Grégoire (1955) or are related to differences between the tissues, remains to be settled. More information on these points is needed in order to permit an evaluation of the possibilities of the application of chemical protectors in the field of radiotherapy.

The number of protective substances has gradually increased. Besides cysteamine a number of other equally active substances has been described in recent years (Table I). In several laboratories various cysteamine derivatives have been investigated, a few of which proved to be as effective as or even slightly better than cysteamine, Doherty and Burnett (1955). From our laboratory the protective activity of a series of dithiocarbamic acid derivatives has been reported, van Bekkum (1956*a*); of these several could be classified as very active. More recently we synthesised the cysteamine derivatives 1-amino-propanethiol-2 and 2-amino-propanethiol-1. They are included in Table I; moreover in preliminary toxicity tests they were found to possess a favourable therapeutic width.

Another group of protective substances has been found among the so-called biological amines. Histamine is one of the best protectors of this group, but various phenylethylamine derivatives also afford considerable protection to mice, Alexander *et al.* (1955). Two other groups of protective substances, namely the sulphhydryl compounds cysteine and glutathione and some respiratory inhibitors, e.g. cyanide and carbon monoxide, are worth mentioning. These compounds afford a clear-cut protection to mice, but are not as effective as the substances listed in Table I.

TABLE I

Activity of some Potent Protectors in Mice *

Compound	Laboratory	Test dose μ M/mouse	% survival 20–30 days		
			675–700r	800r	900–950r
Cysteamine	Liège	40	97	—	50
	Oak Ridge	44	—	100	56
	Rijswijk	40	96	80	50
S,β-aminoethyliso-thiuronium . Br.HBr	Oak Ridge	32	—	80	70
	Rijswijk	18	100	80	80
2-amino pro-panethiol-1	Rijswijk	46	100	80	80
1-amino pro-panethiol-2	Rijswijk	76	100	100	90
Diethyldithiocarb. ...	Liège	40	100	—	40
	Rijswijk	60	70	0	—
Histamine	Liège	40	—	70–100	—
	Rijswijk	66	100	—	15

* Oak Ridge data : Doherty *et al.*, 1955. Liège data : Bacq *et al.*, 1952, Alexander *et al.*, 1955.

With regard to the mechanism of the chemical protection, it is noteworthy that the approach to this problem has recently shown a significant change. For some time there has been a tendency to attribute the protective action of most chemical substances to their postulated interaction with radiation-produced radicals. This attractive hypothesis was strongly advocated by Alexander *et al.* (1955) on the basis of their experiments with a polymer system and with mice. However no satisfactory explanation was provided for the selective action of the chemical protectors in view of the fact that living cells contain numerous substances that are also known to inactivate radicals in model solutions.

In the past year the radical-inactivation postulate has been found to be untenable in the case of several protective compounds. Eldjarn (1956) concluded that only a minor portion of the protective effect of cysteamine can be explained by this mechanism. We have now shown that a number of substances which have a high protective action *in vivo*, e.g. histamine, do not protect isolated cells *in vitro*, van Bekkum and de Groot (1956). It is likely that these substances owe their protective action to specific pharmacological characteristics, which probably cause a lowering of the oxygen tension in some of the radiosensitive tissues. In the case of cysteine and glutathione Patt's original hypothesis that these substances protect by removing oxygen from the cells has recently received additional support, Gray (1956), van Bekkum and de Groot (1956).

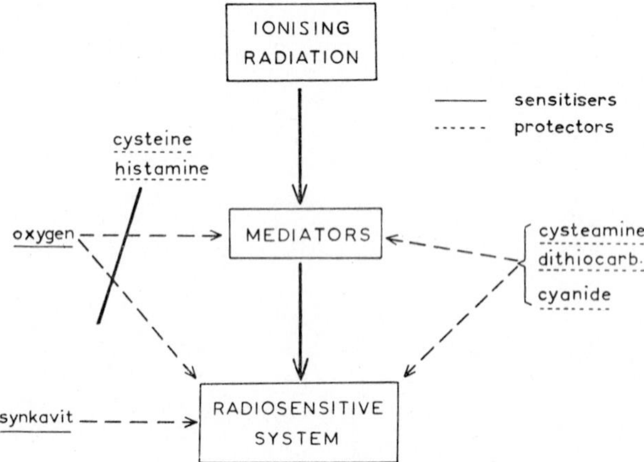

FIG. 1. Postulated mode of action of various radioprotectors and radiosensitisers.

From our results obtained with isolated thymocytes we have previously concluded that this explanation does not hold for the protective action of cysteamine and the dithiocarbamates. In the case of cysteamine, Eldjarn (1956) has reintroduced the hypothesis of an interaction between the protector and the radiosensitive target and this approach seems to be most promising in the case of the dithiocarbamates as well.

These considerations have been visualised in a scheme (Fig. 1) which shows the various ways in which the radiation effect may be modified by chemicals. The term mediator includes all the postulated agents that eventually cause the destruction of the radiosensitive molecules, and these agents may be localised anywhere along the chain of reactions that follows the impact of the ionising event on the organism. The sensitising as well as the protective substances may act either on the mediator or on the radiosensitive site. The most important sensitising agent is of course oxygen and a number of protectors, e.g. cysteine and the biological amines, may simply act by lowering the oxygen tension in the tissues.

The sensitising action of oxygen is generally attributed to its interaction with the mediators, which may lead to the formation of more toxic agents. It is postulated here that oxygen may also cause sensitisation by influencing the radiosensitive site. In the case of some of the protective substances, e.g. cysteamine and the dithiocarbamates, the latter mode of action is also considered most likely. A speculation on the nature of the radiosensitive site has been depicted in Figure 2. It is postulated that this structure is localised somewhere in the oxido-reduction chain of the cellular metabolism and that its sensitivity to radiation depends on the extent to which it is

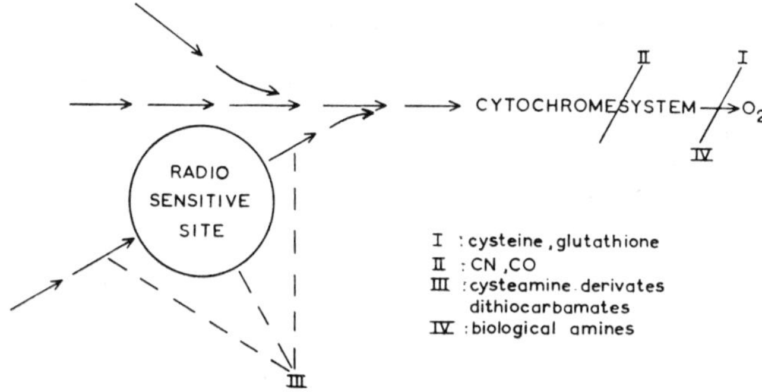

FIG. 2.—Hypothetical scheme of the radiosensitive site.

oxidised. This implies that a high degree of reduction corresponds to radioresistance. In that case all potential electron donors to the system may offer protection in a non-specific way, by increasing the reduction potential at the radiosensitive site. Conversely, electron acceptors (oxygen) increase the radiosensitivity of the system. Groups I and IV of the protective compounds cause a decrease of the electron accepting capacity by interference with the oxygen supply. Group II would act by blocking the electron stream beyond the radiosensitive structure, viz. at the cytochrome oxidase stage, thus inducing a reduced state of the former. The compounds of group III may act directly on the radiosensitive site. On the other hand they might also act by increasing the electron flow towards this structure or by blocking the current of electrons downstream.

At present this scheme constitutes a working hypothesis only. However, it takes into account the observation that radiation causes a disturbance of oxidative phosphorylation in some of the radiosensitive tissues, which is one of the very few biochemical effects that occur within a few hours after irradiation. The recent finding of one of us, van Bekkum (1956*b*), that

Cytochrome *c* may be involved in the radiation-induced uncoupling of oxidative phosphorylation, seems to support the postulate that the respiratory chain is intimately connected with at least one of the radiosensitive sites in living cells.

Bone marrow therapy

The administration of hæmopoietic tissue preparations is only effective after the exposure to radiation. For the past several years the therapeutic effect of these preparations has been generally attributed to the presence of a humoral factor of unknown identity. Only recently the alternative point of view, which has been held by Loutit and co-workers, has been confirmed beyond any doubt by four different groups of workers, Lindsley, Odell and Tausche (1955); Nowell, Cole, Habermayer and Roan (1956); Ford, Hamerton, Barnes and Loutit (1956); Vos, Davids, Weyzen and van Bekkum (1956). We now know that intact viable hæmopoietic cells are required to save the life of irradiated animals. These cells multiply in the host, resulting in the entire or partial substitution of its own hæmopoietic system. The limitations of this form of therapy for the radiation syndrome can now be evaluated more clearly and will probably turn out to be much more severe than was originally believed. On the other hand the whole subject has become of considerable interest from the point of view of the transplantation of living hetero- and homologous tissues. Obviously a large amount of additional information will be needed before a thorough assessment of the possibilities can be made.

With regard to the bone marrow therapy we wanted to know, in the first place, the conditions necessary for the survival of the implanted cells. So far most of the therapeutic results have been expressed on the basis of thirty-day survival recordings. These supply adequate information on the survival and the proliferation of the donated cells during the first few weeks. Figure 3 shows that the method of administration plays a significant role in the transplantation process. When the intraperitoneal route of administration is employed, the number of viable cells required to protect 50% of the irradiated mice has been found to be 75 times greater than in the case of intravenous administration. No appreciable difference was observed between the intravenous injection and the injection into the spleen. It is noteworthy that preparations containing only about 50,000 viable nucleated bone marrow cells are required to provide a significant protection. Figure 4 shows the number of cells from isologous (CBA), homologous (C57 Black) and heterologous (rat) bone marrow that is required to permit a thirty-day survival of the irradiated host (CBA mice). It shows that roughly 20 times more homologous cells are needed to provide the same protection as in the case of an isologous graft. The activity of the heterologous and homologous

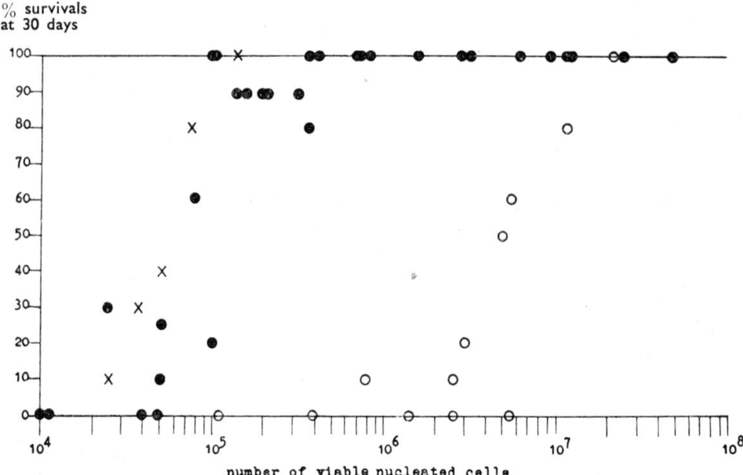

FIG. 3. Therapeutical effect of isologous bone marrow in C57 black mice, using different routes of injection. X-ray dose 700r (LD_{100}). Each point represents the results obtained in a group of 10 animals.

●: intravenous.

○: intraperitoneal.

×: injection into the spleen.

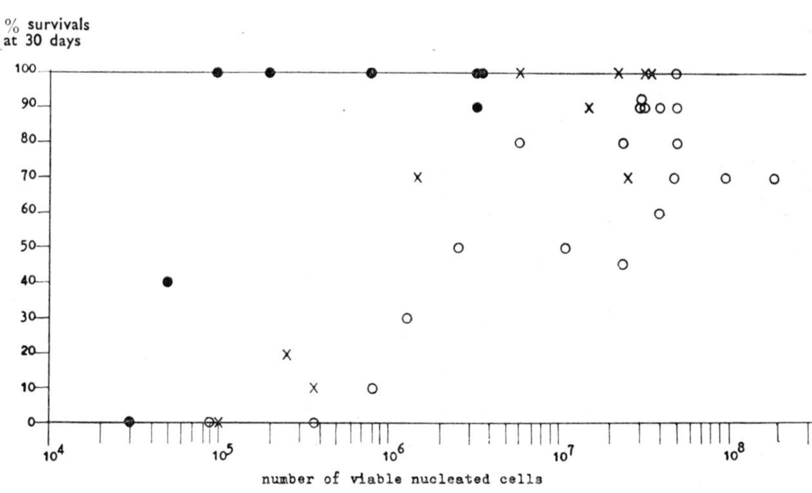

FIG. 4. Therapeutical effect of different kinds of bone marrow in CBA mice. X-ray dose 675r (LD_{100}). Each point represents the results obtained in a group of 10 animals. All cells were injected intravenously.

●: isologous.

×: homologous (C57 Black).

○: heterologous (rat, WAG).

cells does not differ much. Of course these results cannot be generalised. It is likely that different numbers of bone marrow cells will be needed, when different mouse strains are used as donors. The same probably holds for heterologous cells. In a limited number of experiments we did not observe any effect of the injection of bone marrow cells from rabbits and rabbit fœtuses into irradiated mice. Bone marrow cells of hamsters and guinea-pigs were effective although considerably less than rat cells.

TABLE II

The Therapeutic Effect of Heterologous Bone Marrow Cells after Various Doses of Total-Body Irradiation

(Irradiated animals: CBA male mice; 50×10^6 viable nucleated bone marrow cells were injected i.v. per mouse)

Radiation dose	Number of mice	30-day survivals
100r	10	10
200r	10	10
300r	10	10
400r	10	10
500r	10	0
600r	10	1
675r	60	46
800r	20	18

Among the conditions which govern the survival of the graft we have found the dose of total-body irradiation to be of great significance in the case of heterologous transplantation. Table II shows that successful therapy using heterologous bone marrow cells can only be achieved after relatively large doses of irradiation. The curious phenomenon occurs, that after irradiation with a dose of 675r and more, marked protection is obtained with heterologous cells, while no protection is observed after irradiation with lower doses in the lethal range. In the animals irradiated with 500r no rat cells could be found in the peripheral blood, while in the animals irradiated with 600r only in three cases a few rat cells were found. One of these animals has survived 30 days. On the other hand numerous rat cells could be identified in mice after irradiation with a dose of 675r and more.

We are tempted to conclude that with the lower doses some immunological response from the irradiated host remains possible, sufficient to dispose of the graft. Additional support for this hypothesis is gained from an experiment in which lethally irradiated mice received a sufficient amount of heterologous bone marrow cells to protect most of them. In addition we injected about 6 million normal viable isologous lymph gland cells and these doubly treated mice have all died. This demonstrates that only a limited number of lymphoid cells are required to prevent the taking of the graft.

The second point that seems to be of interest is the ultimate fate of the 30-day survivals. Table III shows that 30-day survivals do not die for a long time in the case of isologous treatment, while a marked delayed mortality is observed after treatment with homologous and heterologous cells. This delayed mortality shows a definite peak in the period between 30 and 80 days following the irradiation. The cause of this is not yet known. One of the first questions to be answered is whether this delayed mortality could

TABLE III

Delayed Mortality of CBA Mice after Irradiation with 675r and Treatment with Isologous, Homologous or Heterologous Bone Marrow

Bone marrow injected	Number of surviving animals after			
	30 days	60 days	80 days	100 days
Mouse (CBA) ...	52	51	51	51
Mouse (C57 Black) ...	64	36	27	25
Rat (WAG)	133	78	65	53

be the result of a disturbance of the proliferation of the grafted heterologous cells in the host. It is conceivable that an eventual reappearance of the host's own hæmopoietic cells might enable him to produce antibodies against the transplanted tissue; this could lead to anaphylactic symptoms. One of the most striking clinical findings is the occurrence of severe diarrhœa during this period. This diarrhœa has not been observed after treatment with isologous cells and this would suggest the involvement of immunological factors.

In a number of animals we have followed the relative number of heterologous cells in the peripheral blood after the treatment. Figures 5–7 show the occurrence of host cells and donor cells in the peripheral blood of these animals. In this experiment all mouse erythrocytes and granulocytes have disappeared, being replaced by rat cells. In the observed period no reversal to mouse cells was found. However, in other experiments we observed a few cases in which subsequently a complete return to mouse cells occurred (total reversals). We have found that such total reversals may occur at different times after irradiation, for instance before the 37th day after irradiation and after the 100th day after irradiation. Furthermore we observed some mice that maintained a mixed population of rat and mouse erythrocytes and granulocytes for long periods after the irradiation (about

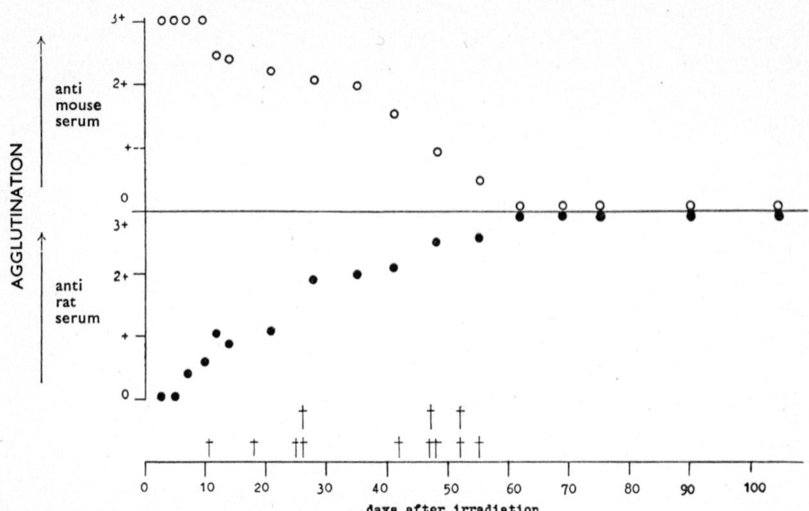

FIG. 5. Identification of erythrocytes in CBA mice, irradiated with 675r (LD_{100}) and treated with rat bone marrow. Twenty mice were irradiated and treated.

†: death of one animal.

●: mean value of agglutination reactions with anti rat serum.

○: mean value of agglutination reactions with anti mouse serum.

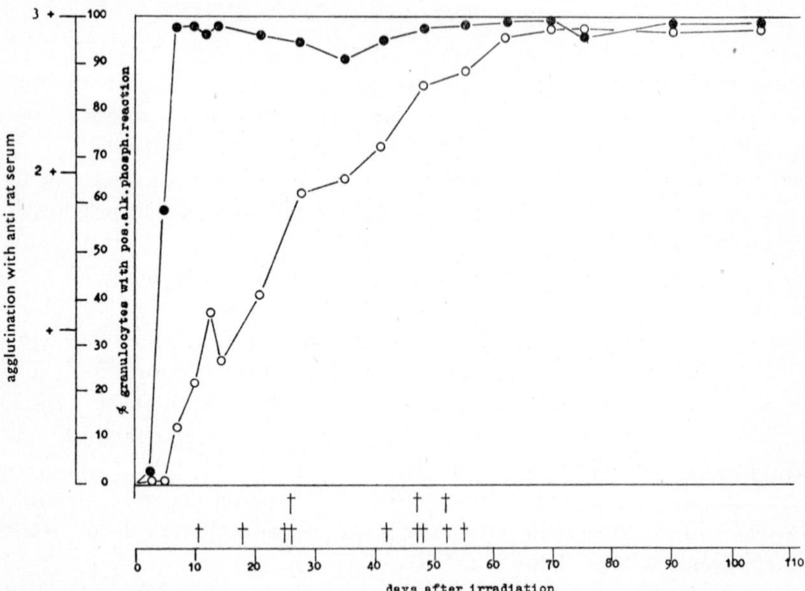

FIG. 6. Identification of rat erythrocytes and rat granulocytes in CBA mice, irradiated with 675r and treated with rat bone marrow. Twenty mice were irradiated and treated.

†: death of one animal.

○——○ : mean value of agglutination reactions with anti rat serum.

●——● : mean percentage of rat granulocytes (positive alkaline phosphatase reaction).

300 days). Our results do not indicate a clear relationship between the reappearance of the host's own blood cells and the delayed mortality.

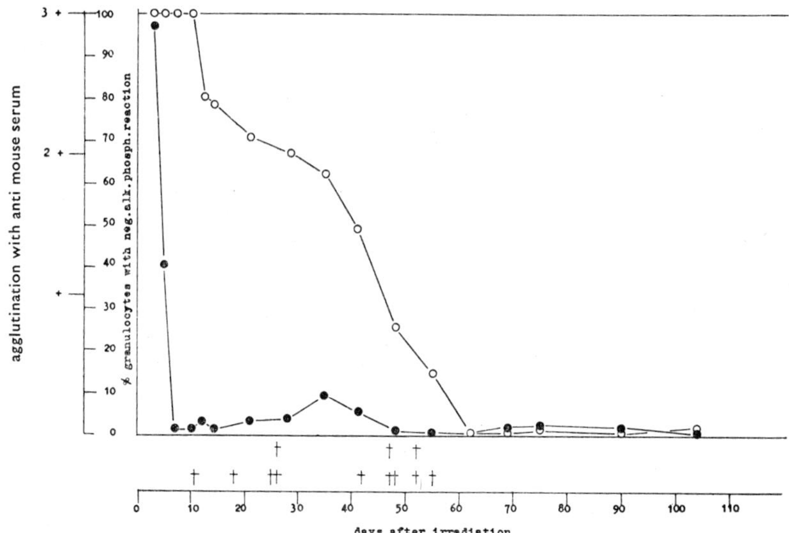

FIG. 7. Identification of mouse erythrocytes and mouse granulocytes in CBA mice, irradiated with 675r and treated with rat bone marrow. Twenty mice were irradiated and treated.

†: death of one animal.
○——○ : mean value of agglutination reactions with anti mouse serum.
●——● : mean percentage of mouse granulocytes (negative alkaline phosphatase reaction).

The possibility should be kept in mind of a delayed lesion caused by radiation doses which kill the animal under normal conditions within 14 days.

TABLE IV

Mortality in CBA Mice Irradiated with Various Doses of Total-Body Irradiation and Treated with Isologous Bone Marrow ($3 \cdot 2 \times 10^6$ Viable Nucleated Cells i.v.)

Radiation dose	Number of mice	Number of survivals after 30 days	60 days
675r	10	10	10
800r	9	9	8
950r	10	10	4
1200r	10	0	—

An observation that supports this hypothesis is that after higher lethal doses of irradiation, therapy with isologous cells is accompanied by a certain amount of delayed mortality (Table IV). It is conceivable that the animals which possess a foreign bone marrow might be less resistant to lesions, resulting in a higher mortality. In fact we have noticed that these animals are less resistant to various animal pests such as mites and lice.

It is obvious that much more data about the condition of the animals treated with homologous and heterologous bone marrow will have to be collected before the value of this therapy can be determined. However, the demonstration of successful transplantation of homologous and heterologous bone marrows and the immunological implications of the experiments described, open unexpected fields of research and bear promise of future clinical applications.

REFERENCES

ALEXANDER, P., BACQ, Z. M., COUSENS, S. F., FOX, M., HERVE, A., and LAZAR, J., 1955. *Radiation Res.*, **2**, 392.

BACQ, Z. M., and HERVE, A., 1952. *Bull. Acad. Roy. Med. Belg.*, **17**, 13.

VAN BEKKUM, D. W., and DE GROOT, J., 1955. *Progress in Radiobiol.*, 1956, p. 243 (Edinburgh, Oliver and Boyd).

VAN BEKKUM, D. W., 1956*a*. *Acta Physiol. Pharmacol. Neerl.*, **4**, 508.

Idem, 1956*b*. *Symp. on Influence of Ionising Radiations on Cell Metabolism* (The Ciba Foundation). To be published.

DOHERTY, D. G., and BURNETT, W. T., 1955. *Proc. Soc. exp. Biol. N.Y.*, **89**, 312.

ELDJARN, L., 1956. *Proc. Int. Conf. Peaceful Uses of Atomic Energy in Geneva*, 1955. *United Nations, New York*, **11**, 335.

FORD, C. E., HAMERTON, J. L., BARNES, D. W. H., and LOUTIT, J. F., 1956. *Nature, Lond.*, **177**, 452.

GRAY, L. H., 1955. *Progress in Radiobiol.*, 1956, p. 267 (Edinburgh, Oliver and Boyd).

LINDSLEY, D. L., ODELL, T. T., and TAUSCHE, F. G., 1955. *Proc. Soc. exp. Biol. N.Y.*, **90**, 512.

NOWELL, P. C., COLE, L. J., HABERMAYER, J. G., and ROAN, P. L., 1956. *Cancer Res.*, **16**, 258.

VERLY, W. G., KOCH, G., and GRÉGOIRE, S., 1954. *Radiobiol. Symp.*, p. 110, (London, Butterworth).

VOS, O., DAVIDS, J. A. G., WEYZEN, W. W. H., and VAN BEKKUM, D. W., 1956. *Acta Physiol. Pharmacol. Neerl.*, **4**, 482.

DISCUSSION

Loutit. I have one question, may I have your opinion on the cause of the delayed death?

Vos. We can only give our provisional opinion because the histological investigations have not yet been completed. We think that immunological factors may play a role, but that they are not the only cause of the delayed death. The delayed death in mice treated with isologous cells after the higher lethal X-ray doses supports this hypothesis. Other irradiation lesions and insufficient proliferation of the grafted hæmatopoietic elements caused by other than immunological factors may be involved.

Sobels. It might be interesting to investigate whether there is a heterosis effect operating. That is, whether implantation of bone marrow cells from a heterozygote between two inbred strains gives better results than cells from either strain by itself.

Vos. We also wondered if bone marrow of a heterozygote of two inbred strains of mice will give better results in one of these inbred strains than the homozygote homologous cells. Experiments of this sort have been started. However, the observation time is as yet too short to allow conclusions to be drawn.

Kanazir. Have you tried to protect with highly polymerised homologous DNA?

Hollaender. We have used highly polymerised nucleic acid solutions but have had no success with regard to bacteria. In contrast to this, Dr. Stapleton had some success with the nucleic acid bacterial extract in the recovery of mice. Here the nucleic acid bacterial extract gives approximately 10–20% recovery after lethal quantities of radiation, but again the results are not very regular.

Muller. Would it be worth while to try to induce tolerance in a group of mice by injecting them in the fœtal stage with homologous cells of a given homozygous strain or from a given animal, and later after these fœtuses become mature, to test whether cells of this same strain of animal would, after post-irradiation injection, protect the irradiated mature mice from the radiation without causing the usual delayed diarrhœa and death?

Vos. We are trying to induce in a group of inbred mice an acquired tolerance against tissue of another inbred strain. We intend to do some of the experiments Professor Muller has suggested with these animals.

Elson. I am very interested in the suggestion of an action on the cytochrome system. Does Professor Cohen have any idea which part of the system may be involved? If, for instance, there is an effect on the succinic dehydrogenase component, this might suggest yet another mechanism for the protective action of cysteinamine as the dehydrogenase is an —SH enzyme, and prevention of inhibition of these enzymes by cysteine, etc. is well known. I do not think, however, there is much evidence for inhibition of the succinoxidase enzyme system by irradiation.

Cohen. No, we did not estimate the succinic dehydrogenase but I agree with you that it has not been found by others to be particularly sensitive to irradiation. As an SH enzyme it should certainly qualify for protection by SH-compounds.

Ford. I was struck by the apparent regularity of the curve for granulocytes and erythrocytes in the heterologously protected mice. This seems to contrast with our cytological observations which indicate great variability in respect of retention or replacement of tissue derived from the donated cells in similarly treated mice. Would Dr. Vos comment on this?

Vos. The regularity in our curve for the replacement of granulocytes and erythrocytes in the heterologous treated mice is only an apparent regularity, because the values shown represent the means of groups of mice. Between the different individuals great differences have been found. A few months after irradiation and treatment with heterologous cells we found three types of mice: (*a*) animals with rat cells only, (*b*) animals with mouse cells only, (*c*) animals with both mouse and rat granulocytes and erythrocytes in the peripheral blood. So in our material there also exists great variability between different individuals. The curve of the mouse and rat cells in any individual mouse, however, is not irregular. Replacement of one cell type by the other occurs very gradually.

II. CHEMICAL METHODS: EFFECTS OF CYSTEAMINE CYSTEIN AND RELATED COMPOUNDS

STUDIES ON THE MECHANISM OF PROTECTION AGAINST IONISING RADIATION BY COMPOUNDS OF THE CYSTEAMINE-CYSTEINE GROUP

ALEXANDER PIHL AND LORENTZ ELDJARN
Norsk Hydro's Institute for Cancer Research,
The Norwegian Radium Hospital, Oslo, Norway

The bulk of the available evidence indicates that the sulphur-containing protective agents act by reducing the immediate biochemical alterations caused by ionising radiation. This holds true, irrespective of the question whether a regeneration factor is protected or the cellular lesions in general are reduced.

When the data on the relative protective ability of different compounds are studied, numerous inconsistencies are found. However, certain trends can be discerned and it seems that protective ability can be correlated with definite structural properties (Koch and Hagen, 1956). Their data indicate that the protective ability of these compounds is strongly influenced by minor changes in the inductive effect exerted on the sulphur function. The widely different protective ability of structurally related compounds must be accounted for by any theory which attempts to explain the mechanism of chemical protection.

It has been generally assumed that the compounds of the cysteamine-cysteine group exert their protective action by reducing in general the indirect radiation effect. Data on the relative ability of the protective agents to capture radicals in solution are so far lacking. In our opinion, the available indirect evidence (Alexander *et al.*, 1955; Fox, 1955) does not support the concept that inherent differences in this respect can account for the different ability of the various compounds to exert protection. Furthermore, evidence has been presented (Eldjarn, Pihl, and Shapiro, 1956) that when cystamine is present in a biological medium, its ability to capture radiation-induced free radicals is too low to explain, by the above mechanism, its protective effect *in vivo*.

Chemical protection obviously requires the presence of the protective agent within the cell. It is, therefore, to be regretted that only limited information is available on the organ distribution and the intracellular localisation of the various protective compounds. The available data indicate that differences in organ distribution can only in part explain differences in protective ability (Patt *et al.*, 1950; Eldjarn and Nygaard, 1954).

The possibility that the protective agents exert their action by reducing the intracellular oxygen tension, has attracted considerable interest. No doubt, sulphhydryl compounds can be oxidised, but it is relevant to point out that a typical disulphide, such as cystamine, is also protective. Furthermore, from a biochemical point of view, it is difficult to see how thiols could consume oxygen to a higher degree than many normal metabolites. Also, the data of Hagen and Koch (1955) show no correlation between protective ability and rate of autoxidation.

When all the facts are taken into consideration, it seems probable that several of the proposed mechanisms of action are operating and contribute, to some extent, to the total protective ability of the sulphur-containing agents. However, it is felt that the main part of their protective effect must be ascribed to other mechanisms being able to account for the observed relationship between chemical structure and protective ability.

A new aspect of the mode of action of the sulphur-containing protective agents was introduced when it was demonstrated that cystamine interacts, *in vitro* as well as *in vivo*, with biological SH groups to form appreciable amounts of mixed disulphides (Eldjarn, Pihl and Shapiro, 1956; Eldjarn and Pihl, 1955 and 1956*a*). On this basis a mechanism was suggested, according to which a temporary formation of such mixed disulphides represents a partial protection of the molecules against the direct as well as the indirect action of ionising radiation.

In the present paper data are presented on the chemical kinetics of thiol-disulphide interactions. It is shown that various disulphides differ greatly with respect to the rate and the extent of interaction with glutathione and cysteine. The data indicate that good protective agents react rapidly with these thiols to form large equilibrium concentrations of mixed disulphides. Cystamine seems to be nearly optimal with regard to the extent of mixed disulphide formation. Furthermore, evidence is presented that, conversely, the protective agents in their thiol form interact similarly with accessible body disulphide groups. The data support the view that the protective effect of agents of this group is associated with the ability to form mixed disulphides.

Experimental

In the present paper the interaction of ^{35}S-labelled cystamine (RS*S*R), cysteamine (RS*H), and N,N'-diacetylcystamine ($(Ac–RS^*)_2$) with glutathione (GSH, GSSG), cysteine (CSH, CSSC), bovine serum albumin, and a commercial Cytochrome *c* preparation, has been studied.

The compounds were incubated in oxygen-free phosphate buffer (0·067 M) under the desired conditions of *p*H and temperature, the reactions were stopped by rapid acidification, and the reaction products separated from the reactants by paper electrophoresis at *p*H 2. The amounts of mixed disulphide formed were measured by scanning the radioactivity on the paper electrophoretograms. The details of the experimental procedure are given elsewhere (Eldjarn and Pihl, 1956*b*).

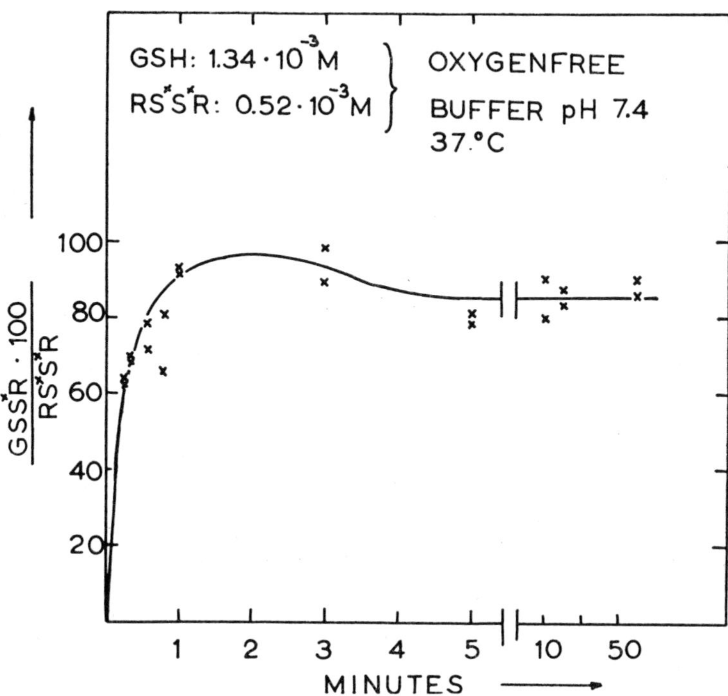

FIG. 1. The rate and extent of formation of the mixed disulphide GSS*R. Glutathione (GSH) and ^{35}S-labelled cystamine (RS*S*R) were incubated, for varying lengths of time, in phosphate buffer, and the mixed disulphide separated by paper electrophoresis.

The reaction between thiols and disulphides is generally formulated as

$$\text{A. } 2\,\text{GSH} + \text{RS*S*R} \underset{}{\overset{k_1}{\rightleftharpoons}} \text{GSSG} + 2\,\text{RS*H}$$

There is ample evidence that the reaction proceeds according to the equations

$$\text{B. I. } \text{GSH} + \text{RS*S*R} \overset{k_2}{\rightleftharpoons} \text{GSS*R} + \text{RS*H}$$

$$\text{II. } \text{GSS*R} + \text{GSH} \overset{k_3}{\rightleftharpoons} \text{GSSG} + \text{RS*H}$$

It has been assumed that the mixed disulphides exist only as intermediates in catalytic amounts (Bersin and Steudel, 1938). However, when glutathione was incubated with labelled cystamine at 37° C., in the absence of oxygen, and at *p*H 7·4, considerable amounts of mixed disulphides were found to be present at equilibrium, as shown in Figure 1. It is also evident that the reaction takes place at a rapid rate, equilibrium values being reached in about one minute.

From the type of data obtained in Figure 1 it is possible to obtain the velocity constants and the equilibrium constants of the system. When the latter constants are known, the equilibrium concentrations of all the five

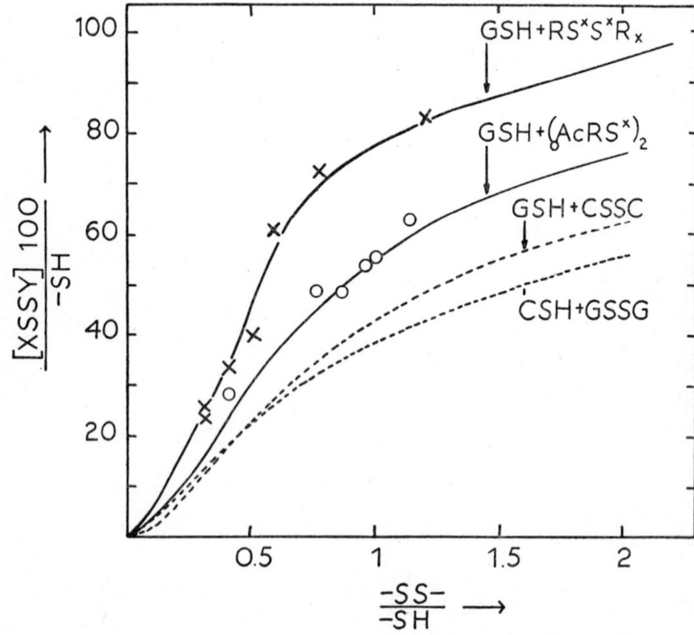

FIG. 2. The equilibrium concentrations of mixed disulphide in various thiol-disulphide systems, expressed as a function of the initial SS/SH ratio.

×——× Drawn on the basis of experimentally determined values.
○——○ The circles are experimentally determined values. The curve has been calculated from the equilibrium constants (Eldjarn and Pihl, 1956*c*).
- - - - - Calculated on the basis of K values given by Kolthoff *et al.* (1955).

participating molecular species can be calculated for any initial ratio of the reactants. The procedure and its application to several thiol-disulphide systems, are described elsewhere (Eldjarn and Pihl, 1956*b*). In the present paper the findings of radiobiological significance will be considered.

The interaction of disulphides with biological thiols

For the purpose of the present discussion it is convenient to consider GSH and CSH cellular thiols. Constant amounts of these are exposed to increasing amounts of the protective agent in the disulphide form. The equilibrium concentrations of mixed disulphide, obtained in the absence of oxygen, are shown in Figure 2. It is immediately apparent that the various disulphides differ widely in their ability to form mixed disulphides. Clearly, there is a qualitative correlation between this ability and the known protective effect *in vivo* of these compounds.

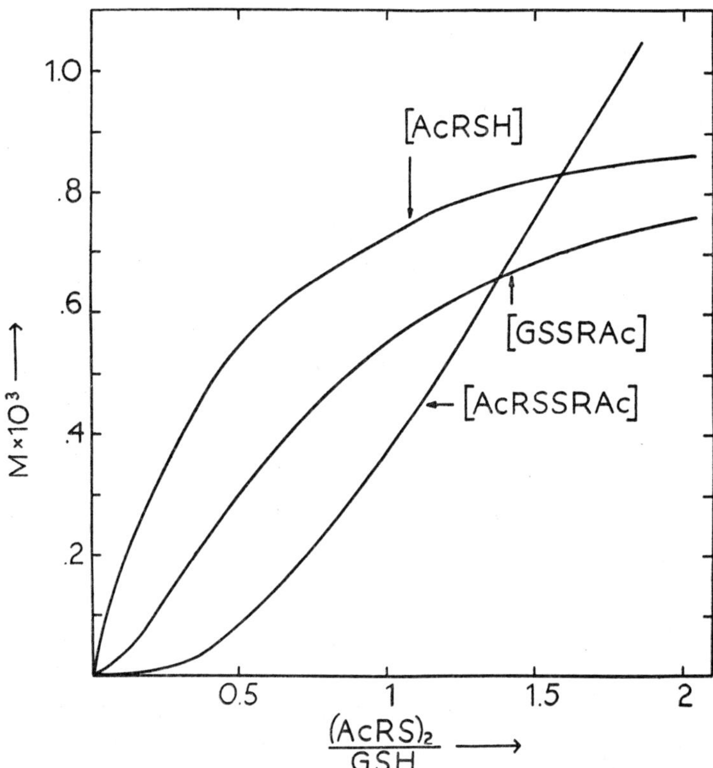

FIG. 3. The equilibrium concentrations of three of the constituents involved in the GSH + $(AcRS)_2$ system, expressed as a function of the initial $\frac{(AcRS)_2}{GSH}$ ratio. The curves have been calculated (Eldjarn and Pihl, 1956*c*) on the assumption that the initial GSH concentration is 1×10^{-3} M.

The equilibrium concentrations of the three molecular forms of the protective agent, obtained when a constant amount of GSH interacts with increasing amounts of N,N′-diacetylcystamine, are shown in Figure 3. The striking fact appears that, when the SS/SH ratio is less than 1 (which presumably is the situation existing in the protected organism), at equilibrium only a small fraction of the protective agent is present in the original disulphide form, whereas considerable amounts are present in the form of the mixed disulphide.

Obviously, the biological significance of a mixed disulphide will be determined, not only by the equilibrium concentrations, but also by the rate of formation. It was pointed out above that cystamine reacts rapidly with glutathione. Measurements of the initial velocities of mixed disulphide formation have revealed that the reactivity of various disulphides is markedly influenced by structural factors. Thus, N,N′-diacetylcystamine reacts 280 times slower with GSH than does cystamine. At 37° C. and *p*H 7·4, equilibrium values are reached after 15 to 20 minutes in the former system (Eldjarn and Pihl, 1956*b*).

It is clear from the reactions depicted in B that, if in this system the oxidised glutathione (GSSG) is continuously reduced (for example by means of the enzyme glutathione reductase), RSSR will likewise be reduced. Recently we have been able to reduce cystamine, N,N′-diacetylcystamine, tetramethyl-(N,N′)-cystamine, tetraethyl-(N,N′)-cystamine, cystine, and homocystine by means of glutathione reductase, in the presence of glutathione (Pihl, Eldjarn and Bremer, 1957). The rate of reduction was found to be proportional to the concentration of GSSG in the system, showing that the mechanism suggested above is indeed operating.

The interaction of thiols with biological disulphides

It is an obvious consequence of the reaction mechanism depicted in B that thiols will react with accessible biological disulphides. Again, the equilibrium concentrations of mixed disulphide will be determined by the equilibrium constants and by the initial amounts of thiol and disulphide. Those compounds, which give rise to high equilibrium values of mixed disulphide when they are incubated in the disulphide form with thiols ($K_2 \gg 1$, $K_3 \ll 1$), will also do so when interacting, in the reduced form, with SS compounds. This was experimentally confirmed in the case of cysteamine + oxidised glutathione. In this case the equilibrium was reached at *p*H 7·4 and 37° C., in about 3 minutes.

Figure 4 shows the equilibrium concentrations of mixed disulphides, obtained when oxidised glutathione or cystine interacts with increasing amounts of various thiols. In this case the disulphides can be looked upon as the body constituents, constant amounts of which interact with increasing amounts of protective agents in the thiol form. Again, large differences in the ability to form mixed disulphides are found, particularly in the concentration range of $SH/SS < 1$. By far the highest equilibrium concentration of mixed disulphide is obtained with cysteamine.

Figure 5 shows the equilibrium concentrations of the three molecular forms of the protective agent, when GSSG is incubated with increasing amounts of AcRSH. Clearly, in the SH/SS range below 1 the mixed disulphide is the dominating form of the protecting agent in this system.

Protein SS groups have so far been assumed to be inert under physiological conditions, thereby contributing to the structural stability of the proteins. However, on the basis of the above findings it would be expected that SH compounds will react, under physiological conditions, with accessible protein SS groups as well. The experimental investigation of this problem is complicated by the fact that our analytical procedure requires acidification and

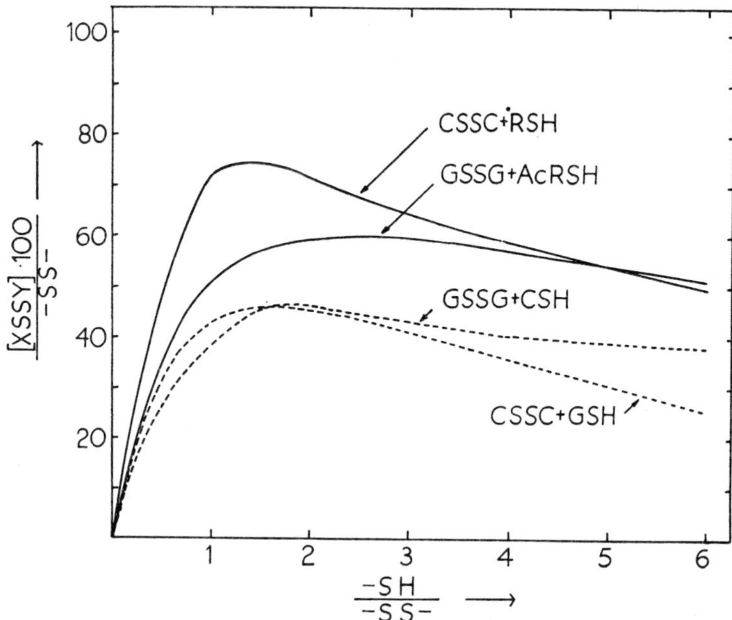

FIG. 4. The equilibrium concentrations of mixed disulphide in various disulphide-thiol systems, expressed as a function of the initial SH/SS ratio.
---- Calculated on the basis of K values published by Kolthoff *et al.* (1955).

electrophoretic separation of the proteins at *p*H 2. So far, the interaction of labelled cysteamine with protein SS groups has been demonstrated to take place in the case of bovine serum albumin, desoxyribonuclease, and a commercial Cytochrome *c* preparation.

The reaction between cysteamine and protein SS groups would be expected to proceed according to the following equations:

$$\boxed{\text{PROT.}}\begin{matrix}-\text{S}\\ |\\ -\text{S}\end{matrix} + \text{RS*H} \rightleftharpoons \boxed{\text{PROT.}}\begin{matrix}-\text{SH}\\ \\ -\text{SS*R}\end{matrix}$$

$$\boxed{\text{PROT.}}\begin{matrix}-\text{SH}\\ \\ -\text{SS*R}\end{matrix} + \text{RS*H} \rightleftharpoons \boxed{\text{PROT.}}\begin{matrix}-\text{SH}\\ \\ -\text{SH}\end{matrix} + \text{RS*S*R}$$

Upon incubation of the Cytochrome *c* preparation with labelled cysteamine, the expected type of curve was indeed obtained (Fig. 6).

In a recent study of a specially purified Cytochrome *c* preparation no disulphide linkage could be demonstrated (Paleus, 1955). However, it is generally agreed that ordinary Cytochrome *c* preparations, even when separated by electrophoresis at different *p*H levels, contain one SS bridge per mole (Paul, 1951).

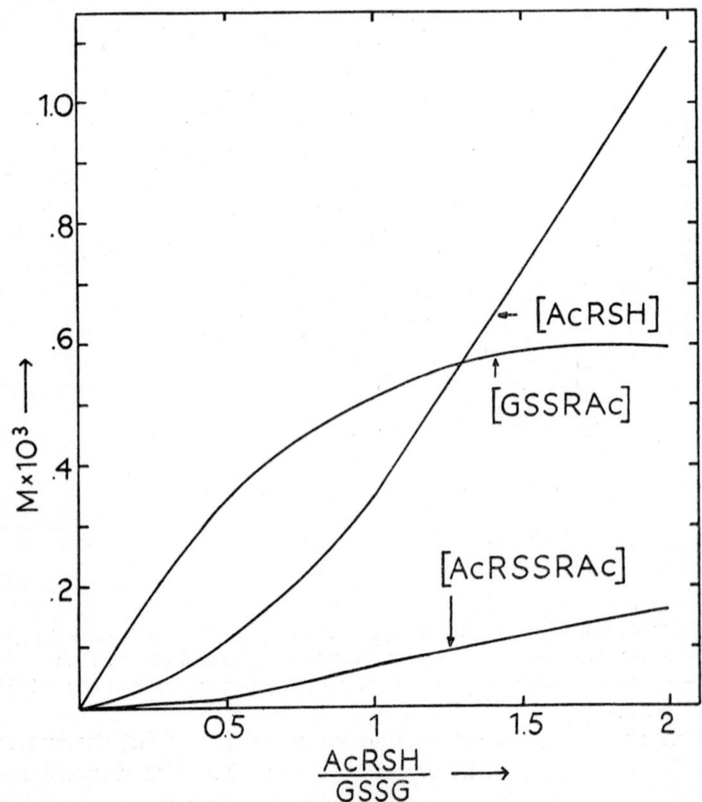

FIG. 5. The equilibrium concentrations of three of the constituents involved in the GSSG + AcRS Hsystem, expressed as a function of the initial $\frac{\text{AcRSH}}{\text{GSSG}}$ ratio. The curves have been calculated (Eldjarn and Pihl, 1956*c*) on the assumption that the initial GSH concentration is 1×10^{-3} M.

The interaction of thiols with aldehydes and ketones

The concept that the protective agents of the cysteamine-cysteine group exert their main protective action after having been bound to body constituents, led us to consider other possible modes of chemical fixation to target molecules. It is a well established fact that thiols may react with aldehydes and ketones to form semimercaptals and mercaptals. We have therefore, by means of procedures similar to those used in the study of thiol-disulphide interactions, measured the rate and extent of the interaction of cysteamine with various aldehydes and ketones (Pihl and Eldjarn, to be published). It is of interest from a radiobiological point of view, that

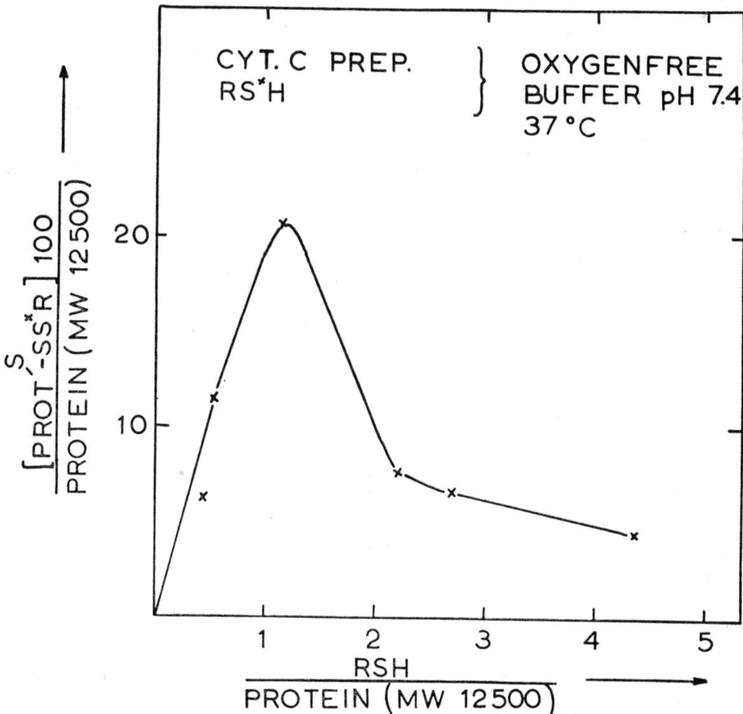

FIG. 6. The binding of ^{35}S-labelled cysteamine (RS*H) to commercial Cytochrome *c*. The Cytochrome *c* preparation was incubated with increasing concentrations of labelled cysteamine, and the protein fixed radioactivity measured after paper electrophoretic separation at *p*H 2.

none of the nucleic acid components tested showed any measurable interaction with cysteamine at *p*H 7·4 and 37° C. (Table). However, certain aldehydes and ketones with adjacent activating groups do react with cysteamine to form semimercaptals. It may be significant that several of these active aldehydes and ketones serve as substrates for sulphhydryl enzymes.

The radiobiological significance of mixed disulphide formation

It follows from the preceding discussion that the sulphur-containing agents react, both in their SH and SS form, with accessible biological disulphide and thiol groups respectively, to form mixed disulphides. Small structural variations in the molecules seem strongly to influence this ability.

The above findings are of interest from a biochemical point of view, since they indicate that the accessible protein SH and SS groups exist *in vivo* in a dynamic equilibrium with the disulphides and thiols of the surrounding medium (Eldjarn and Pihl, 1956*c*). From a radiobiological point of view the significant fact is that when the good protective agents of the cysteamine-cysteine group are administered to animals, these compounds will exist, to a considerable extent, in the form of mixed disulphides with tissue constituents.

TABLE

The Interaction of Cysteamine with Various Aldehydes and Ketones

The aldehydes and ketones (2×10^{-3} M) were incubated, at *p*H 7·4 and 37° C. with cysteamine-S^{35} (2×10^{-3} M), and the semimercaptals isolated by paper electrophoresis at *p*H 2

Carbonyl compounds tested	Incubation time (mins.)	$\frac{\text{Moles RS}^{35}\text{ fixed}}{\text{Mole of CO cpd.}} \times 100$
Thymine, Cytosine, d-ribose, Uric acid, Allantoin	30	<5
Guanylic acid, Uridylic acid, Guanosine	30	<5
p-nitro-benzaldehyde	30	27
α-keto-glutaric acid	30	13
Dihydroxyacetone	30	15
Glyceraldehyde	5	50
Methyl glyoxal	3	50

Radiochemical studies in this laboratory indicate that disulphide bridges are particularly vulnerable to the destructive effect of irradiation. Thus, when cystamine is irradiated in aqueous solution, rupture of the disulphide bridge occurs with the formation of the sulphinic and sulphonic acid oxidation products (Shapiro and Eldjarn, 1955*a*, *b*). Upon irradiation, also a reduction of the disulphide occurs with the formation of cysteamine, the SH compound. On this basis the hypothesis was advanced that the formation of a mixed disulphide on a target SH group may represent a partial protection of this SH group against the indirect action of ionising radiation (Eldjarn, Pihl and Shapiro, 1956; Eldjarn and Pihl, 1955). The assumption was made that approaching free radicals might attack either one of the sulphur atoms, and that, if the cysteamine sulphur were oxidised by the free radicals, the original target SH group would then be reformed.

The hypothesis was also suggested that the formation of a mixed disulphide might offer partial protection against direct hits. It is known that in an ionised organic molecule, the charge may migrate along the carbon chain until rupture of a bond occurs (Wallenstein *et al.*, 1952). If the fission takes place at a susceptible mixed disulphide bond, this event will, in part, represent a non-destructive form of energy dissipation.

Independent evidence that SS bonds are involved when a molecule is struck by a direct hit has recently been brought forward by Gordy *et al.* (1955). On the basis of paramagnetic resonance measurements on proteins and polypeptides, irradiated in the dry state, he concludes that whenever radiation knocks out an electron to create a hole or vacancy at any given point in the protein, this hole or vacancy is quickly filled by an electron borrowed from a cystine group.

Conclusion

Evidence is presented that compounds of the cysteamine-cysteine group react, both in the thiol and in the disulphide form, with accessible disulphide and thiol groups of body constituents. Good protective agents, such as cystamine and cysteamine, react rapidly and extensively with thiols and disulphides. Qualitatively, a correlation was found between the ability of various compounds to form mixed disulphides and their known ability to protect against ionising radiation. The data support our previous hypothesis which provides a mechanism for the protection, both against the direct and the indirect action, of ionising radiation.

Compounds of the cysteamine-cysteine groups will, to a large extent, exist in the tissues in the form of mixed disulphides with body constituents. Any theory which attempts to account for the protective effect of these compounds will have to take this fact into consideration.

Acknowledgment

Supported by grants from The Norwegian Cancer Society (Landsforeningen mot Kreft).

REFERENCES

ALEXANDER, P., BACQ, Z. M., COUSENS, S. F., FOX, M., HERVE, A., and LAZAR, J., 1955. *Radiation Res.*, **2**, 392.

BERSIN, T., and STEUDEL, J., 1938. *Ber.*, **71B**, 1015.

ELDJARN, L., and NYGAARD, O., 1954. *Arch. Int. Physiol.*, **62**, 476.

ELDJARN, L., PIHL, A., and SHAPIRO, B., 1956. *Proc. of the Intern. Conf. on the Peaceful Uses of Atomic Energy*, **11**, 335.

ELDJARN, L., and PIHL, A., 1955*a*. *Progress in Radiobiol.*, 1956, p. 249 (Edinburgh, Oliver and Boyd).

Idem, 1956*a*. *J. Biol. Chem.* (in press).

Idem, 1956*b*. *Ibid.* (in press).

Idem, 1956*c*. *Acta Chem. Scand.* (in press).

Fox, M., 1954. *Radiobiology Symposium*, p. 61 (London, Butterworth).

Gordy, W., Ard, W. B., and Shields, H., 1955. *Proc. Nat. Acad. Sci. Wash.*, **41**, 983.

Hagen, U., and Koch, R., 1955. *Progress in Radiobiol.*, 1956, p. 257 (Edinburgh, Oliver and Boyd).

Koch, R., and Hagen, U., 1955. *Ibid.*, p. 246.

Kolthoff, I. N., Stricks, W., and Kapoor, R. C., 1955. *J. Amer. Chem. Soc.*, **77**, 4733.

Paleus, S., 1955. *Acta Chem. Scand.*, **9**, 335.

Patt, H. M., Straube, R. L., Blackford, M. E., and Smith, D. E., 1950. *Amer. J. Physiol.*, **163**, 740.

Paul, K. G., 1951. *The Enzymes*, **2**, Part I, p. 372 (New York, Academic Press Inc.).

Pihl, A., Eldjarn, L., and Bremer, J., 1957. *J. Biol. Chem.* (in press).

Shapiro, B., and Eldjarn, L., 1955*a*. *Radiation Res.*, **3**, 255.

Idem, 1955*b*. *Ibid.*, **3**, 393.

Wallenstein, M., Wahrhaftig, A. L., Rosenstock, H., and Eyring, H., 1952. *Symposium on Radiobiology*, p. 70 (New York, Wiley).

Discussion

Alexander. Drs. Pihl and Eldjarn have presented very interesting data about the interaction of SH compounds with proteins, but it seems to me that they have given us no reason why this reaction should be of particular importance with regard to the mechanism whereby SH compounds protect. SH substances are highly reactive and can undergo very many different types of reactions. They can act as transfer agents; they compete very rapidly for free radicals; they are reducing agents, etc. The only way in which experiments *in vitro* can provide information about the mode of action of protective agents is by comparing large numbers of substances of varying protective power and seeing whether they have any one property in common, and if this property runs parallel with their capacity to protect. Clearly disulphide inter-change cannot be of this type since very many excellent protectors cannot undergo a reaction of this nature. Experiments conducted along these lines led Professor Bacq and myself to postulate a mechanism for protection which has been described in a series of publications. This was based on a comparison of 100 different protective agents.

We had no indication that these agents act by different mechanisms in the animal and until one has a compelling reason for assuming that this is the case, I think one should not do so. Before one can seriously consider disulphide interchange as a mechanism of protection, Dr. Eldjarn and his colleagues must show that all the non-SH protective agents work by a different mechanism in the animal and that the ability to undergo disulphide exchange runs parallel with protective power in a large series of SH compounds.

Pihl. I agree of course with Dr. Alexander that our hypothesis cannot by any means explain the protective activity of compounds which do *not* contain sulphur. We also agree that this type of experiment should be extended to cover a large number of protective and non-protective thiols and disulphides. Such work is now in progress in our laboratory.

Alexander. We have found that disulphides protect very well against direct degradation of polymers.

Dale. I wanted to stress some facts which in some way supplement Dr. Alexander's point of view and show the difficulties of interpretation of protection experiments. The fact is that sulphur is highly protective as is thiourea and I think in this case all theories of SH-exchange break down. We have worked on the problem that thiourea in contact with other compounds liberates free sulphur. The position is difficult here because we have chain reactions which are oxygen dependent. By the way, have you ever changed *p*H in your experiments?

Pihl. As stressed in the paper, the reaction velocity is strongly dependent on *p*H, but in the systems we have so far studied the equilibria do not seem to be greatly influenced by *p*H changes, at least *p*H between 5 and 7.

Bacq. If the mechanism postulated by Drs. Pihl and Eldjarn for protection with SH or SS compounds is correct, we must find a difference in the biochemical or physiological behaviour of animals irradiated after injection with a sulphurated or a non-sulphurated protector. Such a difference has not so far been seen but it may be because one has not looked carefully for such differences.

Eldjarn. In general it would be expected that the equilibrium concentration of mixed disulphides is *p*H-dependent, since the ionised form of the thiols enters the reaction. However, when the *p*H values of the participating thiols are equal, as is nearly the case for GSH and RSH, this *p*H-effect will not be found.

Dr. Dale mentioned the protective effect of colloidal sulphur. This latter form of sulphur would, of course, be expected to interact with thiols and possibly with disulphides.

I would finally point out that there seems to be a good parallelism between the 'concentration' of the protector and the protection offered. We, therefore, feel that a connection should be expected between protection mechanism and the dominating chemical form of the protector.

RECENT RESEARCH ON THE CHEMICAL PROTECTORS AND PARTICULARLY ON CYSTEAMINE-CYSTAMINE

Z. M. Bacq

Laboratory of General Pathology and Research Laboratory for Protection of the Civil Population, Liège, Belgium

In these last two or three years, the search for new chemical protectors against ionising radiations has slowed down. Langendorff and Koch (1955) and Koch and Hagen (1956) failed to confirm the activity of the thiouronium compound studied by Hollaender *et al.* (1955). From a series of amines, the action of which was demonstrated by Bacq and Herve (1952 and 1954), only tryptamine was found to be effective by them. Likewise, the action of cystamine by oral route in mice (Bacq, 1953) and rats (Bacq, 1956) could not be confirmed by this German group of investigators.

Cystamine $\begin{array}{l} S—CH_2—CH_2NH_2 \\ | \\ S—CH_2—CH_2NH_2 \end{array}$ is definitely less toxic than cysteamine ($HS—CH_2—CH_2—NH_2$) when administered continuously in food (Bacq, 1956 and Bacq and Ponlot, 1955). Rats tolerate 1% of cystamine in their food, but are not protected if submitted to a continuous irradiation from a radium source. At the present time, we prefer cystamine to cysteamine in mammalian experiments because we avoid the complication, when the results have to be interpreted, of the oxidation of cysteamine to cystamine which consumes a small amount of oxygen (Gray, 1956). Cystamine is chemically more stable than cysteamine.

The pharmacological study of cysteamine and cystamine is being completed. These amines have a weak sympathomimetic action; they affect neuromuscular and ganglionic transmission at high doses (Goffart and Della Bella, 1954, and Goffart and Paton, 1955): the liberation of histamine by cystamine is confirmed (Lecomte, 1955): adrenalectomy decreases the resistance of rats to cysteamine (Fischer *et al.*, 1955).

The metabolism of cysteamine and of cystamine has been carefully studied by Verly in Liège, Eldjarn in Oslo, and their observations confirmed (1955 and 1954). There is no question that the substance which is active in the body against radiations is the compound as it is administered. The main metabolites (taurine, SO^{4--}) are inactive. The disappearance of radioprotection follows closely the destruction and excretion of cysteamine. Particular emphasis must be put on the fact that the distribution of cysteamine in the tissues and organs of mammals is not homogeneous. Some organs (bone marrow, thyroid, adrenals) concentrate cysteamine-cystamine, others (testes) have a much smaller content than the serum (Eldjarn and Nygaard, 1954). Since the protection against X-rays is proportional to the concentration of the protector, it is easy to understand why the testis, for instance, is very weakly protected when compared with the bone marrow of the same animal.

New examples of protection have been published. Cysteamine given intravenously before local irradiation to rabbits protects the eye against radiodermatitis and conjunctivitis and slightly against cataract (Francois and Beheyt, 1955). Rugh and Clugston (1956) have shown a very nice protection of the fœtus when pregnant mice are irradiated after an injection of cysteamine; Maisin *et al.* have confirmed in pregnant rats these observations of Rugh and Clugston.

There is no reduction or slowing of mortality if chickens receive an injection of a large dose of cystamine (150 mg./kg.), cysteamine or cysteine a few minutes before irradiation (X-rays, 250 kV, 2,000r) although these animals do react to the protective action of diethyldithiocarbamate, a chelating agent (Beaumariage, 1956) (Figs. 1 and 2). The reason for this peculiar behaviour of birds is unknown.

Some authors have insisted on the point that in the hours following irradiation, there is no difference visible at the microscope, between the tissues of the controls and of the protected animals (Betz, 1956). It has been shown that if one uses quantitative histological methods, it is easy to find the anatomical protection of the spleen, the liver (Gerebtzoff and Bacq, 1954), and the bone marrow (Devik and Lothe, 1955); the intestine (Desaive and Varetto-Denoël, 1955) and the vaginal epithelium (Darcis *et al.*, 1956) are also much less damaged in animals protected by cysteamine.

The mechanism of action of chemical protectors has been much discussed.

(a) Possible hormonal influences

Does cysteamine or cystamine act by a kind of physiological mechanism, for instance through the endocrine reaction of anterior pituitary and suprarenals? Apparently not.

The protective effect of cysteamine is easily obtained in adrenalectomised rats (Bacq *et al.*, 1954).

Cysteamine abolishes in adult rats the 'second' reaction of the suprarenals, the reaction (hypertrophy, drop in cholesterol and ascorbic acid content) which begins 48 hours after 800r and lasts until death. On the contrary, it does not affect the early reaction which is finished at 24 hours (Bacq *et al.*, 1954). A combined treatment with barbiturate and morphine before irradiation does not influence mortality; it abolishes the first reaction, but after two days, the changes in the adrenals of the treated animals are the

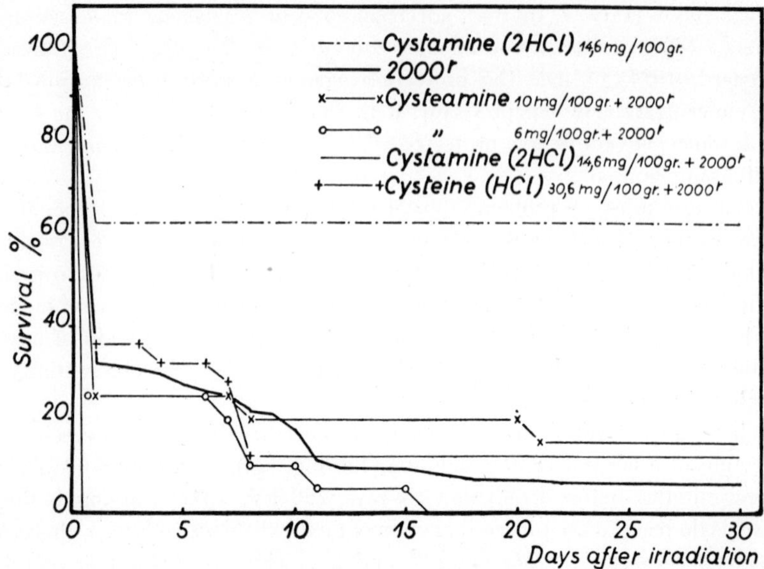

FIG. 1. Survival curves of chickens; controls 2,000r and non-irradiated cystamine injected; irradiated after cystamine, cysteamine or cysteine injection. Absence of protection.

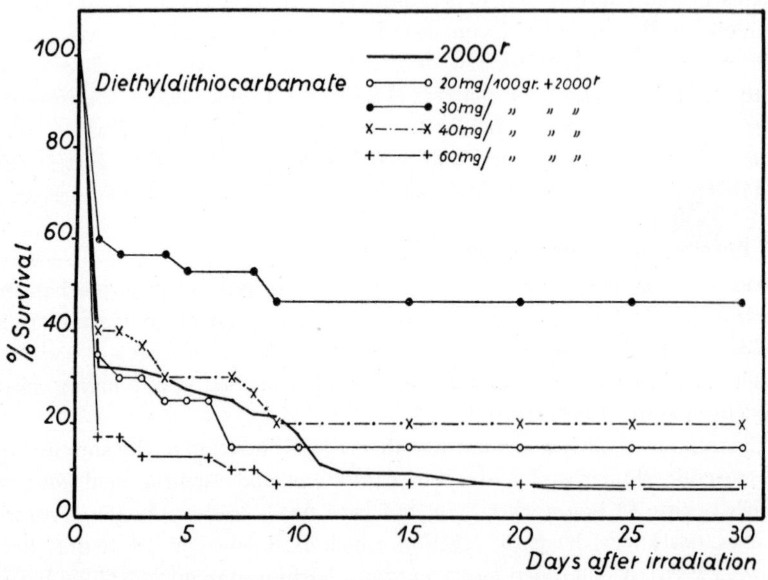

FIG. 2. Survival curves of chickens. Diethyldithiocarbamate at a certain dosage (30 mg./100 g.) injected before 2,000r gives a definite but slight protection.

same as those in the irradiated controls (Bacq and Fischer, 1956). Hypophysectomised rats with pituitary grafts, in a chronic condition of hyposurrenalism, are more sensitive to X-rays than normal controls; the reaction of the adrenals of these grafted animals irradiated on the whole body is very different; for instance, the cholesterol content is increased at the third day (Bacq *et al.*, 1956). New-born rats react very much like hypophysectomised grafted adults: there is no reaction of the suprarenals during the 24 hours following irradiation, and the cholesterol content increases after 3 days; they die earlier than adults when subjected to the same dose of X-rays (Sladić *et al.*, 1956; Bacq and Fischer, 1956).

Thus, in various physiological or pharmacological conditions, it is possible to observe a considerable change in either the early or the second adrenal reaction after irradiation, without changing the mortality. The fact that cysteamine inhibits the second reaction of the suprarenals in rats cannot be interpreted by the assumption that the normal activity of the suprarenals is the reason for the survival; we think that the suprarenals of cysteamine protected rats are normal 3 or 4 days after irradiation, because the animal eats well and has no diarrhœa, or infection.

One must not overemphasise the role of the suprarenals in radiation sickness (Betz, 1956). Obviously, some fundamental physicochemical process is altered during irradiation by cysteamine.

(b) Anoxia produced in mammals by cysteamine-cystamine

Hollaender and L. H. Gray in many meetings (International Conference of Radiobiology; Geneva Conference, August 1955) have independently given numerous arguments in favour of the idea that in mammals, the chemical protectors in general (cysteine in particular) create in the cells or in the tissues of the mammal such a condition of anoxia that the effects of X-rays are reduced. There is no point in repeating these arguments here; we ourselves have gathered a number of facts in favour of this anoxia theory.

Bacq and Fischer have shown that cysteamine mobilises rapidly (in 1 hour) the glycogen store of the rat's liver (Bacq and Fischer, 1953); there is no hyperglycæmia or glycosuria and cysteamine does not increase O_2 consumption. Fischer has shown that after cysteamine, there is a large increase of the concentration in the plasma of organic acids (lactic and pyruvic mainly) and a urinary excretion of these acids is characteristic of anoxic condition in mammal (Fischer, 1956).

Mice injected with cysteamine are much more resistant to oxygen poisoning than controls. At equimolecular doses, cysteamine is about 5 times more active in this respect than cysteine; this factor of 5 has also been obtained for the protection against X-rays. We found that rats injected with large amounts of cystamine (150 mg./kg.) are less resistant to decreased atmospheric pressure (178 mm. Hg.). Not a single injected animal survived more than 60 minutes although 14 out of 18 controls were not killed by a 90 minutes' exposure (Fig. 3). These observations agree with those of Scott (1956). Histamine also decreases resistance to anoxia, but we do not believe that the histamine eventually liberated by cystamine plays a major role because it does not, like cystamine, decrease the oxygen saturation of the venous blood (Bacq *et al.*, 1955).

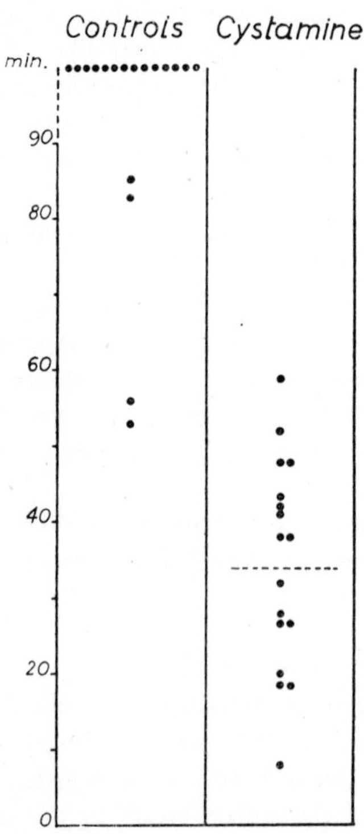

FIG. 3.—Resistance of rats to hypoxia (atmospheric pressure 178 mm. Hg.). Left, normal controls. The experiment lasts 90 minutes. Each point shows the reaction of a single animal. Points above the 90 minutes indicate permanent survival. Right column, rats which have received an intraperitoneal injection of cystamine (2 HCl) 15 mg./100 g.; mean survival time 35 minutes.

The blood taken from the inferior vena cava in rats anaesthetised with a barbiturate, is normally about 70% saturated in O_2; in animals injected with cystamine, the degree of saturation of the venous blood falls to about 40% which is a very low level indeed (Bacq *et al.*, 1955). Oxygen saturation in the arterial blood is not changed. Marked desaturation of vena cava blood also occurs after giving cystamine by stomach tube in protective amounts (Fig. 4).

Since the oxygen consumption is not increased (there is a slight decrease in fact) by cystamine injection, a slowing of the circulation would explain the oxygen desaturation and this has been experimentally verified. The slowing of the circulation lasts about 1 hour.

Cater (1956) has observed a marked and long-lasting fall in oxido-reduction potential of the extracellular fluid in rats after injection of cysteamine. But Synkavit, which sensitises to X-rays, also decreases this oxido-reduction potential.

(c) Arguments against the anoxia theory

Devik and Lothe (1955) have shown by the analysis of chromosome damage and of lethality that the effect of a moderate hypoxia adds itself to that of cysteine, cysteamine or cystamine. For instance, 2 mice only (out of 21) survived when irradiated (1,100r) after an injection of 5 mg. of cystamine; 4 mice out of 20 survived after 1,100r in 8·54% oxygen; but 13 out of 18 were living 30 days after the same irradiation if cystamine injection was combined with low oxygen pressure. Thus one is tempted to argue that cystamine (or cysteamine) does not act by creating an anoxic condition; but the argument is not conclusive because there is some probability that the low oxygen pressure will accentuate the anoxemia of the cystamine-injected animal.

P. Alexander and I, (1955), have made a study of the protective action of more than one hundred substances, in the mouse, on the one hand, on a polymer, on the other hand. The highly polymerised polymethacrylate (PMA) in aqueous solution and in the presence of oxygen is depolymerised by relatively small amounts of X-rays. With two exceptions, the results in mice agree with that on the polymer. For instance, chelating agents are excellent protectors both *in vitro* and *in vivo*, provided that their toxicity is not excessive. In Alexander's polymer system, the partial pressure of O_2 does not change; the great majority of the chemical protectors used were not reducing substances (no —SH substances could be studied on PMA), and were not substances which consume oxygen in aqueous solution. Is it logical to admit that diethyldithiocarbamate, for instance, should have two different modes of action: *in vitro*, on PMA: a competition for free radicals, and *in vivo*: anoxia? Why should a substance lose one of its physicochemical properties because it is introduced into a living organism?

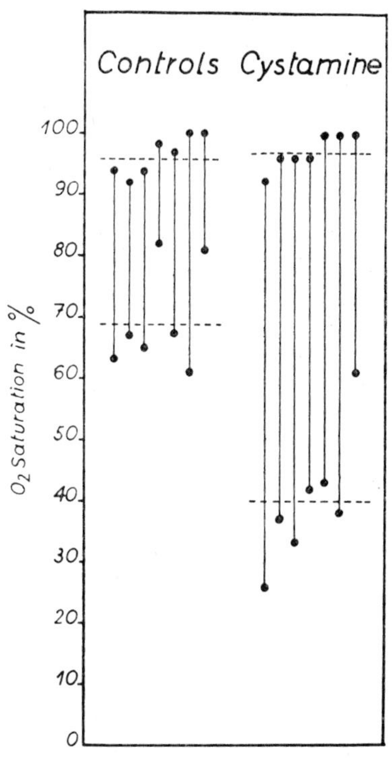

FIG. 4. Oxygen saturation of arterial and inferior vena cava blood in slightly anæsthetised rats. The O_2 saturation of vena cava blood is very low after intraperitoneal injection of 15 mg./100 g. of cystamine (2 HCl).

Dog's reticulocytes *in vivo* are well protected against X-rays by pyruvate and cysteamine when their hæmoglobin content is saturated with O_2 (Nizet *et al.*, 1952).

If, in mammals, anoxia was the main mechanism of action, how can one account for the fact that certain tissues which concentrate cysteamine are better protected than others? Anoxia after cysteamine is due to circulatory troubles and naturally reaches the same level in all tissues.

More important than any other is the argument that in the mammal, it is not difficult to observe a local protection when cysteine or cystamine is locally applied.

Forssberg (1950) made the first demonstration of this fact in 1950. His experiments were confirmed recently by Hoffmann (1955) who used as a test the erythema produced by about 5,000r of soft X-rays (60 kV) on the tail of the rat. Cysteine (5%) was injected subcutaneously or locally applied as an ointment (10%). The erythema was quantitatively measured by a photoelectric method.

Bacq and Herve (unpublished) have observed the regrowth of hair in the previously epilated hind legs of the rat. One leg served as control; on the other, cystamine was introduced by ionophoresis or in a carbowax ointment. The regrowth of hair after a dose of 1,000r to 1,500r is more marked on the protected limb than on the normal side (Fig. 5).

In this case, the anoxia theory cannot provide an explanation. If the chemical protector slows down the local circulation, the result will be an increased oxygen pressure in the tissues where cystamine or cysteine has been introduced, since the general circulation is not affected.

One must also remember that according to classical pharmacology, cyanide injection increases the oxygen concentration in blood because it inhibits oxygen utilisation by the tissues. Cyanide being a good protector against X-rays, this point should be carefully reinvestigated.

(d) Conclusion

There seems to be good reasons to accept an eclectic theory at least as far as mammals are concerned. Anoxia (lowering of O_2 pressure in the tissues) may be contributory, but one cannot explain all the available observations by anoxia alone. Some fundamental physicochemical action must be accepted: (1) competition for free radicals as put forward by Alexander *et al.*, 1955. (2) Eldjarn and Pihl (1956) have developed a hypothesis already suggested by Bacq and Herve (1952). The chemical protector combines with a radiosensitive enzyme or a protein; the combination is radioresistant. After irradiation, the combination is dissociated, and normal activity restored to the enzyme or protein. Unfortunately, this hypothesis cannot give a logical interpretation either of the chemical protection of the PMA *in vitro* or of the protective action in the mammal, of non-SH substances like tryptamine.

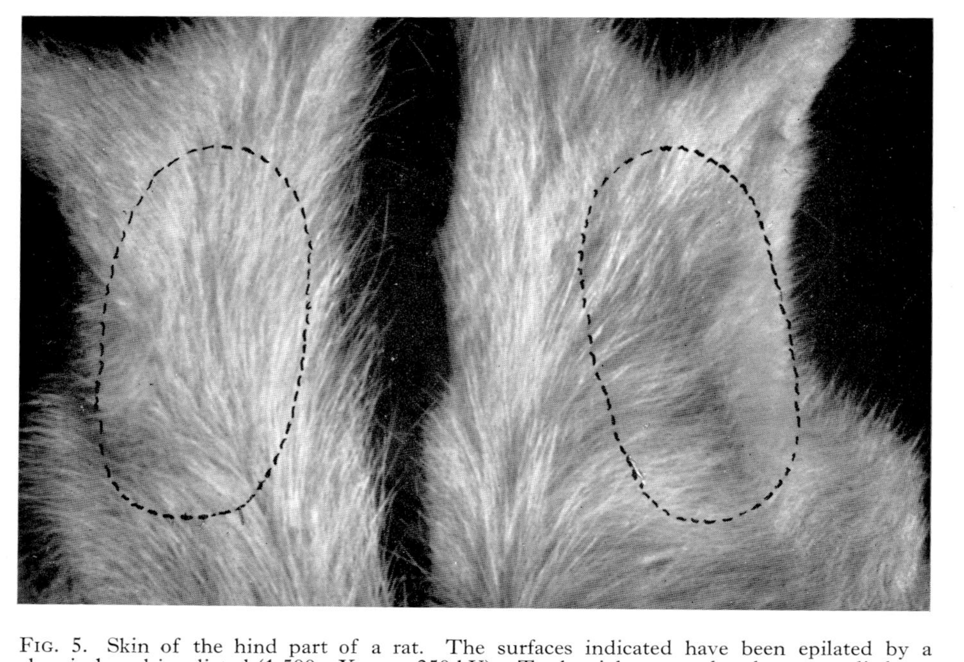

FIG. 5. Skin of the hind part of a rat. The surfaces indicated have been epilated by a chemical, and irradiated (1,500r, X-rays, 250 kV). To the right, control carbowax applied one hour before irradiation; to the left, carbowax +1% cystamine. The rat has been killed 2 months after irradiation. Much of the skin of the back has been removed in order to put the two interesting regions closer together.

REFERENCES

Alexander, P., Bacq, Z. M., Cousens, S. F., Fox, M., Herve, A., and Lazar, J., 1955. *Radiation Res.*, **2**, 392.

Bacq, Z. M., 1953. *Bull. Acad. Roy. Med. Belg.*, **18**, 426.

Idem, 1954. *Acta Radiol.*, **41**, 47.

Idem, 1956, *Bull. Acad. Roy. Med. Belg.*, **21**, 119.

Bacq, Z. M., and Alexander, P., 1955. "Principes de Radiobiologie," *Sciences et Lettres*, Liège, p. 478.

Bacq, Z. M., Cuypers, Y., Evrard, E., and Soetens, R., 1955. *C.R. Soc. Biol.*, **149**, 2014.

Bacq, Z. M., and Fischer, P., 1953. *Arch. Int. Physiol.*, **61**, 417.

Bacq, Z. M., Fischer, P., and Beaumariage, M. L., 1954. *Bull. Acad. Roy. Med., Belg.*, **19**, 399.

Bacq, Z. M., and Fischer, P., 1956. *Arch. Int. Pharmacodyn.*, **107**, 120.

Idem, 1956 *C. Rev. Belge Path. Med. exper.*, **25**, 384.

Bacq, Z. M., and Herve, A., 1952. *Bull. Acad. Roy. Med. Belg.*, **18**, 13.

Idem, 1952. *J. Suisse Med.*, **82**, 160.

Idem, unpublished.

Bacq, Z. M., Martinović, P., Fischer, P., Pavlović, M., and Sladić, G., 1956. *Arch. Int. Physiol. Biochem.*, **64**, 278.

Bacq, Z. M., and Ponlot, R., 1955. *C.R. Soc. Biol.*, **149**, 2012.

Beaumariage, M. L., 1956. Personal communication.

Betz, E. H., 1956. *Contribution à l'étude du syndrome endocrinien provoque par l'irradiation totale de l'organisme*, p. 326 (Paris, Masson).

Cater, D. B., 1955. *Progress in Radiobiol.*, 1956, p. 276 (Edinburgh, Oliver and Boyd).

Darcis, L., Hotterbeex, P., and Onkelinx, C., 1956. *Experientia*, **12**, 286.

Desaive, P., and Varetto-Denoël, J., 1955. *Experientia*, **11**, 242.

Devik, F., and Lothe, F., 1955. *Acta Radiol.*, **44**, 243.

Eldjarn, L., 1954. *Scan. J. Clin. and Lab. Invest.*, **6**, suppl., 13.

Eldjarn, L., and Nygaard, O., 1954. *Arch. Int. Physiol.*, **62**, 476.

Eldjarn, L., and Pihl, A., 1955. *Progress in Radiobiol.*, 1956, p. 249 (Edinburgh, Oliver and Boyd).

Fischer, P., 1956. *Arch. Int. Physiol. Biochem.*, **64**, 130.

Fischer, P., Lecomte, J., and Beaumariage, M. L., 1955. *Arch. Int. Physiol. Biochem.*, **63**, 121.

Forssberg, A., 1950. *Acta Radiol.*, **33**, 296.

Francois, J., and Beheyt, J., 1955. *Ophthalmoligica*, **130**, 397.

Gerebtzoff, M. A., and Bacq, Z. M., 1954. *Radiobiol. Symp.*, p. 290 (London, Butterworth).

Goffart, M., and Della Bella, D., 1954. *Arch. Int. Physiol.*, **62**, 455.

Goffart, M., and Paton, W. D. M., 1955. *Arch. Int. Physiol. Biochem.*, **63**, 477.

Gray, L. H., 1955. *Progress in Radiobiol.*, 1956, p. 267 (Edinburgh, Oliver and Boyd).

Hoffmann, D., 1955. *Strahlentherapie*, **96**, 396.

Hollaender *et al.*, 1955. *Geneva Confer. August.*

Koch, R., and Hagen, U., 1955. *Progress in Radiobiol.*, 1956, pp. 246 and 257 (Edinburgh, Oliver and Boyd).

Langendorff, H., and Koch, R., 1955. *Strahlentherapie*, **98**, 245.

LECOMTE, J., 1955. *Arch. Int. Physiol. Biochem.*, **63**, 291.
NIZET, A., BACQ, Z. M., and HERVE, A., 1952. *Arch. Int. Physiol.*, **60**, 449.
RUGH, R., and CLUGSTON, H., 1956. *Science*, **123**, 28.
SCOTT, O. C. A., 1955. *Progress in Radiobiol.*, 1956, p. 274 (Edinburgh, Oliver and Boyd).
SLADIĆ, D. S., PAVLOVIĆ, M. R., and RADIVOSEVIĆ, D. V., 1956. *Bull. Inat. Neul. Sc. Boris Kidri*, **6**, 199.
VERLY, W. G., 1955. *Bull. Acad. Roy. Med. Belg.*, **20**, 447.

DISCUSSION

Smith. I do not see how you can exclude anoxia as an explanation of the cysteamine effect in your hair regrowth experiments since (1) the cysteamine is constantly available to the cells being irradiated and can thus produce local anoxia and (2) since you have no measurements of local oxygen conditions.

Bacq. The experiment has been done with cystamine, the S-S oxidised compound which does not consume oxygen. It is generally accepted that local vasodilation increases the oxygen tension in the venous blood of the vasodilated area, but I agree that direct measurements might be useful.

Auerbach. Does cysteamine produce anoxia in the chicken?

Oftedal. I can confirm the lack of protection found after injecting cysteamine in chicks. In chick embryos 4 days' old no protection is found either by cystamine, cysteamine or cystein. On the contrary, when a high concentration of cysteine in hypertonic solution is injected, the effect of radiation appears to be enhanced.

Bacq. We have not measured the O_2 desaturation of venous blood in the chicken after injection of cystamine, but we know that the same hypotension occurs as in mammals and eventually the chicken dies after injection of cystamine from cardiovascular collapse.

Swallow. I think it would follow from present ideas that the chemicals which protect biological systems against X-rays would not protect against alpha-particles. I would like to ask whether there is any evidence in the literature or in unpublished experiments whether this is so? If true, it would support the view that alpha-particle radiobiology is a subject with different phenomena and different concepts from X-ray radiobiology.

Bacq. Cysteine is nearly inactive when *Allium* roots are irradiated with alpha-particles; the same chemical protector is half as active in mice against neutrons as against X-rays (Forssberg and Nybom, 1953; Patt, Clark and Vogel, 1953).

Marcovich. As far as bacteria are concerned, when we add to the irradiation medium of irradiation, cysteamine, the protection rises quickly, but it is immediately destroyed by a few seconds' bubbling of oxygen.

Alper. Perhaps Dr. Hollaender would comment on the fact that he has reported protection of micro-organisms by means of cysteamine far beyond that which could be obtained by anoxia alone?

Hollaender. I would mention that reduced oxygen tension by replacing it with nitrogen gives a dose reduction factor in *E. coli* of 1 to 3–4, but cystamine gives a factor as high as 12. It is very difficult to imagine why, but I think cysteamine somehow protects nucleic acids or other important compounds of the cell. Cysteamine is the only compound which shows this high extra protection above that of removal of oxygen.

Scott. Professor Bacq has suggested that radioprotection due to cysteine or cysteamine locally applied could not be explained by tissue anoxia. Is it not possible that the autoxidation mechanism, which Dr. Gray drew attention to last year at Cambridge, could result in a local tissue anoxia after the injection of cystein or cysteamine?

Bacq. We have obtained local protection with cystamine which does not autoxidase.

Scott. The pre-print only mentioned cysteine and cysteamine.

Bacq. I agree, it is an error in the pre-print.

REFERENCES

FORSSBERG, A., and NYBOM, N., 1953. *Physiologia Plantarum*, **6**, 78.

PATT, H. M., CLARKE, J. W., and VOGEL, H. H., Jnr., 1953. *Proc. Soc. exp. Biol. N.Y.*, **84**, 189.

EDITORS' NOTE

It is probable that some of this discussion was based on a misconception arising from an error in the pre-circulated pre-print of Professor Bacq's paper.

THE PROBLEM AND CONSTITUTION OF RADIATION-SENSITISING AGENTS

R. Koch
Biophysics Department, Heiligenberg Institute,
Heiligenberg, Baden, Germany

It has been fairly well established by numerous radiobiological studies on rats and mice with a multitude of sulphhydril compounds (especially cysteamine derivatives) that not all SH-bodies can be regarded as radiation-protecting agents in animals. It has been found that the property of a radiation protective requires a special chemical constitution (4; 6) of the basic configuration:

$$HS—(CH_2)x—N\begin{matrix} R_1 \\ R_2 \end{matrix}$$

(R_1, R_2 = alkyl groups; x = not greater than 3).

All α-thiolamino acids of this configuration (e.g. cysteine, α-homocysteine and in a wider sense, glutathione) have the same reaction. This specific chemical constitution of the SH-protectors also suggests the idea that their biological effect may be specific. It could be specific with regard to metabolic functions, but also with regard to such functions as have been discussed by Eldjarn and Pihl (1955) at the last conference, namely the temporary masking of vital protein groups or other SH-groups. The idea is not too improbable that a specific chemical constitution may also be required for the masking of endogenous sulphhydril groups by other SH or SS compounds. Not all substances blocking SH are capable of sensitising animals for a subsequent irradiation. This is only accomplished by mono-iodo-acetic acid (Langendorff and Koch, 1954). Some time ago the author described the toxicology of mono-iodo-acetic acid and has pointed out definite analogies of histological changes with those produced by X-rays (Koch and Hagen, 1956).

During these studies, two interesting observations were made which will now be reported. As was mentioned earlier, α-thiolamino acids are known to be radiation-protective substances. The author examined the analogous β-thiolamino acids: β-homocysteine and isocysteine. Surprisingly, the β-isomers are radiation-sensitising agents. This not only holds for α-thiolamino acid as against β-thiolamino acid, but also if definite changes are made on the cysteamine molecule. One such change in thioglycole is the substitution of the amino group by an alcoholic OH-group.

TABLE I

Protective Substances	Sensitising Substances
$HS{-}\overset{\gamma}{C}H_2{-}\overset{\beta}{C}H_2{-}\overset{\alpha}{C}H(NH_2){-}COOH$ α-homocysteine	$HS{-}\overset{\gamma}{C}H_2{-}\overset{\beta}{C}H(NH_2){-}\overset{\alpha}{C}H_2{-}COOH$ β-homocysteine
$HS{-}\overset{\beta}{C}H_2{-}\overset{\alpha}{C}H(NH_2){-}COOH$ cysteine	$NH_2{-}\overset{\beta}{C}H_2{-}\overset{\alpha}{C}H(SH){-}COOH$ isocysteine
$HS{-}CH_2{-}CH_2{-}NH_2$ cysteamine	$HS{-}CH_2{-}CH_2{-}OH$ thioglycole

The significance was also studied of chain branching for the observed basic configuration. No effect was found in α-methyl-cysteine, while a moderate sensitisation was observed in penicillamine.

TABLE II

Protective Substances	Sensitising Substances
$HS{-}C(CH_3)_2{-}CH(NH_2){-}COOH$ penicillamine	$HS{-}CH_2{-}C(CH_3)(NH_2){-}COOH$ α-methyl cysteine

The specific protective effect postulated by the author for this basic configuration (cysteine-cysteamine group) was confirmed by these observations. It appears that one is no longer justified in speaking of a radiation-protective effect of the SH-bodies in general, but should only relate this effect to the cysteine-cysteamine group.

One concept would be that the bodies of the cysteine-cysteamine group are metabolites in the general meaning of the word, that the sensitising bodies are antimetabolites, and that the ineffective substances are as indifferent with regard to metabolism as are the *d*-isomers of our amino acids. Yet, one also can make use of the concepts introduced by Eldjarn and Pihl. The protective bodies mask vital endogenous SH-groups while the irradiation is going on. In doing so, they very probably have a specific affinity for definite SH-groups. Sensitising substances here have an effect similar to SH-blockers, i.e. they mask, but persist for so long a time that toxicity occurs. It is obvious that they have the same affinity for definite endogenous SH-groups as have the protective bodies.

The sensitising effects here described must be differentiated from the reactions which are observed when animals are treated with thio-acetamide, barbiturates or thiobarbiturates prior to irradiation. In these cases a sensitising effect is feigned which actually results from the combined action of two different toxic agents. For instance, upon intraperitoneal administration in mice of 0·15 mg. thio-acetamid per g. body weight, liver necroses

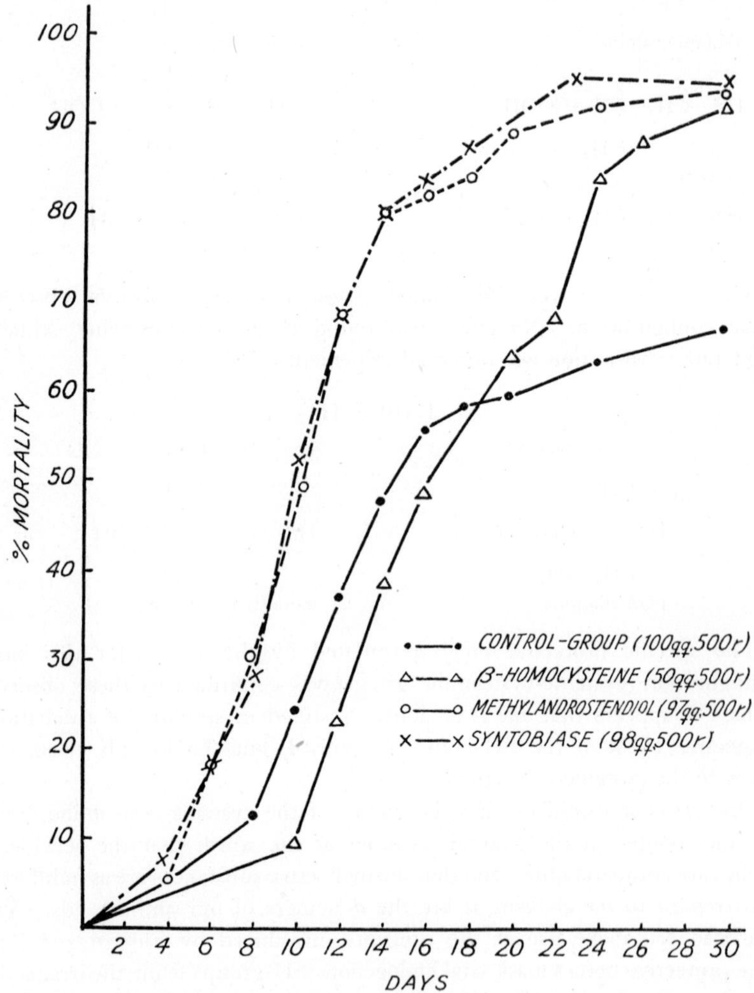

FIG. 1. Experimental findings with β-homocysteine, methylandrostendiol, syntobiase, and a control group.

were always observed. In each case of sensitisation to radiation it must therefore be clarified if the damage caused by the radiation has been enhanced by a substance which *per se* is not toxic, or if a combination of toxic injuries has taken place.

In experiments with rats and mice co-carboxylase, riboflavin (Langendorff *et al.*, 1956), α-alanine and α-alkyl alanines (Langendorff and Koch, 1955) were also found to be—(with a high degree of probability)—true sensitising agents. This effect cannot yet be explained.

Some experiments with methylandrostendiol and 'Syntobiase', a dye of the mono-azo group, are especially noteworthy. Mice and rats were on one occasion given 10 to 25 mg. of methylandrostendiol per animal intramuscularly 10 days before irradiation. In each of 250 animals a distinct sensitisation was observed. The same was found in animals treated with 'Syntobiase'. Some of these findings are shown in Figure 1 in the conventional manner. A numerical evaluation, however, demonstrates the findings much more distinctly.

These findings with methylandrostendiol and 'Syntobiase' point in a direction which in cancer research has been known for about a decade (see Buu-Hoi, 1954 and 1956). Because of peculiar physicochemical properties (K-zones with a high electron density or excitation energies which may originate from Van der Waal forces) some aromatic cyclic compounds appear to be capable of attaching themselves to proteins or to form so-called 'Einschluss' compounds in which energy is transmitted to the protein. As a result, the proteins undergo structural changes which may bring about grave consequences when they occur within the nucleus of a cell. Irradiation may markedly promote these phenomena in methylandrostendiol and the dye.

It has been the object of this paper, to demonstrate with the help of a few examples, that radiation-protective agents and radiation-sensitising agents are important tools for the clarification of the mechanism of radiation protection.

REFERENCES

Buu-Hoi, N. P., 1956. *Arzneimittel-Forsch.*, **6**, 251, and 1954, *Abstr. Pap. Amer. Chem. Soc.*, **14**, 9.

Eldjarn, L., and Pihl, A., 1955. *Progress in Radiobiol.*, 1956, p. 249. (Edinburgh, Oliver and Boyd).

Koch, R., and Hagen, U., 1956. *Arch. exp. Path. Pharmak.*, **228**, 227.

Langendorff, H., and Koch, R., 1954. *Strahlentherapie*, **95**, 535.

Idem, 1955. *Ibid.*, **98**, 245.

Idem, 1956. *Ibid.*, **99**, 567.

Langendorff, H., Koch, R., and Hagen, U., 1956. *Ibid.*, **100**, 137.

Idem, 1956. *Ibid.*, **99**, 375.

Discussion

Alper. In considering certain radiobiological data I have come to the tentative conclusion that radiation acts on the constituents of living cells mainly by direct ionisation of target molecules. It seems plausible to suggest that after ionisation has occurred, the target molecule will be left in a very reactive state, and a chemical reaction will occur which may be of such a nature that the target becomes non-restorable. Or it may be such that restoration is possible, in a manner analogous with that suggested by Charlesby and Alexander for protection against direct effects by SH-compounds. According to this view, oxygen in the neighbourhood of the ionised target molecule will react with it to form a peroxide which will be a non-restorable product. Does Dr. Koch consider it possible that at last some of his sensitising substances may take part in this 'metionic reaction' I have postulated, in an oxygen-like way, so that a non-restorable product is found?

Koch. This is a very difficult question, although Dr. Alper's view seems quite plausible. We are inclined to look at the effects of SH-substances as those of anti-metabolites. In the case of steroids and dyes we rather think they are transferred to a 'metastable' state by the irradiation. These metastable intermediates could then be expected to transfer their energy to, for example, proteins.

Bacq. What is the dose-sensitising factor observed when using your most active substance?

Koch. After 500r we had a dose-sensitising factor of 1·4 in experiments on female mice, independent of the substances assayed. There were, however, some differences with regard to the survival time.

Mitchell. I suggest that one must consider the possibility of a free radical mechanism under physiological conditions, in radiosensitisation produced by a number of compounds of different constitution such as riboflavinphosphate and 2 : 3-dimethyl-1 : 4-naphthohydroquinone diphosphate: the parent quinone of the latter compound does not interact *in vitro* with SH-groups. In this connection reference may be made to the recent work and measurements of the life times of a number of free radicals by Blois (1956).

Höhne. I should like to add to Dr. Koch's paper that we have investigated the influence of a castration on the radiosensitivity of male Wistar rats. The animals were given LD/50/30 doses at various times after castration but we could not find any difference in the survival time as compared with controls.

Koch. We have also worked with castrated male and female rats and had in both cases an increased radioresistance, but it was not so pronounced as in mice. It is important, however, that the irradiations are not performed earlier than 8 weeks after castration.

Eldjarn. I would like to ask Dr. Koch whether he believes that the concentration of the steroid hormones in tissue, days after the i.m. administration—is sufficient to make likely radiochemical reactions involving the steroid hormones as such? If I understood him correctly, he proposed such a mechanism to be possible.

Koch. We find sensitising effects of the steroid hormones when the maximum of physiological action coincides with the irradiation. At this time we should expect an optimal distribution of the hormones in the body; that is, optimal conditions for the steroids to act as energy-transferring substances.

REFERENCE

Blois, R. S., 1956. *Proc. Soc. exp. Biol. N.Y.*

RADIOBIOLOGICAL INVESTIGATIONS ON THE HIBERNATING LOIR (*GLIS GLIS*)

H. A. Künkel, G. Höhne and H. Maass
Universitats-Frauenklinik, Hamburg-Eppendorf, Germany

The influence of temperature on the radiosensitivity of biological systems is of particular interest in radiobiological research. Generally most investigations show that the response of the living tissue to the radiation injury is delayed when the temperature of the organism is maintained below normal (Patt, 1953). For studying the mechanism of both the interaction of ionising radiations with cellular constituents, and of chemical protecting agents, hibernating mammals seem to be particularly suitable, Smith and Greenan (1951) and Doull *et al.* (1952).

In the present investigations we used the loir (*glis glis*), a species which prevalently lives in Southern and Middle Germany and can easily be maintained under laboratory conditions (Fig. 1). For our experiments 92 loirs were available up till now. In the first group whole-body irradiations were carried out in the non-hibernating state with single doses of 700r of 200 kV X-rays (0·5 mm. Cu. dosage rate 85r/min.). The radiosensitivity was found to be about the same as that of a pure inbred strain of *Wister*-rats used by us in other experiments. 83% of the animals died within 30 days after irradiation. In the second group the animals (also in the non-hibernating state) were intraperitoneally injected with cysteine (concentration 500 mg./kg. body weight) 10 minutes before they were irradiated under the same conditions. In this case a protective effect of 100% was observed. No mortality occurred within 30 days after irradiation. In the third group cysteine was applied 3 minutes after irradiation. As expected no difference has been observed between the mortality of this group and that of the animals irradiated without cysteine injection. More than 80% died within 30 days. These results correspond with the numerous findings of many investigators, e.g. Patt *et al.* (1949). Only if applied immediately before irradiation could a considerable protective action of cysteine against ionising radiations be detected, whereas given after irradiation no significant decrease in mortality was found.

For carrying out irradiations during the hibernating state an apparatus was constructed which rendered it possible to keep the animals at a constant temperature of +4° C. This was found to be the optimum temperature of hibernation (Fig. 2). In this box the loirs slept under a sheet of plexiglas

Fig. 1. Loir or *Glis. Glis.*

Fig. 2. Five loirs in circular chamber.

of 5 mm. thickness. It was thus possible to irradiate the animals in the hibernating state without taking them out of the box. By these means any disturbance of the hibernating animals by change of temperature, noise or special preparations for the irradiation was avoided. The loirs fell into the hibernating state within a few days after they had been brought into an environment of 4° C. without being influenced by any drugs. The animals were irradiated while hibernating with the same dose of 700r and maintained in the hibernating state for the following 21 days. During this time all of the animals survived. On the 22nd day after irradiation they were taken out of the refrigerator and brought into a warm room of 20° C., where they

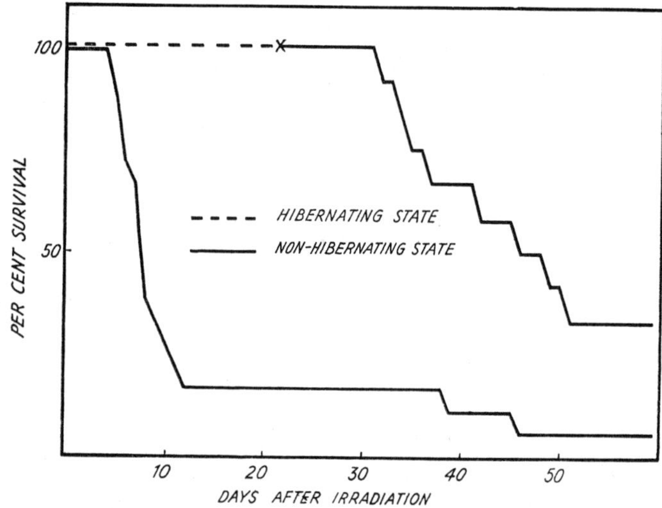

FIG. 3. Survival of loirs (*glis glis*) after whole-body irradiation with 700r.

awoke within about one hour. Only a few days after they were returned to the homeothermic state they developed the typical radiation syndrome as it is known from normal whole-body irradiations. In Figure 3 the survival rate is plotted against the time after irradiation. The broken line shows the period of hibernation. During this period the radiation damage is apparently in a latent phase. The effects of the radiation injury make their appearance first, after the wakening from hibernation, and the slope of the survival curve is then parallel to that of loirs irradiated in the non-hibernating state; as if they had been exposed to radiation on the day they had been returned to room temperature. 67% of the animals died within 51 (= 21 + 30) days of irradiation. The difference between the mortality of the animals irradiated in the homeothermic and in the poikilothermic state is within the range of statistical error. The possibility of a protecting action of the hibernating state with regard to the ultimate mortality is statistically not significant.

These findings correspond with the investigations of Doull and Du Bois (1953), on ground squirrels irradiated in the hibernating state. These authors, too, observed a latent phase of radiation damage during the hibernation.

TABLE

Mortality of Loirs (*glis glis*) after Whole-body Irradiation with 200 kV X-rays

Group	Treatment	Mortality
1	Irradiation with 700r in the non-hibernating state (T = + 20° C.)	83% (in 30 days)
2	Irradiation with 700r in the non-hibernating state (T = + 20° C.). 500 mg./kg. cysteine i.p. 10 min. *before* irradiation	0% (in 30 days)
3	Irradiation with 700r in the non-hibernating state (T = + 20° C.). 500 mg./kg. cysteine i.p. 3 min. *after* irradiation	83% (in 30 days)
4	Irradiation with 700r in the hibernating state (T = + 4° C.). After 21 days return to room temperature (T = + 20° C.)	67% (in 51 = 21 + 30 days)
5	Irradiation with 700r in the hibernating state (T = + 4° C.) 500 mg./kg. cysteine i.p. 15 min. *before* irradiation. After 21 days return to room temperature (T = + 20° C.)	0% (in 51 = 21 + 30 days)
6	Irradiation with 700r in the hibernating state (T = + 4° C.). After 21 days return to room temperature (T = + 20° C.). 50 omg./kg. cysteine i.p. immediately *after* return to room temperature	0% (in 51 = 21 + 30 days)

When we injected cysteine (500 mg./kg. body-weight i.p.) during the hibernating state 15 minutes before irradiation, a full protection of 100% was observed as well.

In the last group, however, cysteine was injected 21 days after irradiation, just at the moment when they were returned from hibernation to room temperature. We were greatly surprised to see that all these animals were still alive after the following 30 days. The results of the 6 experiments are shown once again in the Table. The difference between the two experiments in the hibernating state with injection of cysteine 21 days after irradiation and without cysteine is significant. That means: the probability that the difference between the mortality of 67% in group 4 and the mortality of 0% in group 6 is due to statistical deviations is less than 0·27%.

These investigations prove that cysteine is able to give full protection even 21 days after irradiation, if the animals are in the hibernating state during the irradiation and for the following 21 days. We have intentionally restricted ourselves to giving only experimental facts. Of course, an interpretation of these facts may be very difficult at the present moment. It can be expected, however, that our investigations started in the meantime on both the X-ray induced changes of the serum protein and on the histological and enzymatic processes occurring during the hibernating state may help to elucidate the mechanism involved. It can be stated that the usual theories of action of sulphhydryl-containing agents cannot explain the above findings.

REFERENCES

DOULL, J., and DU BOIS, K. P., 1953. *Proc. Soc. exp. Biol. N.Y.*, **84**, 367.
DOULL, J., PETERSEN, D. F., and DU BOIS, K. P., 1952. *Fed. Proc.*, **11**, 340.
PATT, H. M., SMITH, D. E., TYREE, E. B., and STRAUBE, R. L., 1949. *Science*, **110**, 213.
PATT, H. M., 1953. *Physiol. Rev.*, **33**, 35.
SMITH, F., and GREENAN, M. M., 1951. *Science*, **113**, 686.

DISCUSSION

Cohen. What were the number of animals used in groups 4 and 6?

Künkel. Up till now we have irradiated 21 animals in group 4 during the hibernating state and 15 hibernating animals in group 6 with an intra-peritoneal injection of cysteine 21 days later. This number, it is true, is not yet very large. It is, however, sufficient to give a significant difference between the two groups within the range of 36.

Koch. Dr Künkel suggests that the protection mechanism in his hibernators should differ from that in the case of other cysteine effects. Although we have so far no definite opinion on the protective mechanism of various protecting substances in general, it nevertheless seems possible that his effects and the common cysteine effects may adhere to the same scheme. We have suggested that the substances of the cysteine-cysteamine group could be considered as metabolites in a general sense. If so, the delayed cysteine effect in hibernators may be understood as a consequence of the universal delay in the irradiation reactions.

Künkel. In the present state of our investigation it is difficult to express any opinion on the mechanism. We plan to look into the synthesis of DNA in radiosensitive tissue, histological and serum protein changes and hope to have some information here. There seems to be some resemblance with the findings of Hollaender, i.e. of a certain time interval in the reversibility of the primary irradiation lesions. It may be that the time within which reversibility of the damage can take place is prolonged in hibernators.

THE DOSE-REDUCTION FACTOR FOR CYSTEAMINE AND ISOTHIURONIUM IN THE CASE OF WHOLE-BODY X-IRRADIATED MICE AND RATS

A. Catsch

Biophysics Department, Heiligenberg Institute,
Heiligenberg, Baden, Germany

Quantitatively, the efficiency of a radio-protective substance is best characterised by the so-called dose-reduction factor (DRF) which is defined as the ratio of equi-effective radiation doses. According to Patt *et al.* (1952), the DRF for cysteine amounts to about 1·7 in the case of acute mortality among irradiated mice and is said to be independent of the radiation dosage. As against this, we had obtained the impression in the course of earlier work Catsch *et al.* (1956) that the DRF for cysteamine increased with increasing radiation dosage; as this was not fully borne out statistically, further experiments were made to clarify the problem. The results obtained are described below.

Adult white mice ♂♂ of an inbred strain raised at the Institute were subjected to whole body X-irradiation (150 kV, 20 mA, filter 0·43 mm. Cu, HVL 0·87 mm. Cu, 111 r/min.). An irradiated group served as controls, while the animals in the other groups were given intra-peritoneal injections of 3 mg. cysteamine per animal and 5 mg. β-aminoethyl-isothiuronium-HBr per animal respectively, 5–10 minutes before irradiation. The experimental values obtained are shown in Table I.

The probit-analysis (by logarithmic transformation of the dose coordinate) gave almost identical dose-effect curves for the cysteamine and isothiuronium groups in respect of the LD 50/30 values, as also for the regression coefficients, so that it seemed justifiable to take the two groups together in further analysis. The results of the χ^2 analysis (Table II) show that between the dose-effect curves for the control group on the one hand, and the groups injected with a radio-protective substance on the other hand, there is no parallelism, and this can be proved statistically; in the first case the regression coefficient is approximately 1·7 times greater.

Confirmatory tests with rats (♂♂) gave an analogous result (Table III).

Nevertheless, there is no justification for assuming a simple linear dose relation for mice pre-treated with cysteamine and isothiuronium; the systematic scattering of experimental points from linearity in the direction of a convexity to the right (Fig. 1) is statistically significant

$$(\chi^2_8 = 18{\cdot}20;\ P = 0{\cdot}02\text{–}0{\cdot}01).$$

TABLE I

Dose in r	Control group		Cysteamine		Isothiuronium	
	Ratio of dead animals to number of animals tested	%	Ratio of dead animals to number of animals tested	%	Ratio of dead animals to number of animals tested	%
400	4/101	4	—	—	—	—
450	16/104	15·4	—	—	—	—
500	72/102	70·6	1/59	1·7	—	—
550	70/97	72	—	—	—	—
600	99/105	94·5	4/67	6	6/52	11·5
700	30/30	100	5/50	10	9/68	13·2
800	20/20	100	22/52	42·3	20/62	32·2
900	—	—	22/51	43·2	—	—
950	—	—	—	—	37/66	56
1000	20/20	100	41/54	76	41/49	83·7
1110	—	—	—	—	45/51	88·3
1200	—	—	64/64	100	—	—
1320	—	—	—	—	30/30	100

TABLE II

	χ^2	n	P	χ^2/n	t_{12}	P
Parallelism	30·51	1	<0·001	30·51	2·92	0·02–0·01
Heterogeneity ...	42·93	12	<0·001	3·58	—	—
Total	73·44	13				

TABLE III

Dose in r	Control group		15 mg. per animal cysteamine	
	Ratio of dead animals to total animals tested	%	Ratio of dead animals to total animals tested	%
500	1/60	1·7	2/60	3·3
650	24/58	41·4	8/58	13·8
830	48/50	96	4/50	8

In regard to the fundamental conditions affecting the increasing protective effect with increasing irradiation doses, it could first be imagined the effective principle was not represented by the protective substance as such, but rather by a secondary product and, or, reaction induced in the organism by

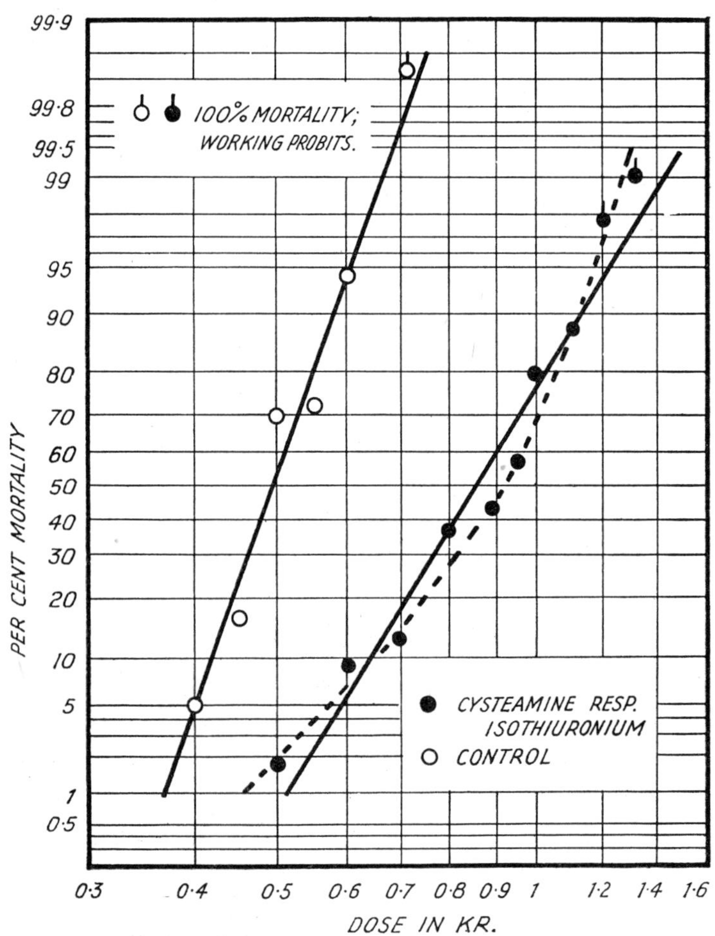

FIG. 1.—Dose-effect curves for the different groups.
——— linear dose relation.
– – – – curvilinear dose relation.
100% mortality; working probits.

irradiation. This assumption seems, however, to be very improbable, in view of research undertaken by Shapiro and Eldjarn (1955) as well as by Hollaender and Doudney (1954), which fails to give any support for this assumption; moreover it is also opposed by the fact, sufficiently proved by experiment, that in the case of cells or tumours irradiated *in vitro* the DRF is independent of the irradiation dose, Hall (1952) and Patt *et al.* (1952).

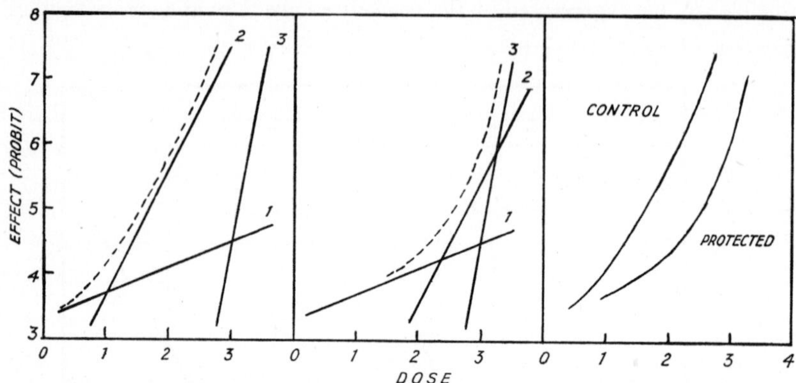

FIG. 2.—Schematical representation of three different mortality components of which only N°2 reduced by a given protective substance.

In our opinion a more accurate interpretation of the results in question would be the following: If one assumes: (1) that the acute irradiation

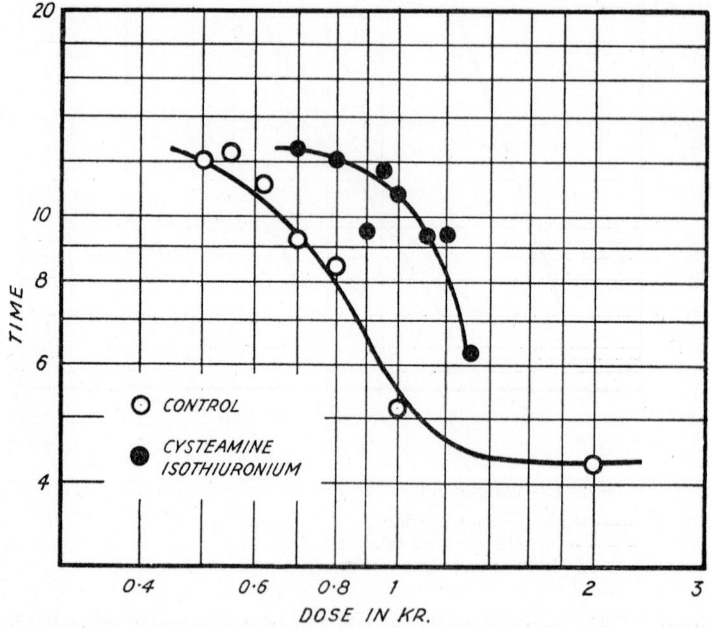

FIG. 3.—Mean survival times for the different groups.

mortality fails to represent any unified phenomenon, but rather that there are several mechanisms independent of each other and characterised by differences in the slopes of the dose-effect curves concerned; (2) that the component characterised by the flat slope is affected to a slight extent only, or not at all, by the protective substances, an apparent divergence of the dose-effect curves must arise with increasing irradiation doses, as shown schematically in Figure 2. This working hypothesis is supported not only by the previously mentioned curvature of the dose-effect curve for the group treated with the protective substances, but also by the fact that we had already been induced by the analysis of time-distribution of mortality Catsch *et al.* (1956), and of the dose-effect curve, Catsch and Langendorff (1956) to postulate various mortality mechanisms. Finally, we have been able to show by a series of tests (Langendorff *et al.*, in press), that by castration of mice (♂♂) only the component characterised by flat slope is reduced, which leads to an apparent reduction in the DRF with increasing irradiation dosage.

An estimate of the DRF for the various mechanisms gives values of about 1·4 and 2·1 respectively. For the third component—the so-called 3·5-day effect, the DRF can be computed as approximately 1·3 on the basis of a comparison of the mean survival time (Fig. 3). Elucidation of the problem of what is the fundamental reason for the various mortality mechanisms and for the differences in the extent to which they are capable of being influenced by protective substances, would be conditional on further investigation. Nevertheless, we may state at this point that the last-named case can be easily explained by the uneven distribution of the protective substance through the various organs.

REFERENCES

Catsch, A., Koch, R., and Langendorff, H., 1956. *Fortschr. Geb. Röntgenstr.*, **84**, 462.

Catsch, A., and Langendorff, M., 1956. *Naturwiss*, **43**, 281.

Hall, B. V., 1952. *Cancer Res.*, **12**, 787.

Hollaender, A., and Doudney, C. O., 1954. *Radiobiol. Symp.*, p. 112 (London, Butterworth).

Langendorff, H. Catsch, A., and Koch, R. In press. *Strahlentherapie.*

Patt, H. M., Blackford, M. E., and Straube, R. L., 1952. *Proc. Soc. exp. Biol. N.Y.*, **80**, 92.

Patt, H. M., Mayer, S. H., Straube, R. L., and Jackson, E. M., 1953. *J. Cellul. Comp. Physiol.*, **42**, 327.

Shapiro, B., and Eldjarn, L., 1955. *Radiation Res.*, **3**, 255 and 393.

Discussion

Cohen. How does Dr. Catsch explain the straight-line relationship on the curve with unprotected animals?

Catsch. The dose curve for the controls is really linear. I think this is due to the fact that we have worked within a rather narrow dose interval, when studying lower doses we obtained non-linear curves in our control material.

Alexander. Can you tell from your extensive data whether the protective agent disturbs the period at which death occurs, because with unprotected animals there seem to be two peaks where death occurs?

Catsch. The various 'mortality components' are characterised by maxima which occur at different times. In the dose interval we have studied, there is no normal distribution but rather bimodal or assymetric distributions. Our protective substances preferably influence the component with a maximum after 8–10 days.

CHEMICAL AND BIOLOGICAL ACTION OF PROTECTIVE SH-COMPOUNDS

ULRICH HAGEN

Biophysical Department of Heiligenberg Institute,
Heiligenberg, Baden, Germany

Three main basic concepts of the mode of action of the different SH-compounds have been developed; these are summarised below:

1. The protective compound cysteine, cysteamine or any similar SH-compound depresses the oxygen partial pressure in the system and with that reduces the effect of irradiation. A similar effect is achieved by the capture of radicals (HO_2 H_2O_2). Gray (1955) discussed this problem at the Cambridge Congress, and according to his measurements cysteine rapidly decreases the oxygen content of a solution, and at the same time oxidises itself. This oxidation is, however, dependent upon the presence of a heavy-metal catalyst whose presence is responsible for the rapid decrease in the oxygen partial pressure in the solution, after the addition of cysteine. Gray finds it desirable "to distinguish between a simple anoxia and a chemical protective effect". It is possible that catalysts of cysteine-oxidation are localised on certain sensitive parts of the cell so that cysteine produces a local reduction of oxygen pressure. This explains the relatively small quantity of cysteine needed to produce a protective effect in mice.

2. It is possible for protective compounds to react with SH-groups in the cell and to mask these by the formation of short-lived mixed disulphides. This suggestion was made by Eldjarn and Pihl (1955) and will be considered later in this paper.

3. The protective compound through its own reactivity may reduce the irradiation oxygenated cell-molecules, or may substitute irreversibly destroyed compounds into their function. This possibility must be considered, although the protective compounds are not effective when given after irradiation. It is probable that after irradiation the protective compounds do not approach the sensitive parts of the cell. Franzen (1956) stated that basic thiols of the type $SH—(CH_2)_n—N—R_2$ can be imagined as models of the enzyme of Glyoxalase I. They catalyse an action in which an energy-rich S-Acyl-bond is formed under physiological conditions.

In 2 and 3 above it is presupposed that the protective compound is effective only on particular sensitive parts of the cell.

In all the possible mechanisms, the protective compound reacts in a certain way with its surroundings. In the first there is an oxidation up to a disulphide, and the precise conditions can be easily studied in model experiments, particularly the heavy metal catalysis. Against this type of reaction occurring there is the formation of mixed disulphides from a totally different type of reaction distinct from the formation of disulphides under oxidation. The ability to reduce other compounds is expressed in the redox potential of the sulphide-disulphide system; it must be emphasised that the redox potential and the oxidation discussed above is not always identified. This physico-chemical problem will not be dealt with here.

Franzen (1956) studied the ability of an SH-compound to act as an enzyme model through the reaction with alpha-Ketoaldehyde:

$$R{-}CO{-}CHO + HS{-}(CH_2)_n{-}NR_2 \rightarrow RCO{-}\underset{\displaystyle OH}{\overset{\displaystyle H}{C}}{-}S{-}(CH_2)_nNR_2$$

A semimercaptale is formed with an S-Acyl-bond. These S-C bonds are often discovered by enzyme reactions, for example, Phosphoglyceraldehyd-dehydrogenase, of the coenzyme A or the above mentioned Glyoxalase I. Another very similar reaction is the one with iodoacetic acid which also gives an S-C bond, Hagen (1956).

$$R{-}SH + J{-}CH_2COOH \rightarrow R{-}S{-}CH_2{-}COOH$$

This reaction suggests that a sulphhydryl-compound can act as a model of an enzyme.

We have, in recent years, investigated quite a number of SH-compounds in connection with the irradiation effect, and we have in the course of these studies found both protective and non-protective compounds. We were especially interested in chemical reactivity and its relationship to protective properties. Differences in reaction may suggest the mechanism of protection in the intact organism.

The following groups will now be discussed and our results briefly summarised:

I. *Redox potential of the SH-compounds*

The redox potential of an SH-compound is difficult to measure by the usual methods, Ender (1954). However, different authors have shown that the effective cysteine and the ineffective thioglycolic acid have nearly the same redox potential.

Redox Potentials of Different Sulphhydryl-compounds

Protective compound Cysteine	Non-protective compound Thioglycolic acid
—0·15 V	—0·15 V (Ryklan and Schmidt, 1944)
—0·220 V	—0·215 V (Fruton and Clarke, 1934)

We did not, therefore, explore further the problem of differences of potential.

II. *Oxidation in the Presence of heavy metals*

We investigated the oxidation of different SH-compounds catalysed by iron and copper. The measured manometrical values show that the protective compounds are oxidised, in part, very fast. But we are not able to say

TABLE I

Oxidation of Different SH-compounds in the Presence of Heavy Metals

Protective compound				Non-protective compound			
Substance	Aut.	Fe	Cu	Substance	Aut.	Fe	Cu
Cysteine	17·3	34	110	Methylacetyl-cysteamine	3·6	0·01	0·8
Cysteamine ...	25·4	4·3	95	Pantetheine ...	2·9	0·05	0·9
Diethyl-cysteamine	13·7	0·05	129	Aminomercapto-heptane	2·0	0·44	1·7
Aminoethyliso-thiuronium	27·6	0·56	36·1	Thioglycolic acid	1·8	11·09	126·0
				Aminophenyl-mercaptane	365	241	224

Aut.: Autoxydation (mm.^{3} O_2/hour)
Fe: Increase of oxydation with iron (mm.^{3} O_2/γ hour)
Cu: Increase of oxydation with copper (mm.^{3} O_2/γ hour)
Methods (Langendorff, Koch and Hagen, 1954)

that all the non-protective compounds consume very little oxygen. It remains to be investigated whether or not certain cell catalysts have a different mode of action from iron or copper ions.

III. *Reaction with iodoacetic acid*

It is again difficult to distinguish between protective and non-protective compounds from this reaction. It is certain that no protective compound

TABLE II

Ability of Different SH-compounds to React with Iodoacetic Acid

Protective compound	k_{37^o}	Non-protective compound	k_{37^o}
Cysteine	2·22	Methylacetylcysteamine ...	0·475
Glutathione	2·08	Aminomercaptoheptane ...	0·033
Homocysteine	0·87	Aminophenylmercaptane ...	90·3
Cysteamine	12·35	Dimethylmercaptopyrimidine ...	1·67
N-alkyl-cysteamines	15–90	Thioglycolic acid	0·185
Aminoethylisothiuronium ...	9·53		

k_{37^o} = Velocity constant k by + 37° C., pH = 7·0

possesses a lower velocity constant k than 0·87, and most non-protective compounds react much more slowly. On the other hand, some non-protectors react rapidly as for example Aminophenylmercaptane. Further work is needed on this subject of protective compounds and enzyme models.

IV. *Ability to activate an inactive enzyme preparation*

So far the best agreement between protective ability and activation of an inactive enzyme has been obtained with inactive cathepsin. Only radio-protective compounds activate in the same manner as cysteine. Other SH-compounds—non-protectors—hardly influence the activity of cathepsin

TABLE III

Activity of Gelatine Splitting Cathepsin after Activation with Different SH-compounds

Protective compound		Non-protective compound	
Cysteine	1·00	Thioglycol	0·37
Homocysteine	1·03	Thioglycolic acid	0·22
Cysteamine	0·96	Mercaptoaminoheptane ...	0·37
N-alkyl-cysteamines ...	0·82–0·89	Dimethylmercaptopyrimidine ...	0·29
Aminoethylisothiuronium ...	0·96	Penicillamine	0·01

at all. The question immediately arises, of what are the details of the reaction? Following the ideas of Gawron and Cheslock (1951) there are two possibilities.

1. A simple reduction of the Enzyme-disulphide by SH-compounds:
   Enzyme—S—S—Enzyme + 2 RSH ⟶ 2 Enzyme—SH + R—S—S—R.
2. The inactive enzyme is present as a mixed disulphide. The enzyme becomes effective by changing with the protective compound, forming at the same moment a new mixed disulphide:
   Enzyme—S—S—R′ + RSH ⟶ Enzyme—SH + R—S—S—R′.

The ability of SH-compounds to activate cathepsin shows their ability to either reduce protein —S—S—bonds or to build mixed disulphides. If the second of these possibilities were confirmed the protective mechanism accepted by Eldjarn would be confirmed. On the other hand, it is possible that protective compounds, by their ability to reduce, with the aid of specific catalysts, retard or prevent certain oxidising effects of irradiation.

REFERENCES

ELDJARN, L., and PIHL, A., 1955. *Progress in Radiobiol.*, 1956 p. 249 (Edinburgh, Oliver and Boyd).

ENDER, E., 1954. *Hand. Physiol. Path. Chem. Anal.*, **1**, 628 (Heidelberg, Spinger-Verlag).

FRANZEN, V., 1956. *Angew. Chem.*, **68**, 381.
FRUTON, J. S., and CLARKE, H. T., 1934. *J. Biol. Chem.*, **106**, 667.
GAWRON, O., and CHESLOCK, K. E., 1951. *Arch. Biochem. Biophysics*, **34**, 38.
GRAY, L. H., 1955. *Progress in Radiobiol.*, 1956, p. 267 (Edinburgh, Oliver and Boyd).
HAGEN, U., 1956. *Arzneimittel-Forschg.*, **6**, 384.
LANGENDORFF, H., KOCH, R., and HAGEN, U., 1954. *Arch. Int. Pharmacodyn.*, **100**, 1.
RYKLAN, L. R., and SCHMIDT, C. L. A., 1944. *Univ. Calif. Publ. Physiol.*, **8**, 257.

DISCUSSION

Bacq. Possibly the protective mechanism of the SH-compounds, cysteine, cysteamine alike, can be understood on the basis of a common chemical property. But how are the effects of, say, tryptamine or polymethacrylic acid to be explained, as they do not have the chemical properties of, for example, cysteine?

Hagen. This remains to be seen but we have deliberately concentrated our work on the SH-compounds because they are easy to reproduce and also because their protective effect is only to be found by substances having the structure $R_1R_2—N(CH_2)_nSH$. It would seem that the mode of action of this latter type of substances is likely to be understood sooner than that of the compounds you mentioned.

Koch. I should like to add to the comments of Professor Bacq that the mechanism of radiation protection is very different in different substances. If we compare cysteamine and tryptamine, it would seem that the action of the latter in some way influences the central nerve system; we also found this to be the case in experiments with benzedrine and oxytryptamine (serotine).

A BACTERIOLOGICAL TEST FOR THE STUDY OF RADIOPROTECTION PROBLEMS IN MAMMALS

H. MARCOVICH AND J. F. DUPLAN
Laboratoire Pasteur de l'Institut du Radium, Paris, France

Studies on the mechanism of radioprotection by substances containing —SH groups have produced several hypotheses which may be summarised as follows:—

1. The substances compete with radiosensitive cellular sites for radicals produced by irradiation, Alexander *et al.* (1955) and Bacq (1955).
2. Eldjarn (1956) thinks, in the case of cysteamine, that unstable —S—S bonds are formed between the —SH group of the protector and the —SH groups of proteins; the latter are considered to be radiosensitive and to be protected if incorporated in an —S—S complex.
3. Experiments performed *in vivo* and *in vitro* by Salerno *et al.* (1952), Patt *et al.* (1952) and Gray (1956) indicate that cysteine acts indirectly by lowering the oxygen tension in the medium. However, Charlier (1954) considers that anoxia is insufficient to account for the protective action of cysteine. Salerno *et al.* (1954) on the other hand believe that this substance acts by pure anoxia while cysteamine produces a cytotoxic anoxia.

A new approach to cysteine protection is described in this paper, which may give additional information on its mode of action. It is known that radioprotection is a general phenomenon conferred on a considerable variety of organisms by the same substance. Cysteine is effective in mammals and bacteria, and it was thought that the change produced in mammals by the injection of cysteine might be reflected in the plasma. The mammalian plasma would then have bacterial radioprotective properties.

Materials and methods

Rabbits weighing 2·5 to 3·0 kg. were used, and were treated as follows :—

1. One thousand I.U. per kg. of heparin was given intravenously.
2. Four to five minutes later 10-15 ml. of blood was taken from the marginal vein of the ear.
3. Immediately afterwards cysteine was injected into the peritoneal cavity. The cysteine was neutralised just before injection and 400 mg. per kg. body weight of a 10% solution was used.

4. Ten minutes later a second blood sample was taken.
5. The plasma was separated by centrifuging for five minutes at 4,000 revs./minute, and then kept in test tubes at 0° C. during subsequent manipulations.

The bacterial suspension was a culture in broth of *Escherichia coli*; this was used at the end of the exponential growth phase, the bacterial density being then 2-3·10^9 cells per ml. For irradiation, 0·5 ml. of culture was added to 4·5 ml. of plasma, or of broth in the case of controls.

Irradiation was carried out in a plexiglas cup, the fluid volume being 0·3 ml. and the thickness on the average 1·0 mm.

The X-ray dose was always 45,000r given at a dose rate of 1,000r/second mean energy 15 KeV.

Bacteria were suspended in the following media for irradiation.

I. Tryptone broth.
II. Tryptone broth + cysteine one part per thousand.
III. Tryptone broth + 1 M ethanol.
IV. Normal plasma.
V. Normal plasma + cysteine added *in vitro* to a dilution of one part per thousand.
VI. Plasma of an animal previously injected with cysteine.

In addition, each of the above suspensions were subjected to the following procedures :—

(*a*) Irradiation immediately after addition of bacteria.
(*b*) As above, but after oxygen bubbling for a few seconds.
(*c*) (*a*) and (*b*) above, but after a storage of the suspensions in the refrigerator for 90-120 minutes.

The surviving fraction of the irradiated bacteria was measured by spreading samples on petri dishes containing nutrient agar medium and comparing with non-irradiated germ cells. The plates were kept for 18 hours at 37° C. before counting.

Results and discussion

The results are summarised in Tables I, II and III. In these experiments ethanol served as the control for 'protectability', and as can be seen ethanol protection is not cancelled by oxygen bubbling. Patterson and Matthews (1951) have shown that ethanol given to mice before irradiation has a protective action. To study this protection of bacteria in plasma a rabbit was injected intravenously with 1 ml. per Kg. of ethanol in solution in 4 ml. of saline. A sample of blood was taken 10 minutes later. At that time the animal presented all the symptoms of acute ethanol intoxication. It can be seen in Table III that this plasma had no protective properties. But if added *in vitro*, into plasma or to broth, ethanol protects at a concentration of 1 mole.

TABLE I

Delay between mixing and irradiation	O_2	Doses of X-rays in r	Bacterial dilution medium (see text)					
			I	II	III	IV	V	VI
0	—	0	376·000	435·000	512·000	479·000	352·000	375·000
		45,000	266	520	7·690	217	6·350	6·500
0	+	0	445·000	544·000	519·000	399·000	339·000	454·000
		45,000	284	182	4·830	671	520	177
90 min.	—	0	498·000	497·000	465·000	273·000	294·000	307·000
		45,000	464	2·730	8·250	993	35·600	3·880
90 min.	+	0	488·000	444·000	491·000	294·000	272·000	278·000
		45,000	257	250	3·480	231	69	1·760

TABLE II

Delay between mixing and irradiation	O_2	Doses of X-rays in r	Bacterial dilution medium (see text)			
			I	IV	V	VI
0	—	0	457·000	466·000	326·000	466·000
		45,000	332	—	1·162	4·830
0	+	0	518·000	510·000	445·000	363·000
		45,000	60	137	833	660
120 min.	—	0	—	525·000	367·000	336·000
		45,000	—	202	14·500	4·450
120 min.	+	0	—	609·000	376·000	427·000
		45,000	—	77	102	170

Three facts have been demonstrated in the cysteine experiments:—

1. Plasma of animals injected with cysteine has a radioprotective action for bacteria.
2. This action is destroyed immediately by oxygen bubbling before irradiation.

3. If cysteine is added *in vitro* and if irradiation is carried out immediately there is no protection. However, the protective action appears if the suspensions are allowed to stand before irradiation; this action is also suppressed by oxygen bubbling.

The phenomenon may be satisfactorily explained by the hypothesis of Gray (1956) and Patt *et al.* (1952).

The oxygen tension of plasma with protective properties (IV, V and VI) has been estimated by P. Joliot in Professor Wurmser's laboratory with an amperometre. The findings agree with the bacterial response, i.e. oxygen tension was low in every case where protection was observed.

Control plasma had no protective action.

TABLE III

Delay between mixing and irradiation	O_2	Doses of X-rays in r	Bacterial dilution medium (see text)				
			I	III	IV	V	VI
0	—	0	366·000	342·000	306·000	290·000	216·000
		45,000	402	8·300	559	6·800	755
0	+	0	—	380·000	288·000	249·000	252·000
			550	6·350	63	4·470	280

It is obvious that one cannot, at present, compare quantitatively the properties of the plasma of an animal injected *in vivo*, to control plasma in which cysteine has been added *in vitro*. The main reason for this is that in the first case there is a time interval of 20-25 minutes between the time when cysteine is injected and when the plasma is tested. Therefore, these results can only be considered in a qualitative way.

The difference between cysteine and ethanol protection with regard to oxygen bubbling clearly indicates that protection by ethanol is not by a pure anoxia.

Similar results have been obtained with rats.

From the above findings, it may be concluded that the protection of bacteria, irradiated in the plasma of an animal which has received an injection of cysteine, is a direct consequence of anoxia. The same conclusion holds if cysteine is added to the system *in vitro*. It seems logical to extrapolate from the plasma to the intercellular medium, and to conclude that mammalian protection by cysteine is mainly, if not entirely, linked to anoxia.

REFERENCES

ALEXANDER, P., BACQ, Z. M., COUSENS, S. F., FOX, M., HERVE, A., and LAZAR, J. 1955. *Radiation Res.*, **2**, 392.

BACQ, Z. M., 1955. *Int. Conf. on the Peaceful Uses of Atomic Energy* (United Nations, N.Y.).

CHARLIER, R., 1954. *Proc. Soc. exp. Biol., N.Y.*, **86**, 290.

ELDJARN, L., 1955. *Progress in Radiobiol.*, 1956, p. 249 (Edinburgh, Oliver and Boyd).

GRAY, L. H., 1955. *Ibid.*, 1956, p. 267 (Edinburgh, Oliver and Boyd).

PATT, H. M., BLACKFORD, M. E., and STRAUBE, R. L., 1952. *Proc. Soc. exp. Biol., N.Y.*, **80**, 92.

PATTERSON, E., and MATTHEWS, J. J., 1951. *Nature, Lond.*, **168**, 1126.

SALERNO, P. R., and FRIEDELL, H. L., 1954. *Radiation Res.*, **1**, 559.

STAPLETON, G. E., BILLEN, D., and HOLLAENDER, A., 1952. *J. Bacteriol.*, **63**, 805.

DISCUSSION

van Bekkum. Have you performed similar experiments with the injection of cysteamine into rabbits?

Marcovich. No, we have not yet performed experiments with cysteamine in animals. However, as far as bacteria are concerned, the protective activity can be cancelled by oxygen bubbling just before exposure to X-rays.

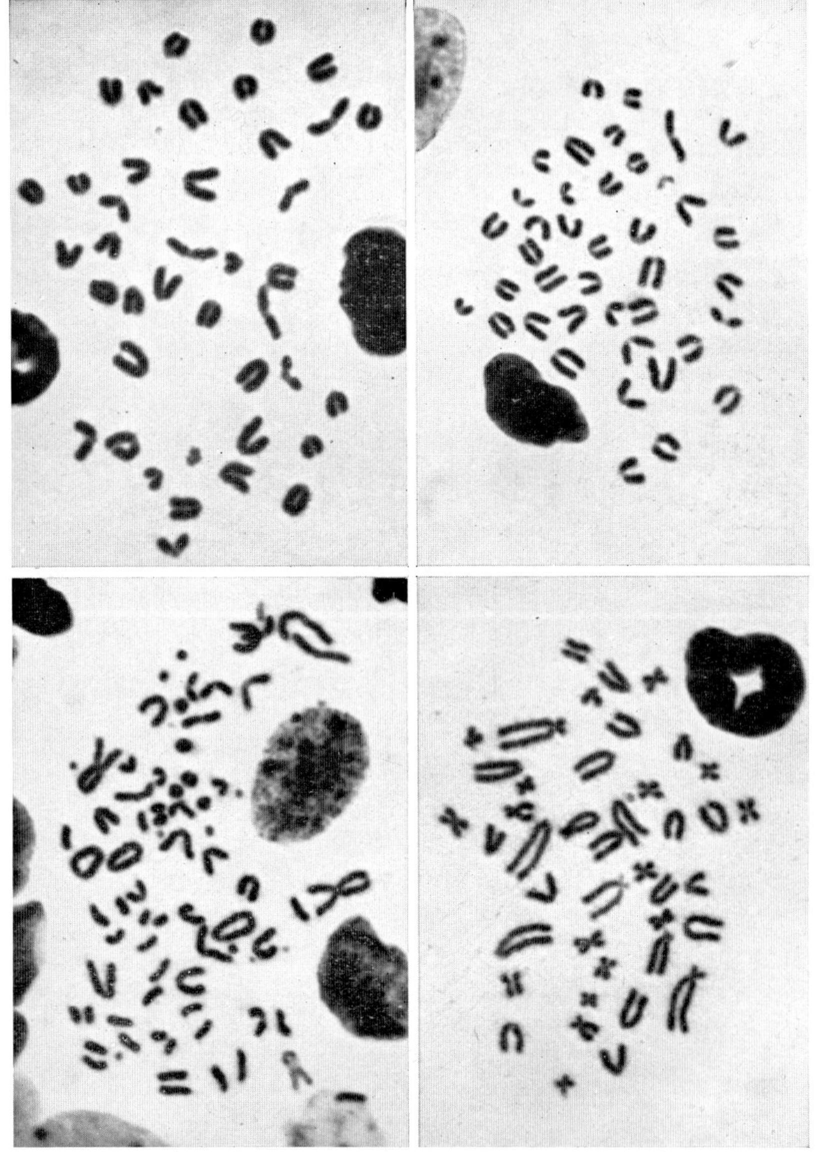

FIG. 1. T6/ + cell in bone-marrow of a CBA mouse 49 days after irradiation and injection with T6/ + cells. The small marker chromosome is at 6 o'clock.

FIG. 2. T6/ + cell in thymus of a CBA mouse 49 days after irradiation and injection with T6/ + cells. The marker chromosome is at 2 o'clock.

FIG. 3. CBA cell with many structural changes in thymus of a CBA mouse 5 days after irradiation and injection with T6/ + cells.

FIG. 4. Rat-cell in bone-marrow of a CBA mouse 19 days after irradiation and injection with cells from rat bone-marrow.

III. TRANSFER OF CELLULAR MATERIAL—IRRADIATION AND IMMUNE REACTIONS

STUDIES OF RADIATION CHIMAERAS BY THE USE OF CHROMOSOME MARKERS

C. E. Ford, J. L. Hamerton, D. W. H. Barnes and J. F. Loutit

Medical Research Council, Radiobiological Research Unit, Atomic Energy Research Establishment, Harwell, Berks, England

It is now well-known that the survival-time of heavily irradiated mice can often be greatly prolonged by injection, after irradiation, of suitable homologous or even heterologous material. By using distinctive chromosomes as markers we have recently been able to demonstrate that cells descended from the injected material appear in the bone marrow, spleen, thymus and lymph nodes of the hosts and establish what amount to grafts, or transplants, of foreign tissue (Ford *et al.*, 1956). Animals, like these, whose tissues are of more than one genetic or antigenic kind have been termed chimaeras. Our first results showed that there was a rapid and virtually complete replacement of dividing host-cells by cells from the donor, which persisted for at least 49 days after treatment. This striking effect is dependent upon two interrelated effects of the irradiation: the destruction of the host's hæmopoietic and lymphatic tissue and the inactivation of its immune response.

In our main series of experiments young adult male CBA/H mice were exposed to 950r of 240 kVp X-rays and then injected intravenously with cell-suspensions prepared from spleens of baby mice heterozygous for the T6 translocation. The donated material is marked by a characteristic small chromosome, and dividing cells containing it are readily identified in good preparations (Figs. 1 and 2). The advantage of this method is that the proportion of native cells to donated cells can be determined directly. Its disadvantages are that it requires the sacrifice of the animal, that it is confined to cells in metaphase of mitosis, and that cross identifications with the cell types of the hæmatologist are not at present possible. The cytological technique employed is a modification of the Feulgen squash method (Ford and Hamerton, 1956). The animals are injected with colchicine beforehand to accumulate metaphases and spread the chromosomes.

Irradiated animals injected with T6 material have been killed for cytological examination at intervals from 5 to 157 days after treatment and the results are presented diagrammatically in Figure 5. By 5 days after treatment nearly all the dividing cells of bone marrow, spleen, and lymph nodes are derived from the injected cells. In the thymus this condition is not attained until later. Replacement by foreign tissue, it seems, must take longer in the thymus. Once the 'transplants' are established they persist for at least 157 days.

The dividing host-cells present in thymus at 5 days were identified by the presence of structural changes in the chromosomes. The production of

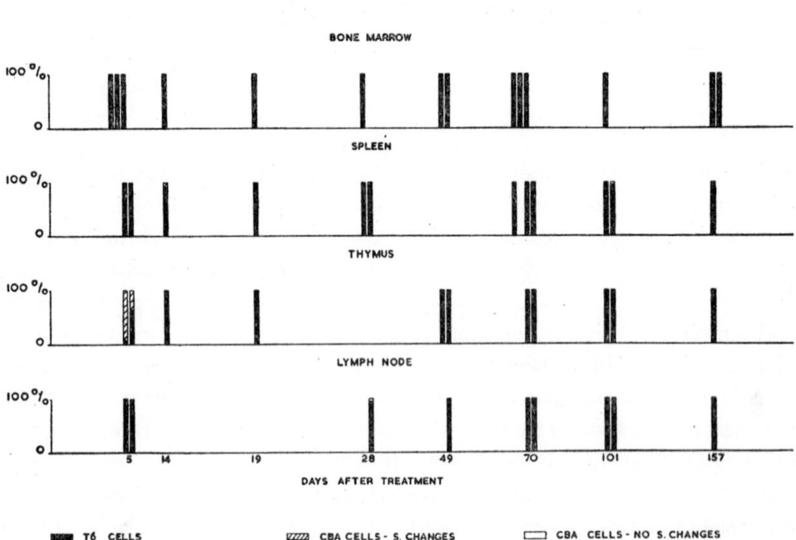

FIG. 5. Percentages of cells in metaphase which are T6, CBA with structural changes, and CBA without structural changes, respectively, in bone-marrow, spleen, thymus, and lymph node of individual mice. The interval in days between treatment and sacrifice is shown on an arbitrary scale.

such changes is a well-known effect of radiation. They are, however, occasionally observed in untreated material, at least in plants and in vertebrates. All the thymus cells concerned contained large numbers of structural changes, and as the marker chromosome was not seen in any of them the identification leaves little room for doubt. An example is shown in Figure 3.

Altogether 80 mice have been injected with T6 spleen-cells after exposure to 950r. Two died almost immediately after injection, as a result, it is believed, of embolism. Thirty-six more have died, 20 in the first 13 days and the others at scattered times up to 84 days after treatment. Of the remaining 42 mice, 20 have been sacrificed and 22 survive. At the present time the longest established chimaeras are 3 mice, grey, but otherwise in very fair condition 275 days after treatment.

A word should perhaps be said about the method of scoring the cytological preparations. These were searched systematically and at first all cells in mitotic metaphase were examined and recorded. This became too time-consuming and in later preparations only those cells in which the chromosomes were reasonably well spread were accepted. If the marker was not identified an attempt was made to count the chromosomes. In most of these cells fewer than 40 were present; in some an exact count was not possible; in 2 cells the full number of 40 normal chromosomes was counted and the absence of the marker confirmed. Since many of the deficient cells had clearly been broken during the making of the preparations, only the last mentioned 2 cells and those with structural changes could be positively identified as the host's own. (Reports in the literature of variation of chromosome number in somatic cells of normal animals are not borne out by our experience of the four tissues concerned.) Altogether 71 host-cells and some 3,000 cells containing the marker have been identified. The diagram is based on these alone, the cells with deficient or uncountable chromosomes being rejected. The latter may have included a higher proportion of host-cells, but since they amount to less than one-sixth of the total the possible bias cannot be very great.

Other similarly irradiated CBA/H mice were injected with suspensions of cells from the femoral bone-marrow of adult albino rats. The chromosome sets of the two species are so different (compare Figs. 1 and 2 with Fig. 4) that the scoring of the preparations is much easier than in the T6/CBA chimaeras. Except in preparation of spleen, which often contain patches of badly fixed cells, all cells in metaphase were scored and only 3% were classified as uncertain. Nearly 15,000 cells were recorded altogether.

On different occasions five groups of animals received this treatment with rat-cells. Each group differed in some degree from the others in respect of survival pattern or cytological behaviour, or both, and it will be convenient to consider them in three series. In series I, 30 mice were treated, 1 died immediately after injection, 10 died between the 7th and 25th day and two more later, and the remaining 17 were killed for cytological examination. The results obtained are summarised in Figure 6. Tissue with dividing cells containing rat-chromosomes is already well established in bone marrow and spleen by 5 days. Thymus at this time contains a high proportion of cells with structural changes, but becomes fully populated with rat-cells later. Lymph node behaves in the same way as thymus.

At 28 days there is an indication that re-appearance of the host's own cells is commencing in both spleen and lymph node. By 49 days this 'recovery' in one of the two animals is almost complete. Thereafter the results are erratic: two out of five animals are partly or largely reverted, but no clear temporal trend is revealed; nor was there any indication of correlation between the cytological picture and the clinical state of the animals at the time of sacrifice.

In an attempt to clarify this situation a further batch of 20 mice were irradiated and injected (Series II). There was a period of heavy mortality between 18 and 28 days during which 7 animals died and many more would have done so had they not been killed for cytological examination. Nine animals were sacrificed during this period and 1 at 42 days after treatment. Several of these were examined rather cursorily but in all four tissues the only dividing cells identified contained rat-chromosomes. Three are still alive, having survived treatment for 126 days.

Two further batches of rat/mouse chimaeras were prepared. These behaved somewhat similarly and will be discussed together as series III. Four of the 40 mice died immediately after injection, 17 died later (15 of them between the 4th and the 15th day after treatment), 16 were killed for cytological examination and 3 survived. The results are summarised in Figure 7. There are two points of difference from series I and II. Firstly, at 5 and 6 days after treatment both bone marrow and spleen contain a few mouse-cells, some of which are normal (i.e. do not show structural changes). Secondly, in thymus and lymph node there is no instance of full replacement of mouse-cells by rat-cells until 67 days after treatment. The cytological results, combined with the large number of early deaths, suggest that the injected cells were less able to establish themselves than in the animals of series I and II.

Brief mention may be made of a further experiment in which the mice received an acute dose of 575r and were then injected with T6 spleen-cells. The preliminary cytological results show that the donated cells were quickly established in all four tissues but that by 28 days after treatment very few of them remained, nearly all the dividing cells observed at this time containing a normal set of chromosomes with no marker. Under these conditions the donated cells evidently gain a temporary hold but are soon replaced by cells of the host.

Discussion

In mice exposed to 950r there is a pronounced difference in behaviour between the 'grafts' derived from injected homologous (T6) and heterologous (rat) material. Once established, T6 cells persist, but rat-cells may be progressively replaced by cells of the host from the sixth week onwards. Although this difference in behaviour is most probably to be associated with the antigenic difference between homologous and heterologous cells, the

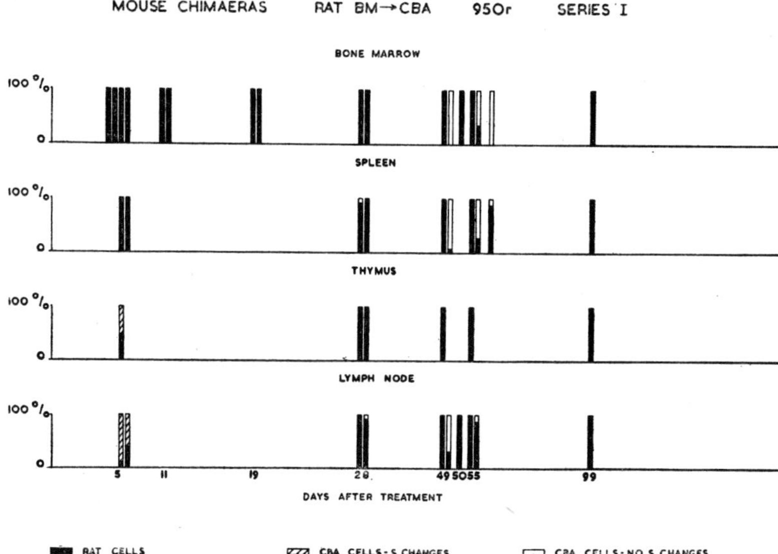

FIG. 6. Percentages of cells in metaphase with rat chromosomes, mouse (CBA) chromosomes showing structural changes, and mouse chromosomes without structural changes, respectively, in bone-marrow, spleen, thymus, and lymph node of individual mice. The interval between treatment and sacrifice is shown on an arbitrary scale.

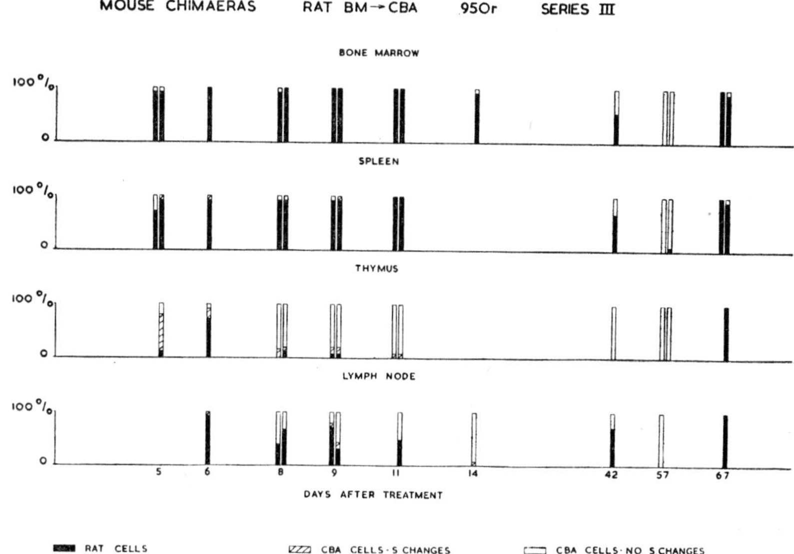

FIG. 7. Percentages of cells in metaphase with rat chromosomes, mouse (CBA) chromosomes showing structural changes, and mouse chromosomes without structural changes, respectively, in bone-marrow, spleen, thymus, and lymph node of individual mice. The interval between treatment and sacrifice is shown on an arbitrary scale.

difference in tissue injected (spleen, bone marrow) or in the age of donors (infant, adult) may also have been important.

In the few animals given 575r the T6 cells appeared transitorily but were rapidly replaced by host-cells. At this lower dose there is presumably a much larger residue of surviving host-cells from which replacement can take place. It may be supposed that these cells have a competitive advantage over the donated cells, and that the latter, either through an initial advantage in numbers, or because the surviving host-cells are still in progress of recovery from the effects of the dose administered, become dominant for a period but then rapidly decline and disappear.

Summary

1. Suitable foreign cells injected intravenously into heavily irradiated mice establish themselves quickly in bone marrow and spleen, probably more slowly in lymph node, and probably more slowly still in thymus.

2. The donated cells multiply in each of these sites and, in effect, form a graft of donated tissue.

3. Dividing host-cells showing large numbers of structural changes in their chromosomes are present in all four tissues, but particularly in thymus and lymph mode, during the first 14 days after treatment. In the same period normal host-cells have only been detected in the exceptional Series III treated with rat material.

4. Several weeks after treatment, replacement of the 'grafted' tissue by the host's own cells may occur in mice which had been injected with bone-marrow cells from adult rats. Similar replacement has not been observed to occur up to 157 days after injection of homologous cells from infant T6 spleens if the dose is high (950r), but takes place rapidly if the dose is lower (575r).

REFERENCES

Ford, C. E., Hamerton, J. L., Barnes, D. W. H., and Loutit, J. F., 1956. *Nature, Lond.*, **177**, 452.

Ford, C. E., and Hamerton, J. L., 1956. *Stain Technology*, **31**, 247.

Discussion

Wolff. I should like to ask Dr. Ford if, with his beautiful technique, he is able to determine whether or not those animals that die after irradiation contain donor cells?

Ford. We have been looking at this in our material. In a series where we had a lot of deaths between the 18th and 21st day one animal which was analysed immediately after death had the rat-cells.

Fahmy. There is no doubt at all that the foreign cells substitute the destroyed ones, but do you think that excludes the possibility that a chemical agent might be a contributing factor also?

Ford. No, we can't exclude that, but it would be very difficult to prove experimentally, because you have to distinguish between substances you inject with the donor material and substances produced by the donor material itself after the injection.

van Bekkum. In your last experiment you mentioned a radiation dose of 575r. What is the percentage of untreated mice that are killed by this dose?

Ford. No untreated CBA mice are killed by 575r.

Hollaender. People in our laboratory—Dr. Congdon, Dr. Makinodan and others—have obtained very similar results to those Dr. Ford mentioned in regard to immunological aspects. Mice that have been given heterologous bone marrow after lethal exposures will survive in very large numbers up to 60 days then we lose about 50% of our mice. Another 20% will be lost in 60 to 80 days after irradiation. Apparently these animals cannot digest the food that they take in but 10–20% will survive for a considerable time. We have mice now that have survived for more than 200 days after irradiation up to 900r when given heterologous bone marrow.

Howard-Flanders. Have you looked for chromosome markers in some other tissues, for example in the basal layer of skin or in the intestinal epithelium?

Ford. We have not examined other tissues for the chromosome marker. Dividing cells in the basal layer of the skin and in the crypts of the intestinal epithelium would be much more difficult to handle technically, but the same principles would, of course, apply.

Chairman. Dr. Loutit, do you have some comments?

Loutit. The only things I will stress are the following points: where you have the heterologous mixture of mouse and rat it is only the minority of the animals which you can bring through to long-term survival. Even in the case of homologous transfer from mouse it is still a matter of chance. We found by experience that certain mice are good donors and by matching such good donors and good recipients we can go through with say, 50% survival up to 1–2 years. But we can also show results where the mice die after about 50 days. Therefore, there are still many unknown immunological or other factors in this procedure.

THE CELL FACTOR IN SPLEEN HOMOGENATES AFTER IRRADIATION

J. Soška, V. Drášil and Z. Karpfel

Biophysical Institute of the Czechoslovak Academy of Sciences, Brno, Czechoslovakia

Using a heterozygous strain of mice for experiments with spleen homogenates, we have observed that the mortality effect is not constant; although a marked recovery of the hæmopoietic tissues has always been observed. The mortality rate, after the use of spleens from adult animals for homogenate treatment and irradiation, has sometimes been even higher than in the control group. Two to four weeks after irradiation, anæmia and spleen atrophy often reappeared together with a luxuriant growth of reticulum. We have tried to explain these results by accepting the assumption of homogenate cell survival in the irradiated animal and of immune reactions occurring between the implanted tissue and the host tissue, as is usual in tissue transplantation.

To verify this hypothesis the following experiments have been performed.

In the first experiment it was assumed that homogenate from the 'own' spleen of mice will be the most effective, effectiveness depending on the identity of the antigens involved.

Partial splenectomies were performed on all the mice, and they were immediately and separately given total-body X-irradiation; the dose being 550r, 180 kV, i.e., LD75. Each homogenate was divided into two parts. One part was injected into the tail vein of the mouse from whose spleen the homogenate had been prepared and that had in the meantime been irradiated, forming a group of 31 mice with 'own' homogenate. The other part of each homogenate was injected into another splenectomised and irradiated mouse, forming a group of mice with 'alien' homogenate. Further mice were injected with homogenates from young (one week old) mice, forming a group of 31 mice with 'young' homogenate. A control group of 29 mice were injected with isotonic saline.

The following points have been evaluated:—survival, body weight, hæmatological changes and intensity of DNA synthesis in the remaining portion of spleen on the third post-irradiation day. In addition, the DNA turnover has been determined in groups of 6–8 mice by determining the specific activity of DNA ^{32}P two hours after $Na_2H^{32}PO_4$ injection. The DNA was isolated from spleen by the Smith-Tannhauser method, purified, counted and the specific activity calculated in relation to the acid soluble phosphorus fraction.

The highest survival rate, 61 per cent., has been observed after the administration of the 'own' homogenate and the lowest, 18%, after the

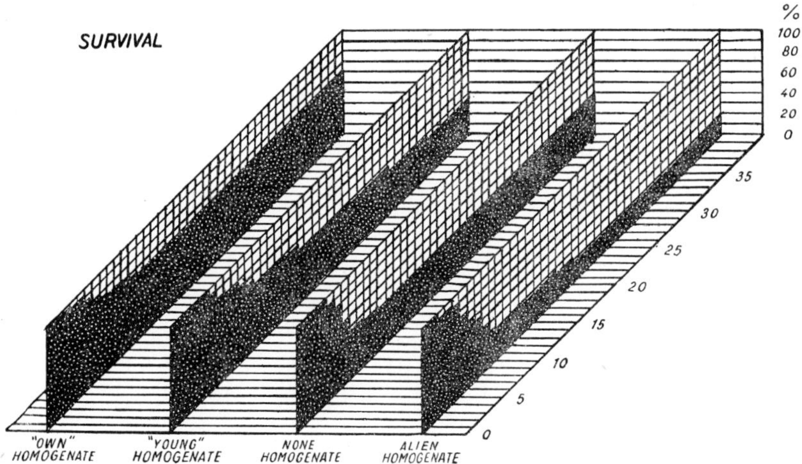

FIG. 1. Mouse mortality after LD75, 550r.

'alien' homogenate (Fig. 1). The difference is statistically significant P = 0·01–0·001. After the administration of 'young' homogenate the mortality was not lower, it was only retarded.

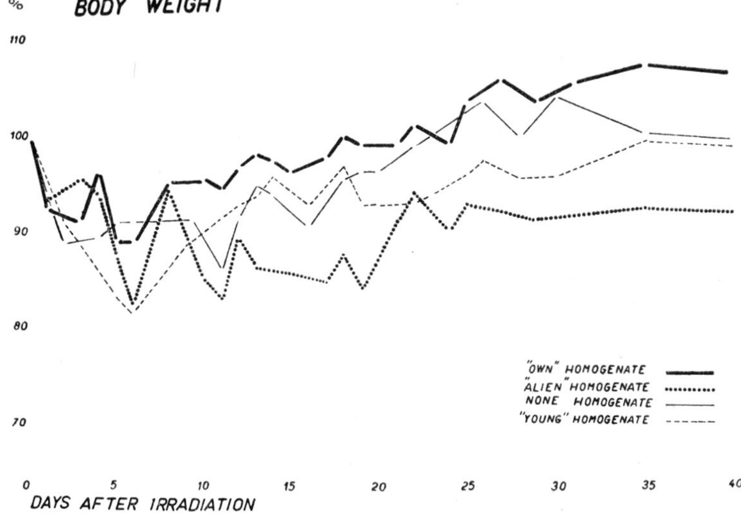

FIG. 2. Mouse body weight changes after LD75, 550r.

The body weight change was favourable (Fig. 2) for survival after 'own' homogenate, and was unfavourable after the 'alien' homogenate.

Peripheral blood response of all components was favourable to all homogenates. The reaction to 'own' homogenate was generally the most permanent. The initial reaction in different components of the blood was usually the most prominent after 'young' homogenate. A marked eosinophilia of 7% occurring in the 'alien' homogenate group, compared with 0·5%–3·0% in the other groups, is worthy of note.

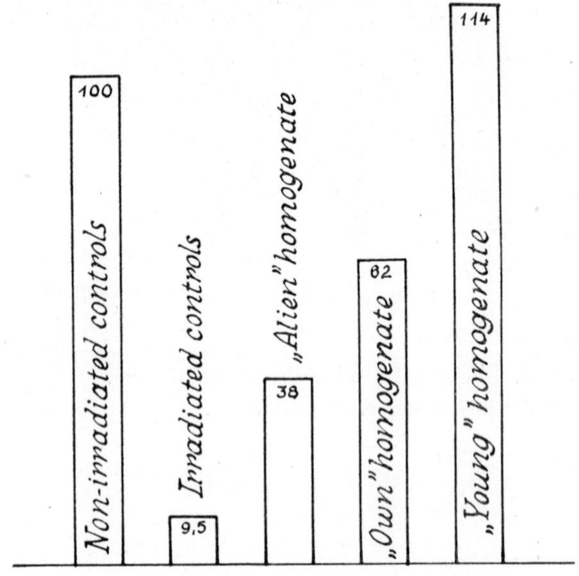

FIG. 3. Relative DNA turnover in spleens.

The relative intensity of DNA synthesis has been expressed in relation to that occurring in normal adult spleens which has been taken as 100. In irradiated mice without homogenate a value of 9·5% was obtained, in mice with 'alien' homogenate a value of 38%, in mice with 'own' homogenate a value of 62%, while in the 'young' homogenate group a value of 114% (Fig. 3). All observations were made on the third post-irradiation day.

These results would seem to be evidence in favour of the reaction to 'young' homogenate being the quickest, but would suggest that the 'own' homogenate while acting more slowly is more permanent in effect. These results tend to confirm the hypothesis that the most effective homogenate is that with the least antigenic difference from the recipient. In addition, these findings are in agreement with the work of Barnes and Loutit (1956) showing that homogenate is permanently effective only when derived from animals of the same inbred strain. Similar laws are valid here to those applicable to tissue transplantation. We, therefore, assume that the effective components of homogenate are living cells, at least in the early part of the reaction. Of course, simultaneous stimulation of the animal's own hæmopoietic tissue by metabolites from the implanted tissue cannot be excluded.

In the second experiment we have attempted to prove the presence of implanted tissue in the spleen after irradiation and spleen homogenate treatment.

About the sixth day after irradiation the spleens of mice injected with spleen homogenate are two to four times as heavy as the spleens of mice

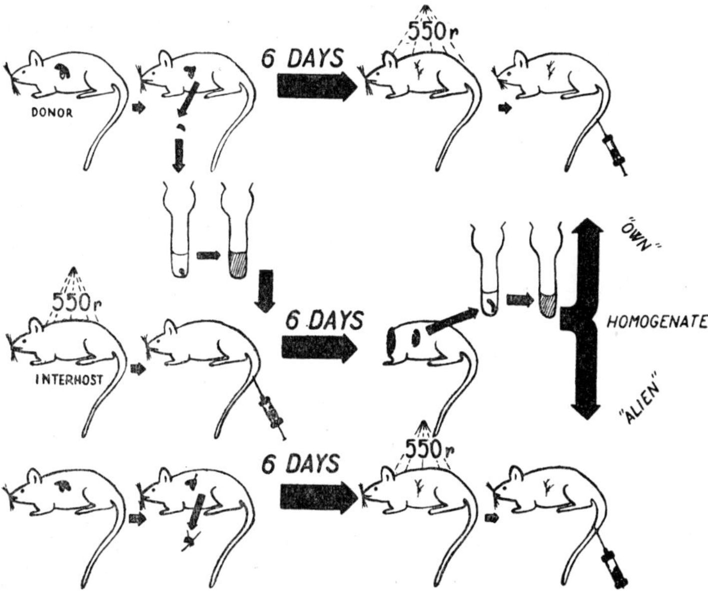

FIG. 4. Scheme of spleen 'interhost' homogenate experiment.

irradiated only, and, moreover, they contain a lot of blood-forming tissue. This raises the question of the origin of the tissue. Is it 'donor tissue' surviving and multiplying in the spleen and marrow of the 'interhost'? If this latter were the case, it would be possible to return the tissue to the original 'donor' which should then react to it as to its 'own' homogenate, as in the preceding experiment. If this is not the case, then it will react to the injection as to an 'alien' homogenate. We then tried to transmit the spleens' cells from the 'donor' through the irradiated 'interhost' partly back to the 'donor' and partly to another mouse. The scheme of the experiment is shown in Figure 4.

Partial splenectomies were first performed in the donor and control mice. Simultaneously, the 'interhosts' and mice planned as the source of irradiated homogenate were irradiated; the dose being 550r, 180 kV, LD75. The homogenates from spleens of the 'donors' were injected intravenously into the 'interhosts'. On the sixth day the 'interhosts' were sacrificed and homogenates prepared from their spleens. The 'donors' and control group mice were then irradiated. Each homogenate from each 'interhost' was divided into two parts; one part was injected into the tail vein of the original 'donor' and called 'own' homogenate, the other part was injected into another irradiated mouse and called 'alien' homogenate. Mice of further control groups were injected with homogenate from normal non-irradiated spleens—'normal' homogenate, with homogenate from mice irradiated six days earlier—'irradiated' homogenate and with Hanks solution—'none' homogenate. Each group contained about 20 mice.

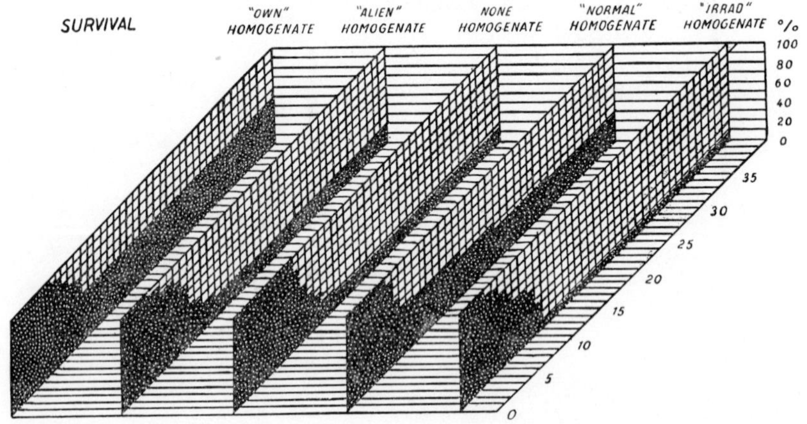

FIG. 5. Mouse mortality after LD75, 550r in 'interhost' experiment.

In the group treated with 'alien' homogenate the survival rate was 19%, and in the group treated with 'own' homogenate, 50% (Fig. 5). The difference is significant $P<0{\cdot}05$. It is of interest to note that the mortality ceased earlier in the 'alien' homogenate group than in the 'normal' homogenate group.

From the seventh day onwards there is a difference in the course of the body-weight curves in both main groups (Fig. 6). The weight of mice with 'own' homogenate rises, while the weight of mice treated with the 'alien' homogenate falls in the same period. Both curves meet again about the twentieth day. In the group with the 'normal' homogenate the course is worse than in groups with homogenates from regenerating spleens.

The recovery of the peripheral leucocyte count was faster after the 'normal' homogenate than after the homogenates from regenerating spleens. The course of recovery after the 'own' homogenate was more favourable than after the 'alien' homogenate, but the difference is not marked. In the red cell count the difference is more marked.

The results thus seem to corroborate the suggestion that on the sixth day, after irradiation and spleen-homogenate injection, the regenerating spleen had an antigenic structure at least partly similar to that of the tissues of the 'homogenate donor'. If this is true then it appears that the mechanism of the homogenate action is in the nature of a substitution of hæmopoietic tissues rather than of their stimulation. Other experiments published recently (Main and Prehn, 1955 ; Ford *et al.*, 1956 ; Lindsley *et al.*, 1955) support this conclusion.

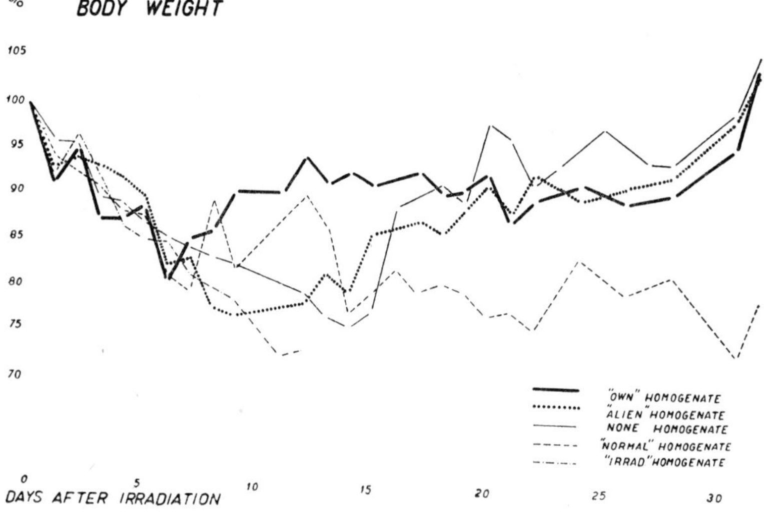

FIG. 6. Mouse body weight changes after LD75, 550r in 'interhost' experiment.

We have also to take into account Cole's hypothesis (1954) that the effective component is the cell nucleus or a nuclear nucleoproteid and that it becomes a constituent of the host's cells and renews their viability. Even this might change their antigenicity, as shown by experiments with nucleus transplantation in amœbæ (Danielli *et al.*, 1955). From this point of view, it is, however, difficult to explain the fact that only homogenates from reticulo-endothelial tissues are effective and that only these tissues react to homogenate treatment. If Cole's hypothesis were correct we would have to accept that the nuclei of the reticulo-endothelial cells contained a specific nucleoprotein component, and this we take to be improbable.

Footnote in Press.

In later experiments we have obtained positive results with homogenates of tissues other than those described in this paper.

REFERENCES

Barnes, D. W. H., and Loutit, J. F., 1955. *Progress in Radiobiol.*, 1956. p. 291, (Edinburgh, Oliver and Boyd).

Cole, L. J., and Ellis, M. E., 1954. *Radiation Res.*, **1**, 347.

Danielli, J. F., Lorch, I. J., Ord, M. J., and Wilson, E. G., 1955. *Nature, Lond.*, **176**, 1114.

Ford, C. E., Hamerton, J. L., Barnes, D. W. H., and Loutit, J. F., 1956. *Ibid.*, **177**, 452.

Lindsley, D. L., Odell, T. T., and Tausche, F. G., 1955. *Proc. Soc. exp. Biol. N.Y.*, **90**, 512.

Main, J. M., and Prehn, R. T., 1955. *J. nat. Cancer Inst.*, **15**, 1023.

Discussion

Vos. The work Dr. Soška has done with non-inbred mice has been done by us previously with 2 inbred strains of mice. CBA bone marrow suspensions were made and injected into lethally irradiated C 57 BL mice. Fourteen days afterwards the cells of the regenerated bone marrow of these mice were injected into lethally irradiated CBA and C 57 BL mice. This regenerated bone marrow was only effective in the CBA mice, not in the C 57 BL mice (suspensions containing 100,000–800,000 viable nucleated cells per injection were used). This is in agreement with the fact that isologous bone marrow is twenty times more effective than homologous bone marrow.

van Bekkum. Did you observe any delayed mortality in your irradiated and homologous spleen-treated mice?

Soška. We have often observed late deaths in experiments with homologous cells, but not in experiments with the animals's own cells or in experiments with 'passed own' cells. With regard to survival, we usually got better reponses to homologous cells after relatively higher doses of radiation, exceeding LD100, than after relatively lower doses, below LD50. After very high doses, there was no effect on mortality. But the favourable effect of hæmopoietic cell treatment—measured by the DNA turnover in the spleen on the 3rd day—has been the same after doses as high as 2,400 or 4,800r as after e.g. 600r.

REFERENCE

Vos, O. *et al.*, 1956. *Acta Physiologica et Pharmacologica Neerlandica*, **4**, 482.

SOME EXPERIMENTS IN FAVOUR OF THE CELLULAR HYPOTHESIS FOR THE SPLEEN CURATIVE FACTOR

D. W. H. BARNES, M. P. ESNOUF AND L. A. STOCKEN
Medical Research Council, Radiobiological Research Unit, Harwell, and the Department of Biochemistry, University of Oxford, England

Two hypotheses have been advanced to explain the ability of splenic preparations to prevent death which follows exposure of small animals to lethal amounts of radiation. On the one hand it has been maintained that survival of the injected cells initiated the cure and on the other, that the agent responsible was a humoral one.

A formidable objection to the cellular hypothesis was the lack of antibody response to the foreign heterologous or homologous preparation. This has now been reconciled by the work of Lindsley, Odell and Tausche (1955), Nowell, Cole, Roan and Habermeyer (1955), and Ford, Hamerton, Barnes and Loutit (1956), which collectively prove the survival of the injected material. One other difficulty was the demonstration by Cole and Ellis (1954) that treatment of the spleen homogenate with DNA-ase destroyed its curative properties, which suggested that nuclear material was the therapeutically active principle. It is extremely difficult to differentiate between intact and slightly damaged cells and nuclei but the recent method of Barer, Esnouf and Joseph (1956) allows a clear distinction to be made. A small sample of the homogenate dispersed in 50% w/v bovine serum albumin solution is observed by positive phase contrast microscopy. Since the cell membrane is impermeable to protein while the nucleus is not, the intact cells appear bright and the nuclei dark and it is a simple matter to carry out cell and nuclear counts and make any necessary adjustments before the bulk of the preparation is used.

In our early experiments using the conventional media and techniques of Dounce, Witter, Monty, Pate and Cottone (1955), Hogeboom, Schneider and Streibech (1952), and Wilbur and Anderson (1951) for isolation of nuclei, the cell count was sometimes as high as 5% of the total. It was then found that by means of a modification of the tissue culture medium described by Pannet and Compton (1924), together with the homogeniser devised by Philpot and Stanier (1956) the cell contamination could be reduced to less than 0·5%. The alterations made in the composition of the tissue culture medium were to reverse the potassium-sodium ratio and to substitute triethanolamine for phosphate buffer. The prime reason for the latter change was to allow enzyme assays to be made on the same preparation.

The medium contained 0·085 M KCl, 0·0085 M NaCl, 0·0025 M $CaCl_2$, 0·0025 M $MgCl_2$, 0·045 M Glucose, and 0·005 M Triethanolamine-hydrochloric acid buffer *p*H 7·2.

Spleens were taken from 5-10-day-old mice and homogenised in 1 ml. sterile medium. The homogenate was diluted to 20 ml. with the medium, filtered through nylon cloth to remove strands of connective tissue and centrifuged at 440×g. for two minutes. The residue was redispersed in the homogeniser, again filtered and centrifuged as before. The sediment

TABLE

30 days' Survival of CBA Mice after 950r X-rays and Intravenous Injection of Spleen Fractions

Nuclear preparation		Cell preparation	Survival
Nuclei $\times 10^6$	Cells $\times 10^3$	Cells $\times 10^3$	
—	—	60	4/4
6·5	60	—	4/4
—	—	50	3/4
1·3	50	—	3/4
—	—	36	3/4
60	36	—	4/4
10	29	—	1/10
—	—	29	0/5
24	24	—	1/4
—	—	12	0/5
2·2	10	—	0/20
36	<0.6	—	0/10
10	<0·2	—	0/10

was finally suspended in fresh rabbit serum and a sample diluted with 50% albumin for cell and nuclear counts. When a preparation of whole cells was required, spleens were disintegrated with a mechanical cutting device using rabbit serum as a suspending medium. This usually gave 20–30% nuclei and damaged cells, with the remainder as intact cells. Intravenous injections were made into 3-month-old CBA male mice which had been exposed to 950r whole body X-radiation about two hours previously (for technique see Barnes and Loutit, 1955). The mice have been kept for long-term observation but survivivors were scored at 30 days. It will be seen (Table) that when the number of whole cells was reduced to below about 30,000 the 30-day survival was markedly affected, and that the therapeutic activity of the preparations was independent of the number of nuclei. Injection of 10×10^6 nuclei on days 1 and 4 after irradiation gave no increase in survivors when compared with mice given one injection of 10×10^6 nuclei on day 1.

It is of interest to compare these results with those of Jacobson, Marks and Gaston (1954), who found that injection of between $0 \cdot 5 \times 10^6$ and 10×10^6 nucleated spleen cells caused the survival of 70% of LAF mice after 900r X-rays. The fact that so many cells were required and that the survival rate did not rise with the number injected, might be explained by the difficulty of differentiating between intact and slightly damaged cells.

None of these experiments, however, exclude the possibility that combinations of cell fractions might have therapeutic value, but nuclei alone isolated by our procedure cannot prevent death following a lethal exposure to X-rays.

REFERENCES

Barnes, D. W. H., and Loutit, J. F., 1955. *J. Nat. Cancer Inst.*, **15**, 901.

Barer, R., Esnouf, M. P., and Joseph, S., 1956. *Science*, **123**, 24.

Cole, L. J., and Ellis, M. E., 1954. *Radiation Res.*, **1**, 347.

Dounce, A. L., Witter, R. F., Monty, K. J., Pate, S., and Cottone, M. A., 1955. *J. Biophys. Biochem. Cyt.*, **1**, 139.

Ford, C. E., Hamerton, J. L., Barnes, D. W. H., and Loutit, J. F., 1956. *Nature, Lond.*, **177**, 452.

Hogeboom, G. H., Schneider, W. C., and Streibech, M. T., 1952. *J. Biol. Chem.*, **196**, 111.

Jacobson, L. O., Marks, E. K., and Gaston, E. O., 1955. *Radiobiol. Symp.*, 1954, p. 122 (London, Butterworth).

Lindsley, D. L., Odell, T. T., and Tausche, F. G., 1955. *Proc. Soc. exp. Biol., N.Y.*, **90**, 512.

Nowell, P. C., Cole, L. J., Habermeyer, J. G., and Roan, P. L., 1955. *USNRDL-TR*-59.

Philpot, J. St L., and Stanier, J. E., 1956. *Biochem. J.*, **63**, 214.

Wilbur, K. N., and Anderson, N. G., 1951. *Exp. Cell Res.*, **2**, 47.

Discussion

Loutit. You quote the observation of Cole and Ellis (1954) that DNA-ase abolishes the recuperative properties of 'nuclear' preparations, but you have not explained why this should be so, if the contaminating whole cells are the effective fraction of the preparation.

Esnouf. The results which we have obtained with DNA-ase are ambiguous. We have found that DNA-ase destroys the therapeutic activity of dilute cell suspension whereas other workers have shown that DNA-ase has no effect on the curative properties of concentrated cell suspensions.

Cohen. Did the albumin method differentiate between living and dead cells?

Esnouf. Yes. Since only the living cell membrane is impermeable to protein; only intact cells appear bright under positive phase contrast. The term 'whole cells' in our paper refers only to these reversed cells.

STUDIES ON THE MECHANISM OF POST-IRRADIATION PROTECTION

ERIC L. SIMMONS, LEON O. JACOBSON AND JOANNE DENKO
Argonne Cancer Research Hospital, The University of Chicago, Chicago 37, Illinois, U.S.A.

Within the past two years, an exciting array of independent experimental findings in the field of post-irradiation protection has led to a re-evaluation of the entire concept of how partial body shielding and cell injections bring about recovery of the lethally irradiated animal.

It is now well established by the work of Jacobson and his associates (1949 and 1952) that lead-shielding of the spleen, spleen transplants, and injections of spleen and embryo cell suspensions save the lives of irradiated mice following X-ray doses far above the lethal range. Lorenz *et al.* (1951) expanded these techniques by the use of cell suspensions of bone marrow. Subsequent findings that success could be obtained with combinations as diverse as guinea-pig or rat marrow to mice were made possible by these techniques.

In the early days of protection studies it was difficult to believe, on general immunological grounds, that cells from one genus of mammal could survive in the body of an animal of another genus, since it was assumed that species-specific antibodies would destroy them. And so for years the search went on for a factor or factors contained in reticulo-endothelial tissues that presumably hastened the regeneration of undamaged primitive cells that were present originally in the irradiated animal's body. Jacobson *et al.* (1950) demonstrated that the capacity to produce antibodies to an injected particulate antigen is retained by irradiated animals if they are spleen-shielded during the exposure, and Dixon and Maurer (1955) showed that when irradiated animals are stressed with a foreign protein during the refractive immune state which follows irradiation, the animals develop a tolerance to this antigen and thereafter will not produce antibodies against it. These findings made it quite possible, then, to accept the fact that injected cells successfully invade the immunologically altered host and, failing to encounter any resistance, thrive and re-colonise the tissue spaces that have been depopulated by death of the host cells.

In 1954, Barnes and Loutit suggested that the evidence for direct colonisation by injured cells should be re-evaluated. They reported that when two different strains of mice were used, and one strain was pre-immunised against the other, cell injections failed to produce survival, suggesting that living cells were destroyed rather than the failure of a factor to gain entry. They also reported that when mice are given lethal exposures of X-radiation but are kept alive with injections of spleen cells from strains other than their own, survival results at the usual 30-day period are misleading since the mice die subsequently before 200 days. They postulated that the injected cells colonise the immunologically-weakened host animals, but that when the antibody-forming capacity of the irradiated mouse recovers, the foreign marrow is destroyed, resulting in death (which would not agree with the Dixon and Maurer (1955) hypothesis).

In any event, the inability of homologous cells to confer long-lasting protection has been substantiated in our own laboratory using a variety of strains of mice and cell types, and has also been reported by Trentin (1956), Congdon and Urso (1956), and others. To investigate the nature of these late deaths following successful recovery beyond the acute 30-day period, irradiated CF No. 1 female mice were given DBA/2 bone marrow cells, and mice that survived 24 days or more were killed for histological study when they appeared moribund.

At the time of death the bone marrow fell into one of three classifications: (1) all cells appeared very immature; (2) some cells appeared very immature, while others were differentiated in the myeloid direction; or (3) there was differentiation in both myeloid and erythroid directions. This may mean either that one or both series are not maturing, or that they are maturing but are being released into the peripheral circulation very rapidly. In addition, all spleens showed at least minimal erythropoiesis as well as myelopoiesis. Other organs showed a variety of pathological conditions, so that no single common cause of death was identified throughout the series. Rather, the mice seemed weakened and died with a variety of conditions, including pneumonia, infarcts, liver lesions, and parasites. In many cases, the cause of death was not apparent.

A variety of experiments using rodents now lend physiological proof to the hypothesis that foreign cells may invade the host body and function in a fashion typical of the donor animal. Lindsley and his associates (1955) have injected bone marrow from rats of one blood type into irradiated rats of another type, and have found that in the peripheral blood, the donor cell type makes its appearance and eventually predominates. Nowell *et al.* (1956) used the alkaline phosphatase test to show that the bone marrow of the mouse, which is typically negative, shows the positive reaction characteristic of the rat after mice are irradiated and treated with rat marrow. Other lines of evidence have included the work of Merwin and Congdon (1956) using non-vascularised tumour implants as the test object, and of

Russell (unpublished) who is able to alleviate a genetically derived macrocytic anæmia in mice by irradiating them and giving marrow cells from normal donors.

In experiments in our laboratory, we have used sensitivity to lymphatic leukæmia P–1534 to which the DBA/2 mouse is specifically susceptible but to which the CF No. 1 mouse is resistant, as a test of altered immunity pattern following irradiation and cellular protection. We have used the CF No. 1 mouse extensively in experiments with this leukæmia, and have never observed a take. On the other hand, as Kirschbaum (1951) has summarised in his review on mouse leukæmia, if one inoculates the F_1 hybrid of a susceptible strain, the genetic constitution that it has inherited results in susceptibility. Thus, 53 (CF No. 1 × DBA/2)F_1 hybrid mice showed 100% mortality when injected with leukæmic cells.

In view of the increasing evidence of cellular colonisation when protective cell injections are given to irradiated animals, it seemed of interest to irradiate CF No. 1 mice, which are normally resistant to leukæmia P–1534, protect them with cell injections from leukæmia-sensitive mice and, following their recovery from the acute radiation phase, stress the survivors 28 days later with an inoculation of leukæmic cells in order to see whether such mice, which were previously resistant, would now acquire the immune response pattern of the DBA/2, accept the leukæmic cells, and die.

The plan of experiment was as follows: Groups of CF No. 1 mice were given 750r and were injected intravenously either with 5 million DBA/2 bone marrow cells in 0·5 ml. Locke's solution or with 5 million (CF No. 1 × DBA/2)F_1 hybrid bone-marrow cells. Because of the high initial mortality encountered with the DBA/2 cells at this dosage, additional series were set up using a lower dosage of 600r. One month after irradiation, surviving animals were divided into two groups and given an intraperitoneal inoculation of DBA/2 leukæmic spleen cells, or a control injection of normal DBA/2 spleen cells. Since leukæmia kills well within the first 30 days after inoculation, and irradiated animals given homologous cells will continue to die sporadically as Barnes and Loutit (1955) have shown, it was decided to compare survival results at the end of 30 days following the inoculation.

The results to date are shown in the Table. Cells of DBA/2 or hybrids that are introduced into resistant CF No. 1 mice following irradiation result in increased susceptibility to inoculation with DBA/2 leukæmic cells, suggesting that the altered immune response pattern is due to colonisation by the foreign cells that confer leukæmic receptivity. Experiments now in progress show that this transfer can be effected when hybrid baby liver or spleen cells are used for the initial protection, as well as by adult bone marrow. The continued presence of leukæmic sensitivity has been observed as late as 68 days after the initial 750r and homologous injection, suggesting the continued presence of colonised cells.

We have also used the transfer of leukæmic sensitivity as a tool to find out whether, once an animal has been X-irradiated and colonised by foreign cells, further X-irradiation of the surviving animals and the injection of cells of its own strain will destroy the immune properties originally conferred by the foreign cells. CF No. 1 mice were given DBA/2 bone-marrow cells to keep them alive after irradiation. Surviving mice were

TABLE

30-day Mortality in CF No. 1 Female Mice following Leukæmic Stress

Original injection	X-ray exposure (r)	Stress	Mice (no.)	Mortality (%)
CF No. 1 bone marrow ...	750	Leukæmia	23	17·4
DBA/2 bone marrow ...	750	Leukæmia	8	75
DBA/2 bone marrow ...	750	Normal spleen	19	26·3
Hybrid bone marrow ...	750	Leukæmia	10	100
Hybrid bone marrow ...	750	Normal spleen	7	14·3
CF No. 1 bone marrow ...	600	Leukæmia	23	0
DBA/2 bone marrow ...	600	Leukæmia	28	78·6
DBA/2 bone marrow ...	600	Normal spleen	20	30
DBA/2 adult spleen cells ...	600	Leukæmia	17	70·6
DBA/2 adult spleen cells ...	600	Normal spleen	18	11·1

again X-irradiated 21 days later, and this time, injected with cells of their own strain, namely, CF No. 1 bone marrow. Twenty-eight days after this attempt to flush out the DBA/2 cells, survivors were divided into two groups and either inoculated with leukæmic cells or given a sham injection of normal DBA/2 spleen. By 30 days, all of the animals given leukæmic cells were dead, while 86% of the control animals were still alive, suggesting either that the original colonisation by DBA/2 cells had not been destroyed completely or that whatever immune receptivity had been originally conferred by the DBA/2 cells was still in effect.

These results, using sensitivity to a specific leukæmia as a test of colonised cell function, are in agreement with the work of Main and Prehn (1955) who used skin grafting following irradiation and cell injection as their test. BALB/cAnN skin grafts will normally not be retained by DBA/2JN mice. However, when DBA/2JN mice were given 800r and injected with (BALB/cAnN × DBA/2JN)F_1 hybrid cells, subsequent grafts took and were retained. Their feeling was that, in addition to saving the animal's life, the injected cells served some further function in changing the immunity response pattern of the host and allowing the skin graft to be retained.

Perhaps the most beautiful visible proof of cellular colonisation is to be found in the work of Ford *et al.* (1956*a* and *b*).* It should be kept in mind from the recent work of Askonas and White (1956), however, that even bone marrow may have antibody-forming properties.

If Dixon and Maurer (1955) are correct in stating that the irradiated host is immunologically unresponsive thereafter to the antigens of the injected cells; then the possibility exists that the colonised foreign tissue produces antibodies against its host's antigens. This may well explain the secondary pathological syndrome that develops when homologous cells are used. Experiments will have to be designed to determine the extent of immunity interactions between host and colonised tissue.

At this stage it is not known whether cellular colonisation is involved initially, or is maintained when widely different animals are used, for example mouse cells to rabbit. Rabbits whose lives have been protected in this fashion in our laboratory, have now lived for over one year, so do not show the post-acute failure seen in mouse, rat, and guinea-pig work. Despite the fact that such injections of mouse cells to rabbits significantly enhance their survival, it is interesting that they have no appreciable effect upon hæmatological recovery as observed in the peripheral blood; as is seen when mouse cells are injected into irradiated mice. In our search for a test that might indicate whether the rabbit retains its own antibody-forming ability or whether the injected cells assume dominance in antibody formation, use was made of the fact that the rat does not form antibodies (precipitins) to soluble antigens such as bovine serum albumin, although the rabbit will respond to soluble antigens. Rabbits were given 900r and injected with liver cells of newborn baby rats. Survivors were injected with soluble antigen, including a booster dose one month later. When their serum was tested by the antigen-antibody ring test for reaction to such soluble antigen, a positive reaction typical of the rabbit indicated that antibody had been formed. Thus it would seem that whether or not the rat cells have colonised the rabbit, at least they have not taken over the control of antibody production sites in the body of the rabbit.

A variety of cell types from chicken embryos as well as from young chicks has been injected into CF No. 1 mice following 750r. Animals usually died by the 30th day, except for an occasional survivor. Blood smears have been made from such surviving animals between two months and one year in order to see whether hæmatopoietic stem cells of avian origin had colonised the mouse marrow and whether nucleated erythrocytes typical of the chicken were present. None has ever been seen. It would appear from these

* This volume p. 197

preliminary investigations that attempts to bring about radiation protection as far apart as different phyla is too distant a relationship for cell growth and colonisation.

Thus we have witnessed the filling of one gap in our knowledge of the pathways of recovery from irradiation damage in the rodent with the realisation that supportive cells invade and colonise the recipient animal, at least temporarily. Additional investigations will be needed to clarify the immunity reactions involved, the possible secretory role of such cells, and their transformations into other cell types. Regeneration of lymphoid elements, especially the thymus, does not seem to keep pace with that of the granulocyte series, and may represent a different mechanism. Jaroslow and Taliaferro (1956) have demonstrated the restoration of antibody-forming capacity in the irradiated rabbit when antigen is mixed with cell-free preparations of splenic tissue, yeast autolysate, etc., prior to injection, suggesting that some substance derived from living cells may enter into the reaction. Living cells produce an effect by work that they do; the challenge that now faces the radiobiologist is to elucidate the functional role of the colonised cell.

REFERENCES

Askonas, B. A., and White, R. G., 1956. *Brit. J. exp. Path.*, **37**, 61.

Barnes, D. W. H., and Loutit, J. F., 1955. *Radiobiol. Symp.* 1954, p. 134 (Academic Press, N.Y., and London, Butterworth).

Congdon, C. C., and Urso, I., 1956. *Radiation Res.*, **5**, 474.

Dixon, F. J., and Maurer, P. H., 1955. *J. exp. Med.*, **101**, 245.

Ford, C. E., Hamerton, H. L., Barnes, D. W. H., and Loutit, J. F., 1956. *Nature, Lond.*, **177**, 452.

Idem, 1956. This volume, p. 197 (Edinburgh, Oliver and Boyd).

Jacobson, L. O., 1952. *Cancer Res.*, **12**, 315.

Jacobson, L. O., Marks, E. K., Gaston, E. O., Robson, M. J., and Zirkle, R. E., 1949. *Proc. Soc. exp. Biol. N.Y.*, **70**, 740.

Jacobson, L. O., Robson, M. J., and Marks, E. K., 1950. *Proc. Soc. exp. Biol. N.Y.*, **75**, 145.

Jaroslow, B. N., and Taliaferro, W. H., 1956. *J. infect. Dis.*, **98**, 75.

Kirschbaum, A., 1951. *Cancer Res.*, **11**, 741.

Lindsley, D. L., Odell, T. T., and Tausche, F. G., 1955. *Proc. Soc. exp. Biol. N.Y.*, **90**, 512.

Lorenz, E., Uphoff, D., Reid, T. R., and Shelton, E., 1951. *J. Nat. Cancer Inst.*, **12**, 197.

Loutit, F. J., 1954. *J. Nuclear Energy*, **1**, 87.

Main, J. M., and Prehn, R. T., 1955. *J. Nat. Cancer Inst.*, **15**, 1023.

Merwin, R. M., and Congdon, C. C., 1956. *Fed. Proc.*, **15**, 129.

Nowell, P. C., Cole, L. J., Habermeyer, J. G., and Roan, P. L., 1956. *Cancer Res.*, **16**, 258.

Russell, E., unpublished data, E. Roscoe B. Jackson Memorial Lab., Bar Harbour, Maine.

Trentin, J., 1956. *J. Proc. Amer. Assn. Cancer Res.*, **2**, 153.

Discussion

Klein. Have you tried to inoculate the irradiated CF1 mice given DBA/2 cells with any mouse leukæmia of non-DBA/2 origin?

Simmons. No, we have not. The normal CF1 mouse is resistant, and irradiated CF1 mice that are spleen shielded or given cell injections are restored to normal. It would be wise for us in the future to test other known mouse leukæmias to see whether the DBA cells have conferred sensitivity only to P–1534, or to mouse leukæmia in general.

Klein. Radiation is known to abolish resistance to homotransplantation of leukæmia and other tumours. The fact that the cell-injected animals are restored to normal in some other respects does not necessarily mean that their capacity to develop a homograft reaction has been restored. It would be of great interest to investigate the specificity of this phenomenon.

THE INFLUENCE OF IONISING RADIATION UPON THE NATURAL IMMUNITY

Data on the Mechanism of the Derangements of Natural Immunity under the Influence of Irradiation

V. L. Troitsky, M. A. Tumanjan and A. J. Friedenstein
Gamaleya Institute of Epidemiology and Microbiology, Academy of Medical Sciences of the U.S.S.R., Moscow, U.S.S.R.

It is impossible at present to fully explain the whole complex mechanism of decreasing the natural immunity under the influence of irradiation. Changes in the permeability and derangements in the barrier properties of the tissues and lymphatic nodes play an important part in this mechanism. It may be, however, that the most important part in decreasing the natural immunity is played by derangements in the clearing mechanism, with the macrophages being very important elements of this process.

This can be demonstrated, for example, by an experiment in which rabbits were inoculated with dysentery bacilli per os.

As is known, rabbits, as well as some other laboratory animals, are naturally resistant to dysentery infection. The dysentery bacilli even when introduced into the organism in a large dose are very soon destroyed in the organism of such animals. Only very large doses introduced directly into the blood or peritoneum can cause the death of the animals. This is a result of toxicosis developing in the animals under the influence of large amounts of microbes which, having been destroyed in the organism, release large quantities of endotoxin. The oral inoculation of rabbits with a massive dose of dysentery bacilli does not cause the appearance of any clinical symptoms. The inoculated animal does not become a carrier either. Only on the day following the inoculation are dysentery bacilli sometimes detected in the fæces.

The oral inoculation of irradiated animals brought quite different results. The rabbits were X-irradiated with 800r. On the 1st to the 21st day after irradiation they were inoculated orally by 50,000 million bodies of *Flexner's bacillus*. The animals remained carriers for a long time. Figure 1 shows the results obtained in one such experiment. The positive fæces cultures are marked with black circles. The double circles marked off the cases when positive blood cultures for the dysentery bacilli were obtained simultaneously. Pathological intestinal phenomena (stools containing mucus, sometimes blood) are marked off with rectangular blocks. As can be seen irradiated rabbits remained carriers for a long period after the inoculation. In the surviving animals the bacilli in the fæces were detectable for 5 weeks. The autopsy results of the dead rabbits proved that the dysentery bacilli could be found both in the internal organs and in the intestine, of the irradiated animals.

Thus in the irradiated rabbits, which are naturally resistant to dysentery, the oral inoculation not only produced some clinical symptoms of dysentery

The dose 800r.

NN	DATES OF INFECTION AFTER IRRADIATION	DAYS AFTER IRRADIATION																											
		1	2	3	4	5	6	7	8	9	10	11	12	13	14	15	16	17	18	19	20	21	22	23	24	25	26	27	28
1	3 DAYS	▬	▬	▬	▬	▬	▬	▬	▬	◉ ● ▬		●	†																
2	"	● ▬	◉ ● ▬	▬	†																								
3	"	▬				▬	▬		●				●		●					●	†								
4	5 DAYS	● ▬	●		● ▬	▬	▬	▬	● ▬	◉ ● ▬	†																		
5	"	● ▬				● ▬	†																						
6	"	● ▬		●	▬	● ▬	◉ ● ▬	▬	▬	● ▬	†																		
7	15 DAYS	▬	▬	▬	▬	● ▬	◉ ● ▬	▬	▬								†												
8	"	▬					▬																	●	▬	▬	● ▬	▬	● ▬
9	"	▬			▬			●	●	● ▬			●	●	●	●	●								●				●

DESIGNATIONS

▬ - *DIARRHOEA, MUCOUS, BLOOD;* ● - *B. FLEXNER IN FAECES;* ◉ *B. FLEXNER IN BLOOD;* † - *DEATH*

FIG. 1. Table.

but even made them carriers for a long period. That is a proof of a long-term persistence of the dysentery bacilli in the organism of the irradiated animals.

This leads to a conclusion that the natural immunity of the rabbits to dysentery first of all depends upon a rapid clearing from the organism of the dysentery bacilli.

Bacteræmia, which appears under the influence of the ionising irradiation, is a clearer example of the derangements in natural immunity.

In contrast to the transitory bacteræmia, which sometimes appears in the normal organism, radiation bacteræmia is a cause of the autoinfection which exerts a great and very often decisive influence upon the course and outcome of radiation sickness. This can be proved by the positive results of administering antibiotics to the animals affected by ionising radiation.

Our experiments on monkeys proved that rationally administered chemotherapy, in combination with the application of antibiotics and with the use

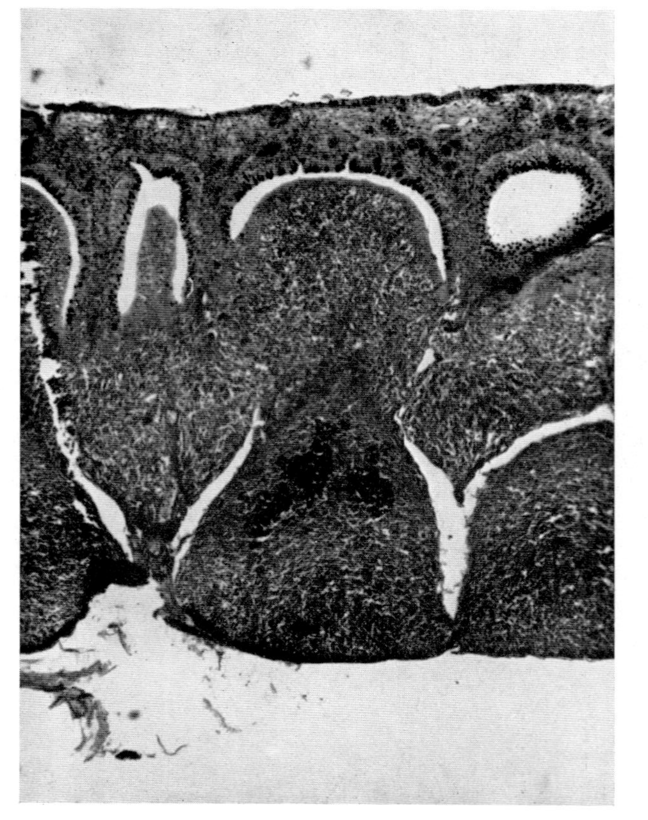

FIG. 2. Microphotograph of the appendix wall of a normal rabbit.

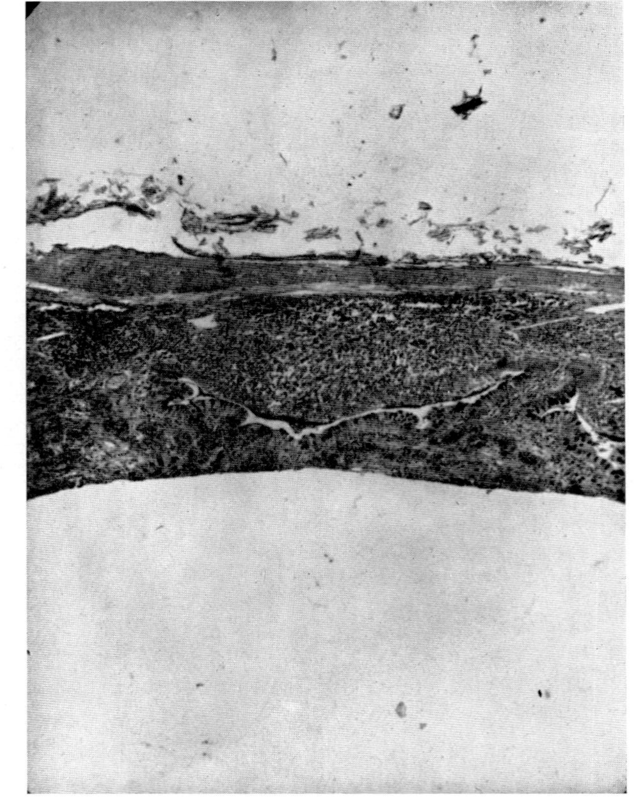

FIG. 3. Microphotograph of the appendix wall of a rabbit 4–5 days after irradiation.

of various routes for their introduction, prevents the monkeys from dying, even in the cases irradiated with lethal doses. This confirms the part played by the bacteræmia both in the course of radiation sickness and in its outcome.

According to the majority of workers, radiation bacteræmia is caused by the migration of bacteria from the intestine into the blood. The possibility of the migration of bacteria from the intestine into the intestinal wall is established without any doubt. But the mechanism of these phenomena, their significance and even frequency under normal conditions and after irradiation have not been determined. The main problem is to establish the factors which control the migration of the bacteria into the lymphatic and then into the blood vessels of irradiated animals. Do the bacteria migrate due to the changes in the permeability of the tissues under the influence of the irradiation, or is this process conditioned by damage inflicted upon the defensive mechanism, with the macrophages in the intestinal wall playing the main part in it? It is quite possible that it is derangements in this latter mechanism that enable the bacteria to migrate from the intestine into the lymphatic and blood systems, causing radiation bacteræmia.

The study of this problem by ordinary histological methods and by the use of microbiological methods for staining microscopic slices, do not reveal any specific change in the morphological picture and, consequently do not permit a solution of the problem. New possibilities appeared when histochemical methods of staining polysaccharides were introduced by Hotchkiss (1948) and Schabadasch (1947). These methods made it possible to establish that, in the normal adult rabbit, the wall of the appendix possesses a special mechanism for detaining and destroying a huge number of the bacteria, which are migrating from the intestine even through the undamaged epithelium of the crypts. These micro-organisms, which give a positive reaction for the polysaccharides, are morphologically of very many forms. There are large and small bacilli, cocci and spore-forming bacilli among them.

Figure 2 represents a microscopic section of the appendix of a normal rabbit. Small accumulations of phagocytosed bacteria can be seen under the epithelium of the crypts in the upper part of a follicle. Compact masses of polysaccharides which are stained red are situated in the depth of the intestinal wall in the centre of the lower part of a lymphoid follicle. They represent chiefly the products of phagocytosis of the micro-organisms which have penetrated into the intestinal wall. These products are accumulated in the reticular cells.

In the appendix of irradiated rabbits 4-5 days after the irradiation, Figure 3, a sharp decrease in the amount of polysaccharides can be observed, alongside a characteristic reduction of the lymphoid tissue.

Let us give some additional details of phagocytosis in the wall of the appendix of irradiated and non-irradiated animals.

Figure 4 demonstrates phagocytosis of the microbes which penetrated through the epithelium of the crypts, by the reticular cells-macrophages in the appendix of a normal rabbit. This phagocytosis is complete. Digestive vacuoles can be observed in the cells and the microbes are destroyed, with polysaccharides being released.

Figure 5 shows the macrophages in the appendix of an irradiated rabbit. The process of phagocytosis can be observed in this preparation too. However, in the majority of cases this process is not complete. The destruction of the macrophages leads to the release of bacteria which have not been destroyed.

As a result there appears a huge mass of polysaccharides in the lower part of a follicle, in the reticular cells of the normal rabbit (see Fig. 6). They were formed as a result of intracellular digestion of the bacteria when free and unchanged bacteria are absent.

Simultaneously, in an irradiated animal, Figure 7, free bacteria can be observed alongside with some amount of the same polysaccharide. These bacteria are capable of further penetration into the organism.

The chemical nature of these polysaccharides has not yet been determined. Some attempts to solve this problem are now being undertaken. It is possible to state, however, that on the one hand, there exists a definite morphological relationship between these polysaccharides and the substances which are formed as a result of intracellular digestion of bacteria and, on the other hand, that the polysaccharides can be stained if the slices are previously treated with amylase and methanol-chloroform. This leads to a conclusion that this polysaccharide is neither glycogen nor a glycolipid.

The above histological data completely corresponds with the picture of changes in the morphology of the appendix after irradiation as it was described by de Brynu and other authors. The use of a special method of staining has enabled us to establish that there is a special defensive mechanism. It is based on phagocytosis of bacteria by the macrophages, on the digestion of these bacteria and on depositing the polysaccharides into the intestinal wall. The latter is a result of the intracellular digestion of the microbes. The irradiation causes derangements in the activity of this mechanism enabling the bacteria to penetrate into the organism.

Thus, radiation bacteræmia, which is often observed as one of the expressions of the suppression of the natural immunity after irradiation, is mainly a result of the inhibition of the clearing mechanism, the basic part of which is the phagocytic activity of the macrophages in the intestinal wall.

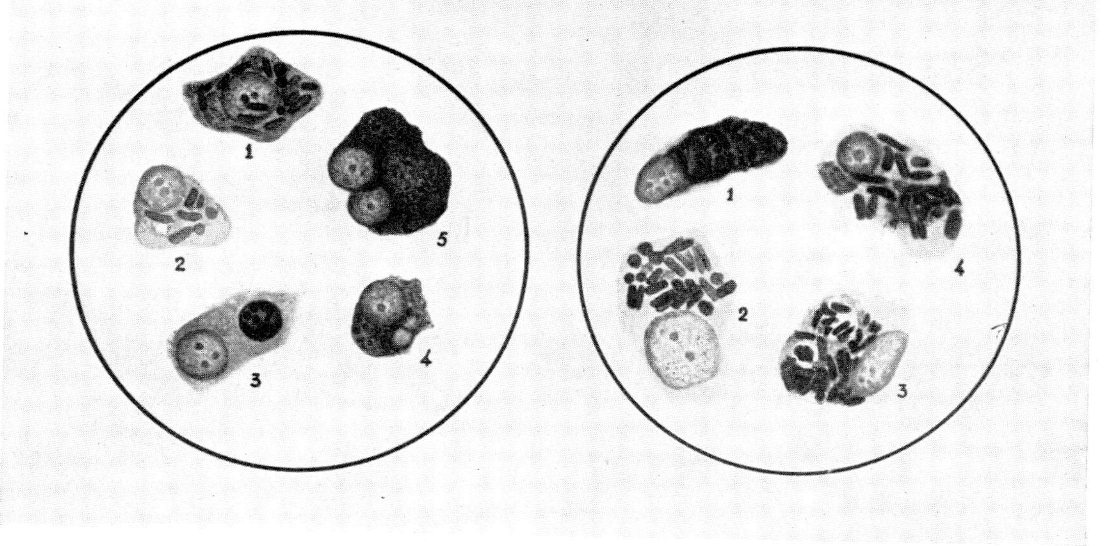

FIG. 4. Phagocytosis in the appendix of a normal rabbit.

FIG. 5. Phagocytosis in the appendix of an irradiated rabbit.

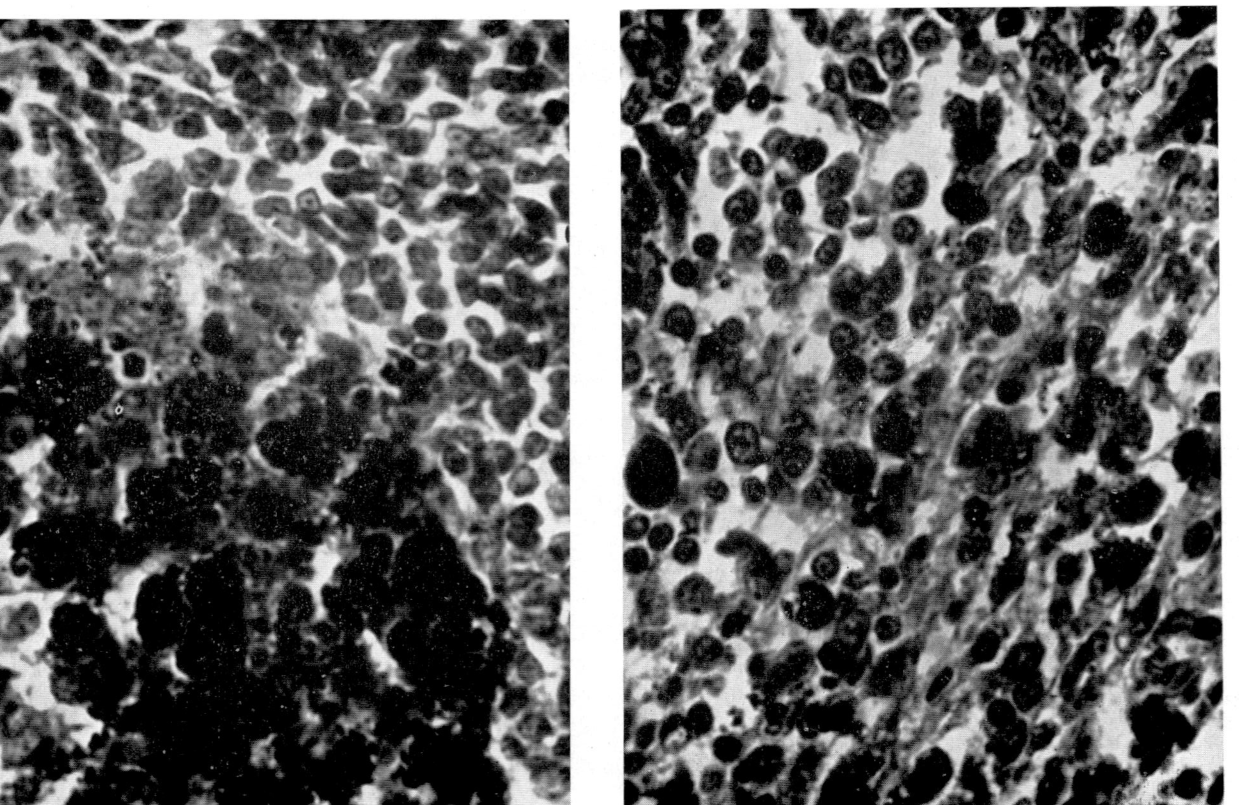

FIG. 6. Microphotograph of lower part of normal appendix

FIG. 7. Microphotograph of lower part of appendix follicle

REFERENCES

HOTCHKISS, R., 1948. *Arch. Biochem.*, **16**, 131.
SCHABADASCH, A., 1947. *Isvest Acad. Nauk U.S.S.R., biol. serv. No.* 6, p. 745.

DISCUSSION

Lajtha. Relatively large doses of radiation do not seem to inhibit the phagocytosis of colloids (colloid gold, india ink) by the reticulo-endothelial system. I wonder whether the damage in your experiment is rather to the digestion of the ingested material than to the process of ingestion by the cells?

Troitsky. We think that the damage is manifold, firstly to the permeability of the intestinal wall and secondly a disarrangement of the macrophages.

IV. OTHER APPROACHES TO THE MODIFICATION OF IRRADIATION EFFECTS

MODIFICATION OF THE X-RAY REACTION IN THE SKIN OF MICE BY SHIELDING OF MINUTE AREAS OF THE SKIN

FINN DEVIK
Institute of General and Experimental Pathology,
University of Oslo, Norway

It is well-established that the course of reactions to ionising radiations in living organisms can be influenced by interference with the cells or tissues prior to, or during irradiation. Some findings indicate that it is also possible to alter the reactions by interference after irradiation (*Brit. J. Radiol.*, 1954).

An interaction between irradiated and non-irradiated cells is likely to occur after irradiation. The use of sieves and grids in radiotherapy (Jolles, 1953), and the results of experiments with shielding of minor amounts of hæmopoietic tissue during irradiation demonstrate that this effect can be striking (Jacobson, 1952).

The purpose of the present experiments was to study the interaction between normal and irradiated cells of the skin; and to see how small an area of the skin, when shielded, would be effective in modifying the skin reaction after irradiation.

Material and methods

Mice from a recessive hairless strain were used, and on their backs a flap of the loose skin was temporarily fastened to a frame, and the skin irradiated with a dose of 2,700r in 16 mice, and 4,000r in 4 mice. (Philips Metalix unit, 50 kV, 2 mA, 1 mm. Al filter, HVL 1·1 mm. Al, FSD 27 mm., dose rate 677r/min.).

Across the field of irradiation, 10 mm. × 11 mm., four platinum wires of 0·05 mm. diameter were placed in contact with the skin. The wires were arranged parallel to each other at intervals of 2 mm. The shielding effect was restricted to a thin layer of that part of the skin which faced the X-ray tube. Because of scattering, and the geometry of the set-up, the shielding of the thin wires was ineffective on the back side of the skin flap, although the flap was only 0·5–0·8 mm. thick. The back side of the skin flap therefore served as a control to the effects observed on the shielded side.

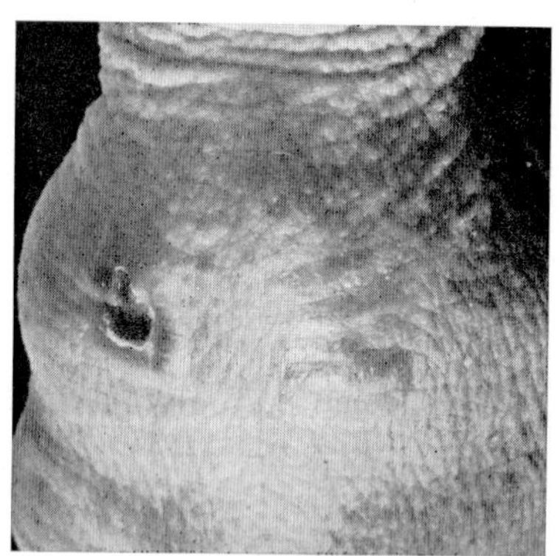

FIG. 1. Mouse 11 days after 2,700r. Right: test side, skin in contact with grid during irradiation. Left: control, back side of flap, with typical roentgen reaction of the skin.

The reactions were inspected daily. In 7 animals 22 intraperitoneal injections of fluorescein were given. Histological sections were made during the acute reaction of skin from 11 mice sacrificed 4-14 days after irradiation, and the late effects were studied in sections from 9 mice killed 3-5 months later.

The typical X-ray reaction on the control side started about the 4th day, and later on, desquamation, exudation and occasionally necrosis were observed (Fig. 1). The reaction was most pronounced 8-10 days after irradiation, and then healed gradually. A temporary hyperplasia of the epithelium started at the periphery and proceeded towards the centre. Except for the first days of the visible reaction, when both sides had the same appearance, the ensuing acute skin reaction was consistently profoundly altered by the application of the platinum wire 'grid'. Desquamation, exudation and necrosis were not observed. Instead, a hyperplasia of the epithelium developed and became pronounced, in contrast to the control side where the loss of epithelium at the same time was nearly complete.

After the acute reaction, the more permanent effects which remained differed less obviously on the two sides. Four tiny pale stripes remained on the test side; they were easily identified in sections as the only places where sebaceous glands had survived. Outside the stripes, no difference between the test and the control side was evident on histological examination. Some difference could be demonstrated, however, when the size of the area of the affected skin was used as a criterion. The control, or back side, which in fact had received a slightly lower dose than the test side, consistently showed a greater retraction than the latter. The size was 70% of the test side, as an average measured in 9 mice. In mice irradiated without the platinum wires, the corresponding figure was larger than 100. The shielding effect of the platinum wires can therefore also be discerned in the permanent effects, although it is less pronounced than during the acute reaction.

The distinct reduction in degree of the skin reaction after irradiation, described above, must be regarded as a modification which is due to factors that act after irradiation. Such factors apparently do not come into action until after the beginning of the acute reaction, because no difference can be seen during the first days of the reaction; neither by inspection nor by the use of fluorescein which accumulates in the irradiated field. It is only when the reaction has been visible for a few days that the dye injection also reveals the presence of thin stripes of shielded tissue (Fig. 2 (*a*) and (*b*)).

The nearly complete absence of epithelial destruction, and the early and extensive hyperplasia of epithelial cells on the test side, can be explained: either as a result of cell restoration after irradiation, or by migration and multiplication of normal epithelial cells which have been shielded during irradiation, or by a combination of both factors.

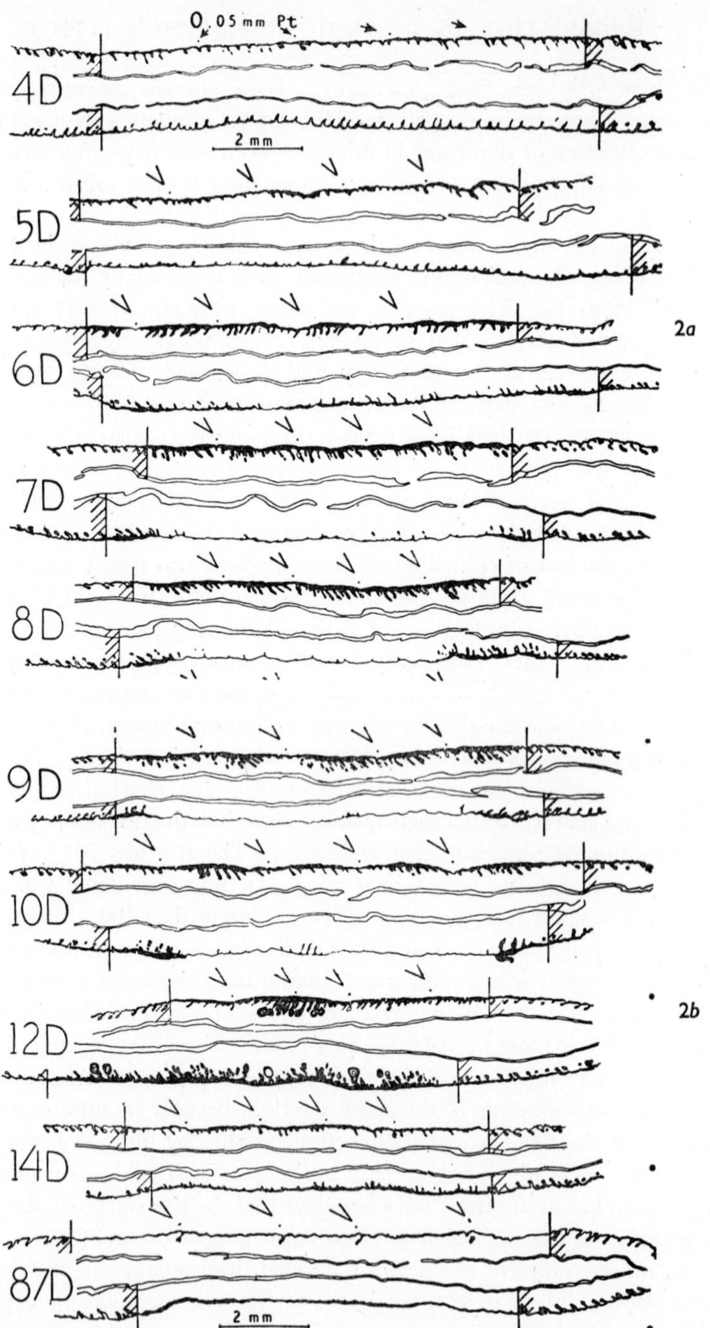

FIG. 2, *a* and *b*. Camera lucida drawings of epithelium and skin muscle from sections of irradiated skin at intervals of 4 to 87 days after 2,700r. Test sides face upward, control sides from the same mice face downward. The position of the platinum wires during irradiation is indicated. The wires are drawn to scale. The border between irradiated and non-irradiated skin is pointed out.

Another possibility is that the shielded cells might act by detoxication of a hypothetical toxic product of the irradiated cells. This is regarded as improbable because the shielded cells are of normal appearance during the reaction, and we have no clear evidence of the existence of diffusible radiation-induced toxic substances which appear after the acute reaction has started with the sudden appearance of abnormal cell divisions.

The shielding of skin apparently presents close analogies to hæmopoietic tissues, which show a great capacity for recovery or restoration when a small part of the tissue has been shielded during whole-body irradiation. Jacobson has ascribed this effect to the existence of humoral factor(s) which induce regeneration of hæmopoietic tissues (1952). Lately, Ford *et al.* (1956) have been able to demonstrate that non-irradiated, homologous or heterologous hæmopoietic cells can almost completely repopulate the bone marrow in mice after whole-body irradiation. They have thus weakened the arguments for humoral factors. By analogy, migration of non-irradiated epithelial cells in the skin would seem to afford a plausible explanation for the skin effects which have been described.

In the hyperplasia of the skin, however, migration is considered improbable for the following reasons: (1) No loss of epithelium has occurred when the hyperplasia starts. From experiments in general pathology it seems that complete loss of epithelium is a prerequisite for migration of epithelial cells. (2) The hyperplastic epithelium contains numerous cells with micro-nuclei, which shows that these cells have undergone an abnormal division after irradiation. Increased mitotic activity of the shielded cells was not observed. (3) It has also been found that hyperplasia can be induced in the middle of a field of irradiation before the hyperplasia of the periphery has developed, e.g. by intracutaneous injection of the dye Sudan IV (Scharlachrot) suspended in olive oil (Devik and Osnes, unpublished). This has been found after irradiation of rabbits' ears with doses up to 3,000r, and occasionally in mice after 2,700r.

The assumed recovery of the epithelial cells, however, is not complete. The sebaceous glands do not escape destruction, and the remaining epithelium is unable to form new glands. It is also conceivable that the late radiation effects in the epithelium may to some extent be influenced by radiation-induced changes in the connective tissue of the skin.

The experiments show that shielding of very narrow stripes (0·05 mm.) of skin in mice during irradiation, may have a pronounced effect on the acute skin reaction. It is regarded as a process of partial recovery due to the proximity to non-irradiated cells, which indicates that the normal cells of the skin contain some factor(s) of importance to the recovery of epithelial cells that have been injured by irradiation (Devik, 1955).

Acknowledgment

This work has been made possible through a research grant from the Norwegian Cancer Society LANDSFORENINGEN MOT KREFT.

REFERENCES

Brit. J. Radiol., 1954, **27**, 36.

DEVIK, F., 1955. *Acta radiol., Stockh. Suppl.* 119.

FORD, C. E., HAMERTON, J. L., BARNES, D. W. H., and LOUTIT, J. F., 1956. *This Volume*, p. 197, and 1956, *Nature, London*, **177**, 452.

JACOBSON, L. O., 1952. *Cancer Res.*, **12**, 315.

JOLLES, B., 1953. *X-ray Sieve Therapy in Cancer* (London, Lewis).

DISCUSSION

Carter. Dr. Devik was using mice which were homozygous for the 'hairless' mutant. These mice have skin which is genetically abnormal. I would like to ask if he thinks this has any effect on the interpretation of his results?

Devik. I do not believe that the genetic constitution of the hairless mouse can explain the results, because the skin reaction is similar to that of other laboratory animals. The hyperplasia has been described by Jolles, and can also be seen in illustrations from early papers on radiation effects.

Howard-Flanders. Have you any data using even smaller wires or wires at different spacings?

Devik. For technical reasons it would be difficult to use smaller wires, they might not absorb enough radiation, and there would be too much scattering, so that the shielding effect could be ineffective. If the wires are further apart than about 2 mm., desquamation and exudation occur.

Alexander. You mentioned the results of Jolles, but does he not claim the opposite result to yours?

Devik. The findings are similar to some of those described by Jolles, but contrary to his interpretation a recovery or restoration factor has been postulated.

CHANGES IN THE REACTIVITY OF THE ORGANISM UNDER THE INFLUENCE OF IONISING RADIATION

O. Bogomolets, A. Boico, G. Diadiucha, Z. Zekhova,
V. Lavrick and G. Levtchouk
Research Institute of Public Health, Kiev, Ukraine, U.S.S.R.

Numerous experiments have shown that a profound inhibition of the organism's reactivity accompanied by suppression of the physiological functions of the connective tissue, constitutes one of the consequences of radiation injury. This system takes part in the defensive, plastic and trophic processes, and any derangement of its activity, as a part of the general decrease in reactivity, leads to decreasing the organism's immunobiological resistance.

A. A. Bogomolets followed Mechnikov's ideas on cytotoxins and produced an antireticular cytotoxic serum (ACS). Small doses of the serum stimulate the physiological system, whereas large doses inhibit its activity. Many workers have shown that the best curative effects of ACS are in patients with diseases which are accompanied by a suppression of the connective tissue activity.

These findings led the authors to study the changes in reactivity of the physiological system of the connective tissue in irradiated animals treated with A. A. Bogomolets' antireticular cytotoxic serum.

It must be remembered that the administration of ACS can only be successful in cases when changes in the elements of the connective tissue are still reversible.

The experiments were performed on dogs, rabbits and rats which were X-irradiated with single doses. The doses were—for dogs, 400 and 500r; for rabbits, 700r; for rats, 400r. The best results were obtained when dogs and rabbits were inoculated with antireticular cytotoxic serum beginning on the fourth day after the irradiation, and in rats half an hour after irradiation. Stimulating doses of the serum were selected after taking into account its titre. Of the three doses tested on dogs (titre—1 : 800) the best results were obtained after six injections of ACS, with a single dose being 0·00125 ml. of the serum. For rabbits, an effective dose (titre—1 : 200) was 0·0005 ml. of ACS and for rats, 0·0001 (titre—1 : 400). In all cases, ACS, before being introduced into the organism, was dissolved in physiological saline. The injections were made subcutaneously at three-day intervals.

The following factors were taken into account in evaluating the results of the serum's action; changes in opsonocytophagic indices and in the coefficient of a skin test by trypan blue, morphological picture of the wound exudate (macrophages, lymphocytes, histiocytes—M. P. Poprovskaya's method), changes in interrelations of the protein fractions, the degree of oxygenation of the blood and in its morphological character, as well as the clinical picture of the course of the radiation syndrome and the data from pathological investigations.

Our experiments demonstrated that there is a distinct difference between the changes in the reactivity of the irradiated animals which received ACS on the one hand, and of the control animals or those which were given normal blood on the other.

Let us cite some general findings:

As can be seen in Figure 1 for rats, the lowest number of the histiocytes, lymphocytes and macrophages was observed in all irradiated animals on the eighth day after irradiation. However, the degree of the decrease was considerably less in the animals injected with ACS than in the animals which were not so injected.

On the 30th day after irradiation there appeared a sharp difference in the counts of lymphocytes and macrophages in the wound exudates of the rats which received ACS and of those from the control group. In the former, the lymphocyte count reached the initial level and the number of the macrophages came near to it. The number of histiocytes was higher than at the beginning of the experiments. In all the animals which were not injected with ACS the number of lymphocytes, macrophages and histiocytes remained at a very low level. This experiment clearly demonstrates the stimulating effect of small doses of the antireticular cytotoxic serum upon the physiological system of the connective tissue. Similar results were obtained in the experiments on dogs.

As is shown in Figure 2 for dogs, the phagocytic index of the ACS-treated dogs, despite considerable fluctuations, was constantly higher than in the animals which received no serum.

The experiments demonstrated that the phagocytic function of the blood leucocytes in the irradiated animals could be stimulated by small doses of ACS introduced repeatedly.

As is shown in Figure 3 the X-irradiation decreased the index of the skin test in dogs. The decrease of the index (QD) was considerably less in the dogs which received the serum than in the control ones. The phenomena were well shown in the dogs which were irradiated both with 400r and 500r.

Figure 4 shows changes in blood protein levels for dogs. During the experiment, a progressive hypoalbuminæmia and a decrease in the protein

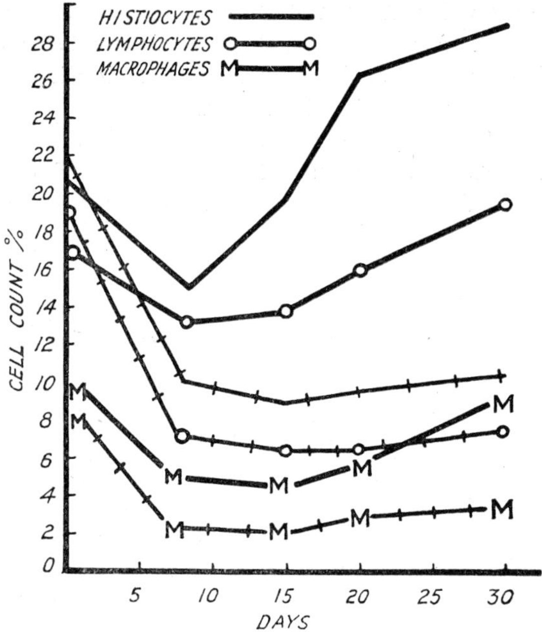

FIG. 1. Changes in the number of lymphocytes, macrophages and histiocytes in the exudates of 20 irradiated rats (average figures); 10 received ACS (solid lines), the remaining were used as controls (dotted lines). The number of lymphocytes are drawn ○ - - ○ - - ○, histiocytes ———, macrophages M———M. The number of each form was counted per 1,000 cellular elements.

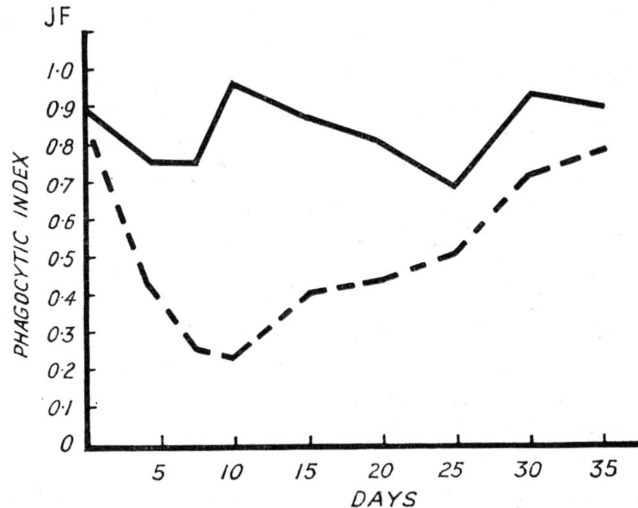

FIG. 2. Changes in the phagocytic index of 20 irradiated dogs (average data). Ten of them were inoculated with ACS. The solid line shows the changes in the phagocytic index in the dogs which received serum, the dotted line represents the index for the control animals. The experiments were conducted with *Staphylococcus albus*.

content were observed in the control animals. In the animals injected with ACS in the month after irradiation, the interrelations between the serum protein fractions were restored almost to normal after hypoalbuminæmia and hyperglobulinæmia caused by irradiation.

These data and other observations lead us to state that the phagocytic activity of the leucocytes is stimulated by small doses of the antireticular cytotoxic serum, which intensifies a reactive inflammatory process in lesions,

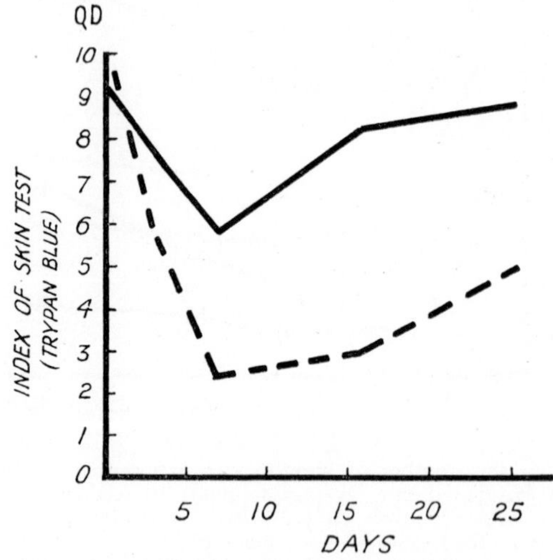

FIG. 3. The curves which reflect the fluctuations of the index of a skin test with trypan blue in 16 dogs which received ACS (solid line) and in 8 control dogs (dotted line).

activates the trophic functions, brings a rapid restoration of the interrelationship of the protein fractions and hastens the normalisation of the morphological contents of the peripheral blood. The general condition of the irradiated animals improves, the latent period of the disease becomes longer and visible hæmorrhagic phenomena decrease.

Pathological studies of the few dead animals which received the serum showed that the atrophy in their bone marrow was not so marked as in the bone marrow of the control animals. The folliculi of the spleen were not necrotic as was always the case with the spleen of the control animals. The number of hæmorrhages into the tissues was considerably less in the ACS-treated animals than in the controls.

The above-mentioned facts demonstrate that the administration of ACS in cases exposed to ionising irradiation, sometimes helps to maintain the functions of the physiological system of the connective tissue, to raise its reactive properties and even to limit the development of destructive phenomena. Results which were almost similar have been obtained in the laboratories of P. D. Horizontov, A. A. Gorodetsky and A. I. Smirnova-Zamkova.

We did not expect to cure radiation sickness by means of cytotoxinotherapy. The aim of our investigations is formulated in the title of the paper. It is, nevertheless, possible to state that A. A. Bogomolets' conception has been fully confirmed by our experiments. This conception was formulated by him in the following way:

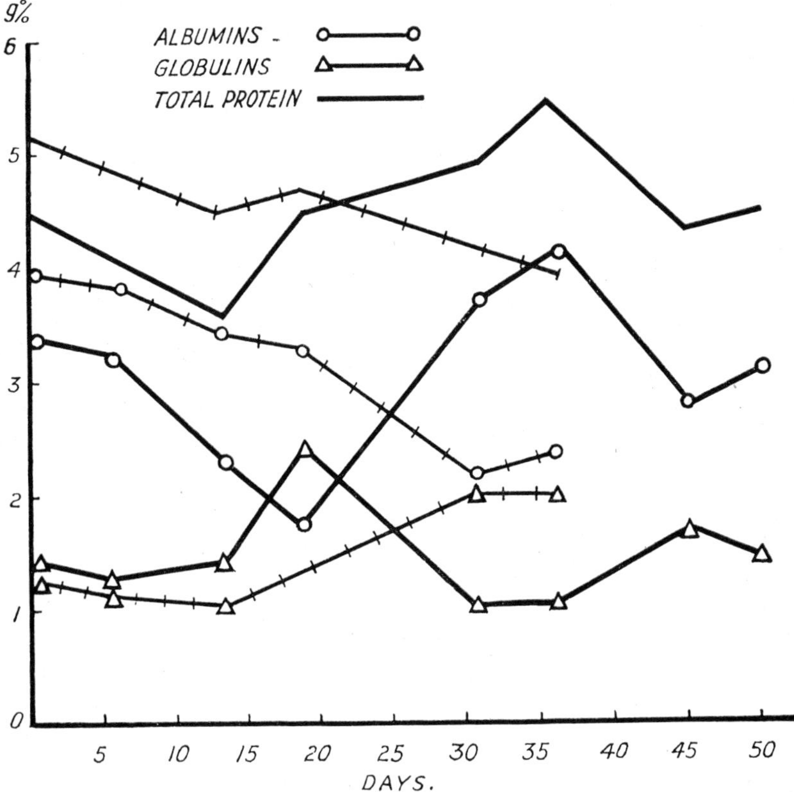

FIG. 4. The relations between albumins (○ -- ○ -- ○), globulins (△ -- △ -- △) and total protein in blood serum (———) of 10 irradiated dogs which received ACS (solid lines) and 5 control dogs (dotted lines).

"The clinical therapy (chemical, physical, surgical) directed at the elimination of the cause of the disease, should always be a pathogenic therapy, that is, it should be directed at supporting and intensifying those reactions of the organism which are the best obstacles to the development of the disease."

Summary

These experiments have led to the following conclusions:

(*a*) A considerable decrease in reactivity is observed in irradiated animals. It is accompanied by a profound inhibition of the physiological system of the connective tissue. The restoration of its reactivity proceeds very slowly.

(*b*) An increase in the reactivity of the irradiated animals can be obtained by administering the antireticular cytotoxic serum in small doses. Its mode of action is a relatively rapid restoration of the functions of the physiological system of the connective tissue.

(*c*) The antireticular cytotoxic serum is one means of intensifying the reactivity in the irradiated animals.

REFERENCE

Cytotoxin in Present Medicine, 1956 (U.S.S.R., Kiev, Gosmedisdat).

Discussion

Kanazir. Have you some biochemical data on the composition of ACS and some idea about the mechanism of its action on connective tissue? Can you also tell us what is your method of injection to obtain stimulating effects in rats?

Bogomolets. It seems that we are dealing with an agent which normalises the abnormal processes. The nature of this action has been studied by Professor Medvedova. It is preferably a gamma-globulin, but in some animals a beta-globulin.

As to the method of injection: the first injection should be performed on the 4th day followed by two injections at three-day intervals.

Bacq. Do small doses of ACS decrease mortality of irradiated animals. And do large doses increase the mortality?

Bogomolets. Yes, and mortality is rising markedly after high doses.

Bacq. It is well known that the physiology of the connective tissue is normally regulated by the secretion of the adrenal cortex. Have you tried to combine injections of cortisone with those of ACS?

Bogomolets. No, that was not our task. We paid our full attention to the study of the effectiveness of ACS. We believe that the functions of the connective tissue are first of all regulated by the nervous system.

IRRADIATION AND ADRENAL AND PITUITARY FUNCTION

Adrenal Changes after Whole-Body Irradiation in the Hypophysectomised Rat Grafted with Pituitary Tissue in the Anterior Chamber of the Eye and of the Adrenalectomised Rat Grafted with Adrenal Tissue in the Eye

Z. M. BACQ, P. MARTINOVIĆ, P. FISCHER, M. PAVLOVIĆ AND G. SLADIĆ

Laboratory of Pathology and Therapeutics, Liège,
the Research Laboratory for Protection of the Civil Population, Belgium and the Boris Kidrić Institute, Belgrade, Yugoslavia

Irradiation produces important changes in the suprarenals of rats which have been discussed at some length (Bacq *et al.*, 1954, 1955). These changes are dependent on the anterior pituitary gland. In order to study the part of the central nerve system in these reactions, one of us has used barbiturate preparations which partially suppress the control which the hypothalamus exercises on the pituitary (Fischer, 1955). Nembutal anæsthetised rats do not show the characteristic fall of cholesterol in the suprarenals 3 hours after irradiation, although the concentration changes in ascorbic acid are similar to those occurring in normal animals (Fischer, 1955).

The hypophysectomised rats which had been grafted with pituitary tissue in the eye, show a good physiological condition in spite of a chronic insufficiency of the suprarenals (the amount of ascorbic acid and cholesterol is low, see Table). The connection with the central nerve system is suppressed. Twenty-five hypophysectomised rats were used in these experiments at the Boris Kidrić Institute; the details of the technique have been published elsewhere (Martinović, 1951; Bacq *et al.*, 1954).

It appears from the table that hypophysectomised and grafted rats do not show a cholesterol fall, but an increase, having a maximum 72 hours after irradiation. However, a decrease in ascorbic acid concentration takes place 3 hours after irradiation, but normal values are found again at 24 hours and thereafter.

Adrenalectomised and grafted animals show the same irradiation effects as normal ones, except that the ascorbic acid and cholesterol concentrations are lower shortly after irradiation.

TABLE

Time after irradiation (hours)	Normal rats					Adrenalectomised and grafted rats					Hypophysectomised and grafted rats				
	No.	Wt.	Weight of supra-renals mg./100 g. body weight	Ascorbic acid, mg./100 g.	Choles-terol mg./100 g.	No.	Wt.	Weight of supra-renals mg./100 g. body weight	Ascorbic acid, mg./100 g.	Choles-terol mg./100 g.	No.	Wt.	Weight of supra-renals mg./100 g. body weight	Ascorbic acid, mg./100 g.	Choles-terol mg./100 g.
Controls	39	127 ± 5·1	17 ± 0·5	397 ± 10	6·3 ± 0·2	10	193 ± 16	9	260 ± 20	3·4 ± 0·45	6	121 ± 6	6 ± 0·4	275 ± 16	3·9 ± 0·31
3	21	115 ± 7	19·3 ± 1	289 ± 13	4·6 ± 2·3	6	210 ± 13	8·5	188 ± 18·6	2·6 ± 0·5	5	110 ± 6·1	8·1 ± 0·67	186 ± 28	4·5 ± 0·65
24	15	117 ± 5·1	18·7 ± 0·8	407 ± 9·7	5·5 ± 0·3	6	197 ± 18	6·6	263 ± 35	2·5 ± 0·35	5	116 ±4·5	7·9 ± 0·77	258 ± 18·5	4·9 ± 0·3
48	10	130 ± 19	20·9 ±1·8	441 ± 13	3·8 ± 0·3	4	221 ± 22	10·5	258 ± 14	1·75 ± 0·3	4	113 ± 7	8·3 ± 0·7	242 ± 70	5·68 ± 0·38
72	10	105 ± 6·2	21 ± 1	287 ± 23		6	190 ± 12	10	171 ± 23	1·06 ± 0·1	5	116 ± 3·5	9 ± 0·55	256 ± 25	4·5 ± 0·46
96	17	98 ± 5·7	29·6 ± 2·5	394 ± 27·7	1·7 ± 0·36										

Conclusions. The suprarenals grafted in the eyes of adrenalectomised rats react to irradiation just like normal glands. In contrast to this, the adrenals of hypophysectomised grafted rats show an increase in cholesterol although a transient decrease in ascorbic acid takes place.

REFERENCES

Bacq, Z. M., Fischer, P., and Beaumariage, M. L., 1954. *Bull. Acad. Roy. Méd. Belg.*, **19**, 399.

Bacq, Z. M., and Alexander, P., 1955. "Principes de Radiobiologie," *Sciences et Lettres*, Liège.

Fischer, P., 1955. *Arch. Int. Physiol. Biochem.*, **63**, 134.

Martinović, P. N., 1951. *Methods in Medical Research*, **4**, 240 (The Year Book Publishers Inc.).

Discussion

Curtis. How does this picture of adrenal response following radiation differ from other forms of stress such as cold?

Bacq. The first adrenal response which occurs 1 to 8 hours after irradiation is similar to the reaction which occurs after a short stress (cold, histamine or adrenaline injection). The changes which occurs on the second day following irradiation have a different character and cannot be considered as a simple reaction.

Kanazir. In connection with Dr. Bacq's paper I should like to mention some results of Dr. Hajducovich's work on the role of the adrenal gland cortex in irradiation damage, particularly with respect to hæmatological changes. New-born rats respond to irradiation stress only after the 9th day. He gave 700r to (1) groups of 8–9 days old mice and (2) groups of 14, 21, and 30 days old mice. The hæmatological response was followed for 7 days after irradiation. The increase of reticulocytes was considerable 5 minutes after irradiation of group (1) mice as well in those 14 days old, but not so pronounced in the other groups. A decrease of leucocytes occurred in all animals shortly after irradiation and 48 hours after the dose this decrease was more pronounced in the second group. Likewise lymphocyte counts showed a pronounced fall following immediately upon irradiation, but with no differences between the groups, nor did we find any differences on the bone-marrow smears. It would seem then from our data that the adrenal gland cortex does not play an important role in the immediate blood changes following X-radiation.

Bacq. We have confirmed our observations on hypophysectomised grafted animals by the study of adrenal behaviour after irradiation of new-born rats, which are known not to react to stress before about the 13th day, because the hypothalamic control of the anterior pituitary is not yet established. Fischer and myself in Liège have shown that the first significant drop in ascorbic acid or cholesterol content two hours after 800r total-body dose occurs in rats 13 days old. At 72 hours after irradiation of rats aged up to 14 days, there is no drop in ascorbic acid value of the suprarenals and a significant increase in cholesterol content.

Our results on new-born rats are very similar, if not identical to those obtained independently at the Boris Kidrić Institute of Belgrade by Sladić and Radivojević.

ALKOXYGLYCEROL-ESTERS IN IRRADIATION TREATMENT

ASTRID BROHULT
Radiumhemmet, Stockholm, Sweden

The effect of alkoxyglycerols and alkoxyglycerol-esters obtained from bone marrow and shark liver oil respectively has been studied in cases of leucopenia caused by irradiation. The preparation has also been tried in thrombocytopenia. Experiments with alkoxyglycerol-esters in leukæmia are in progress. A preliminary communication was published in *Nature* (Brohult and Holmberg, 1954). Simultaneously, but independently, Edlund (1954) obtained a beneficial protective effect on irradiation in mice using synthetic batyl alcohol.

In 1952, experiments were started in cases of children's leukæmia in co-operation with Holmberg, using unsaponifiable fat from cattle bone marrow. Since there was reason to assume that the active component of this unsaponifiable fraction consisted of alkoxyglycerols, primarily of batyl alcohol, preparations containing both batyl alcohol and other alkoxyglycerols were then used instead for the treatments. On the basis of experience gained in this connection, the irradiation treatment experiments have been developed.

It is not a new idea to administer extracts from blood-forming tissue in blood disturbances. In 1930 Watkins (Watkins and Giffin, 1933) for the first time reported promising results from oral administration of cattle bone marrow in granylocytopenia. However, it proved that the medication was unsuitable since the marrow had to be taken in such large quantities that it caused digestive disturbances. Marberg and Wiles (1937 and 1938) showed that the same effect on the blood could be obtained by using solely the unsaponifiable fraction of the fat. This fraction constitutes only about 2% of the bone marrow. Holmes *et al.* (1941) found that batyl alcohol is a component of this fraction. Sandler (1949) published a paper on the stimulating effect on blood of both unsaponifiable bone marrow fat and of pure batyl alcohol.

In all natural sources the alkoxyglycerols are found esterified with fatty acids. In the bone marrow about 2% of the fat of the gelatinous tissue and about 0·6% of the spongy tissue consists of alkoxyglycerol-esters. The alkoxyglycerols have the following general formula:

$$R—OCH_2—CHOH—CH_2OH.$$

The alcohols commonly found are: batyl alcohol, chimyl alcohol and selachyl alcohol (Fig. 1). The main portion of the alkoxyglycerols of bone marrow consists of batyl alcohol (approximately 60%). Unpublished investigations by Holmberg prove that alkoxyglycerol-esters are present in many kinds of human and animal fats. Besides being in bone marrow, they are also found in, for example, the spleen, liver and the red cells.

$$\begin{array}{l} CH_2OH \\ \quad | \\ CH\ OH \\ \quad | \\ CH_2.O.R \end{array} \qquad \textit{Alkoxyglycerol}$$

$$\begin{array}{lll} R = & (CH_2)_{15}.CH_3 & \textit{chimyl alcohol} \\ \ \ = & (CH_2)_{17}.CH_3 & \textit{batyl alcohol} \\ \ \ = & (CH_2)_8.CH = CH.(CH_2)_7.CH_3 & \textit{selachyl alcohol} \end{array}$$

	Alkoxyglycerols	
	in bone marrow	*in shark liver oil*
	0·2–0·7%	16–17%
(*as diesters*	0·6–2%	46–50%)
Distribution:		
chimyl alcohol	?	5
batyl alcohol	*approx.* 60	43
selachyl alcohol	?	52
polyunsaturated alcohols	?	<1

FIG. 1. Alkoxyglycerols and their distribution.

When it proved difficult to obtain sufficient quantities of a concentrated preparation from bone marrow two methods of progress were open: to try a synthetic preparation or to find another source of alkoxyglycerols. Large quantities of alkoxyglycerols are found in some sharks. For example, the liver oil of *Somniosus microcephalus*, Greenland shark, contains up to 50% of alkoxyglycerol-esters. The approximate proportions of the alkoxyglycerol components are: selachyl alcohol 52%, batyl alcohol 43% and chimyl alcohol 5%. The alkoxyglycerols are esterified mainly with polyunsaturated fatty acids. So far it is chiefly this type of 40-50% shark liver oil which has been used in the investigations. Lately it has been possible to achieve a concentrate yielding practically pure alkoxyglycerol-esters (95%). The preparation has been administered orally in the form of sugar-coated tablets, capsules or as an emulsion, the dosage being 1-2 g. of oil per day for adults.

Alkoxyglycerol-esters have been given to about a hundred patients suffering from irradiation leucopenia. Of these, 75% responded with an increased leucocyte count in spite of continued irradiation treatment. In 15% of the cases the leucocyte level was stabilised and only 10% showed a further decrease.

Alkoxyglycerol-esters were given to a nurse who had been occupationally exposed to radium, and who for more than one year had a leucocyte count of approximately 2,000. After 3 days' treatment the count had risen to 3,600. This level has been maintained for more than one year by five treatments, the treatment periods being of 3-5 days' duration (Fig. 2).

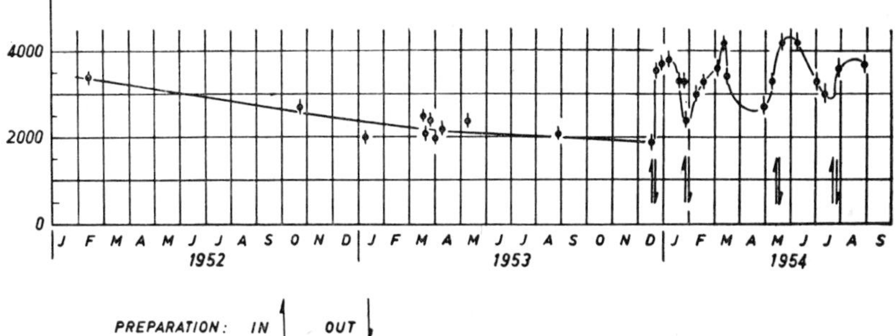

FIG. 2. The white cell count of a radium nurse before and after the administration of alkoxyglycerol-esters.

To obtain directly comparable results from treated and non-treated cases, alkoxyglycerol-esters were administered to patients suffering from cancer of the uterine cervix. These patients form the largest homogeneous group of irradiation treatment cases at Radiumhemmet. They are generally treated by two series of intracavitary radium applications in which radium cylinders are introduced into the corpus and packed against the cervix. The treatments are carried out with an interval of about three weeks. After 3 more weeks, X-ray treatment is given for 3–4 weeks. Alkoxyglycerol-esters have been administered to every second patient, each day during the treatment. Each alternate case has been used as a control and 125 cases of each kind have been studied and followed by blood examinations.

The results are summarised in Figure 3.

The initial leucocyte count for both groups lies at 6,000. The controls at the beginning of the X-ray treatments have a white cell count (V_F of 4,000, the corresponding value for the group which has received alkoxyglycerol-esters is 4,700. The white cell count at the end of the treatment (V_S) is 3,200 for the control group and 3,900 for the prophylactic or treated group. The mean value of the number of white cells during the X-ray treatment (V_M) is 3,450 for the control and 4,000 for the group which received alkoxyglycerol-esters. The difference between the two groups is highly significant. ($P = 0{\cdot}0001$.)

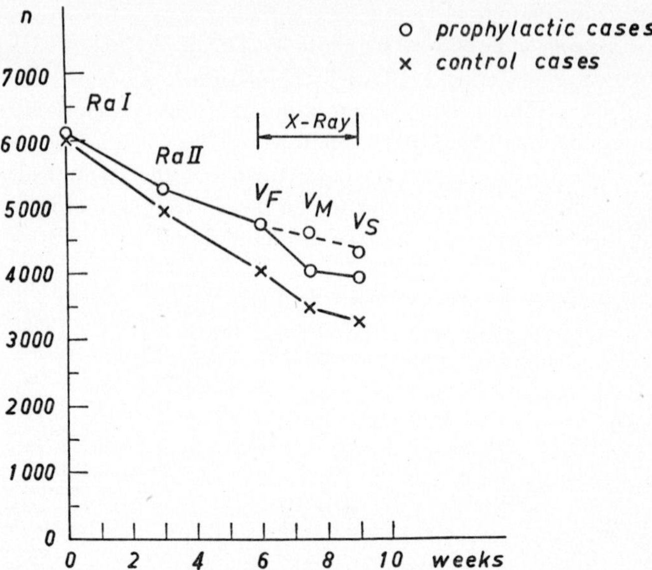

FIG. 3. Cancer of the uterine cervix: white cell count during the radiation treatment.

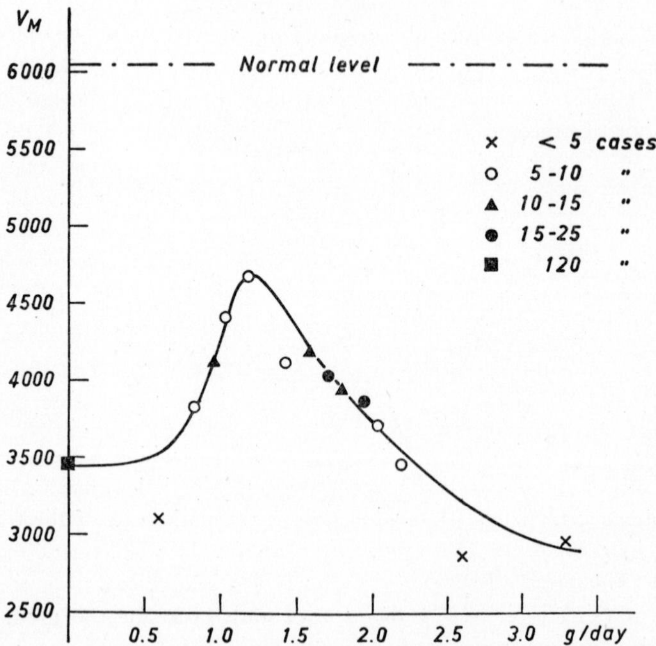

FIG. 4. White cell count (V_M) as a function of the amount of oil taken. V_M: Mean value of the number of white cells during the whole X-ray treatment.

As already stated, these averages relate to all cases. If the 'prophylactic' group is subdivided according to the amounts of alkoxyglycerols administered, a correlation between effect and dose becomes apparent. This is shown by the dotted line. There seems to be an optimal dosage during the X-ray treatment period amounting to 1·2 g. of oil per day. With this dosage V_M is 4,650. Both larger and smaller doses give lower values. Doses higher than 2·5 g. seem to give lower counts than in the controls (Fig. 4).

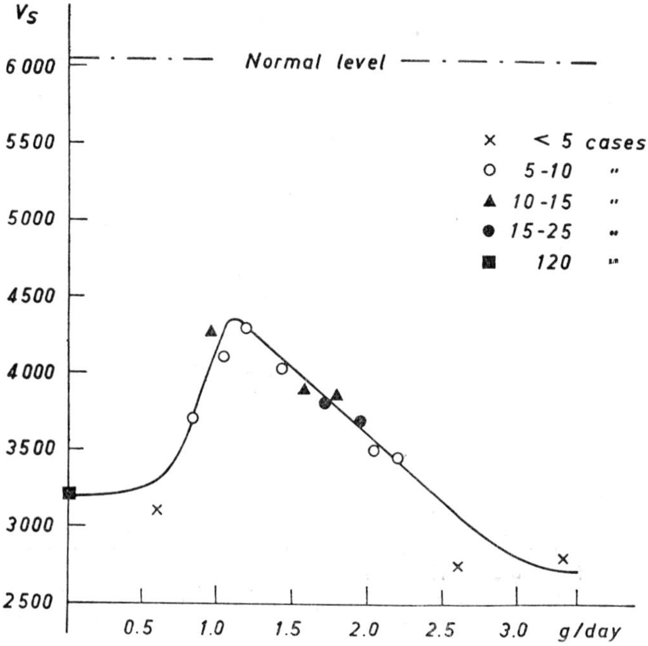

FIG. 5. White cell count (V_S) as a function of the amount of oil taken. V_S: Mean value of the number of white cells at the end of the X-ray treatment.

Exactly the same relationship was also found if the blood count at the end of the course of X-ray treatment (V_S) is plotted against the amount of oil taken (Fig. 5).

As regards the average of the absolute number of lymphocytes during the irradiation treatment the optimal dose of oil seems to be the same.

The thrombocyte average count during the X-ray treatment at this optimal dosage is 190,000 and the corresponding figure for the controls is 155,000 (Fig. 6).

It has not been possible to demonstrate any definite effect on the erythrocytes.

The results have varied somewhat with different preparations of alkoxyglycerol-esters. The effect produced by the concentrated preparation is lower than expected and it cannot at the present state of this investigation be decided if the activity is due to the whole complex of alkoxyglycerol-esters in the shark liver oil, or to some special alkoxyglycerol-esters or even to some other unknown component of the oil.

We hope that experiments in progress with pure alkoxyglycerols and alkoxyglycerol-esters will provide more information about the active principle.

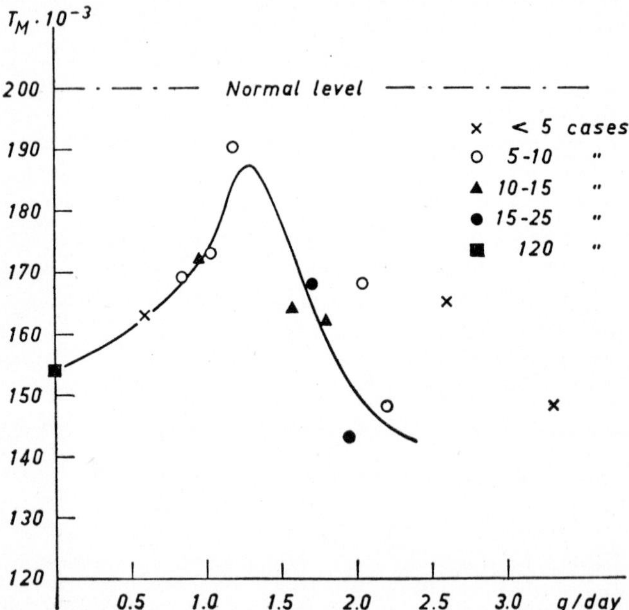

FIG. 6. Thrombocyte count (T_M) as a function of the amount of oil taken. T_M: Mean value of the number of thrombocytes during the whole X-ray treatment.

Not enough time has passed to be certain whether the alkoxyglycerol-esters influence the effectiveness of the radiation therapy, but all the indications are that the effect on the tumours is in no way reduced. Moreover, in some cases it may be possible by giving the oil to continue radiation treatment of patients whose blood picture would otherwise have made it impossible to do so.

Acknowledgments

These investigations have been supported by grants from The Swedish Cancer Society and The Cancer Society of Stockholm; and valuable help has been given by Forskningslaboratoriet LKB and A/B Kabi, Stockholm.

The author is greatly indebted to Professor Sven Hultberg and Dr. Hans-Ludwig Kottmeier for their help and interest.

REFERENCES

Brohult, A., and Holmberg, J., 1954. *Nature, Lond.*, **174**, 1102.
Edlund, T., 1954. *Ibid.*, **174**, 1102.
Holmes, H. N., Corbet, R. E., Geiger, W. B., Kornblum, N., and Alexander, W., 1941. *J. Amer. Chem. Soc.*, **63**, 2607.
Marberg, C. M., and Wiles, O. H., 1937. *J. Amer. med. Ass.*, **109**, 1965.
Idem, 1938. *Arch. intern. med.*, **61**, 408.
Sandler, O. E., 1949. *Acta med. scand. Suppl.*, p. 225.
Watkins, C. W., and Giffin, H. Z., 1933. *Amer. Med. Ass. Sci. progm.*

Discussion

Elson. Is the effect on the lymphocytes or on the neutrophils?

Loutit. Are there any observations as to the mechanism of this effect? Is it a stimulation of the bone marrow or is it a mobilisation of lymphocyte from other sources?

Brohult. The figures given concern the effect on the total number of white cells. The effect on the lymphocytes alone seems to be less than on the neutrophils which suggests that the effect may be a stimulation of bone marrow.

Elson. The fact that the platelets are increased does perhaps also suggest the effect may be a stimulation of bone marrow.

Fliedner. Have you any information on the effects of alkoxyglycerols on the marrow-picture? We often find a hyperplastic marrow in patients with a peripheral leucopenia after radiation treatment for cancer. We have tried *Prednisolon* in these patients and have seen in some cases an increase of the granulocytes after 2–3 days.

Brohult. A few marrow-punctures have been carried out on patients suffering from radiation leukopenia who had responded to alkoxyglycerol-esters with an increased white-cell count, in spite of continued radiation treatment. In these cases the marrow did not show any important changes.

Hulse. In view of the somatic mutation theory for the origin of post irradiation leukæmia (Loutit, this volume, page 388) do you think that by artificially increasing the leucocyte content of the marrow during a course of radiotherapy you are increasing the population in which such a mutation may occur, and so increase the likelihood of a post-irradiation leukæmia?

It might be interesting to follow your series to see if the leukæmia incidence amongst them is greater than amongst those patients treated by radiotherapy alone.

Brohult. Since the treatment has been used for a comparatively short time (about two years), it is impossible to answer this question. However, no change has been observed in the marrow-function of the few patients investigated. In this connection it might be of interest to mention that some cases of leukæmia have responded favourably to treatment with alkoxyglycerol-esters.

THE PROTECTIVE EFFECT OF THE REDUCTION OF THE BODY TEMPERATURE ON ADULT AND YOUNG MICE AFTER WHOLE-BODY IRRADIATION

SHIRLEY HORNSEY

Medical Research Council, Experimental Radiopathology Research Unit, Hammersmith Hospital, London, W.12, England

It has been shown that the reduction of the body temperature of newborn rats (Hempelmann *et al.*, 1949) and mice (Lacassagne, 1942, and Storer and Hempelmann, 1952) increases their survival rates after whole-body irradiation: it has also been shown that a reduction in the body temperature of adult rats to 18° or 19° C. reduces their sensitivity to whole-body irradiation (Baclesse and Marios, 1954), whereas the chilling of adult mice to 20–29° C. has shown no reduction in sensitivity (Storer and Hempelmann, 1952). Considerable protection from the effects of whole-body irradiation of adult mice may be obtained by irradiating the animals while their body temperature is between 0 and 0·5° C. (Hornsey, 1956). It is suggested that this protection is due largely to the anoxic state of the mice and some corroborative evidence for this suggestion is given using young mice.

Method

All the mice used were of the 'T' strain bred at The National Institute for Medical Research, London. The adult mice were 3–5 months old and weighed 23–31 grams. Irradiations of normal mice were done in a cardboard box, 6–8 mice at a time. Cooled mice were irradiated singly and were packed in ice throughout the irradiation. All the irradiations were with 190 kV X-rays, 6 mA, filtered through 1·5 mm. of Cu and 1 mm. of Al. The body temperature of the mice was lowered using the technique of Goldzveig and Smith (1956). The temperature was measured with a thermometer in the colon. Breathing stopped at approximately 5° C. and the heart stopped beating at approximately 3° C. The irradiations were carried out 10 minutes after breathing had stopped and the mice were revived after 30 minutes.

The irradiation of young mice was carried out when they were one or two days old. Litter-mates were assigned to the various types of treatment by weight on the day of irradiation. They were placed in a stream of N_2, and after approximately 10 minutes, when breathing had ceased, were either irradiated, or packed in ice and irradiated 5 minutes later. A comparison of the growth of animals in the same litter was made for 30 days following the irradiation.

Results

In these experiments the adult mice not cold-treated were found to have an LD_{50} at 30 days after the irradiation of approximately 620r of X-rays. No mice of this strain survived a dose of 900r. Mice irradiated while their colonic temperature was between 0°C. and 0·5° C. were found to have an LD50/30 of approximately 1,760r. There was 100% survival in the groups given 900, 1,200 and 1,500r; (14, 15 and 13 mice respectively in each group). After a dose of 1,800r (15 mice), approximately 40% survived at 30 days. The first deaths occur earlier after 1,800r. whole-body irradiation given to cooled mice than after doses giving comparable survival at 30 days (e.g. 700r) given to animals at normal body temperatures (Fig. 1).

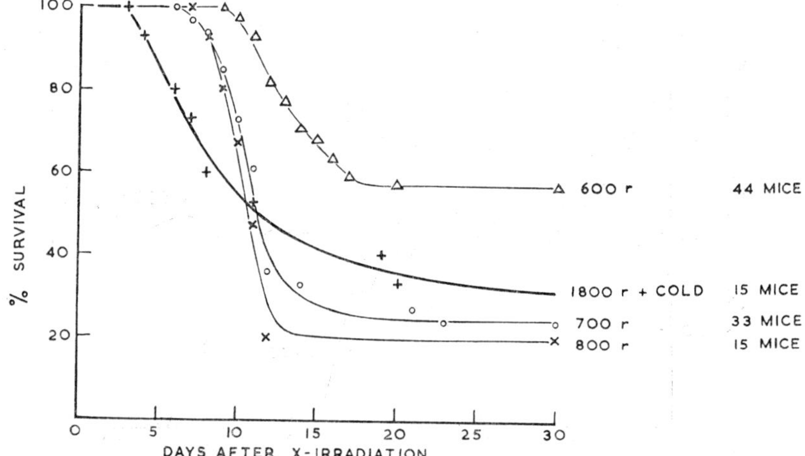

FIG. 1. The survival of male mice over a 30-day period after whole-body X-irradiation of 600r, 700r or 800r at normal body temperatures and of 1,800r given to mice at reduced body temperatures.

Cooled animals lost more weight during the first day after irradiation than animals irradiated at normal body temperature (Fig. 2), owing to the combined trauma produced by the technique of hypothermia and by the irradiation. Mice lost between 1 and 4 gm. body-weight in the two days following hypothermia (Hornsey, 1956, and Goldzveig and Smith, 1956). The weight of all irradiated animals fell steadily, but, in those animals which showed subsequent recovery, the minimum body weight appeared to be reached earlier in cooled animals than in animals irradiated at normal body temperatures.

When one-or two-day-old litter-mates were irradiated with 1,500r X-rays while they were made anoxic by breathing nitrogen, the growth rate was the same when the irradiation was given at normal body temperature as when the body temperature was reduced by packing them in ice (Fig. 3).

A similar result has been obtained after an irradiation of 900r under these conditions. No allowance for the sex of the animal was made in assigning them to the various groups, as no difference, or only very slight differences, in sensitivity between male and female mice have been reported (Abrams, 1951 and Chapman, 1955) and it can be seen that in these two litters there is no detectable difference.

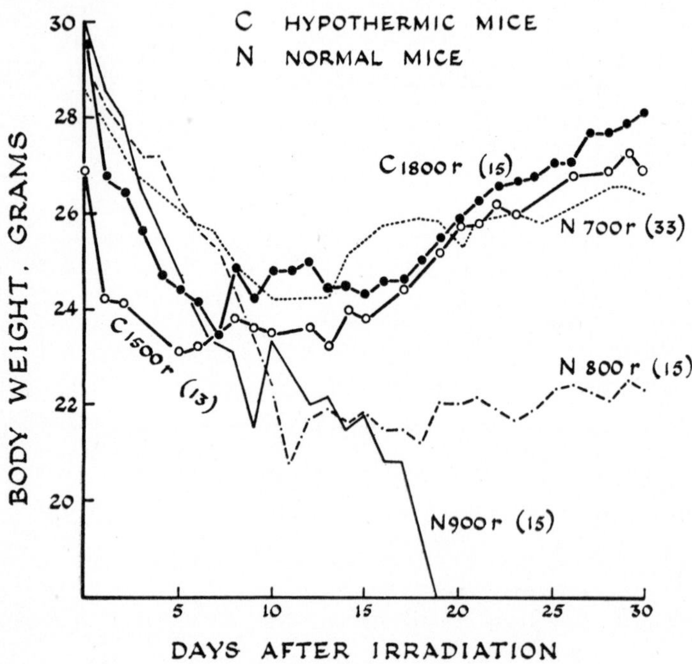

FIG. 2. The weight changes in male mice over 30 days after whole-body X-irradiation at normal or reduced body temperatures.

Discussion

The shape of the survival curve obtained after 1,800r whole-body irradiation given to hypothermic mice differs from that obtained after the irradiation of normal mice with slightly less than LD 100% (Fig. 1). After 1,800r X-irradiation to cold animals a number of them die at 3–5 days, showing all the signs of 'intestinal death' (Quastler, 1956). It would be expected that this type of death would occur in unprotected animals after doses of 1,000–1,200r X-rays (Quastler, 1956 and Rajewsky, 1955), and it is therefore clear that the hypothermia has not given the same degree of protection to the mice against this post-irradiation 'intestinal death' as it has against damage which results in death at a later time, e.g. to the hæmopoietic tissues. The earlier minimum in the loss in body weight reached after 1,800r to hypothermic animals compared with that of animals having a similar survival value at 30 days after irradiation at normal body temperature (e.g. 700r X-rays) also suggests that damage showing effects later than the first week after irradiation is less in the hypothermic animals. It

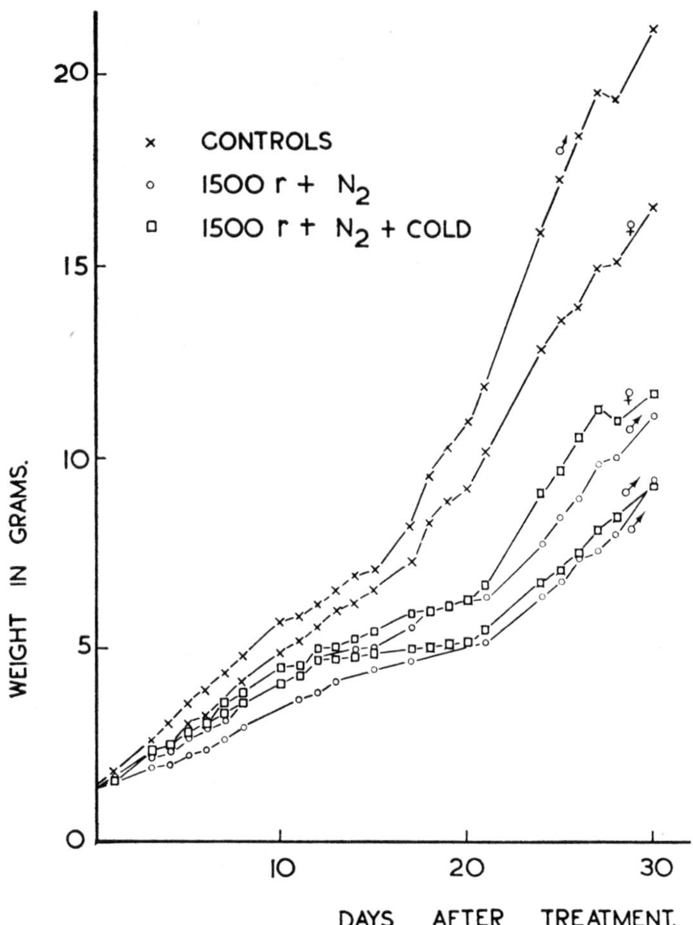

FIG. 3. The increase in weight over 30 days of individual mice either untreated controls or after similar doses of whole-body X-irradiation in N_2 at room temperature or in N_2 at the temperature of crushed ice.

may be, however, that the hypothermic treatment allows recovery from the post-irradiation syndrome to start earlier and this is reflected in the earlier minimum in weight-loss of animals after irradiation and the proportionately fewer deaths seen at later times.

The reduction of the body temperature of anoxic young mice during irradiation has no effect on the subsequent growth of the animal. Thus the reduction in body temperature gives no increased protection to the animals additional to that already conferred by anoxia during the irradiation. Storer and Hempelmann (1952) concluded that the degree of anoxia is a most important factor in the increased survival of infant mice irradiated at body temperatures of 3–7° C. They found that the addition of N_2, after reducing the body temperature, produced no additional protection, whilst the addition of O_2 reduced the survival considerably.

The body temperature of the adult mice used in these experiments is lowered by first placing the animal in an atmosphere of decreasing oxygen tension and, when breathing has almost ceased, packing the animal in ice, when, after gasping irregularly for a few minutes, the mouse stops breathing altogether. It seems likely, therefore, that at the time of irradiation these animals are under a considerable degree of anoxia. It is not possible to say whether the reduction in temperature itself resulted in some protection against the radiation in addition to that which is given by the anoxic state of the animals. However, the comparisons of anoxic and cold young mice suggest that the considerable protection afforded these adult animals by the reduced body temperature is an indirect one due to the higher degree of anoxia tolerated by cold animals than by mice at normal body temperature.

REFERENCES

ABRAMS, H. L., 1951. *Proc. Soc. exp. Biol. N.Y.*, **76**, 729.
BACLESSE, F., and MARIOS, M., 1954. *C.R. Acad. Sci., Paris*, **238**, 1926.
CHAPMAN, W. H., 1955. *Radiation Res.*, **2**, 502.
GOLDZVEIG, S. A., and SMITH, A., 1956. *J. Physiol.*, **132**, 406.
HEMPELMANN, L. H., TRUJILLO, T. T., and KNOWLTON, N. P., 1949. *A.E.C.U.*, 239.
HORNSEY, S., 1956. *Nature, Lond.*, **178**, 87.
LACASSAGNE, A., 1942. *C.R. Acad. Sci., Paris*, **215**, 231.
QUASTLER, H., 1956. *Radiation Res.*, **4**, 303.
RAJEWSKY, B., 1955. *Radiobiol. Symp.*, 1954, p. 81 (London, Butterworth).
STORER, J. B., and HEMPELMANN, L. H., 1952. *Amer. J. Physiol.*, **171**, 341.

DISCUSSION

Wolff. I noticed that the body weight of the animals irradiated under cold conditions reached a minimum sooner than did that of the controls and also that those irradiated cold died sooner than did those irradiated under normal conditions. Do you think there is any correlation between these two phenomena?

Hornsey. I believe the considerable weight loss seen in the first few days after the irradiation of cooled mice is due largely to the method of producing hypothermia. Control mice which are cooled to a body temperature of 0–0·5° C. but which receive no irradiation, lose between 1 g. and 4 g. in the first two days after cooling. You will recall that the loss in weight of hypothermic mice which had received 1,500r whole-body X-irradiation, shown in Figure 2, was similar over the first few days to that of hypothermic mice which had received 1,800r; but after 1,500r we have obtained 100% survival.

Wolff. Since your hypothesis is that the protection derived from cold is caused simply by anoxia, I wonder also if you have considered testing this by utilising the phenomenon described by Russell of no oxygen effect occurring for induced mutations in mouse testes. I wonder whether or not you would get protection by cold in this case?

Hornsey. We have not yet done any tests for mutations in these mice.

van Bekkum. In the cooled-irradiated series you observed a number of deaths occurring before the 6th day after irradiation. These animals, you mentioned, died from the intestinal syndrome. I should like to know the criteria you used to diagnose the intestinal syndrome? I should point out that early death is not limited to the intestinal syndrome, but may also occur as a consequence of the bone-marrow syndrome, when the mice become infected with some particularly infectious micro-organism.

Hornsey. I have used the criteria suggested by Quastler (1956) for 'intestinal death'. Death could, however, have been due to some other cause. The mice died 3–5 days after the irradiation, which is earlier than with mice irradiated at normal body temperatures with doses of less than 900r. The intestine also shows considerable damage.

Carter. I should like to ask Mrs. Hornsey if she has found any protective action against radiation-induced sterility?

TABLE

Fertility in Male Mice, 7–10 Weeks of Age when First Tested, after Irradiation within Two Days of Birth

Treatment	Males		Matings	Litters	No. of young born	No. still-born	Still-birth rate (%)	Sex drive
	Tested	Fertile						
600r in air ...	7	5	16	9	54	22	40	Absent in 2 males
900r in air ...	1	0	3	0	0	0	0	Normal
900r in nitrogen	3	3	14	11	68	0	0	Normal
1,500r in nitrogen	11	8	28	11	55	22	40	Reduced
1,800r in nitrogen	1	1	1	1	5	0	0	Absent by and after 120 days old

Hornsey. Some of the mice irradiated within two days of birth were tested for fertility at 7–10 weeks old. The tests were carried out at the National Institute for Medical Research, London, N.W.7, by Miss H. M. Bruce.

In the male after 600r in air (7 males) or 1,500r in nitrogen (11 males) fertility was reduced and the viability of the offspring impaired. At the higher dose sex drive was also diminished. The fertility of the 3 males examined after 900r in nitrogen appeared to be normal and the young were viable. Further details are given in the Table. The interpretation of these results needs caution because of the small number of mice.

There was little evidence of ovarian activity. Most of the 23 females examined were acyclic by the time the tests were started and few matings took place. After 600r in air, one female (out of 16) produced a stillborn litter of one; two others mated but failed to become pregnant. After 900r in nitrogen one of the 2 females tested became pregnant but died during gestation, probably from an infection. Implantations were present in both horns. None of the 5 females tested after 1,500r in nitrogen mated.

All the irradiated mice, both male and female, were extremely stunted in growth.

SECTION VI

THE TIME FACTOR IN RADIOBIOLOGY

THE EFFECT OF REPEATED SMALL DOSES ON THE FERTILITY OF THE WHITE MOUSE *

H. AND M. LANGENDORFF
Institute of Radiology, University of Freiburg,
Freiburg, Germany

The effects of repeated small doses of irradiation have until now mostly been scored as influences on the hæmatological system and on the incidence of tumours in mammals. We have during the last few years studied the influence of such irradiation conditions on the fertility and on some properties of the descendants of the irradiated animals.

When male mice were given 2·5r/day and were afterwards mated to un-irradiated females we found that the number of successful pairings decreased only after doses of 400r and above. The number of descendants in each litter, however, even after a total dose of 800r was the same as in non-irradiated animals namely 5·02 per litter as an average.

In the irradiation experiments on females we first determined the number of litters and the number of animals per litter in successive pairings of the same non-irradiated males and females. It was found that the percentage of litters was independent of the order in which the pairings were undertaken (Table I).

TABLE I

	Number of pairings	Number of litters	% of litters	Probability P
1. Litter	295	281	94·4	
2. Litter	293	265	90·6	0·3–0·2
3. Litter	195	180	92·5	

The number of F_1-descendants per litter, however, was significantly higher in the second pairing as compared with the first one, whereas no difference was found between the second and the third pairing (Table II). These results are in accordance with similar observations of L. B. Russell. The increase of animals per litter in the second pairing was explained by the fact that in these groups we had an over-representation of high litter numbers, whereas in the first group the size of the litters, fairly well, followed a normal distribution curve.

* This paper, due to Professor Langendorff's illness, was read by R. Koch.

The experiments giving females fractionated doses were performed in the following way. After irradiation at a rate of 2·5r/day for 0–40 days

TABLE II

	Average litter number in controls
1. Litter	5·78 ± 0·138
2. Litter	6·61 ± 0·141
3. Litter	6·30 ± 0·165

the females were repeatedly mated to non-irradiated males. Sterility occurred after a total dose of 200r and increased at a rapid rate with dose.

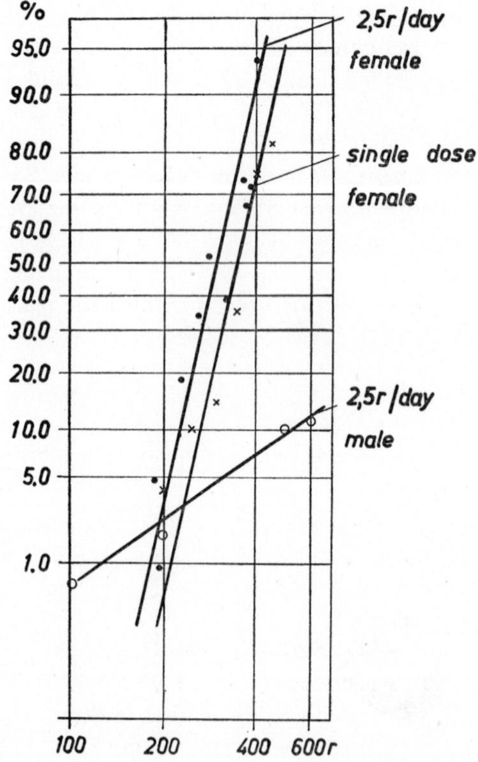

Fig. 1. Mouse sterility and irradiation.

In order to see if there was any influence of the time factor, we gave at the same time, single doses in the interval 50 to 450r to 596 females (Fig. 1). Apparently there is no pronounced difference in regard to sterility whether we irradiated with single or fractionated doses. The slight shift to the left shown by the curve for animals having fractionated doses possibly may be ascribed to a slightly stronger effect of the latter treatment, as suggested by Regaud.

TABLE III

♀ single dose				♀ fractionated dose			
Dose (r)	No. of pairings	No. of F_1-animals	Relative litter size Contr. = 100	Dose (r)	No. of pairings	No. of F_1-animals	Relative litter size Contr. = 100
0 = contr.	60	271	100	0 = contr.	787	4,479	100
50	60	259	91·0				
				59	59	349	103·0
				82	60	318	93·0
100	60	269	100				
				124	61	381	96·5
				136	62	344	96·5
150	60	251	93·5	149	56	301	94·8
				194	119	666	98·1
200	60	231	86·6				
				228	60	225	65·0
250	59	158	57·7				
				270	113	302	40·8
300	59	142	53·3				
				310	34	98	43·9
350	59	86	33·3				
				359	57	58	17·5
				380	81	74	15·8
400	59	38	15·6				
450	60	20	6·7				

The corresponding sterility curves when males are irradiated with 2·5r/day have quite a different slope (Fig. 1). Only by giving very high doses do we find any significant decrease in the fertility. The explanation is probably that the sperms which are not heavily damaged by the irradiation stand a better chance to fertilise. This is also substantiated by the fact that the litter size is the same within the whole dose range up to 800r.

Considering the limited number of egg cells in the mouse ovaries one would expect that the litter size would tend to decrease with rather small doses. This, however, is not the case (Table III). Independent of the method of irradiation we find the litter size remains constant up to a dose of 200r. Similarly the sterility curve begins to show a decrease at this dose and in studies of the viability of the sucklings we also found the same trend, although the last mentioned studies have not so far given statistically significant figures.

To summarise we can state: when males are given fractionated small doses the sterility of the mated females increases only slowly with the doses, probably due to sperm competition. When females are irradiated, whether with single doses or with fractionated doses, we find already at low dose levels pronounced sterility. In both cases the dose effect is completely additive. An influence on the litter size is, independently from the method of irradiation, only visible after rather high doses. It would seem that 200r will start to give irreparable egg cell damage. The rather slow decrease of the litter size with the increase in total dose on the females may indicate that only the damage to the egg cells and not that to the surrounding tissues is the responsible factor, whereas on the contrary the irradiation effect on the males suggests a co-operation of constitutional factors which add to the decreased fertility. We believe these conclusions to be supported by the fact that the fertility is significantly higher in the dose range 200–400r when given as a fractionated dose as well as the fact that the litter size is little influenced by doses higher than 400r (up to 800r) although sterility increases.

Considering the genetical importance of repeated small radiation doses we believe it to be of particular interest to analyse the offspring of females, which, although having obtained rather high total doses, nevertheless do not show an appreciable decrease in fertility. In this group we should expect to find suitable material for studies of genetic damage.

EFFECTS OF RADIATION ON AGEING

H. J. Curtis and Ruth Healey
Biology Department, Brookhaven National Laboratory,
Upton, New York, U.S.A.

It was demonstrated several years ago that the life span of animals subjected to an ionising radiation such as X-rays is markedly shortened (Henshaw *et al.*, 1947). It was demonstrated that the animals developed various forms of cancer at a relatively early age and died as a direct result of these neoplasms. These observations have been confirmed several times since then, but there is a subjective impression on the part of most investigators that the increased death rate in the irradiated population is not due entirely to the increased tumour incidence, but is due to an acceleration of the entire ageing process.

In addition, it has long been postulated that a severe disease will shorten the life expectancy of the individual. On the basis of a statistical analysis of human populations Jones (1956) has concluded that this is true not only for X-ray exposures but for all illness.

The present experiments were designed to test, in mice, whether X-ray exposures is indeed the same as other stresses with respect to shortening of the life span; and to determine whether or not the increased death rate in irradiated animals could be explained entirely by an increased tumour induction rate.

Methods

Strain CF1 female mice were used in these experiments. They were kept in air-conditioned quarters in plastic boxes, and fed a standard laboratory diet. When they were 8 weeks of age they were divided into several groups as shown in the Table.

The X-ray was 250 kVP; 1 mm. Al + $\frac{1}{4}$ mm. Cu; doses were measured in air. The typhoid toxin was given as undiluted typhoid vaccine (Lederle) intraperitoneally. The animals became very acutely ill and, as seen from the Table, many of them died. Controls were injected with physiological saline. Nitrogen mustard was given intravenously in the tail vein in physiological saline, and the controls were injected with physiological saline. These mice became acutely ill and did not recover for many days. When they did, there was often almost complete degeneration of the tail. Three of these mice became blind about a year later.

Approximately 12 months after the treatments, mice were selected from Groups 1A and 4A and given further treatments as follows. Twenty mice from each of these two groups were given a dose of 500r of X-ray. In addition, a group of 10 mice from Group 1A were given $\frac{1}{2}$ ml. of typhoid toxin, and 30% of them died. In addition, a group of 10 mice from Group 4A were given 1 ml. of typhoid toxin, but none died. All survivors were kept for life-span studies.

In addition to the stress experiments, an attempt has been made to measure the efficiency of the capillary circulation in these animals as a function of age and previous treatment. This was attempted by two techniques. The first was to inject a radioactive compound in the tail vein, and measure the rate of build-up of the material in a distant segment of skin as a function of time by means of a Geiger counter. The isotopes ^{131}I, ^{32}P, and ^{45}Ca have

Group no.	Treatment	Number of animals	% mortality in 30 days
1A	525r-X-ray	300	30
1B	420r-X-ray/week for 4 weeks	70	98
1C	262r-X-ray/week for 6 weeks	70	64
1D	Specific controls for Group 1A	50	0
1E	Specific controls for Groups 1B and 1C	20	0
2A	1 c.c. typhoid/day for 5 days	70	34
2B	1/2 c.c. typhoid/day for 5 days	30	0
2C	Specific controls for Groups 2A and 2B	20	0
3A	0·25 mg. nitrogen mustard	70	87
3B	0·125 mg. nitrogen mustard	30	20
3C	Specific controls for Groups 3A and 3B	20	0
4A	General controls	106	0

been used for this. The second method is to allow the mice to inhale ^{131}Xe for a short period of time, and thereafter follow the excretion of Xe from the lungs by means of a total-body crystal scintillation counter. Unfortunately neither of these methods has shown a significant difference between young, old or X-rayed mice. This may simply be an indication of the fact that the error of the method is too large to detect the relatively small changes involved, but seems to indicate that there are no gross changes as measured by these methods.

Results

The results of the life-span studies are shown in Figure 1. There was no apparent difference between the various control groups, so they were all grouped together. It will be seen that the animals given typhoid and nitrogen mustard are not significantly different from the control animals. It must be stated that there was an unknown infection in the colony when the mice were 7 to 10 months old, and this increased the experimental error somewhat. Thus at the present time all that can be said is that if there is an effect on life span of a severe toxic stress, the effect is not large.

On the other hand, an X-ray stress of equal acute severity alters the life span markedly. Furthermore, as has been shown previously (Henshaw *et al.*, 1947), this stress is additive, that is, there is a component of the X-ray damage from which the animal does not recover, and these irreparable increments of injury can add up to give almost any desired degree of reduction of the life span. Thus the repetition of a sub-lethal dose can lead to marked shortening of the life span.

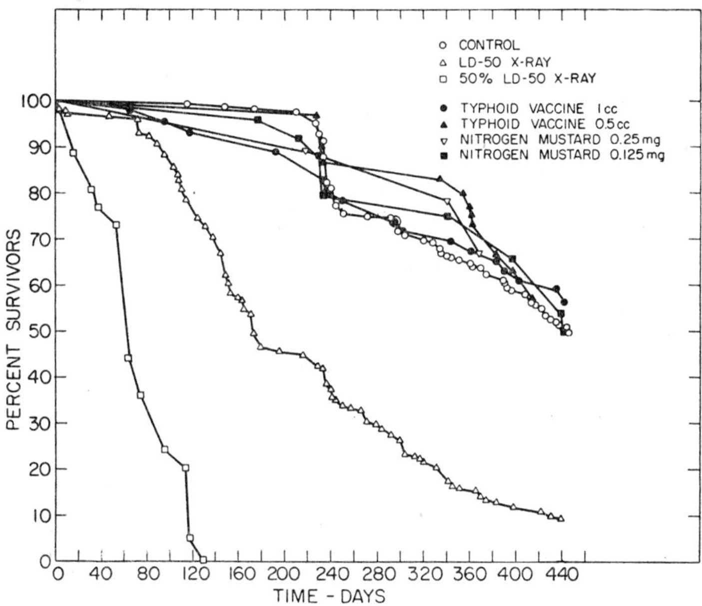

FIG. 1. Survival curves for mice who have survived the acute (30 day) effects of various stresses. X-rays doses were 525r (LD50) and 262r per week for 6 weeks (50% LD50).

These differences are also reflected in the weight curves of Figure 2. Here it will be seen that the animals given a toxic stress lost weight during the stress period, but regained it soon after the stress was stopped. The X-rayed animals lost weight during the treatment period also, but they never regained all of it so they always stayed below the controls. Again the effect of the infection can be seen in the weight curves, and it is also apparent that it was more severe in the control group than in the experimental group.

The animals given an X-ray stress when they were quite old, died faster if they had received a dose of X-ray when young, and this is shown in Figure 3. The control animals appear to be unaffected by the radiation,

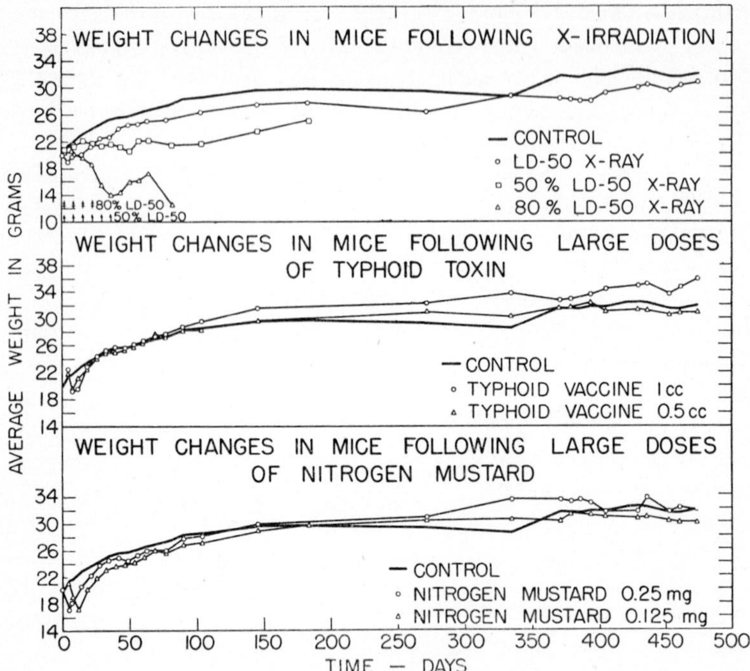

Fig. 2. Weight curves for mice given various treatments as indicated. X-ray doses were 252r (LD50); 420r per week for 4 weeks (80% LD50); and 262r per week for 6 weeks (50% LD50). All treatments given at time zero except the divided X-ray doses which were given as indicated by the arrows.

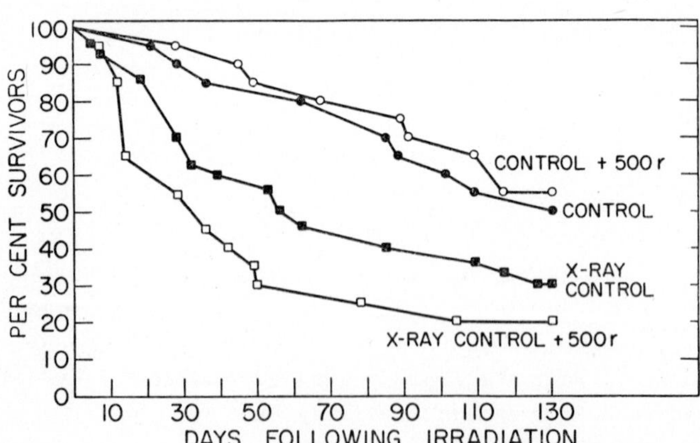

Fig. 3.—Survival of mice irradiated when 13 months of age. The control group had had no previous treatment and the experimental group had received a dose of 500r when 2 months old.

but this is probably a combination of too small a dose and relatively small numbers of animals. Animals receiving an X-ray stress when young are also much less able to withstand a toxic stress when they are old since the control animals could withstand more than twice the dose of typhoid toxin.

Discussion and conclusions

It is unfortunate that the infection developed in the colony, and that there were not greater numbers of mice in several of the groups. However, in spite of these errors two conclusions seem inescapable. The first is that mice that receive a severe toxic stress, but from which they recover, will have a virtually normal life expectancy. This is to be contrasted with mice receiving an X-ray stress of comparable acute severity in which the life-expectancy is markedly shortened. These results do not substantiate the ideas of Jones (1956) obtained on the basis of a statistical analysis of human populations. However, his study included diseases such as rheumatic fever from which there is essentially no recovery, so a lack of agreement does not seem surprising. Since the error of the experiment is quite large, it does not seem justified to conclude that a toxic stress causes no shortening of the life span, but the contrast with X-ray stress is marked.

The second conclusion is that animals receiving a dose of X-rays as young adults undergo premature ageing quite apart from that associated with increased tumour induction. This was demonstrated by the inability of the mice to withstand either a toxic or an X-ray stress as well as the controls.

Acknowledgments

This research was carried out at Brookhaven National Laboratory under the auspices of the U.S. Atomic Energy Commission.

REFERENCES

HENSHAW, P. S., STAPLETON, G. E., and RILEY, E. R., 1947. *Radiology*, **49**, 349.

JONES, H. B., 1956. *Advances in Biological and Medical Physics IV* (New York, Academic Press).

DISCUSSION

Alexander. Have you had an opportunity to examine the physico-chemical proportion of the collagen of the irradiated and non-irradiated animals, since this is claimed to give a quantitative measure of age?

Curtis. I have not tried the ageing methods of Verzar.

Rajewsky. May I ask you, Dr. Curtis, have you given the irradiations both as short-term and long-term doses.

Curtis. The dose rate is very important for the ageing phenomenon. There are two aspects to radiation damage, the acute and the chronic. The former is very sensitive to dose rate and the latter very insensitive.

Hulse. I have been wondering if there is an increase in the incidence of leukæmia and of tumours after irradiation amongst your mice. If the increase is of any appreciable size, may it not mean the degree of 'post-irradiation ageing' is not as great as your graphs suggest ?

Curtis. The tumour spectrum in these animals is approximately the same as for normal animals, but they develop the tumours sooner. It is not possible to explain the increased death rate of the irradiated animals on the basis of increased tumour incidence alone.

Lajtha. These results seem to suggest that apparently there is no true recovery from radiation. Certain organs, however, like bone marrow, intestinal epithelium, and regenerating liver, do seem to recover from radiation damage—I emphasise 'seem' to recover. Also, in tissue culture work, it appears that those cells which survive irradiation do so without apparent loss of viability.

I wonder therefore, whether these long-term effects of irradiation of the whole body are not due to the damage of a specific organ with a slow turnover of cells, such as liver, pancreas or central nervous system. I wonder whether the pathological findings give any indication as to whether there is damage of a specific organ in your experimental animals?

Curtis. Several different histological manifestations of ageing have been described but none of them seem definite. It is my own feeling that all these can be related to the capillary circulation, and this is why we have made an effort to study this system.

Elson. I suggest that you try the other type of so-called radiomimetic substances, e.g. 'myleran'. The nitrogen mustards imitate only part of the effects of X-radiation and myleran imitates others, mainly the myeloid ones; and it may be these effects which are concerned in the ageing effects of radiation.

STUDIES ON THE TIME-INTENSITY FACTOR AFTER WHOLE-BODY X-IRRADIATION

B. Rajewsky, K. Aurand and I. Wolf
Max Planck Institut für Biophysik, Frankfurt am Main, Germany

When studying radiation protection problems, it is important to consider how the effect of equal radiation doses depends on the time distribution of their administration. Experience on long-continued and fractionated whole-body irradiation in man are so scarce, that it is impossible to estimate the value of the 'time-intensity factor' when irradiating human beings. It is therefore necessary to make experiments on animals in order to study the sensitivity of mammals after long-continued and fractionated whole-body irradiation (Rajewsky, 1947; Sacher, 1953; Betz, 1950; Ellinger and Barnett, 1950; Mole, 1955; Paterson *et al.*, 1952).

From 1938 to 1945 Rajewsky and co-workers (1947) studied the effect of long-continued X-irradiation on mice. The results of these experiments and others are summarised in Figure 1 on a double logarithmic scale with values obtained by Rajewsky and Dorneich (1954) and by Sacher (1953).

In this graph there are also plotted our data obtained with higher dose rates. Male mice of the inbred strain of our institute (19–24 g.) were exposed to long-continued X-irradiation until death (200 kV, 10–20 mA tube current, 2 mm. Al-filter). The focus-object distance varied between 30 and 200 cm.

The curve shows the mean survival time of the mice (d) plotted against the dose rate (kr/d). The mean survival time decreases with increasing dose rate up to 1,000r/d. Between 1,000r/d and about 10,000r/d the animals die at a constant rate after 3·5 to 4 days. This plateau corresponds with the '3·5-Tage-Effekt', which has been explored after single whole-body X-irradiation of mice by Rajewsky (1947 and 1953). A further rise in the dose rate results in a rapid decrease in the mean survival time.

In Figure 2 we have tried to display the mortality curves for different dose rates in the range we have explored in order to analyse their shapes.

We examined 16 dose rates and plotted the mortality against irradiation time (or total dose received by the mice). The time interval, after which the animals die when exposed to long-continued X-irradiation, depends markedly on the dose rate (e.g. at 2r/min. (2,880r/d) the mice die between 85 and 130 hours, at 150r/min. (216,000r/d) between 3·5 and 4·8 hours). We therefore normalised these time intervals in order to compare the shape of the curves.

Figure 2 shows some characteristic curves. We can distinguish 3 different shapes, type A was observed with high dose rates, type B with low dose rates. Type C is a composite curve of these two types.

Figure 3 demonstrates the dose (kr) which is necessary to kill 50% of the animals during long-continued irradiation (LD50) as a function of the dose rate (r/min. respectively kr/d).

The LD50 increases rapidly with increasing dose rate and reaches a maximum at 12r/min. (17,200r/d). Then the curve shows a decrease up to 40r/min. (57,600r/d), a further slight increase up to 90r/min. (129,000r/d). The LD50 is 40,000r at this dose rate and 34,000r at a dose rate of 150r/min.

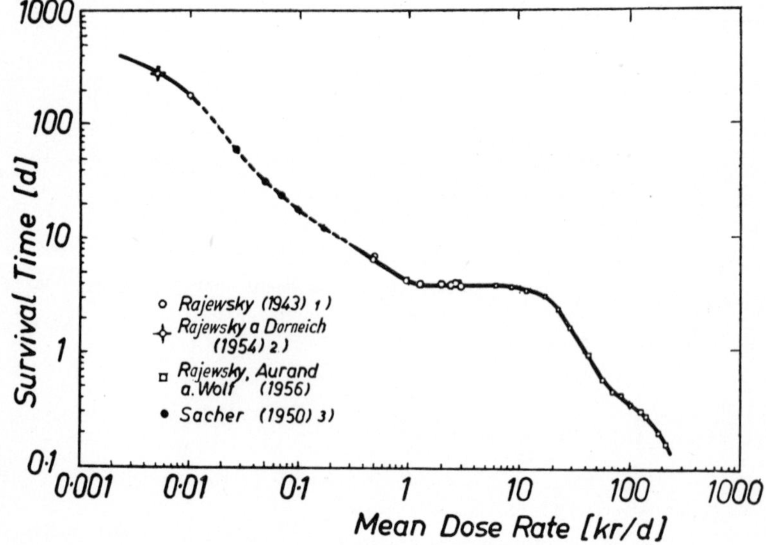

FIG. 1. Long-continued X-irradiation (white mice).

In the following experiments we studied the influence of pre-irradiation on the mortality curve during long-continued irradiation. We chose a dose rate of 90r/min. for long-continued irradiation, because this mortality curve corresponds to the Type C and changes in this ' labile' range might be most easily detected.

First we studied the different effects of irradiation on male and female mice. Figure 4 shows the mortality curves for male and female mice during long-continued irradiation with 90r/min.

The curve for the male mice was obtained with 9 series of a total of 360 animals, that for the female mice with 120 animals. There is a significant difference between the two sexes. For the following experiments only male mice were used.

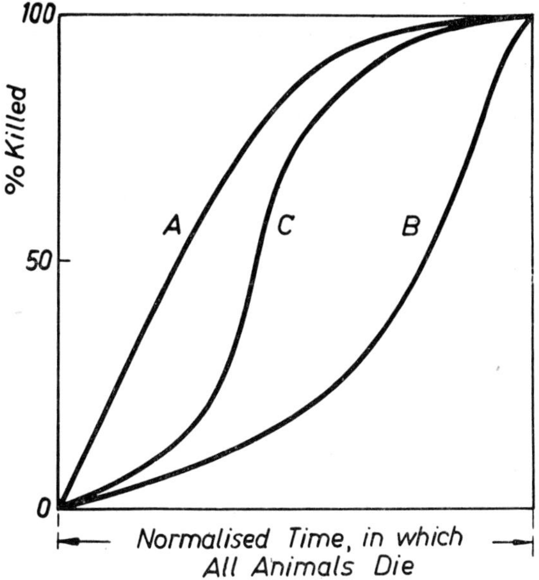

FIG. 2. Mortality curves during long-continued X-irradiation with different dose rates.

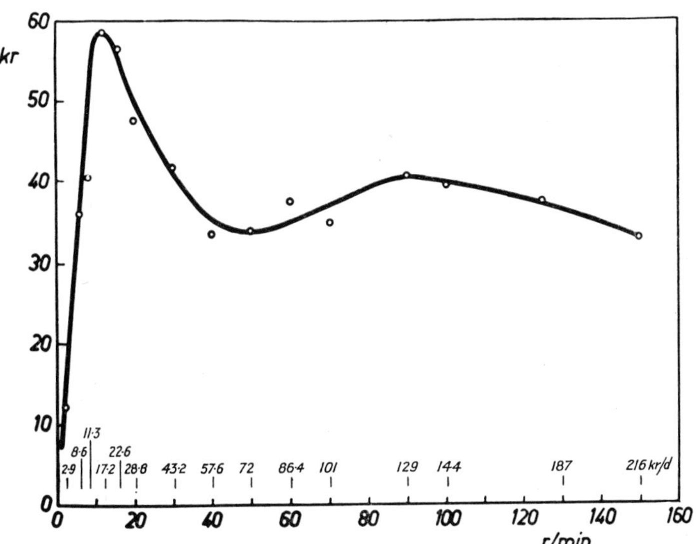

FIG. 3. Long-continued X-irradiation (white mice). (LD50 as a function of the dose rate).

Figure 5 demonstrates the results of earlier experiments on fractionated whole-body irradiation (10) (Aurand, 1954). The mice were exposed to a total X-ray dose of 12,000r, which was applied in two partial doses of 350r and 850r at different time intervals. In Figure 5 the mean survival time is plotted for each group as a function of the time between the first and the second irradiation. The hatched regions correspond to the mean survival time, with standard deviation, for single whole-body irradiation with 1,200r and 850r. These data indicate that the mean survival time varies with the time interval between the two fractionated irradiations.

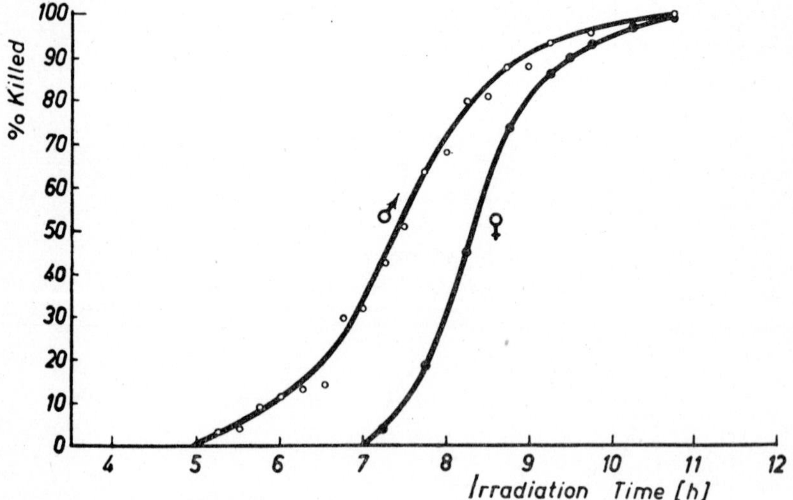

Fig. 4. Long-continued X-irradiation (90r/min.). (White mice ♂ and ♀).

It is important to remember that the second irradiation acts upon an organism which has been altered by the first irradiation. There are the following possibilities of the effects of fractionated irradiation:

1. The effects of the two irradiations are additive.
2. The second irradiation acts on an organism which has been sensitised by the first irradiation. The effect can be smaller than in the case of total accumulation of the doses.
3. The second irradiation is applied at a time, when the first irradiation has no more influence.
4. The first irradiation causes alterations which can diminish the effects of the second dose of irradiation. The animals live longer than after irradiation with only 850r.

This periodical shape does not correspond with the theoretical and experimental recovery functions which were obtained for simple biological systems by Hug and Wolf (1955 and 1956).

In further experiments we studied the influence of a pre-irradiation dose of 150r X-rays on the mortality curves obtained with long-continued whole-body X-irradiation at 90r/min. In Figure 6 the mortality curves are

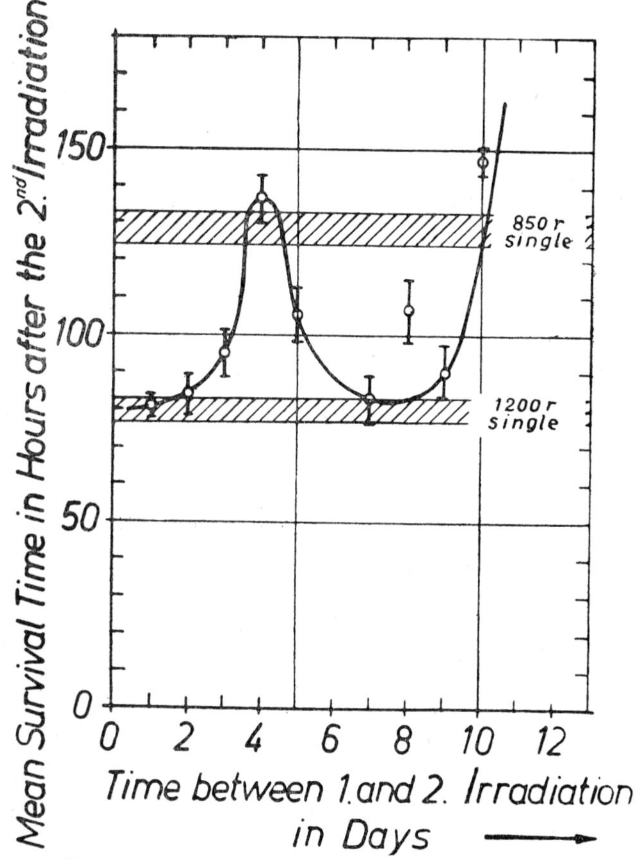

FIG. 5. Fractionated whole-body irradiation of white mice. (350+850*r*).

summarised for different time intervals between pre-irradiation and long-continued irradiation ($\frac{1}{2}$–14 days). The mortality during long-continued irradiation with 90r/min. (without pre-irradiation) is plotted in weak lines for comparison.

The significant influence of pre-irradiation (the animals die earlier) decreases steadily, if the time interval between pre-irradiation and long-continued irradiation is increased. There is a marked decrease after 3 days, only a small influence after 7 days and practically no influence after 14 days.

These experiments demonstrate the possibility of steadily decreasing damage in irradiated animals. These data are dissimilar to the findings on fractionated irradiation with 350r and 850r.

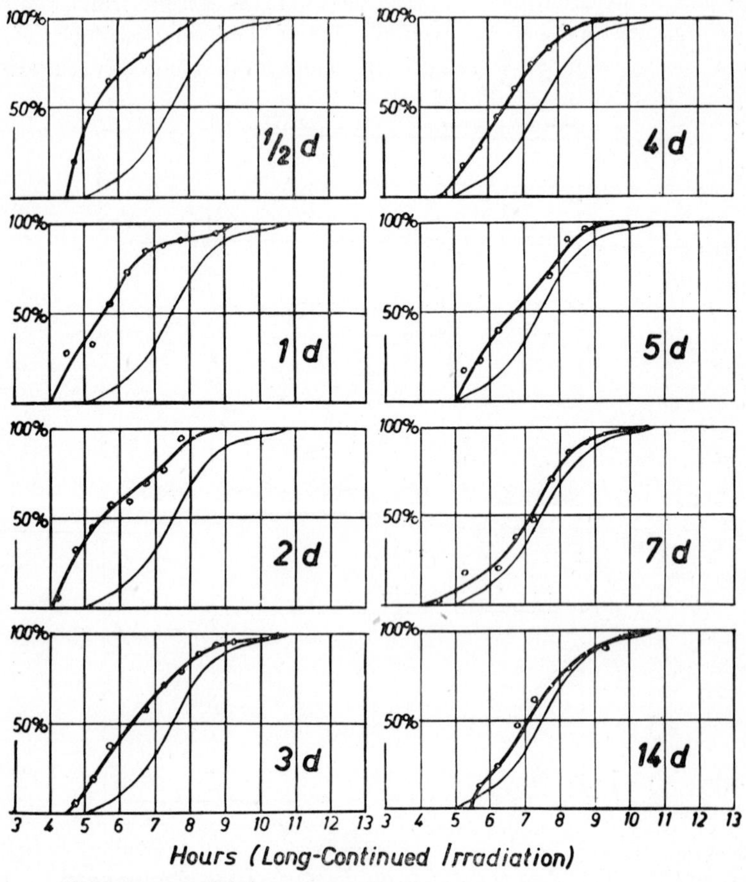

FIG. 6. Long-continued X-irradiation (90r/min.). Pre-irradiation with 150r at different time intervals.

Numerous experiments are necessary to study the problem of time-intensity factor after whole-body irradiation, in order to elucidate this important question, and to obtain a correlation with the experiments on simple biological systems and with theoretical considerations.

REFERENCES

AURAND, K., 1954. *Roentgen Congress, Wiesbaden.*

BETZ, E. H., 1950. *C.R. Soc. Biol., Paris*, **144**, 1437.

ELLINGER, F., and BARNETT, J. C. 1950. *Radiology*, **54**, 90.

Hug, O., and Wolf, I., 1955. *Progress in Radiobiol.*, 1956, p. 23 (Edinburgh, Oliver and Boyd); *Strahlentherapie*, 1956, **35**, 209.

Mole, R. H., 1955. *J. nat. Cancer Inst.*, **15**, 907.

Paterson, E., Gilbert, C. W., and Matthews, J. J., 1952. *Brit. J. Radiol.*, **25**, 427.

Rajewsky, B., 1947. *Fiat Review of German Science*, Bd. 21, *Biophysics Inst., Wiesbaden.*

Rajewsky, B., and Dorneich, M., 1954. *Unveröffentlicht.*

Rajewsky, B., Heuse, O., and Aurand, K., 1953. *Z. Naturforsch.*, **8**b, 157.

Sacher, G., 1953. *Amer. J. Roentgenol.* (J. F. Thomson and W. W. Tourtelotte), **69**, 830.

RESTORATION OF PRIMORDIAL FOLLICLES IN THE IRRADIATED OVARY

P. DESAIVE

Laboratory of General Surgical Pathology,
Liège University, Belgium

The object of this paper is to determine the part played by restoration and by variation of radio-sensitivity, in the final effect on the follicular system of the ovary, after a given amount of X-rays, either as a single dose or as the same total dose equally fractionated.

Experimental methods

1. We used virgin doe rabbits weighing about 2 kg. Irradiation was given at the lumbar level, and included both ovaries ; these were taken out after various lengths of time, fixed in Bouin's solution, and cut into series of sections 6 μ in thickness. In each ovary all the existing follicles were numbered by a technique we have described elsewhere.

2. Three methods of irradiation were used.

(*a*) A single dose of 2,500r (which, according to our previous experiments, sterilises the ovary and completely destroys the follicular system).

(*b*) A single dose of 1,050r (which does not destroy or sterilise but is enough to cause much damage, partially restorable, to the follicular system).

(*c*) The same dose of 1,050r, but given as five doses of 210r, at intervals of 48 hours.

3. We have considered, in the follicular system : (*a*) Follicles—including stages A, B, C and D, corresponding to the growing period of the oocyte, and stages E, F and G, characterised by the formation of the antrum and by the preparation of the cell and egg its envelopes for ovulation; (*b*) primordial follicles P of the cortex storage, numbering about 150,000 in the prepubescent female rabbit. These represent the generative layer of the follicular system, and have a cyclical activity, the rhythm of which we have described at length elsewhere, but which is essentially based on successive peaks separated by 33-hour intervals.

4. Sterilisation of an ovary depends on destruction of all follicles P existing in this organ at the time of irradiation; unless we admit the existence

in the female rabbit of postpuberal ovogenesis (in which case, sterilisation of the ovary depends on complete arrest of the germinative epithelium). For this reason, we are specially interested in these germinative elements and for them we drew up curves showing progressive diminution with time, for each of the three experimental methods above.

5. To draw these graphs for comparison we had to have some reference point from which to begin. The existing follicles in an ovary, however, vary widely from one rabbit to another. This variation may even be as much as a mean variation of $\pm 11{\cdot}15\%$ between one ovary and its twin. To overcome this difficulty we calculated first of all the percentage of variation in the number of follicles P in an irradiated ovary, as compared with its twin removed as a control before irradiation. We then established the percentage of variation in the number of follicles P in an irradiated ovary, as compared with the other ovary of the same pair, but taken out 2, 4, 6, 8, 10 or 12 days previously. By modifying the intervals between removals from one pair to another, we were able to establish for all ovaries in each of our experiments, the successive relations $\frac{\mathrm{N}t}{\mathrm{N}o}$ of the numbers of follicles P existing at any time, *t*, with the numbers of these elements assumed to be existing at time *o* immediately before irradiation. For example, in the experiments when we gave one single dose of 1,050r, we found in a female rabbit (no. 159) that the right ovary, taken out two days after irradiation, still contained a number of follicles P representing 26·8% of those counted in the control left ovary. Moreover, in another female rabbit (no. 155) of the same experimental series, we found that the right ovary, excised four days after irradiation, contained a number of follicles P equal to 36·9% of the number found in the left ovary examined two days after the same irradiation. By extrapolation, the number of follicles P existing in right ovary no. 155 four days after irradiation with 1,050r, contained, as compared with its state before irradiation, a percentage of follicles P equal to 9·89% of the initial figure N_o; which gives:

$$\mathrm{N}_{4j\cdot} = \mathrm{N}_{2j\cdot} \times \frac{36{\cdot}9}{100} = \mathrm{N}_o \times \frac{26{\cdot}8}{100} \times \frac{36{\cdot}9}{100} = 0{\cdot}0989\ \mathrm{N}_o.$$

6. The experiment where we used one single sterilising dose of 2,500r, included 10 female rabbits, the ovaries of which were taken out at time *o*, immediately before irradiation, or between the 2nd and the 12th day after the end of the irradiation. The experiment with one single non-sterilising dose of 1,050r, included 8 female rabbits, the 16 ovaries of which were taken out alternately from the 2nd to the 92nd day after irradiation. Finally, the experiment with the fractionated non-sterilising dose (5×210r at 48-hour intervals), was carried out on 10 female rabbits, the 20 ovaries of which

were taken out from the 2nd to the 112th day after the end of the irradiation (partial or total).

Results of the Counting of Primordial Follicles in the 146 Ovaries studied

Table and Figure 1 summarise these results given as successive values of $\frac{Nt}{No}$ relations.

We find that:

1. After a single sterilising dose, disappearance of evoluting follicles is rapid and definite, while that of primordial follicles follows a simple exponential law specifically influenced by the factor of radiological destruction.

TABLE

Successive Values of $\frac{Nt}{No}$ relations

Days after irradiation	Doses of X-irradiation		
	2,500r (single)	1,050r (single)	1,050r (fractionated)
2	0·285	0·268	0·610 (after 210r)
4	0·045	0·089	0·417 (after 420r)
6	0·012	0·071	0·333 (after 630r)
8	0·003	0·066	0·269 (after 840r)
10	0·000	0·062	0·242 (2 days after 1,050r)
12	—	0·054	0·198 (4 days after 1,050r)
20	—	—	0·159 (12 days after 1,050r)
34-35	—	0·040	0·125 (27 days after 1,050r)
69	—	—	0·039 (61 days after 1,050r)
92	—	0·029	—
112	—	—	0·011 (104 days after 1,050r)

2. After a single non-sterilising dose, the curve of primordial follicles decreases first, in a similar manner to that of the preceding experiment; then from the 2nd day, this proportion grows considerably smaller owing to restoration.

3. After a fractionated non-sterilising dose, the ovaries, when examined ten days later, are found to be less damaged than after the same lapse of time and when the same dose had been given in a single irradiation; but the damage is more extensive than on the 2nd day following the single dose irradiation. Afterwards the number of follicles P in this third experiment decrease so that on about the 80th day the effect is greater than was obtained in experiment 2 above.

Short-term study of the phenomena observed in irradiated ovaries

The decrease in the number of primordial follicles in the irradiated ovary is expressed by the following equation:—

$$\frac{Nt}{No} = e^{-Qt} \qquad \text{......(1)}$$

where Q represents the algebraic sum of factors of disappearance (due to the action (q) of X-rays, to that (a) of atresia, and to that (p) of the passage

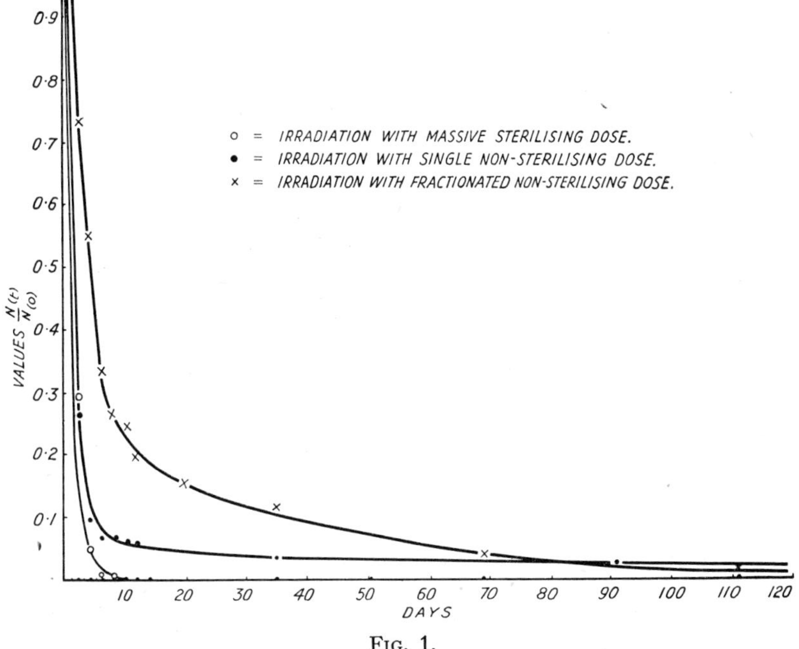

FIG. 1.

from stage P to the following stage) and the factors of recovery (due to post-radiological restoration (r) and to the eventual regeneration (g) by the germinative epitehlium).

As factors a, p and g interfere either in an accessory manner or not at all, we then have the following expression $\phi = -q + r$.

In experiment 1 (2,500r in one irradiation), $\phi = -q$ and equals $-0{\cdot}72 \pm 0{\cdot}06$.

In experiment 2 (1,050r in one irradiation), restoration is non-existent from zero time until the 2nd day, it becomes positive between the 2nd and 4th day, and acquires a constant value of 0·022 between the 4th and the 10th day.

In experiment 3 (1,050r in 5 fractions of 210r at two-day intervals), restoration is also non-existent from zero time until the 2nd day, equals 0·049 between the 2nd and the 4th day, and afterwards acquires a constant rate of 0·0738 until the 10th day.

We have checked analytically this latter value, and if we assume that the follicles P restored after one irradiation may again experience the effects of the following one, we have for (r), the following expression:

$$r = \frac{\dfrac{Nn}{No} - e^{-nqt}}{\dfrac{(n - e^{-(n+1)qt} - e^{-qt}) \cdot e^{-qt}}{e^{-qt} - 1}}$$

which, when applied to the results of the 3rd experiment, confers to r the value — 0·0723, close to the experimental value.

Long-term efficiency of massive non-sterilising irradiation or fractionated irradiation

If we express (1) as follows: $n_t = \frac{Nt}{No} e^{-\phi t}$ we obtain $\phi t = -\frac{1}{n_t} \cdot \frac{dn}{dt}$

$$dt = 2 \text{ days}, \; dn = -q + r \text{ and } \phi t = -\frac{1}{n_t} \cdot \frac{(-q + r)_{t,\,t+2}}{2}.$$

If we consider that factors relating to destruction and recuperation (q^{rel} and r^{rel}) equal half the ratio of the number of destroyed or recuperated follicles in an interval of say, two days, to the number of whole follicles at the beginning of this interval, we finally obtain:

$$\phi_t = -q^{rel}{}_{t=1} + r^{rel}{}_{t=1}$$

Our calculations show, as can be seen from Figures 2 and 3, that ϕ, in experiments 2 and 3, tends towards a balance between values 0·078 and 0·029 respectively, thus showing the greater efficiency of fractionated doses over an equal single dose.

Experimentally, with two groups of 8 ovaries irradiated with 1,050r and taken out from the 21st to the 241st day (for massive irradiation), and from the 20th to the 293rd day (for fractionated irradiation), we obtain the following mean figures for various follicular types:

1,050r (massive irradiation): 5,136 foll. P, 271 evoluting foll. (223A, 23B, 10C, 6D, 4E, 4F, 1G) and 7 special atresic forms GT and GH.

1,050r (fractionated irradiation): 3,274 foll. P, 115 evoluting foll. (91A, 10B, 5C, 3D, 2E, 2F, 2G) and 2 special atresic forms.

Thus it seems that in the ovary of a female rabbit, fractionation of a given non-sterilising dose of X-rays causes, in the long run, a greater destruction of primordial follicles.

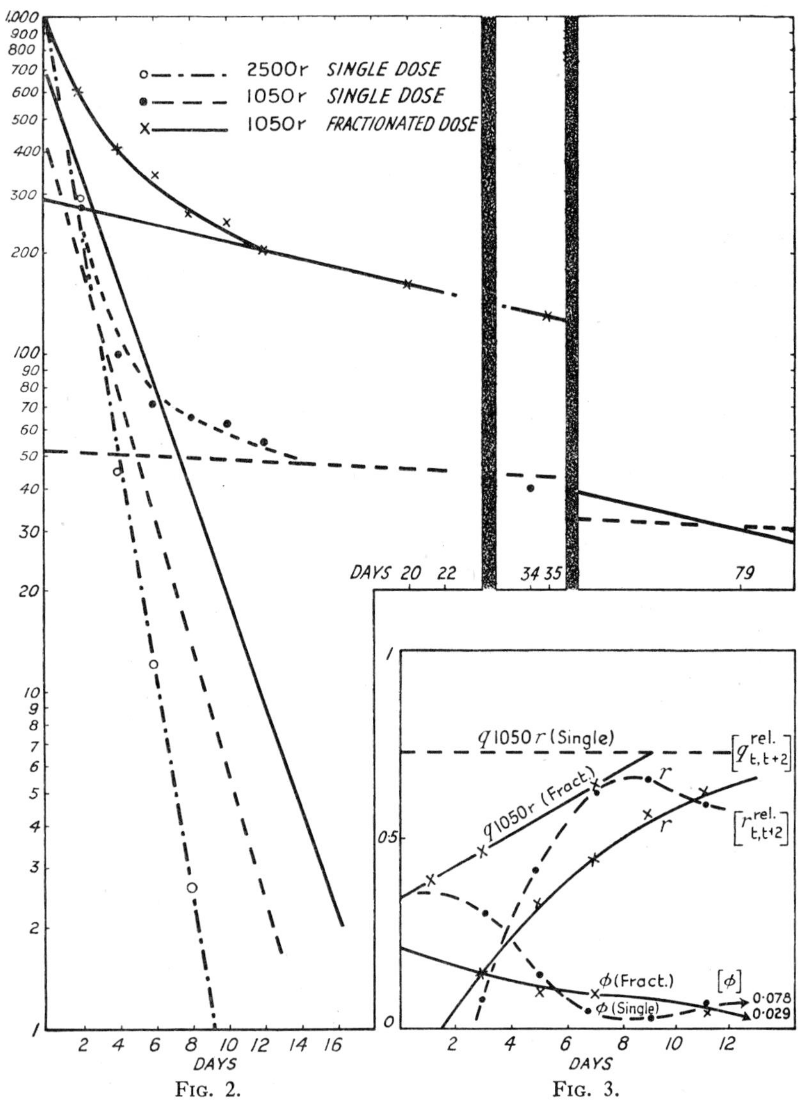

FIG. 2. FIG. 3.

Conclusions

As far as the ovary of the female rabbit is concerned, there exists an irradiation dose of 2,500r, by which the follicular apparatus of this organ is completely and definitely destroyed.

A non-sterilising dose of X-rays, 1,050r, is followed by restoration, the effects of which, in the long run, are undoubtedly more important than those following the same application of 1,050r, in 5 successive doses of

210r, at 48-hour intervals. This is the result of a considerable increase in radio-sensitivity of the primordial follicles in the days immediately following the first fractionated irradiation, an increase which markedly enhances the efficiency of the succeeding fractions.

Discussion

Ebert. Have you observed any hormonal or cytological effects in your irradiated ovaries and did the recovered follicles show any changes compared with normal ones?

Desaive. I have not studied the hormonal behaviour of the ovaries. On the cytological level, I have not noticed any important differences between normal and recovered follicles. But I have the impression that the number of Graafian follicles which are developed in the irradiated ovaries is slightly subnormal.

Kanazir. What is the probable explanation for the increased radio-sensitivity of primordial follicles following the first irradiation?

Desaive. I think that is due to the periodicity in the development of follicles. As I have shown in previous communications (*Arch. de Biologie*, 1940, 1948) this takes place at intervals of about 32 hours. Probably the radio-sensitivity increases when the evolution begins. Fractionated doses with intervals of 24-48 hours are probably 'hitting' a larger number of P follicles than a single dose, which may reach them in a stage of relative resistance.

'INDUCTION' BY LONG-TERM IRRADIATION IN LYSOGENIC BACTERIA

H. Marcovich and R. Latarjet
Laboratoire Pasteur de l'Institut du Radium, Paris, France

The study of the biological effects of ionising radiations delivered at low dose rates presents numerous difficulties. It is essential that during the long exposures required, the material is able to retain unchanged its biological properties, and to accumulate without loss the modifications brought about by the radiation. In this respect, bacteria are in some ways an ideal material since they do not alter their properties when kept over prolonged periods at low temperatures.

The results presented in this paper concern the irradiation over a period of several weeks of the lysogenic strain of bacteria *E. coli K* 12 (λ) kept in the frozen state.

When exposed to ultra-violet rays (Weigle and Delbruck, 1951), this strain of bacteria exhibits the phenomenon of induction discovered by Lwoff *et al.* (1950). Induction, which has been shown (Latarjet, 1951) also to occur under X-radiation, consists in the development of bacteriophage within the bacteria. This development is linked to the presence of a provirus in the genetic pattern of the cell. After a definite latent period, the bacteria undergo lysis, and the phage are liberated into the culture medium. In this condition, the number of free phage may be determined. As the burst-size (the number of phage liberated by a bacterium undergoing lysis) and the latent period, are independent of dose (Marcovich, 1956*a*) and of the energy of the absorbed photon (Marcovich, 1956*b*), the number of free phage in the medium is proportional to the number of lysed bacteria.

The sensitivity of this system as a biological detector of ionising radiation, is limited by the background of free phage liberated by the spontaneous lysis of the bacteria. This background can be decreased and the sensitivity of the system thus increased by the use of specific antiphage serum. If this serum is added during the latent period, and its activity suppressed just before bacterial lysis, one specifically inactivates those phage which are outside the bacteria before the induced lysis. Two other factors make it convenient to use this system as a 'biological dosimeter'. Firstly, the existence of a linear relation between radiation dose and biological response for doses under 1,000r. Secondly, the independence of the induction phenomenon to the presence in the medium of dissolved oxygen or of diverse protective agents.

If the system is kept at 0° C. in broth or inorganic buffer for several days, there is a significant decrease in the aptitude of the bacteria to induction. Long-term experiments in liquid medium are therefore unsatisfactory.

If the suspension of bacteria is frozen, in addition to the decrease in aptitude, a variable fraction of the bacteria is killed. If, however, prior to freezing a suspension of *E. coli K* 12 in tryptone broth or in synthetic medium, one adds glycerol at a concentration of 15 to 20% by volume, the bacteria may be kept at low temperatures (under — 20° C.) for a considerable period without harmful effects.

Method

A culture of *K* 12 (λ) in tryptone broth at the end of the exponential growth period and giving a titre of 3×10^9 bacteria per ml. was brought to 0° C. A quantity of glycerol equal to one-fifth of the volume total was then added and the temperature of the suspension slowly brought to — 40° C. The frozen bacteria were subsequently preserved in a refrigerator at — 25° C. Tubes containing such suspensions were placed (at — 25° C.) in the presence of one milligram of radium in conditions such that the dose rate received by the bacteria was 13r/day. After various periods of time the tubes were removed and their contents allowed to melt at 0° C. The phage was then titrated by the above-mentioned technique. It was found that the presence of glycerol lowered the sensitivity of the test by a factor of 2 or 3. This appears to be mainly due to modifications brought about in the latent period and lysis. Each irradiated sample was compared to a non-irradiated one tested at the same time.

Results

The relation obtained between the radiation dose and the biological effect is plotted in Figure 1 on log-log coordinates. The curve is linear with a slope slightly less than one. If, however, the experiment is carried out with short exposures in a liquid medium, one obtains a curve of slope equal to unity (Marcovich, 1954 and 1956*b*). The difference between these two results is at the limit of significance. That this difference is not due to the loss of a certain fraction of the induced bacteria is shown by the fact that a non-irradiated sample kept at —25° C. for 48 days, instead of being estimated immediately, gives the same result as a sample titrated without delay.

The induction of *K* 12 (λ) in the frozen state is independent of temperature between — 25° C. and — 76° C.

In this system physiological conditions appear to be brought to a standstill, and the radiochemical effects integrated.

The sensitivity of the induction test, however, appears to be insufficient to make it possible to detect within a reasonable period of time, the inducing effects of naturally occurring ionising radiations; and in particular of cosmic rays at sea level, even under conditions where the production of cascade showers are optimum.

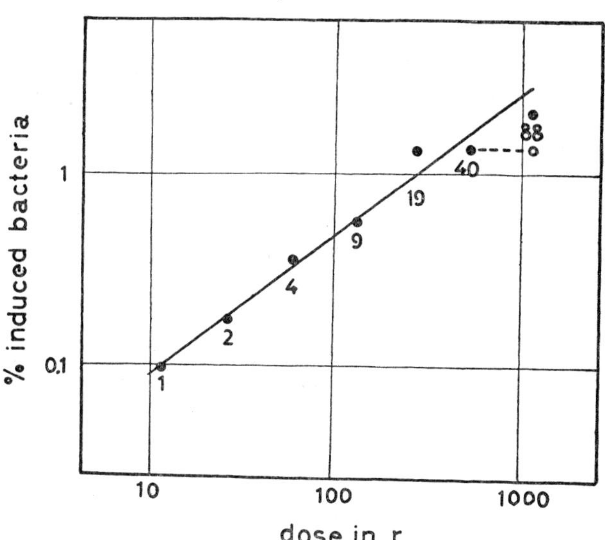

FIG. 1. Percentage of induced bacteria as a function of the dose. The open circle corresponds to a sample which has been isolated from the radium source and kept for 48 days at —25° C. before titration. Determinations at 1–88 days.

In spite of this, and due to the very large number of publications concerning the biological effects of cosmic rays it seems justifiable to carry out an experiment, with all controls, to attempt the detection of their action.

REFERENCES

LATARJET, R., 1951. *Ann. Inst. Pasteur*, **81**, 389.
LWOFF, A., SIMINOVITCH, L., and KJELDGAARD, N., 1950. *Ibid.*, **79**, 815.
MARCOVICH, H., 1954. *Nature, Lond.*, **174**, 796.
Idem, 1956*a*. *Ann. Inst. Pasteur*, **90**, 303.
Idem, 1956*b*. *Ibid.*, **90**, 458.
WEIGLE, J., and DELBRUCK, M., 1951. *J. Bacteriol.*, **62**, 301.

DISCUSSION

Hutchinson. How does the observed spontaneous rate of lysis in your material correlate with the rate which the natural radiation background would be expected to induce?

Marcovich. Spontaneous lysis of this lysogenic strain cannot be accounted for by natural occurring radiation by a factor of 100 or 1,000.

Troitsky. Is there any limit in the low-level doses of irradiation causing mutation in a single microbe cell?

Marcovich. It is possible to demonstrate that, with a dose of 10r, the probability that two electrons will hit the same cell is very low. Hence, in this particular case, one may say that there cannot be a threshold for the lowest possible dose.

Auerbach. How do you interpret the relationship between dose and effect? Do you think that the prophage represents a target to be hit?

Marcovich. I cannot tell what a target can be in the case of induction. The only thing that I feel able to say is that induction by X-ray results from the occurrence of a single event.

Sobels. It seems as if there is a certain correspondence between the induction of lysis in bacteria and the induction of a lethal mutation in higher organisms. Is it correct to make such a comparison?

Marcovich. From a formal point of view, induction of lysis in lysogenic bacteria can be considered as a lethal mutation since it is a lethal effect linked to the presence in the genetic pattern of a structure, the prophage, which is perpetuated from generation to generation. However, I don't think that the correspondence between lethals in higher organisms and induction is more than a matter of definition.

SECTION VII

STUDIES ON THE DISTRIBUTION OF ISOTOPES IN TISSUES

CHANGES IN THE RABBIT TIBIA AND AN ESTIMATION OF THE DOSE RECEIVED BY THE BONE TISSUE FOLLOWING A SINGLE INJECTION OF STRONTIUM

MAUREEN OWEN AND JANET VAUGHAN
Nuffield Department of Medicine, Oxford, England

Introduction

A preliminary report on lesions in the skeleton of rabbits given a single intravenous injection of ^{90}Sr (500–1,000 μc./kg.) at the age of 5–6 weeks and killed 6 months later, was given by Janet Vaughan at this meeting last year. The present report is a more detailed study of the retention of ^{90}Sr and the sites of radiation damage in the proximal half of the tibia of the same animals. As previously described (Vaughan and Jowsey, 1956) the majority of the strontium retained in the tibia six months after injection was associated with an abnormally thick bar of epiphyseal bone in the anterior wall at the level of the plate at the time of injection. It is shown how the presence of this bar can be accounted for in terms of (1) the pattern of ^{90}Sr uptake, (2) the pattern of bone growth and remodelling, and (3) the effects of radiation.

(1) The pattern of ^{90}Sr uptake

In these young animals the strontium is taken up rapidly in the actively calcifying tissues of the epiphyseal plate and to a lesser extent in the endosteal and periosteal tissues of the metaphysis and diaphysis. Figure 1*a* is a diagrammatic representation of a longitudinal section through the middle of the anterior wall and the opposite corner of the proximal half of the tibia of a 5–6-weeks-old rabbit. Figures 1*b* and 1*c* are diagrams of cross sections through the levels *x* and *y*. In both the longitudinal and cross sections areas of primary spongiosa (i.e. calcified cartilage matrix) secondary spongiosa (i.e. remnants of calcified cartilage matrix surrounded by bone), non-calcified cartilage and epiphysis are clearly marked. The anterior, posterior and internal walls are also indicated.

Autoradiographs of three representative sections corresponding to those in Figure 1, from a rabbit killed a few days after injection showed uptake of strontium in all areas of primary and secondary spongiosa, the heaviest reaction being in the primary spongiosa adjacent to the plate and decreasing with distance from the plate. Because of the shape of the epiphyseal plate the uptake of strontium at level *x* was confined almost entirely to the anterior wall;

at level y strontium was taken up throughout the three walls, the heaviest deposition being in the portion of primary spongiosa at corner IP, but the largest area of uptake was the anterior wall. To a lesser extent there was uptake in the endosteal and periosteal tissues of the metaphysis and diaphysis.

(2) The pattern of bone growth and remodelling

The shape of the normal rabbit tibia is such that the anterior surface is relatively flat and straight compared with the more curved posterior and internal surfaces. A previous study (Owen, Jowsey and Vaughan, 1955) of the growth of the tibia showed that this characteristic shape is maintained throughout the period of growth by resorption taking place unevenly on

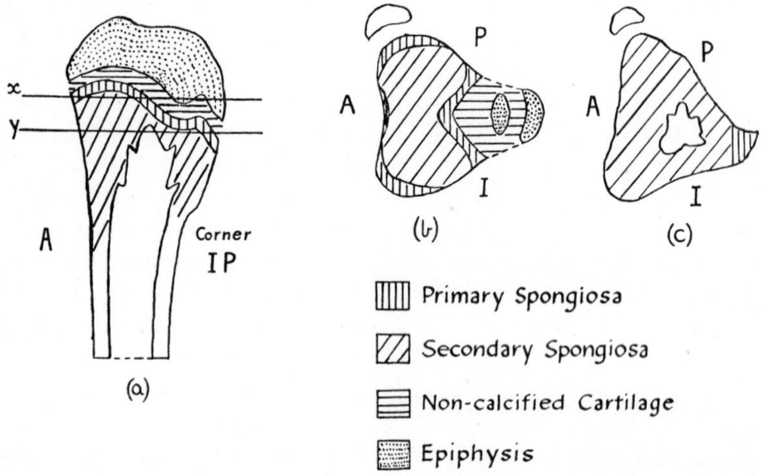

FIG. 1. (*a*) Longitudinal section through the middle of the anterior wall and the opposite corner of the proximal half of a rabbit tibia aged 5–6 weeks. (*b*) Cross section through x. (*c*) Cross section through *y*. A = anterior, P = posterior, I = internal.

the bone surfaces. In fact it was found that in the remodelling of the metaphysis, resorption of trabeculæ and deposition of lamellar bone on the endosteal surface are much more rapid on the posterior and internal walls than on the anterior wall. This is illustrated in Figure 8 which is a diagram of a longitudinal section, through the middle of the anterior wall and the opposite corner, of a 7-months-old rabbit superimposed on a similar section of a 5–6-weeks-old rabbit. Clearly there has been considerable resorption of the corner IP and the posterior and internal surfaces during the period of growth from 6 weeks to 7 months and less resorption of the anterior wall. Consequently in the normal 7-months-old rabbit tibia there were remains of epiphyseal bone in the anterior wall at the level xy (Fig. 8) whereas no epiphyseal remnants were found at this level in the posterior and internal walls. In rabbits injected at the age of 6 weeks and killed 6 months later,

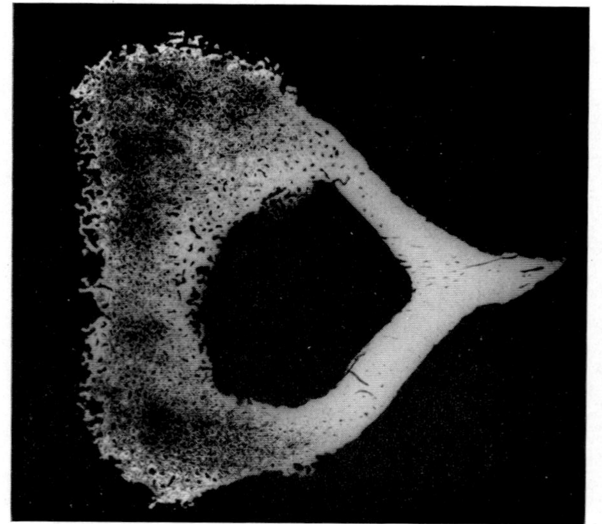

FIG. 2. Microradiograph of cross section of tibia from rabbit injected aged 5–6 weeks, killed 6 months later. Section is from level of plate at time of injection (region xy, Fig. 8). Note thickened anterior wall containing many remnants of calcified cartilage. $\times 8{\cdot}5$.

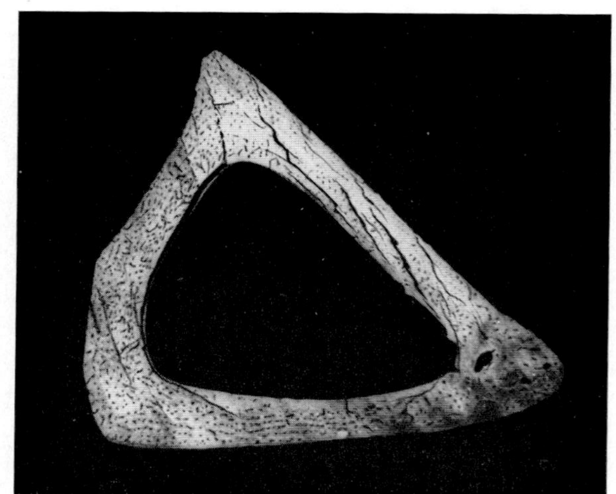

FIG. 3. Microradiograph of cross section of tibia from control rabbit aged 7 months. Section is from same level as that in Fig. 2. Compare appearance and shape of normal bone here with that in Fig. 2. $\times 8{\cdot}5$.

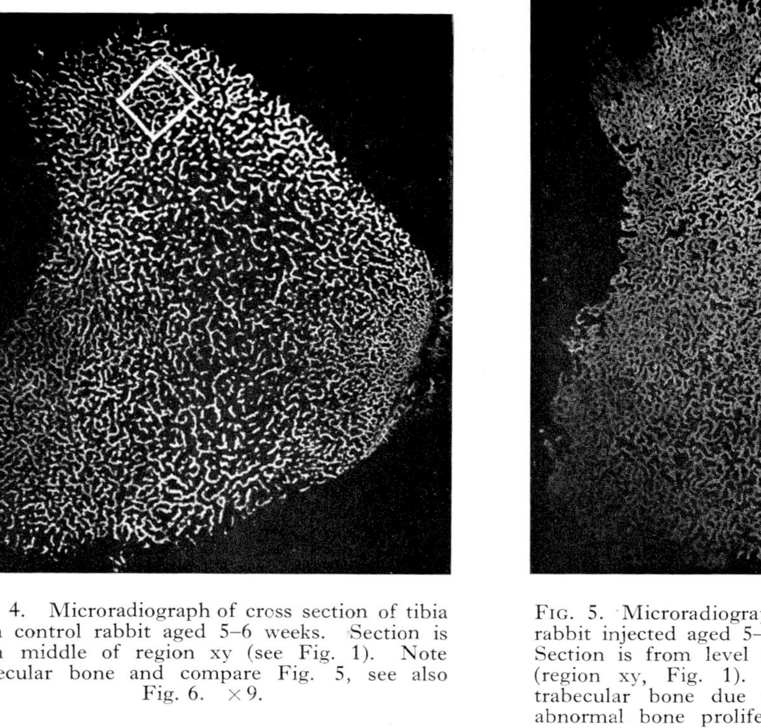

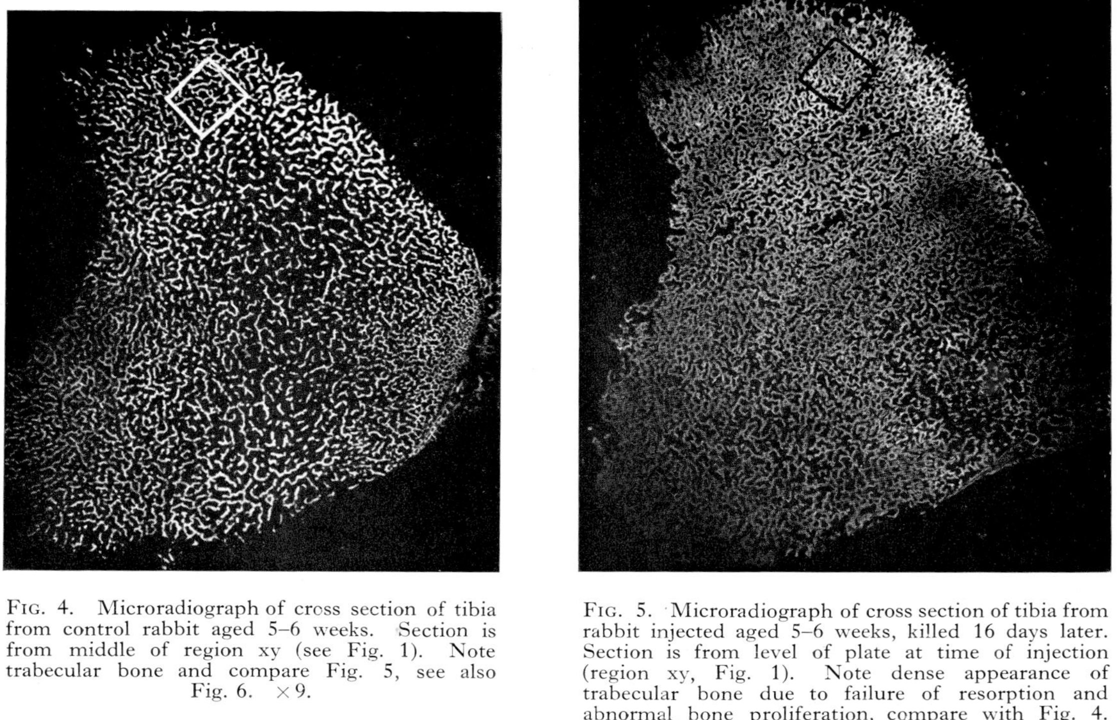

FIG. 4. Microradiograph of cross section of tibia from control rabbit aged 5–6 weeks. Section is from middle of region xy (see Fig. 1). Note trabecular bone and compare Fig. 5, see also Fig. 6. ×9.

FIG. 5. Microradiograph of cross section of tibia from rabbit injected aged 5–6 weeks, killed 16 days later. Section is from level of plate at time of injection (region xy, Fig. 1). Note dense appearance of trabecular bone due to failure of resorption and abnormal bone proliferation, compare with Fig. 4. see also Fig. 7. ×9.

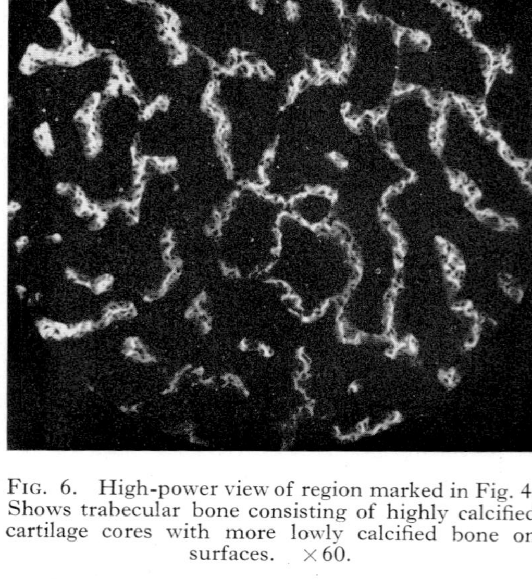

FIG. 6. High-power view of region marked in Fig. 4. Shows trabecular bone consisting of highly calcified cartilage cores with more lowly calcified bone on surfaces. ×60.

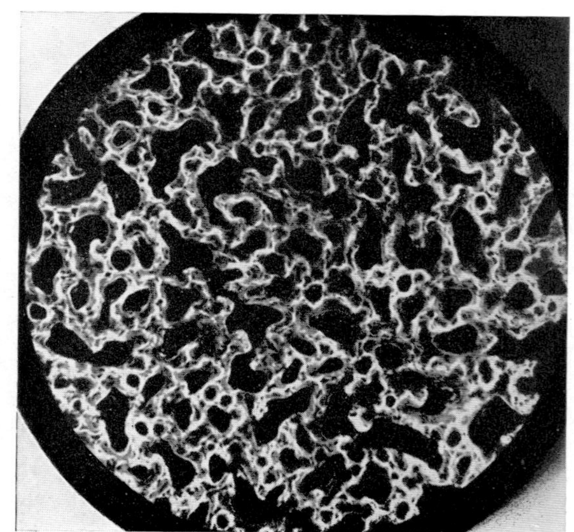

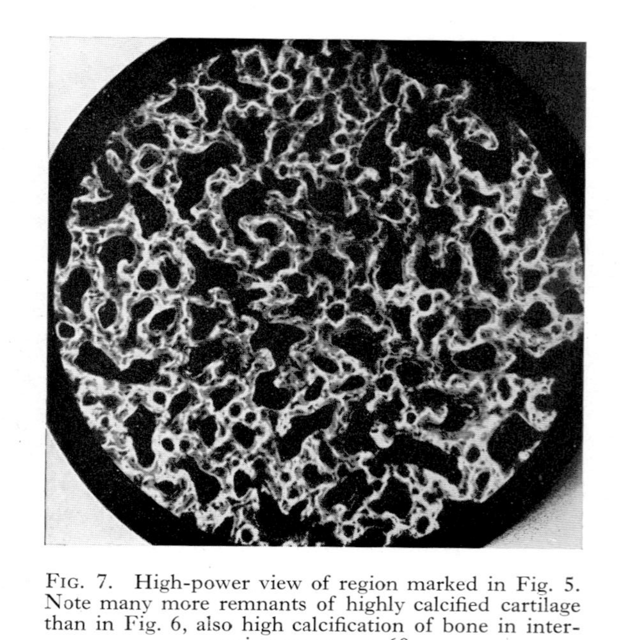

FIG. 7. High-power view of region marked in Fig. 5. Note many more remnants of highly calcified cartilage than in Fig. 6, also high calcification of bone in intervening spaces. ×60.

it is not surprising then, that remnants of epiphyseal bone containing strontium should be found only in the anterior wall at the level of the plate at the time of injection.

(3) The effects of radiation

Our results obtained with 8 rabbits given a single high dose (500–1,000 μc./kg.) at the age of 5–6 weeks and killed 6 months later showed, that in all cases though to a varying degree, there was a gross excess of cartilage remnants and epiphyseal bone containing strontium, constituting a thickened bar in the anterior wall at the level of the plate at the time of injection. Few remains of epiphyseal bone were found in the posterior and internal walls though the shape of the bone was deformed to some extent. A typical example is shown in Figure 2 which can be compared with a control section from the same level, Figure 3. In addition to the excess cartilage remnants in the anterior wall (indicating failure of resorption in this wall) there was intense proliferation bone on the surfaces of the unresorbed trabeculæ. We have evidence that this abnormal proliferation takes place very soon after the uptake of the strontium. A microradiograph of a cross section at about level *x* (see Fig. 1) of a 5–6-weeks-old control rabbit is shown in Figure 4. Figure 5 is a microradiograph of a cross section from the same level of a rabbit given 1,000 μc./kg. and killed 16 days later. High-power photographs of the regions marked are shown in Figures 6 and 7. The irradiated bone (Fig. 7), retains many more remnants of calcified cartilage and the intervening spaces are filled with abnormal type Haversian bone which often has highly calcified rims. The type of bone shown in Figure 7 is similar to that in the bar of the anterior wall of an animal killed 6 months after injection (Fig. 3), and it is this bone which is the site of tumour development in several of the animals.

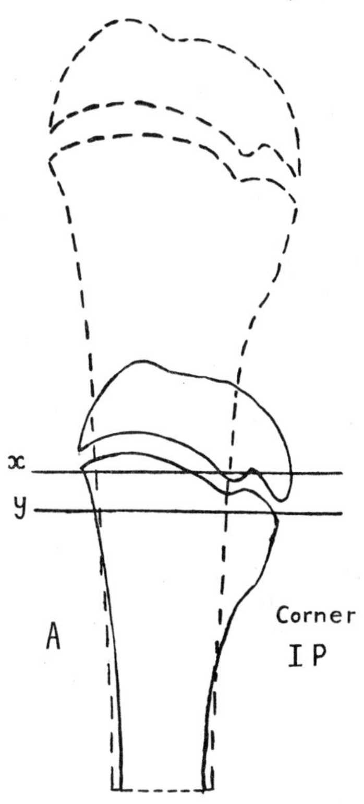

FIG. 8. Longitudinal section through the middle of the anterior wall and the opposite corner of the proximal half of a rabbit tibia, – – – – aged 7 months, ——— aged 5–6 weeks.

T

To summarise: The selective retention of strontium in association with a thickened bar in the anterior wall, can be attributed to the following: the pattern of normal growth is such that more strontium is originally taken up in the anterior wall and resorption is normally slower there. Rapid resorption on the posterior and internal walls removes the strontium before gross damage is done. On the anterior wall, where deposition is anyhow heavy, the slow resorption allows radiation damage to occur, as evidenced by failure of resorption and abnormal bone proliferation resulting in the bar described above.

It should, however, be mentioned that rabbits given a smaller dose, i.e. 100 μc./kg., did not show the same signs of gross radiation damage.

Dosage measurements

An estimate of the dose received by the tissues in various parts of the tibia has been made using a technique similar to that developed by Sinclair and others for the measurement of dosage distribution due to uptake of iodine by the thyroid (Sinclair, Abbatt, Farran, Harriss and Lamerton, 1956). Contact autoradiographs were obtained of plane surfaces cut transversely through the tibia in the region where it was desired to measure the tissue dosage. The thickness of the bone specimen was greater than the range of the β-particles from the ^{90}Strontium or its daughter product Yttrium so that the distribution of the blackening in the autoradiograph corresponded to the total dose from both isotopes received by tissues in the plane of the film. The radiation dose was estimated by comparing the blackening of the autoradiograph with that from a known ^{90}Strontium standard.

Measurements of the dose received by different parts of the tibia have been made for two rabbits injected at the age of 5–6 weeks and killed 16 days later, the injected doses given were 1,000 μc./kg. and 100 μc./kg. respectively. The results are shown in the Table.

TABLE

Maximum Doses to Parts of Tibia in Rabbits Injected at the Age of 5–6 Weeks and Killed 16 Days Later

Initial injection to rabbit	Maximum dose, rads/hour, in region of plate at time of injection	Maximum dose, rads/hour, in mid-shaft due to uptake on periosteal surface
1,000 μc./kg.	220 rads/hour	44 rads/hour
100 μc./kg.	3 rads/hour	5·3 rads/hour

It will be noted that the maximum doses received in the mid-shaft due to uptake on the periosteal surface are approximately in the ratio of the injected doses. The maximum doses received by the tissues in the region of the plate at the time of injection are relatively much higher in the animal given 1,000 μc./kg. This is a consequence of the great failure of resorption of the primary spongiosa in this animal. It is not surprising, therefore, that in rabbits given these high doses extensive damage is done in this region.

In a previous paper (Jowsey, Rayner, Tutt and Vaughan, 1953) radiochemical measurements of the ^{90}Strontium retained at different times after injection in the 'ends' and diaphysis of the tibias of 5–6-week-old rabbits have been reported. Using this data and assuming a uniform distribution of the isotope per gram of bone the average dose received by the 'ends' and diaphysis in rabbits killed 16 days after injection of 100 μc./kg., have been calculated to be 1·7 rads/hour and 2·2 rads/hour respectively. The results reported here show that the localised dose to parts of the bone is somewhat greater than this.

Summary

Rabbits were given a single intravenous injection of strontium at the age of 5–6 weeks and killed 6 months later. The main site of retention of the ^{90}Strontium and radiation damage in the proximal half of the tibia was found to be the anterior wall at the level of the plate at the time of injection. There was gross excess of unresorbed epiphyseal bone in this wall with the highest doses, though not in the other two walls. This can be explained in terms of the pattern of ^{90}Sr uptake, the pattern of growth of the tibia and the effects of radiation.

Measurements of the maximum doses received in various parts of the bone show that they are greater than those calculated assuming a uniform distribution of the isotope per gram of bone.

Acknowledgments

This work was begun on behalf of the Protection Subcommittee of the Medical Research Council's Committee on Medical and Biological Applications of Nuclear Physics. Expenses were met from a personal grant to Janet Vaughan.

REFERENCES

Jowsey, J., Rayner, B., Tutt, M., and Vaughan, J., 1953. *Brit. J. exp. Path.*, **34**, 384.

Owen, M., Jowsey, J., and Vaughan, J., 1955. *J. Bone Jt. Surg.*, **37**-B, 324.

Sinclair, W. K., Abbatt, J. D., Farran, H. E. A., Harriss, E. B., and Lamerton, L. F., 1956. *Brit. J. Radiol.*, **29**, 36.

Vaughan, J., and Jowsey, J., 1955. *Progress in Radiobiol.*, 1956, p. 429 (Edinburgh, Oliver and Boyd).

Discussion

Williams. I should like to ask Dr. Owen if she can tell us anything of the pattern of uptake of strontium and of the bone growth in the flat bones, as I believe that multiple skull tumours have been seen in weanling rabbits previously injected with high doses of strontium? Could she tell us whether the pattern of uptake is similar when adult rabbits are injected and whether there is any difference in the physiological behaviour of old bone?

Owen. It is true that multiple skull tumours have been seen in the rabbits referred to in the present experiments. Unfortunately we do not know anything about the pattern of bone growth and hence the pattern of uptake of strontium in the skulls of rabbits of this age. The only bone whose pattern of growth we have studied in detail is the upper half of the tibia. Here the pattern of strontium uptake in adult bone is quite different to that of the young actively growing bone.

Lajtha. I am very interested in this problem of bone proliferation due to irradiation from the radioactive strontium. Are you sure that the answer is an active proliferation and not only a faulty resorption of cartilage with consequent bone deposit on it? The tumour seems to grow in the abnormal bone, far away from the dense strontium deposit. I find it very difficult to believe that there can be active cellular proliferation under a shower of 1,000 rads per hour, lasting for weeks. Can you present evidence that active growth occurs under such conditions?

Owen. We have no direct evidence that there is active proliferation of bone (that is, osteoblastic activity in association with bone deposition) occurring in conjunction with these heavy doses of radiation. There is certainly failure of resorption of the cartilage remnants in this region, and on the surfaces of this unresorbed cartilage there is what we have called 'an abnormal amount of bone proliferation'. It would be perhaps more accurate at this stage to call this 'an abnormal amount of calcification'. Our only evidence at present comes from microradiograpps. On these, regions of abnormal calcification are clearly seen, and in these regions the cells are arranged in the pattern usual for bone, concentrically around the vessels. A further point is that in the rabbits which died at 6 months, there has been considerable growth in length after injection. In the near future we do, however, plan to do detailed histological studies on similar material.

Smith. It is surprising that the remodelling process in the posterior portion of the head of the bone proceeds normally, since in this region of the plate large amounts of strontium are deposited and the dose delivered (1,000 rads per hour) is very high.

Owen. The remodelling of the posterior and internal walls is not completely normal. There is, for example, considerable deformation of the bone shape. One can say, however, that the process of resorption of the cartilage remnants is less inhibited in these walls than in the anterior wall. This is consistent with the normal pattern of growth of the tibia where resorption takes place more rapidly on the posterior and internal walls than on the anterior wall.

Mackay. Has Dr. Owen any information on the histology of the tumours produced? In particular were all the tumours of one uniform histological type or were tumours of different histological types produced in different bones?

Owen. Many of the tumours have been examined histologically by Dr. Sissons. In all except one they were osteosarcomas. One skull tumour also showed formation of tumour cartilage.

Abbatt. Do you have any information about the response of the hæmopoietic system during the period of your observations and when the tumour developed? If so, what is the nature of these changes?

Owen. Shortly before death all the animals developed considerable anæmia and leucopenia. There were no hæmorrhagic manifestations and no evidence of leukæmia. The marrow was almost invariably gelatinous.

English. With regard to recording the growth of bone as distinct from remodelling and absorption I would call attention to the technique of Björk of Copenhagen, Tandläkarhögskolan. Small stainless steel pegs about 0·2 mm. diameter, cut in 0·8 mm. lengths, are driven into bone at various positions at the beginning of the experiment. A series of roentgenograms will reveal the relative distances between markers and show where maximum growth has occurred in terms of deposition of new bone as distinct from remodelling changes. These pegs are driven into the bone by means of a cylindrical tube containing a plunger slightly larger than the wire at its tip, though more solid throughout its length, thereby enabling it to stand the blow of a mallet required to drive the peg into the bone.

BONE AND RADIOSTRONTIUM

An Attempt to Evaluate the Dose Distribution

R. BJÖRNERSTEDT, C.-J. CLEMEDSON, A. ENGSTRÖM AND A. NELSON

Department of Medical Physics, Karolinska Institute, Stockholm, and Research Institute of National Defence, Stockholm, Sweden

The hydroxyapatite in the osseous tissues has marked ionic exchange properties. Elements belonging to the same category as calcium are especially easily incorporated into bone. For example, the strontium isotopes have been shown to localise in certain parts of bone tissue which is of great importance when considering the radioactive Sr-isotopes (^{90}Sr, ^{89}Sr) from fission products.

X-ray microscopic investigations of bone tissue have shown that the mineral salts are distributed in an uneven manner. This non-uniform distribution of mineral salts seems to be a result of the continuous rebuilding going on in bone tissue. Thus, in spongy bone we find a great percentage of bones tructures (e.g. Haversian systems) which have a relatively lower content of mineral salts than are found in the compact bone, although the lowly mineralised structures are frequent there too. Autoradiographic studies using ^{45}Ca, ^{32}P, $^{35}SO_4$, ^{90}Sr and others have shown that the radioactive substances become localised to these lowly mineralised structures which thus may act as some sort of ionic exchange centres in the bone tissue. When an isotope like ^{90}Sr, therefore, is introduced into the organism it becomes localised in the skeleton in these lowly mineralised centres possessing ion exchange properties.

In order to clarify the molecular structure of the bone salt, a series of investigations has been undertaken. Thus, the scattering of X-rays at high and low angles has been extensively studied. It has been shown that the bone salt can, without any doubt be classified as hydroxyapatite. Already the wide angles diffraction pattern indicates that the apatite crystallites are elongated and that the C axes of the crystallographic unit cells are arranged parallel to the collagen fibres. The sharpening of reflections having index 001 indicates that the crystallite is elongated in the C direction and profile analyses of the 002 line reflection gives a length of about 200 Å for the crystallite. The low angle particle scatter from oriented bone specimens also shows that the long dimension of the apatite particles lies parallel with the collagen and from that type of scatter a particle dimension of 200×50 Å has been obtained.

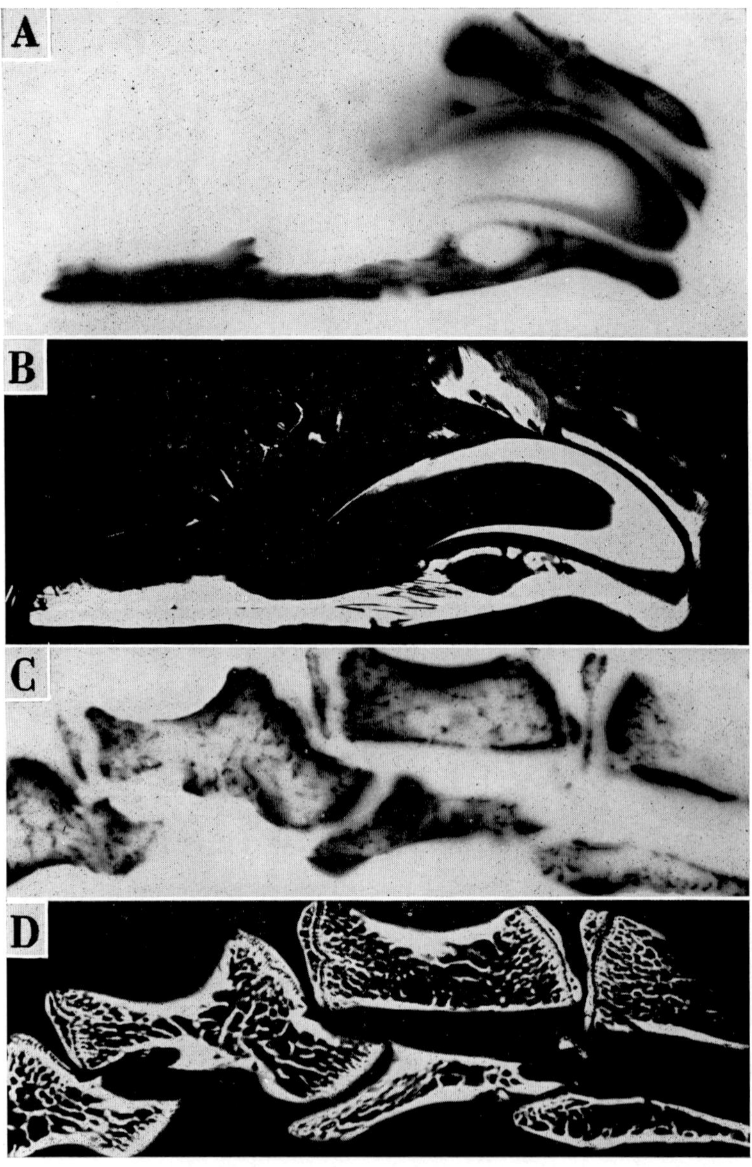

Fig. 1. A and C: Autoradiograms of thin ground sections showing the distribution of ^{90}Sr. A, incisor; C, vertebræ.

Fig. 1. B and D: Microradiograms of the same sections showing the distribution of minerals. B, incisor ; D, vertebræ.

Recent electron-microscopic investigations have also shown that in mature bone the apatite has the shape of rods or filaments, precisely aligned along the collagen fibres. Thus, there seems to be a good agreement between the results of electron-microscopy and X-ray diffraction studies at high and low angles.

The small apatite crystallites have a large surface area which is about 130 square metres per gram for the mature particles. These particles can thus exhibit a large surface area having different patterns of charge, suitable for various types of exchange processes. Although no significant differences

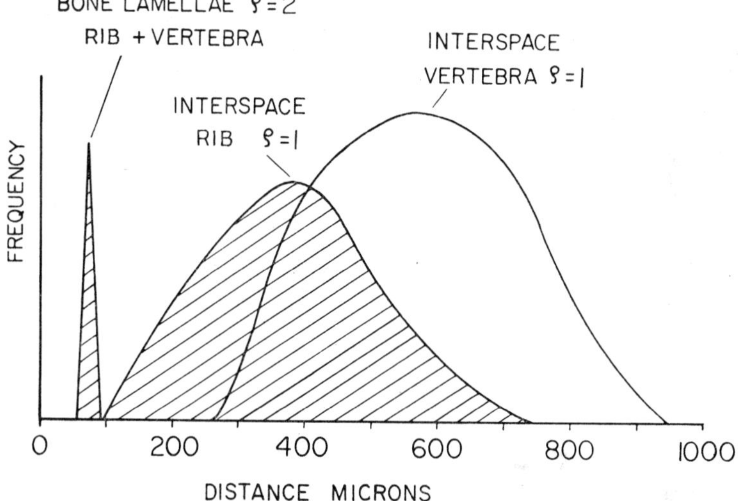

FIG. 2. Dimensions of bone lamellæ and marrow interspaces of human ribs and vertebræ.

have yet been obtained between the X-ray scattering from lowly and fully mineralised areas, it may be supposed that in the lowly mineralised areas the great majority of particles have sizes still smaller than those just mentioned. This has been clearly shown in electron-microscopy of embryonic bone, where small spots of apatite particles precisely arranged on the collagen fibres are found.

The osseous distribution of ^{90}Sr administered to rats has been studied by combined autoradiographic and X-ray microscopic techniques. Figure 1 shows autoradiograms and microradiograms of dental and osseous tissues. A very high uptake of radiostrontium is seen in the rapidly growing incisor (Fig. 1A). In general, structures with a low degree of mineralisation show a high uptake of ^{90}Sr. Thus, the radioactivity has an uneven distribution and this has been correlated with a histological three-dimensional reconstruction of bone tissue and marrow, in order to calculate the dose distribution.

For these calculations we have assumed models of spongy bone (human), consisting of 70 microns thick bone lamellæ of infinite lateral extension, interspersed with bone marrow of thicknesses 140, 280, 420, 560, 700 and 840 microns. Density of bone and marrow was chosen as 2 g./cm.$^{-3}$ and 1 g./cm.$^{-3}$, respectively. A beta-emitting isotope with maximum energy

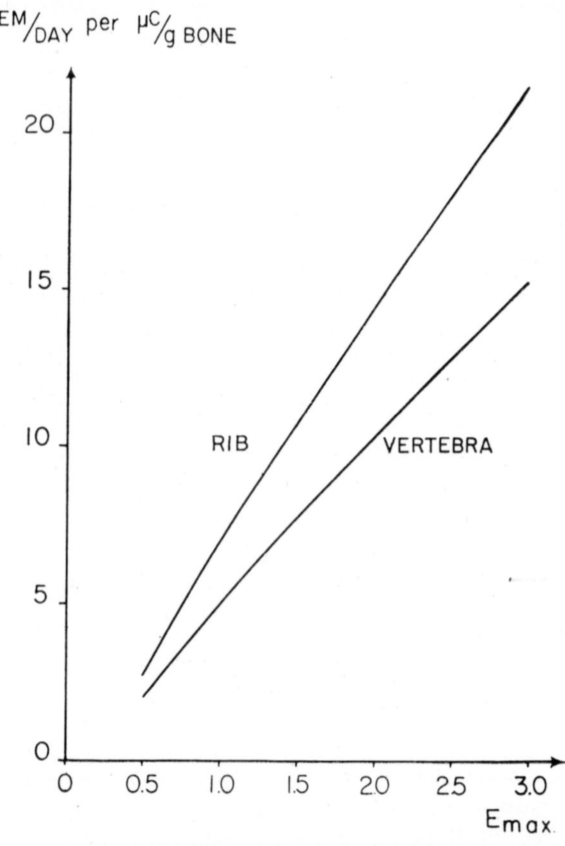

FIG. 3. Radiation dose in bone marrow.

between 0 and 3·0 MeV is considered to be homogeneously distributed in these microscopic bone lamellæ. Taken as a whole therefore, the bone tissue has an uneven distribution of radioisotopes. In the microscopic structures of the order of 10-100 microns, however, the isotope was found to be relatively homogeneously distributed. The dose distribution in both bone and marrow is calculated by successive integration over monoenergetic electrons with constant range, range distribution and having a spectrum

$$P(E) = E(E_{max} - E)^2 . F(E),$$

where F(E) has been approximated to F(E) = kE. The contributions from all bone lamellæ within the maximum range give the dose distribution for the various models chosen, for example, 70 microns bone interspersed with 560 microns of marrow, etc. The dose distribution has been calculated for 6 models and with the following values of E_{max}: 0·5, 1·0, 2·0, 2·5 and 3·0 MeV. These six models have been correlated with the histological investigations (Fig. 2).

A graphical integration gives the radiation dose in the bone marrow as a function of E_{max} as presented in Figure 3. The diagram is valid also for isotopes other than ^{90}Sr and also for mixtures of isotopes. The results have also been correlated with the autoradiographic and X-ray microscopical investigations.

This study is being continued and it is hoped that it will be possible to obtain information on the actual magnitude of the radiation dose received by the bone marrow and to establish the body burden and the maximum permissible concentration of bone-seeking isotopes.

REFERENCES

Carlström, D., 1955. *Acta Radiol. Stockh.*, Suppl. 121.

Carlström, D., and Engström, A., 1956. *Biochemistry and Physiology of Bone*, Chap. VI.

Engfeldt, B., Björnerstedt, R., Clemedson, C.-J., and Engström, A., 1954, *Acta Orthopæd. Scand.*, **24**, 101.

Fernández-Morán, H., and Engström, A. *Biochim. Biophys. Acta*, in press.

DECONTAMINATION STUDIES

A Simple *in vitro* System to Study the Interaction of Radioactive Metals with Proteins from Body Fluids and Tissues

RUTH LEWIN, BETTY ROSOFF, HIRAM E. HART, KURT G. STERN * AND DANIEL LASZLO
Division of Neoplastic Diseases, Montefiore Hospital, New York, N.Y., U.S.A.

The increasing utilisation of radioactive materials for industrial as well as medical purposes calls for a consideration of the hazards due to accidental contamination. It has become essential to direct our efforts towards the problem of removing hazardous radioactive metals from the body. With this end in view the metabolism of various radioactive metals, including strontium, is being investigated in our laboratory. Since we have proposed to use ^{90}Y therapeutically for intrapleural and intraperitoneal application in patients with cancerous effusions, Lewin *et al.* (1954), it was of primary importance to provide means for removing this material from the body should it be necessary to do so, after the injection.

The search for effective means of decontamination can be approached in several ways. One consists in screening a large number of compounds which are expected to combine with the contaminating metal, compounds which should not be toxic to the body and which can be excreted easily. The other approach involves the investigation of the fate of the metal in the body on a molecular level, and the elucidation of the physico-chemical form in which the body transports and finally fixes it in the organs of localisation. Such studies may lead to the design of more effective decontaminating agents.

We have been pursuing both lines of research and the purpose of this paper is (1) to present data on the interaction of yttrium with a purified serum protein, (2) to illustrate a simple *in vitro* system for testing the relative metal-binding effectiveness of various compounds in the presence of body constituents which are also metal-binders, and (3) to compare these results with the decontaminating effectiveness of a chelating agent in a human subject.

Methods

The equilibrium dialysis system has been described by Klotz (1953). The protein employed was crystalline bovine albumin of which a 3% solution in physiological saline was prepared. One ml. of the albumin

* Deceased February 1956.

solution was placed inside a cellophane casing and was immersed in a fixed volume of $^{90}YCl_3$ solution. After attainment of equilibrium the amount of yttrium bound by albumin was determined.

The filtration method here adopted has been described by McFarland and Hein (1953). In this case the $^{90}YCl_3$ solution was added to the protein solution. Five ml. of the mixture were transferred into cellophane casing which was fastened on to a stopper. This assembly was placed into a conical

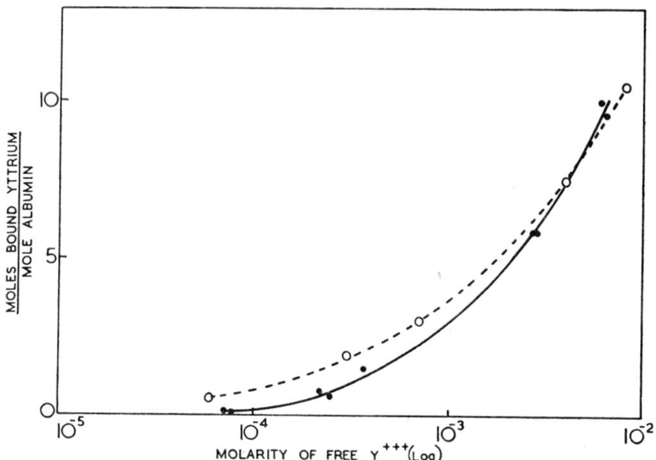

FIG. 1. Binding of yttrium ions by bovine serum albumin as studied by pressure filtration (●——●) and equilibrium dialysis (○ - - - ○) at p H6.

tube and centrifuged at 600 g. for 10 minutes. From the ^{90}Y activity in the mixture and in the filtrate the amount of diffusible and bound yttrium could be calculated.

Results

1. The interaction of serum albumin with ^{90}Y when present as the chloride, was studied at various metal ion concentrations and at two pH levels. The molecular weight of albumin was assumed to be 69,000 and the results were expressed in terms of moles of yttrium bound per mole of albumin. These quantities were plotted against the molar concentrations of free yttrium ion at equilibrium. As can be seen from Figure 1 in the concentration range studied, the greater the concentration of free yttrium, the more yttrium is bound per mole of albumin. At pH 4 less metal was bound than at pH 6 (Fig. 2) which may be attributed to the ionisation of different metal-binding groups on the albumin molecule at the different pH levels. At concentrations of 10^{-4} M this binding was reversible if the concentration of free yttrium ion in the system was decreased.

Although albumin is capable of combining with considerable amounts of yttrium this binding is governed by the amount of free ion in the system and seems subject to the law of mass action. When we express these results as the percentage of yttrium which is unbound we find that about 50–70% of the total metal remains diffusible in the presence of albumin.

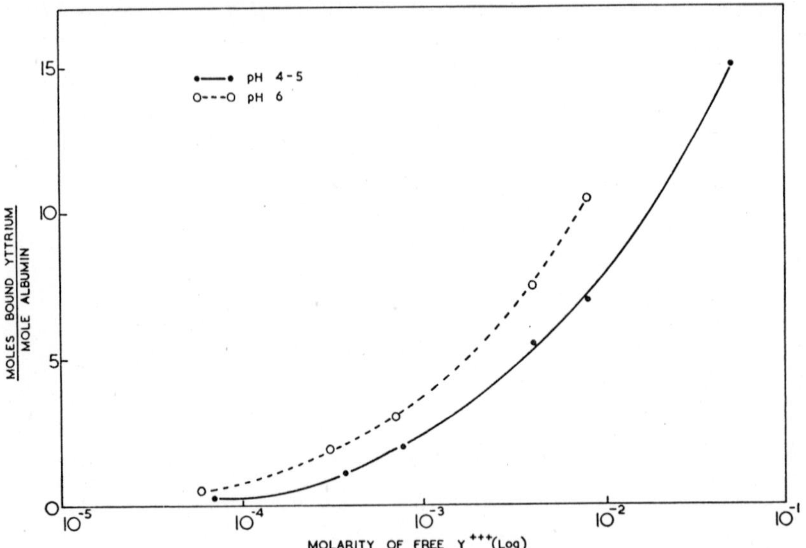

FIG. 2. Binding of yttrium ions by bovine serum albumin. (Equilibrium dialysis).

2. In contrast to the results found with albumin, $^{90}YCl_3$ when added to serum becomes almost completely non-filterable at salt concentrations from 0·0001 to 0·01 M (Table I). It can therefore be concluded that the yttrium binding in serum is due to components other than albumin. Whether we are dealing with a metal-binding protein, such as the iron and copper-binding components of serum, remains to be investigated further. Preliminary paper electrophoresis experiments with mixtures of carrier-free $^{91}YCl_3$ and serum were performed and the paper strips were assayed for ^{91}Y. In addition to a certain amount of radioactivity that remained at the point of application and a distant fraction that moved faster than the albumin component, ^{91}Y was found at a distance corresponding to the general range of globulin mobilities, and was not associated with a particular globulin component. Control strips showed activity only at the point of application and at the distance of a fast moving fraction.

Although the mechanism of interaction of YCl_3 and serum constituents has to be studied further, whole serum was found useful for testing the relative strength of various yttrium-chelates, and these experiments will be considered next.

The following yttrium-chelates were tested for their capacity to keep the metal in a diffusible state: isopropylenediaminetetraacetate (IPDTA),

TABLE I

$^{90}Y\ Cl_3$ interaction with Serum

Experiment no.	Yttrium conc. M	Per cent. filterable
	SERUM	
1	10^{-4}	0
2		0
3	10^{-3}	1·5
4		2·5
5	10^{-2}	1·3
6		2·6
	SALINE CONTROL	
7	10^{-4}	93
8	10^{-3}	96
9		99

TABLE II

Effectiveness of ^{90}Y-Chelates

Compound	Stability constant (Log K_1)	Per cent. filterable
Y-Chelate		
Y-IPDTA	14–5	95·6
Y-EDTA	14·2	88·7
Y-EDTA-OL	10–11	44·2
Y-NTA	10·9	1·7
Y-EDG	8·6	1·0
Y-IDA	6·8	0·1
Y-Fe^3 spec.	5·2	0·2
Ionised yttrium		
YCl_3		0·1

ethylenediaminetetraacetate (EDTA), N-hydroxyethylethylenediaminetri-acetate (EDTA-OL), nitrilotriacetate (NTA), beta-hydroxyethyliminodi-acetate (EDG), iminodiacetate (IDA) and N,N′-dihydroxyethylglycine (Fe^3 specific). The stability constants of these chelates were determined and can be compared with the filterability of the compounds (Table II). As expected, compounds with high stability constants showed the greatest filterability, while compounds with stability constants of less than 10 were practically non-filterable. However, two compounds with similar stability constants, namely Y-EDTA-OL and Y-NTA, behaved differently with regard to filterability.

A factor affecting the binding of a metal under these conditions is the sequence in which the various reactants are added. If $^{90}YCL_3$ is mixed with serum and thereafter IPDTA is added in a tenfold excess to the amount required stoichiometrically, only 40% of the metal becomes filterable as compared to 95% when ^{90}Y is added as Y-IPDTA.

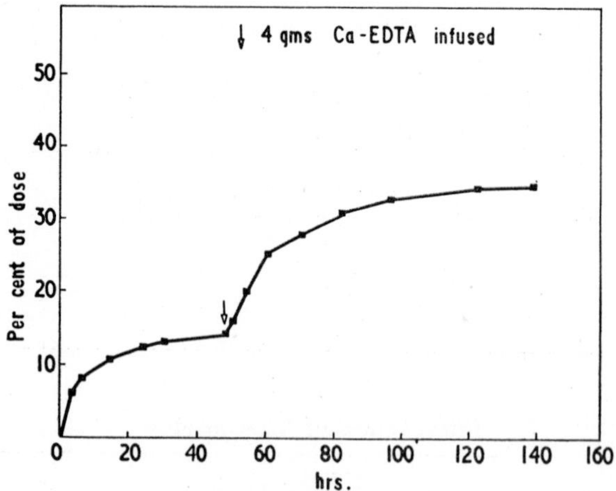

FIG. 3. Cumulative urinary excretion following intravenous injection of $^{90}YCl_3$

3. Finally, one of the effective chelates, EDTA, which had been proven to be non-toxic especially when injected as the calcium salt, di-Na-Ca-EDTA, was tested in man. A patient was injected intravenously with a tracer dose of $^{90}YCl_3$. As can be seen from Figure 3 the total cumulative urinary excretion of ^{90}Y 48 hours after the injection was about 15% of the dose. The rate of excretion had levelled off and the amount on the blood was practically zero. Blood levels after $^{90}YCl_3$ injections have been reported previously, Hart *et al.* (1955). At 48 hours a total of 4 gm. Ca-EDTA was infused over a 4-hour period. A rapid increase in the rate of excretion of ^{90}Y was observed and the total cumulative urinary excretion was brought up to 35% during the next 48-hour period.

It is hoped that these data will encourage further work on decontamination procedures. As more compounds will be made available to us by the chemists and as metals other than yttrium will have to be further investigated in regards to their removability from the body, a simple system as described above may be of help to decide which compounds deserve further and more time-consuming investigation. It should also be interesting to extend these experiments to tissue fractions isolated from organs which fix the metals such as liver and bone.

Acknowledgment

This work was carried out under Contract No. AT(30–1)–1551 with the United States Atomic Energy Commission.

REFERENCES

HART, H. E., GREENBERG, J., LEWIN, R., SPENCER, H., STERN, K. G., and LASZLO, D., 1955. *J. Lab. clin. Med.*, **46**, 182.

KLOTZ, I. M., 1953. *Protein Interactions in the Proteins* (New York, Academic Press, Inc.), Vol. I.

LEWIN, R., HART, H. E., GREENBERG, J., SPENCER, H., STERN, K. G., and LASZLO, D., 1954, *J. nat. Cancer Inst.*, **15**, 131.

MCFARLAND, R. H., and HEIN, R. E., *U.S.A.E.C. Report*, April 1953, TID-5098, p. 27.

DISCUSSION

Carlqvist. Dr. Lewin, have you investigated other fission metals, especially divalent ones such as Sr ? In general divalent ions are less strongly bound by chelating agents than metals of higher valency. Furthermore, the complexes of Sr and Ba are weaker than the corresponding Ca-complexes.

Lewin. Dr. Laszlo and his co-workers are actively engaged in studying radio-strontium metabolism in man. With Sr they are using a different approach. It has been possible to enhance the excretion of carrier-free strontium by giving large intravenous doses of Ca as Ca-gluconate. Another possibility which is under investigation is the use of catabolic hormones.

Elson. How specific are these chelating agents for yttrium, do they remove any other metals, e.g. calcium?

Lewin. When EDTA is injected as the Ca-salt it will not interfere with the calcium metabolism of the body. As has been shown in our laboratory, this Ca-chelate is rapidly excreted in the urine without removing any endogenous calcium. As far as specificity is concerned, EDTA is not specific for any one metal, but the stability constant of Y-EDTA is much higher than that of Ca-EDTA. Thus the calcium ions in the complex will be exchanged for yttrium ions, provided that the chelate will meet the yttrium on its passage through the body.

Lamerton. I wonder if Dr. Lewin has any experiences of removal of yttrium from bone?

Lewin. We do not yet have experimental data to answer this question, since tissue distribution studies could not be made in this case. However, since no Y was found in the circulation at the start of the infusion of the chelate it is assumed that the yttrium appearing in the urine subsequent to the infusion came from stores in bone and possibly liver.

Williams. These *in vitro* studies are of the greatest interest since the use of chelating agents offers some therapeutic promise for the removal of these radioactive substances deposited in the body, particularly the bone seekers. The physical state may be of some importance, whether colloidal or in true solution; this could have some bearing on the renal clearance from the blood stream. If protein complexes or colloidal aggregates are formed, these would not be expected to dialyse. If, however, a metal chelate is found to dialyse readily *in vitro*, this would probably be favourable *in vivo*. It would be interesting to know whether Dr. Lewin has any observations on this.

Lewin. In answer to this point, I should like to mention the following data. The blood levels after intravenous injection of $^{90}YCl_3$ remain high, e.g. after 3 hours about 50% of the dose is still in the serum. Consequently, the urinary excretion is low, as we have seen in the case reported here. On the other hand, when Y^{90}-EDTA is given intravenously, the blood level falls very rapidly and the rate of the excretion is high. On the basis of these results we see, as Dr. Williams has suggested, that a metal chelate which is filterable *in vitro* can be expected to be readily removed from the blood stream and excreted by the kidneys.

IODINATED TYROSINES AND RADIOSENSITIVITY IN THYROTOXICOSIS

A Preliminary Report of Results obtained by the Use of Low Activity Chromatography

JOHN D. ABBATT * AND HELEN E. A. FARRAN
Medical Research Council, Department of Medicine, Hammersmith Hospital, London, W.12 ; New End Hospital, Hampstead, London, England

Introduction

It has been known for many years that one of the most satisfactory methods of treating thyrotoxicosis is to reduce the activity of the thyroid gland either by partial surgical removal or by irradiation. During the last 6–8 years internal irradiation with radioiodine has become increasingly popular. Experience has shown that there is a considerable variation in the radiosensitivity of thyrotoxic glands, as judged by the absorbed mean radiation dose to the thyroid, needed to produce a euthyroid state. The majority of patients require between 8,000 and 15,000 rads mean dose to the thyroid following a single dose of ^{131}I while a few may only become euthyroid after multiple treatments and very high total mean rad doses of 50,000-60,000 rads.

We report here the preliminary findings on a number of thyrotoxic patients who have had serial blood samples subjected to radioactive chromatographic analysis during their investigation with tracer doses of radioiodine and their subsequent treatment with ^{131}I.

The main interest of these preliminary results is centred on the presence or absence in the plasma, as identified by chromatography, of iodinated tyrosines, and the apparent correlation of their presence with relative radioresistance of the thyroid tissue.

Materials and methods

Blood samples were taken into heparin as part of the routine procedure employed in the management of 32 thyrotoxic patients who were being treated with oral radioactive iodine. The activities administered ranged from 6·0 to 60·0 millicuries for therapeutic doses and 50–100 microcuries for 'tracer' doses. A routine sample was taken 48 hours after ^{131}I administration, and where possible, additional samples were obtained over a period of several days.

* Present address: The General Electric Co., Ltd., Atomic Energy Department, Erith, Kent, England.

Butanol extracts were prepared from the acidified plasma (*p*H 3·0). The extract was then concentrated by evaporation under reduced pressure so that aliquots corresponding to 0·5–1·0 ml. of plasma were available for two-dimensional ascending paper chromatography. Whatmans no. 4 paper was used, with butanol-acetic acid and butanol-dioxan-ammonia as solvents.

All chromatograms were run in duplicate. A standard solution was made up containing 100·0 mg. of thyroxine and of 3 : 5 : 3 triiodothyronine and 20 mg. of tyrosine, diiodotyrosine, monoiodotyrosine and histidine in 10·0 ml. of 75% ethanol. 0·01 ml. of this solution was applied to each duplicate chromatogram and also to a blank sheet, together with $Na^{131}I$, to obtain standards and points of identification.

After drying, the paper chromatogram was cut into strips which were counted between two scintillation counters connected through a ratemeter to a recording ammeter. The sensitivity of the counting arrangement was such that 1·0 microcurie on the paper gave 3,000 counts per second with a background of about 10 counts per second. The recording ammeter was adjusted so that the chart was fed through at the same rate as the paper strips. Increased counting accuracy can be obtained by stopping the motor and taking prolonged readings at any desired point on the chromatogram.

After counting, the chromatogram was sprayed with Pauly's diazo reagent. The counting record was then matched with the appropriate strips of chromatogram paper and areas of increased radioactivity were identified from the known spots due to standard substances and from the standard map.

Since the patients studied were suffering from thyrotoxicosis, plasma activities of the order of 1% of the administered dose per litre were usual. It was found necessary, where the administered activity was of the order of 50–100 microcuries, to concentrate the plasma extract by passing it through a column containing an ion exchange resin (ZeoKarb). Using this method it was found possible to run the equivalent of 2–4 ml. plasma, so that the various radioactive components could be identified after 'tracer' doses (Abbatt *et al.*, in press).

Autoradiographs were obtained where there was sufficient radioactivity on the chromatogram.

Results and discussion

Plasma samples have been taken at varying times from 1 hour to 9 days after the administration of the radioiodine label. Since all the patients in this study were undergoing routine investigation or treatment, 48-hour samples have always been obtained, and it is primarily the results of chromatography of these 48-hour samples that we report here.

Iodide in the plasma is present only in small amounts at 48 hours, and

(b)

THY.

D.I.T.

M.I.T.

T.

H.

X ORIGIN

(a)

→ DIOXAN-AMMONIA

FIG. 1. (*a*) Sprayed paper chromatogram after counting and reconstitution, and (*b*) the radioactive counting record of this chromatogram for comparison. The positions of the known standards are marked on the chromatogram (THY = Thyroxine and 3 : 5 : 3 triiodothyronine, D.I.T. = Diiodotyrosine, M.I.T. = Monoiodotyrosine, T. = Tyrosine, H. = Histidine.

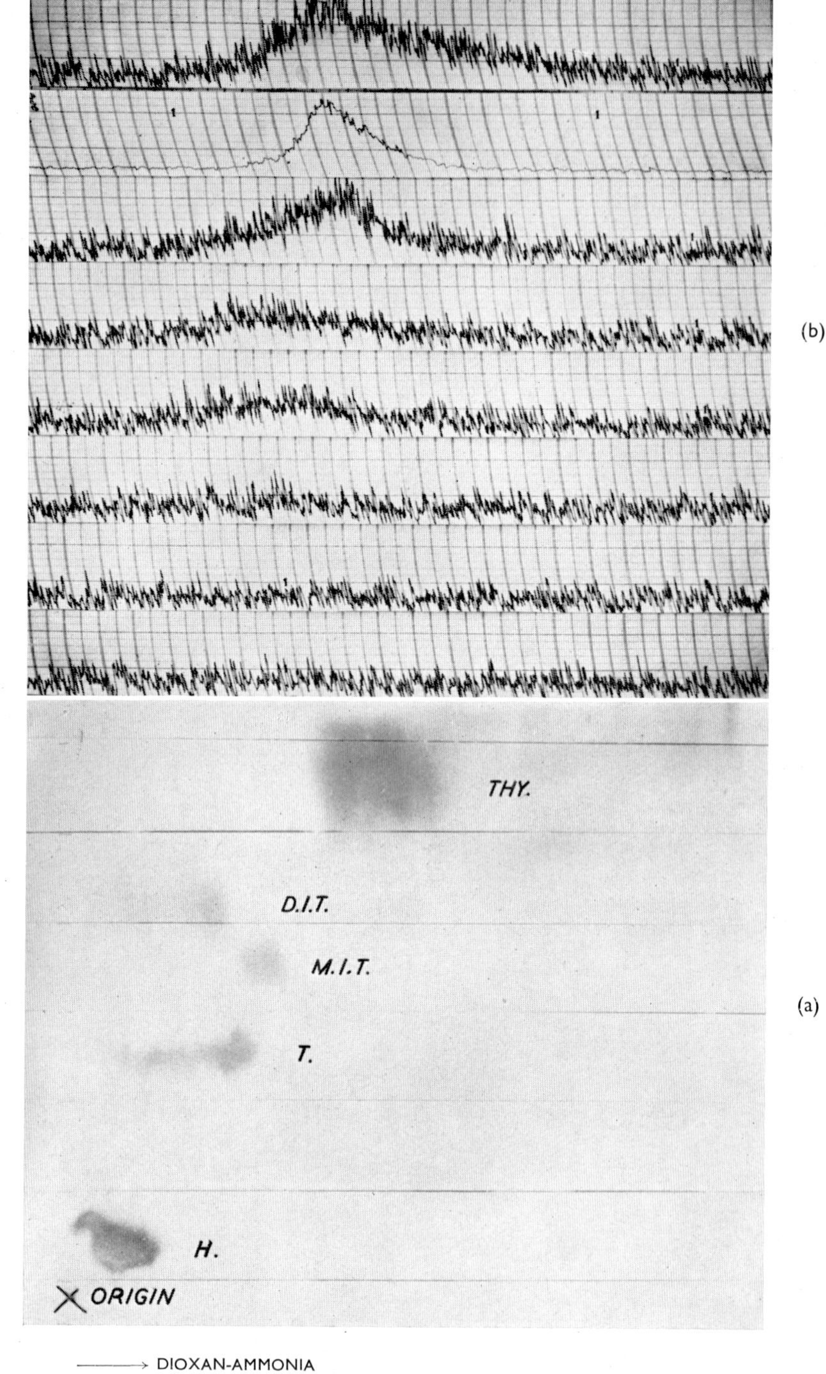

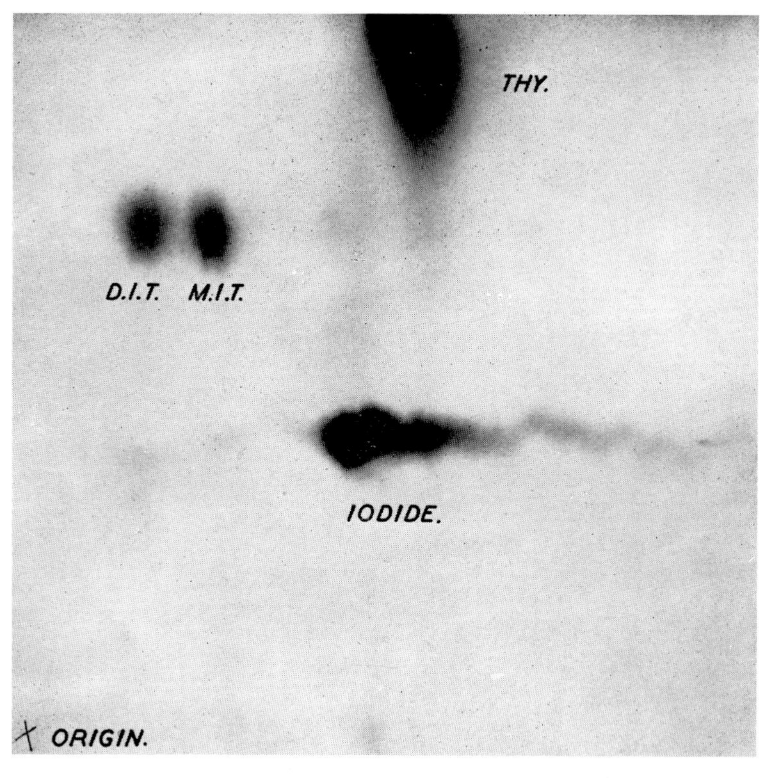

BUTANOL-ACETIC ⟵

⟶ DIOXAN-AMMONIA

Fig. 2. Autoradiograph of chromatogram prepared from plasma of a patient who had received a therapeutic dose of radioiodine; showing iodide (streaked), thyroxine and mono- and diiodotyrosine.

the technique of butanol extraction employed still further reduces this amount by evaporation, so that only some of the 48-hour chromatograms show the presence of traces of radioiodide. These 'spots' have readily been identifiable from standards.

Thyroxine is consistently present in all chromatograms, and 3 : 5 : 3 triiodothyronine has been identified in about half the cases studied. The separation of thyroxine and 3 : 5 : 3 triiodothyronine obtained with the butanol-dioxan-ammonia solvent was not uniformly good; probably because of varying plasma fat contents and the high degree of plasma concentration needed to obtain identifiable spots in the low activity plasmas. For this reason, it is probable that 3 : 5 : 3 triiodothyronine was present more often than has been recorded, and that it was merged in the thyroxine spot and, therefore, inadequately separated for positive identification.

In 17 of our patients definite spots tentatively identified as either mono-iodotyrosine or diiodotyrosine have been observed, these are distinct from the easily identifiable iodide spots (Fig. 1 (*a*) and (*b*) and Fig. 2). The identification of these iodinated tyrosines is at present tentative only, and more work is needed to exclude any possibility that they are artefacts occurring as a result of the particular technique employed. This possibility is somewhat unlikely since these substances have already been demonstrated in gradually increasing and decreasing amounts and activities in serial samples from the same patient.

The significance of these iodinated tyrosines in man has been discussed by a number of workers, but always after millicurie or multimillicurie doses of radioiodine had been given to thyrotoxic patients or to individuals in whom it was desired to ablate all functioning thyroid tissue for the treatment of either carcinoma or heart diseases (Brown and Jackson, 1954, and Robbins *et al.*, 1952 and 1955). In our experience, both monoiodotyrosine and diiodotyrosine can occur together or separately.

Both monoiodotyrosine and diiodotyrosine have been reported as being present in plasma, either together or separately, by a number of workers (Benna and Dobyns, 1955; Benna *et al.*, 1955; Albert and Keating, 1951; and Stanbury *et al.*, 1955). Other workers have been unable to identify these substances in plasma (Gross and Leblond, 1951; Rosenberg, 1951).

Iodinated tyrosines are usually considered to be intermediary compounds in thyroid hormone formation and destruction. It is not known whether they ever occur in the plasma of euthyroid individuals.

Thirty-two patients have so far been studied in detail, but only 22 of these have been followed up for at least 3 months.

In 17 of these 22 patients iodinated tyrosines have been identified (Table) and all these 17 have needed a mean dose of 15,800 rads or more before they became euthyroid. This must be compared with the mean dose of

8–15,000 rads needed to produce an 85–90% remission rate in unselected thyrotoxic patients.

Of the 17 patients in whom tyrosines have been identified, in four, examination of plasma had been possible following a 50–100 microcurie pre-therapy tracer dose. Iodinated tyrosines were present in the plasma of these four patients before treatment. This, while not being conclusive, would suggest that the presence of iodinated tyrosines in the plasma of thyrotoxic patients, either before or after treatment, indicates relative radioresistance to radiation, and, incidentally, an increased liability to recurrence of thyrotoxicosis whatever form of treatment is employed.

TABLE

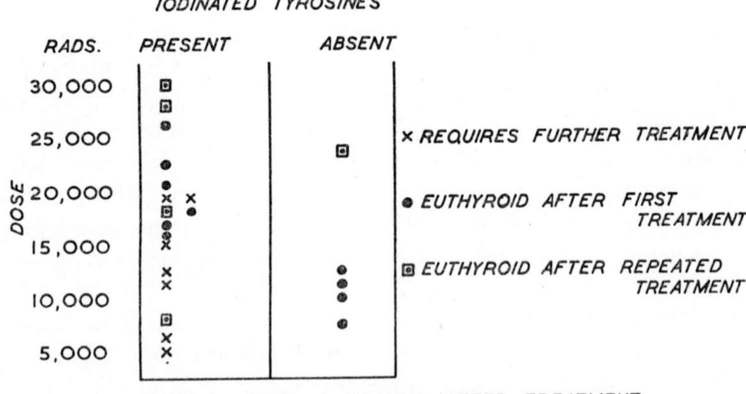

The biochemical significance of the presence of tyrosines in the plasma of thyrotoxic individuals with relatively radioresistant thyroid glands remains a matter for speculation.

Acknowledgments

We would like to thank our clinical colleagues at New End Hospital for their co-operation, and we are particularly grateful to Mrs Peter Jackson and Mrs V. Parker for their valuable technical assistance.

REFERENCES

ABBATT, J. D., FARRAN, H. E. A., and PARKER, V. *Brit. J. Radiol.*, in the press.
ALBERT, A., and KEATING, F. R., 1951. *J. Clin. Endocrinol. and Met.*, **11**, 996.
BENNA, R. S., and DOBYNS, B. M., 1955. *Ibid.*, **15**, 118.
BENNA, R. S., DOBYNS, B. M., and NIMER, A. J., 1955. *Ibid.*, **15**, 1367.
BROWN, F., and JACKSON, H., 1954. *Biochem. J.*, **56**, 399.
GROSS, J., and LEBLOND, C. P., 1951. *Endocrinology*, **48**, 714.

ROBBINS, J., RALL, J. E., BECKER, D. V., and RAWSON, R. W., 1952. *J. Clin. Endocrinol. and Met.*, **12**, 856.
ROBBINS, J., RALL, J. E., and RAWSON, R. W., 1955. *Ibid.*, **15**, 1315.
ROSENBERG, I. N., 1951. *J. clin. Invest.*, **30**, 1.
STANBURY, J. B., KASSENHAAR, A. A., MEIJER, J. W. A., and TERPSTRA, J. J., 1955. *J. Clin. Endocrinol. and Met.*, **15**, 1216.

DISCUSSION

L.-G. Larsson. Have you performed any electrophoretic studies on the plasma? It would be very interesting to know if the iodinated tyrosines demonstrated in your patients are bound to the thyroxin-binding protein or to some other fraction of the plasma proteins. I am also thinking a little about the abnormal iodinated protein studied in cases with thyroid carcinoma by Robbins *et al.*

Abbatt. No, we have done no electrophoretic studies on the patients referred to in this report. I agree that the next logical step is to explore the nature of the protein binding of the iodinated tyrosines in the plasma of hyperthyroid patients.

THE RADIATION EXPOSURE OF THE ORGANISM BY INHALATION OF NATURALLY RADIOACTIVE AEROSOLS

W. JACOBI, K. AURAND AND A. SCHRAUB
Max Planck Institut fur Biophysik, Frankfurt am Main, Germany

The decay products of Radon and Thoron in air attach themselves within a short time to the aerosol particles. Recent studies (Bale, 1955; Schraub *et al.*, 1955, and Aurand *et al.*, 1956) have shown that the α-dosage to the respiratory tract by direct inhalation of these decay products is much greater than that from Radon and Thoron and their daughter products built up in the organism.

This new knowledge is of importance for industrial hygiene in Uranium mines, in the Uranium processing industry and for radon therapy; it is of especial importance in the problem of natural radiation dosage to human beings. With regard to this last problem some radiobiological studies are reported here. They concern long-term inhalation of Radon- and Thoron-daughter products.

1. Accumulation of inhaled radioactive substance in the repiratory tract

The accumulation of activity in the respiratory tract (A) after inhalation of radioactive aerosols follows the function:

$$A = \frac{aRV}{\lambda_{eff}} \cdot (1 - e^{-\lambda_{eff} t})$$

where a = activity per unit volume of inhaled air, V = breathing volume, R = retention factor of the present aerosol, λ_{eff} = effective decay constant of the isotope in the respiratory tract.

The activity approaches a saturation value when the number of radioactive atoms retained is equal to that of the eliminated atoms in the respiratory tract measured at the same time.

As for the inhalation of the daughter products of Radon and Thoron the mechanism of accumulation in the respiratory tract is much more difficult, because the decay products decay again into radioactive daughter substances. In this way every retained Ra A-atom causes several decays in the organism. We have in this case a steady state equilibrium, see Figure 1.

The physical decay processes are passing in a horizontal direction—characterised by the physical decay constants—in a vertical direction the biological elimination takes place—symbolised by the 'biological decay constant'. A combination of the two reactions leads to the effective elimination rate ($\lambda_{eff} = \lambda_{phys} + \lambda_{biol}$).

In the case of the naturally occurring radioactive aerosols, the irradiation of the respiratory tract is chiefly given by the α-active, short-life daughter products, this means Ra A and Ra C′, Th A, Th C and Th C′. For every decay product we get a saturation activity, which will be reached within about three times the effective half-life after the beginning of inhalation.

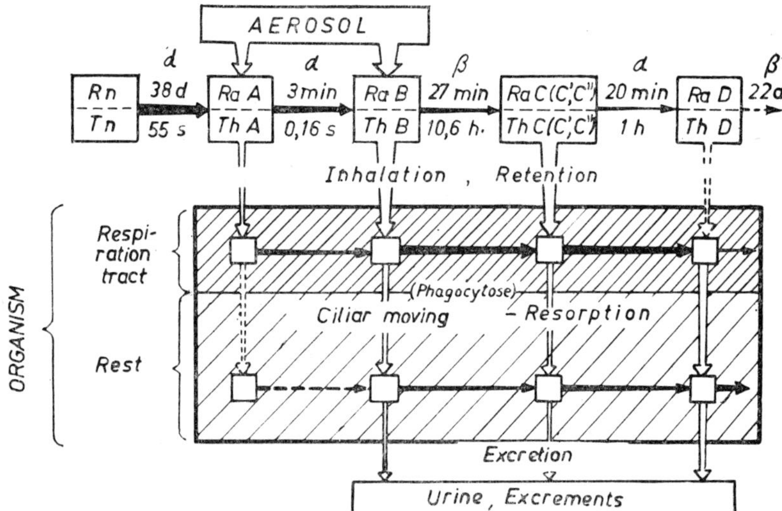

FIG. 1. Scheme of incorporation and distribution of Radon- and Thoron-decay products after inhalation.

The radiation dosage to the respiratory tract depends therefore on the following physical and biological processes:

(*a*) The concentration of decay products in the inhaled air, i.e. the rate of equilibration of Radon and Thoron with their daughter products.
(*b*) The retention of the carrier aerosol in the respiratory tract.
(*c*) The biological elimination processes in the respiratory tract (λ_{biol}); this means especially the transport by ciliary movement and resorption.
(*d*) The distribution of activity in the respiratory tract.

Our measurements of the concentration of Radon- and Thoron-daughter products in normal air, the properties of the carrier aerosol and its retention will not be discussed here because they are of more physical interest and also have been partly published (Schraub *et al.*, 1955; Aurand *et al.*, 1955 and 1956). Therefore only our studies about distribution and elimination of inhaled naturally radioactive aerosols in the organism will be reported here.

2. Elimination of inhaled aerosols in the respiratory tract

Transport of radioactive material takes place in the trachea and bronchii by the action of ciliary movement, and in the alveolar tissue by resorption and phagocytosis. The first process depends upon the place of retention and the particle size, the latter one chiefly on the chemical properties of the radioactive substance and its carrier aerosol. If we take into account the

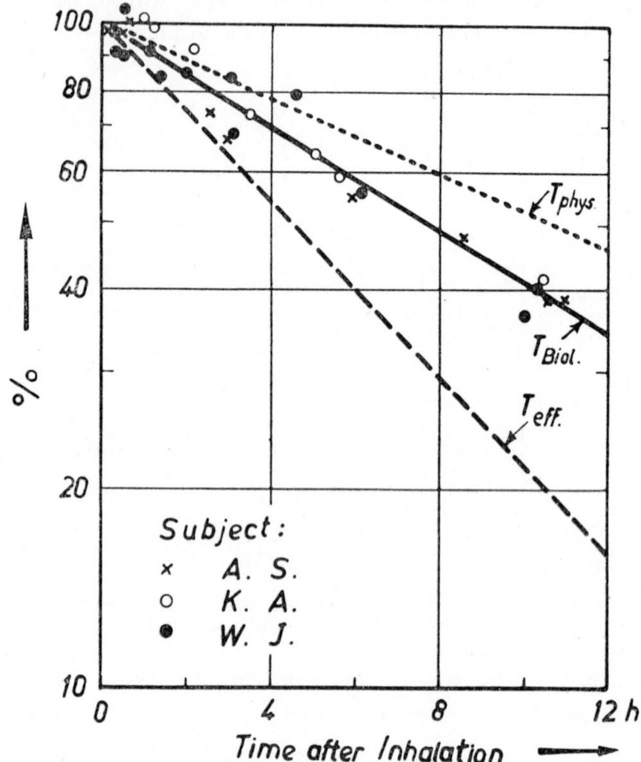

FIG. 2. Decay of lung activity of man after short-term inhalation of naturally occurring aerosol, with attached Th B.

heterogeneous sizes, forms and chemical composition of the naturally occurring aerosols, it will be evident how complicated is the process of elimination.

In spite of this our inhalation experiments with Thorium B, attached to naturally occurring aerosol, gave some interesting results. Figure 2 shows the decay curve of lung- and bronchial-activity of man after short-term inhalation.

The measurements were carried out with a directional γ-sensitive scintillation counter which was put on the surface of the chest. Independent of the inhaling person, we got the same decay curve, with a biological half-life of about 8 hours.

Similar experiments of Albert and Arnett (1955) with rough kaolin dust gave a value of about 60 hours; Stokinger *et al.* (1951) also obtained 60 hours with rats inhaling UO_2-fumes.

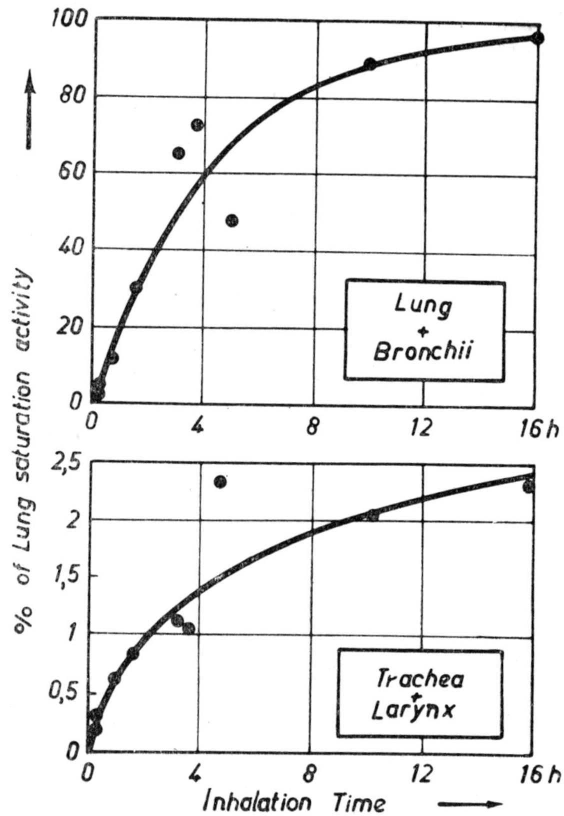

FIG. 3. γ-activity of the respiratory tract of rats after inhalation of aerosol, charged with Th B.

This value of 8 hours for the naturally occurring aerosols charged with Th B, is also much lower than the elimination rate from the whole body. This must be taken into account in estimation of the radiation dose to the respiratory tract. In all cases where $\lambda_{biol} \leqslant \lambda_{phys}$, the respiratory tract will be the 'critical organ' for estimation of radiation damage. We got similar results in experiments with rats and mice. Immediately after inhalation the animals were sacrificed and the activities of definite sections were measured. Figures 3 and 4 show the increasing activity with inhalation time in the respiratory tract and in muscle tissue of rats. In accordance with theory the accumulation can be described by a function

$$(1 - e^{-\lambda_{eff} t})$$

with an effective half-life of about 4 hours. 95% of the activity retained in the respiratory tract was deposited in the lung and bronchii and only 5% in the trachea and larynx. The specific saturation activity in muscle is

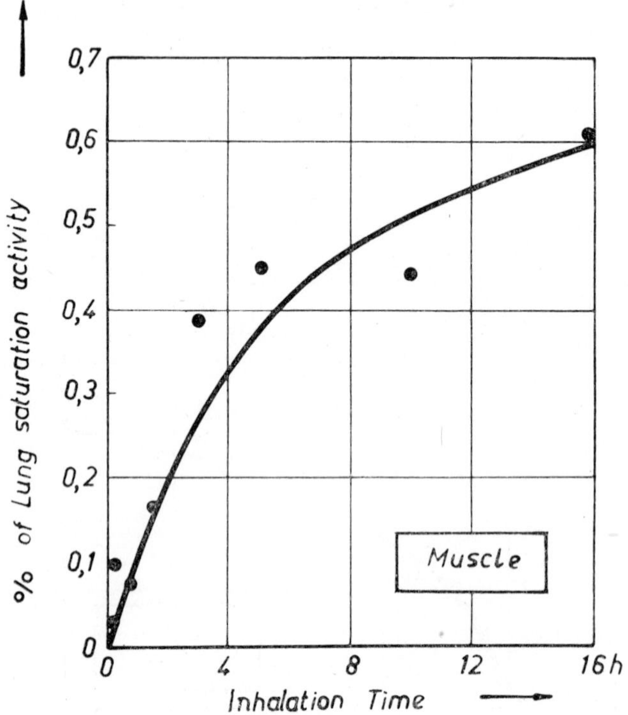

FIG. 4. Specific γ-activity of muscle tissue of rats after inhalation of aerosol, charged with Th B.

about 0·1% of that in the lung. The total resorbed activity (without the kidneys) is therefore only 10% of the retained activity.

The accumulation of Th B (+ Th C) in the kidneys is important for radiation dosage to the organism following inhalation, see Figure 5, a value of about 20% of the whole lung activity is reached. The accumulation curve of the kidneys has the opposite curvature (in the first period after inhalation) to that for lung and muscle in accordance with the excretion function of the kidneys. Considering the relatively high radiation resistance of the lung tissue it might be necessary in some cases to denote the kidneys as 'critical organ'.

The question arose, is resorption in lung or in the gastro-intestinal tract (after swallowing the activity) dominant? The discharging of the radioactive substance from the surface of the carrier aerosol is here the deciding factor. To answer this question we prevented the activity from being swallowed by tying the œsophagus of the rats and studying the distribution

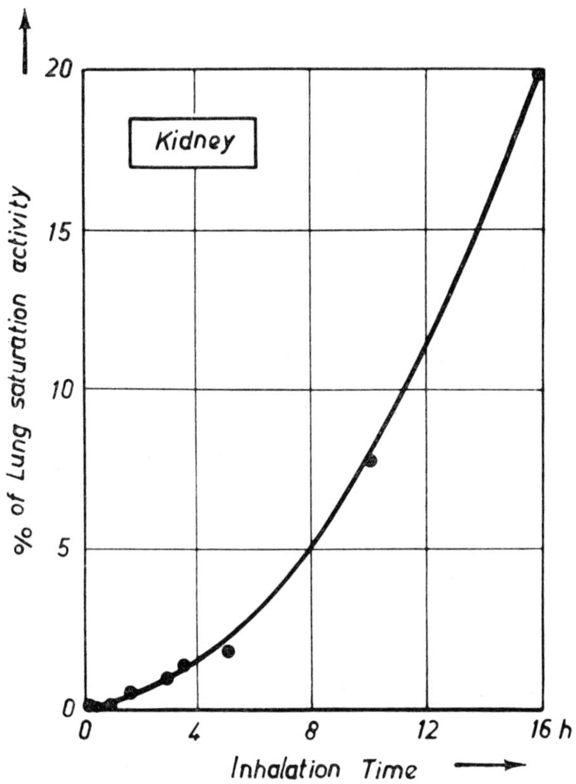

FIG. 5. γ-activity of the kidneys of rats after inhalation of aerosol, charged with Th B.

of activity in the body. The results are shown in Figure 6. After short-term inhalation (1·5 hours) the resorbed activity is much lower than that of the control animals with an open œsophagus. But after an inhalation time of 15 hours there is no significant difference between operated rats and controls. These results seem to prove that the velocity of resorption in the gastro-intestinal tract exceeds the direct resorption in the alveolar tissue. This means that the activity in tissue and organs has reached the saturation value before lung resorption becomes efficient.

3. Distribution of activity in the lung

For estimating the real radiation dosage to lung tissue after inhalation of radioactive aerosols, the knowledge of the non-uniformity of the distribution in the lung is of importance. In the literature, a non-homogeneity factor of two to tenfold the median concentration is mentioned (Bale, 1955;

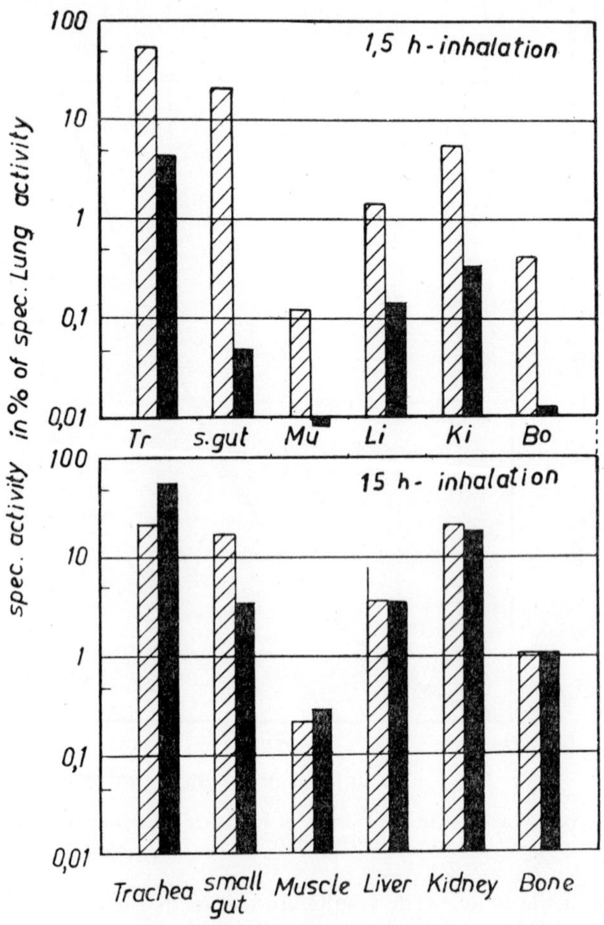

FIG. 6. Distribution of activity in rats with closed œsophagus after inhalation of radioactive aerosol.
■ œsophagus closed
/// controls with open œsophagus

Stokinger *et al.*, 1951; Chamberlain and Dyson, 1956, and Eisenbud, 1952). This value does not seem to be well supported by experiments.

By measuring the specific activity of 22 different sections of the rat lung we got the distribution shown in Table I. It is astonishingly homogeneous;

TABLE I

Specific Activity of Several Sections of the Rat Lung after Inhalation of Radioactive Aerosol

Lung section	Weight (mg.)	No. of sections	Spec. activity (rel. units) Median	Min.	Max.
Right lung :					
1. Lobe	145	4	108	97	115
2. Lobe	115	4	92	52	123
3. Lobe	303	6	106	73	134
4. Lobe	121	4	62	38	87
Left lung	408	4	104	92	114
Whole lung	1092	22	95 ± 22		

TABLE II

Median Radiation Dosage to the Human Lung after Inhalation of the Daughter Products of Radon and Thoron (naturally occurring aerosol; retention 30%; RBW = 10; lung weight 800 g.)

Decay product	$T_{1/2}$	Equilibrium rate F	R	MRC	T_{biol}	Radiation dosage $\left[\frac{mrem}{week}\right]$ F	R	MRC
Ra A	3·05 min.	1·0	1·0	1·0	} ∞	0·30	3·6	880
Ra B	26·8 min.	0·3	0·8	1·0				
Ra C (+ C′)	19·7 min.	0·3	0·8	1·0	} 8 hr.	0·28	3·4	810
Th A	0·16 sec.	1·0	1·0	—	} ∞	0·07	1·0	—
Th B	10·6 hr.	0·03	0·15	—				
Th C (+ C′)	60·5 min.	0·03	0·15	—	} 8 hr.	0·03	0·35	—

F = free air; Rn-activity 10^{-16} c./cm.^{3}; Tn-activity $2 \cdot 10^{-17}$ c./cm.^{3}
R = room air; Rn-activity $5 \cdot 10^{-16}$ c./cm.^{3}; Tn-activity $5 \cdot 10^{-17}$ c./cm.^{3}
MPC = max. permissible Rn-activity in air = 10^{-13} c./cm.^{3}

and the median deviation is only 20–30%. This result could be confirmed by autoradiographs of lung tissue.

4. Radiation dosage in the human lung following inhalation of naturally radioactive aerosols

Using the results on resorption and distribution in the respiratory tract mentioned above, we estimated the radiation exposure of the human lung in the case of naturally occurring concentrations of Radon, Thoron and their daughter products in free air and in rooms (Hultqvist, 1956). The natural radiation exposure following inhalation of Radon- and Thoron-daughter products exceeds the dosage of all the other components of natural radiation exposure. For example, the dosage caused by cosmic rays is about 35 mr/year (Libby, 1955).

In addition, the dose arising from inhalation of the maximum permissible dose of Radon and Thoron in equilibrium with their daughters far exceeds the tolerance value of 300 mrem/week. The reason for this great difference is:

(*a*) The daughter products are only in a few cases in equilibrium with Radon and Thoron.

(*b*) The accumulation of inhaled radioactive aerosols in the respiratory tract was not adequately taken into account.

The investigations above, on this problem should help to estimate the influence of the Radon- and Thoron-daughter products on dosage to the respiratory tract.

REFERENCES

ALBERT, R. E., and ARNETT, L. C., 1955. *Arch. Indust. Health*, **12**, 99.
AURAND, K., JACOBI, W., and SCHRAUB, A., 1955. *Naturwissenschaften*, **42**, 398.
Idem, 1956. *Report 2nd Int. Sym. Use of Radioisotopes, Badgastein, Austria*, and *Sonderaband, Strahlentherapie*, **36**, 266.
BALE, W. F., 1955. *Conf. Peaceful Uses of Atomic Energy, Geneva*, Report No. 76.
CHAMBERLAIN, A. C., and DYSON, E. D., 1956. *Brit. J. Radiol.*, **29**, 317.
EISENBUD, M., 1952. *Arch. Indust. Hyg.*, **6**, 214.
HULTQVIST, B., 1956. *Studies on Naturally Occurring Ionising Radiations*, Stockholm.
LIBBY, W. F., 1955. *Science*, **122**, 57.
SCHRAUB, A., AURAND, K., and JACOBI, W., 1955. *Phys. Ther.*, **7**, 437.
STOKINGER, H. E., STEADMAN, L. T., WILSON, W. B., SYLVESTER, G. E., DZIUBA, B. S., and LA BELLE, C. W., 1951. *A.M.A. Arch. Indust. Hyg.*, **4**, 347.

SECTION VIII

IRRADIATION EFFECTS ON THE HÆMOPOIETIC SYSTEM

I. ERYTHROPOIESIS

EFFECT OF WHOLE-BODY IRRADIATION AND VARIOUS DRUGS ON ERYTHROPOIETIC FUNCTION IN THE RAT. STUDIES WITH RADIOACTIVE IRON

L. F. LAMERTON AND E. H. BELCHER
Physics Department, Institute of Cancer Research,
Royal Cancer Hospital, London, S.W.3, England

Introduction

It has been shown by various workers (Hennessy and Huff, 1950; Belcher *et al.*, 1954) that the uptake of radioactive iron into the peripheral blood is profoundly depressed by prior whole-body radiation, but the precise mechanism by which these changes are brought about is by no means fully understood. It has been found (Lajtha and Suit, 1955; Bonnichsen and Hevesy, 1955; Suit *et al.*, 1955) that the incorporation of iron into the hæmoglobin of the red cells is not a radiosensitive process, and an explanation of the effects observed must be looked for in terms of changes in the cellular population. In this paper we would like to present, in a somewhat compressed form, some recent experimental data we have obtained which has a bearing on this subject.

One approach to the problem of the mechanisms involved is to follow the uptake of radioactive iron in the peripheral blood when administered at various times relative to whole-body radiation or administration of drugs. We have carried out such experiments with various doses of whole-body radiation and also with certain anti-leukæmic drugs, including a nitrogen mustard (CB1348), Myleran and Aminopterin.

In our studies two highly inbred strains of rat have been used, the 'August' and the 'Marshall', and their behaviour, with regard to iron metabolism, has been found to be very similar. Details of the experimental techniques employed are given in previous publications (Belcher *et al.*, 1954; Baxter *et al.*, 1955; Belcher *et al.*, 1955).

Before discussion of the radiation and drug experiments it will be useful to review certain aspects of iron uptake and metabolism in the control animal.

Radioactive iron uptake and turnover in normal rats

Radioactive iron administered parenterally will label the metabolic pool, part of which is taken up by the erythropoietic tissues for hæmoglobin synthesis in the red cell precursors and part of which goes to the liver and other sites in the body. Figure 1 shows the rate at which intravenously administered ^{59}Fe appears in the peripheral blood. It will be seen that 44% of the injected activity has emerged into the peripheral blood by 2 days, and that thereafter the rise in blood activity is slow, indicating that by far the greatest amount of iron entering the bone marrow is taken up by the maturing red cells within two days of their emergence into

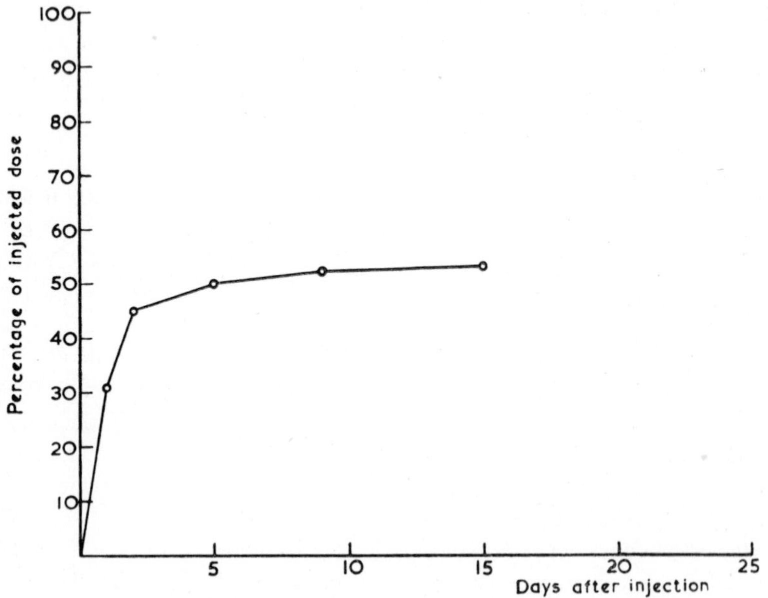

FIG. 1.—Rate of uptake in blood of intravenously injected ^{59}Fe. Control Marshall rat.

the peripheral blood. The extent to which the radioactive iron appearing in the blood is taken up directly by the bone marrow following the injection can be determined from the turnover rate of administered radioactive iron in bone marrow and other tissues of the body. Figure 2 shows the turnover curves of ^{59}Fe in various tissues of the August rat following intravenous injection. From measurement of the distribution of ^{59}Fe in various parts of the skeleton (Baxter *et al.*, 1955) we know that the active bone marrow of each hind limb represents 14% of that in the whole animal. In addition some extramedullary erythropoiesis occurs in the spleen. From these data one may draw up a balance sheet, as follows (Table).

From this it appears that all, or almost all, of the ^{59}Fe appearing in the peripheral blood within two days of injection was taken up directly into the bone marrow following the injection. The release of radioactive iron from tissues other than those engaged in erythropoiesis is slow in the normal animal, presumably because of isotope dilution.

At the cellular level the uptake of radioactive iron can be studied using high resolution autoradiography techniques. Miss Harriss is reporting on this work, and we will quote only her general finding that, in the rat, all the stages from early normoblast to reticulocyte take up iron at a fairly constant rate.* The stage at which the rate of iron incorporation falls is uncertain, but the reticulocyte of the peripheral blood still has an appreciable iron-incorporating capacity.

TABLE

August rat. Intravenous injection of ^{59}Fe

Appearance of ^{59}Fe in peripheral blood	
By 1st day	27% of injected dose
2nd day	44% ,,
5th day	50% ,,
Uptake of ^{59}Fe by erythropoietic tissue	
Content of bone marrow 6 hours after injection	32·1%
Content of spleen 6 hours after injection	2·3%
Direct uptake into reticulocytes of peripheral blood ...	3·0%
Further uptake by erythropoietic tissues estimated from plasma activity at 6 hours	4·0%
Total	41·4%

Studies with X-radiation

Experiments have been carried out with 100r, 200r and 400r whole-body radiation, the ^{59}Fe injections being given 24, 12 and 5 hours before irradiation, at the same time as the irradiation and at 6, 12, 24, 48, 72 and 96 hours afterwards. For each interval five rats in each of the irradiated groups have been used and also five control rats. Blood samples have, in most experiments, been taken by cardiac puncture at 1, 2, 5, 9 and 15 days following the ^{59}Fe injection. In the present communication it is not possible to present all the results, but certain of the data can be shown. In Figure 3 are shown the mean curves of uptake in blood when the ^{59}Fe is given at the same time as irradiation. It will be seen that the initial rate of appearance of activity is less in the irradiated than in the control animals, but that after 100r and 200r the level of activity in the blood by 15 days is the same as that in the control animals. This suggests that a dose of 200r or less will delay the emergence of red cells into the peripheral blood, but will not actually destroy cells which are taking up appreciable amounts of iron, unless it is assumed that the stock of iron of a cell destroyed by radiation can be handed on directly to another precursor in the bone marrow, which is unlikely. It is possible that with dosage levels of 200r and less some cell destruction

* Harriss: this volume, p. 333.

may occur in the early normoblast stage which would not be apparent in the blood uptake curve, but the results suggest no appreciable cell destruction in the later stages. On the other hand, the results after 400r indicate that an appreciable proportion of the cells incorporating iron at the time of irradiation fail to appear as mature red cells.

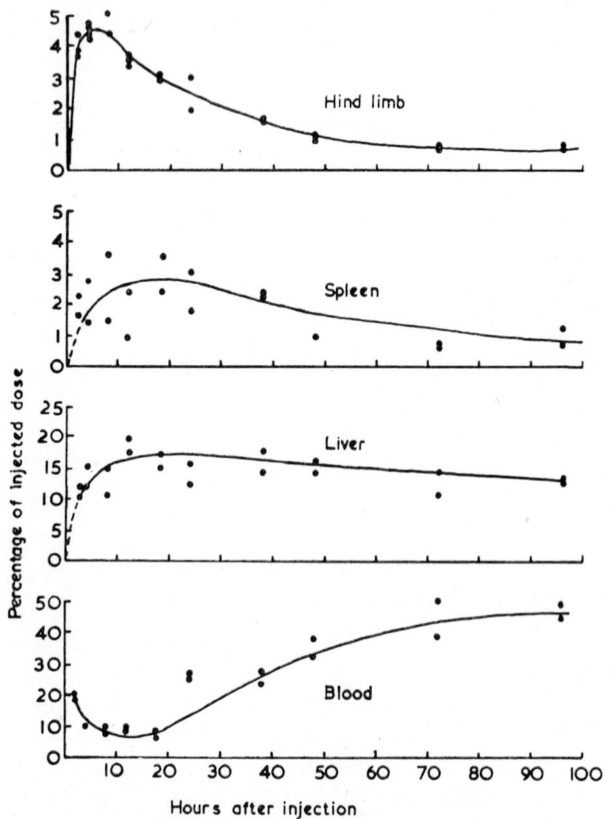

FIG. 2. Tissue turnover of intravenously injected ^{59}Fe. Control August rat.

At 2 days after radiation the depression of iron uptake is very marked for each of the three levels of dosage used, and in Figure 4 are shown the blood uptake curves for this interval. The depression of 24-hour uptake is profound for 200r and 400r and severe for 100r, and there is some depression of the 15-day uptake for each of the dosage levels, but much less than in the case of the 24-hour uptake.

The depression in 15 day uptake indicates a reduction in the iron-incorporating capacity of the marrow at the time of injection. Expressed in

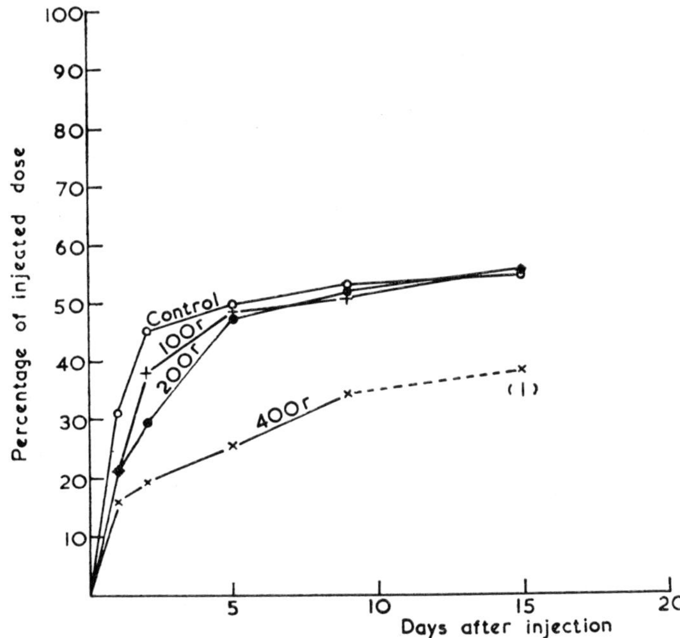

FIG. 3. Rate of uptake in blood cf intravenously injected ^{59}Fe given at same time as whole-body irradiation. Marshall rat.

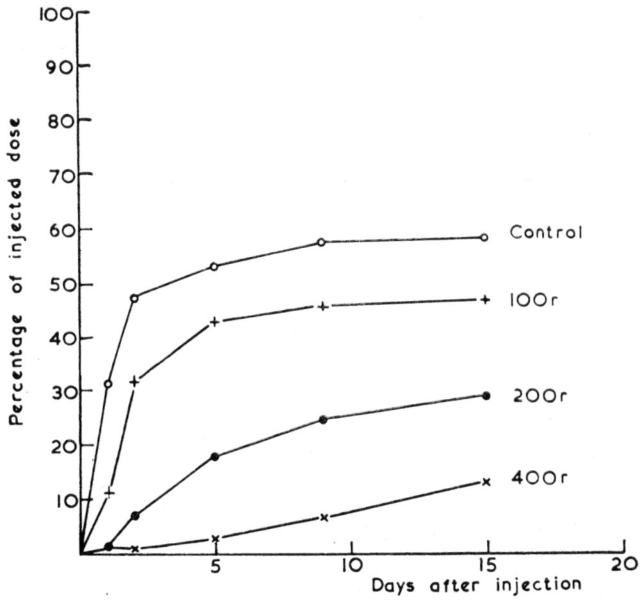

FIG. 4. Rate of uptake in blood of intravenously injected ^{59}Fe given 2 days after whole-body irradiation. Marshall rat.

terms of control values, the value of 15-day uptake is 80% for 100r, 50% for 200r and 22% for 400r, but the extent to which these figures can be taken as a quantitative measure of bone-marrow depletion depends on a number of considerations. The 15-day uptake will be influenced by the extent to which iron is withdrawn from the labelled stores, by any change in the rate at which a given precursor takes up iron as well as by changes that may have been produced by radiation in the values of plasma iron concentration and rate of plasma iron clearance at the time of injection.

It is probable that the 15-day uptake in irradiated animals will be more influenced by withdrawal of labelled iron from stores than in the case of the normal animals, if the period of 15 days after injection covers a period of regeneration of the irradiated marrow. We are at the present time attempting to estimate the degree of withdrawal from stores by measuring the ^{59}Fe uptake in rats given a course of daily doses of an iron-dextran preparation ('Imferon'), started after the injection of the radioactive iron, which will maintain the plasma iron concentration at a high level, and thus limit the utilisation of radioactive iron stores for erythropoiesis. Results to date indicate that for ^{59}Fe injected 2 days after 200r, an appreciable fraction of the iron appearing in the red cells at 15 days after injection has come from stores, so that the amount of iron initially taken up directly by the bone marrow 2 days after 200r is less than 50% of that of the controls. Nevertheless, the depression of the 15-day uptake is still much less severe than that at 24 hours, indicating a delay in the emergence of the injected radioactive iron entering the marrow compared with the control animals.

Relevant to this discussion are tissue turnover data for ^{59}Fe in normal and irradiated animals. The curves of Figure 5 show that, at 2 days after 200r there is still a considerable uptake of ^{59}Fe into the marrow, though it is released more slowly than in the case of the controls. In this experiment the iron was given subcutaneously, and this will cause some reduction in the uptake of iron by the tissues, but will not materially affect the comparison between normal and irradiated animals. Carrying out the same type of analysis as in the Table, the initial uptake of ^{59}Fe in the erythropoietic tissues of the control animal was approximately 37% of the injected dose, and 17% in the irradiated animal. Thus from these data the initial uptake of ^{59}Fe in the erythropoietic tissues of the rat given 200r two days previously was about 45% of that in the control animal.

The delay in release of radioactive iron in the irradiated animal cannot be due to lack of late forms in the marrow at this stage since the results with ^{59}Fe given at the same time as radiation indicate a release of red cells into the circulation between 2 and 5 days after 200r. However, the release of red cells between 2 and 5 days after 200r is apparently the result of a hold up of late forms in the marrow. If this hold up were essentially a delay in release rather than in maturation, so that iron-incorporation continues, the cells may become fully, or almost fully, hæmoglobinised some time before emergence so that their capacity for iron uptake will be much reduced. We have tested this hypothesis by measuring the *in vitro* uptake of reticulocytes emerging 2 days after 200r and find it to be considerably less than in the case of the control animals, in fact, almost zero. Histologically it can be seen that the reticulocytes emerging at this time are all very late forms.

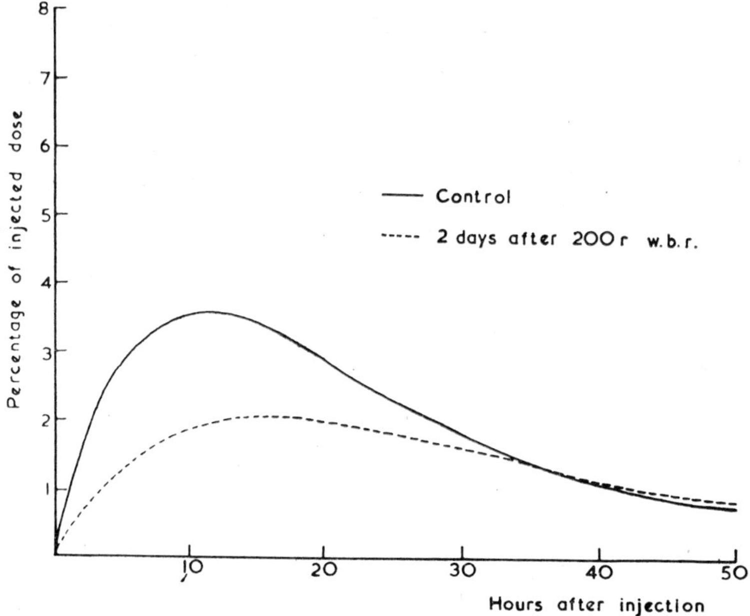

FIG. 5. Turnover of subcutaneously injected ^{59}Fe in hind limb, in control rat and in rat given 200r whole-body irradiation 2 days previously. August rat.

The uptake in bone marrow of ^{59}Fe given 2 days after 200r would be explained by the demands of early and intermediate forms, regenerating after the radiation damage, and such iron would not appear in the peripheral blood until these cells have come through to maturity.

On this basis the marked depression in 24-hour uptake 2 days after whole-body radiation would be the result mainly of two factors :—

(*a*) a depletion of late forms in the marrow,

(*b*) a delay in release of late forms which would cause their iron-incorporating capacity to be reduced during the latter period of their stay in the marrow.

With regard to (*a*), this presumably arises as a result of radiation affecting the division of the earlier precursors, though not, with doses of 200r and less, by actual destruction of appreciable numbers of the iron-incorporating forms, at least from the intermediate normoblast up to the reticulocyte stages.

With regard to (*b*), at the moment only a guess can be made of the mechanism by which the release of late forms is delayed. It may be that their emergence is dependent on their being followed by a full population of earlier cells.

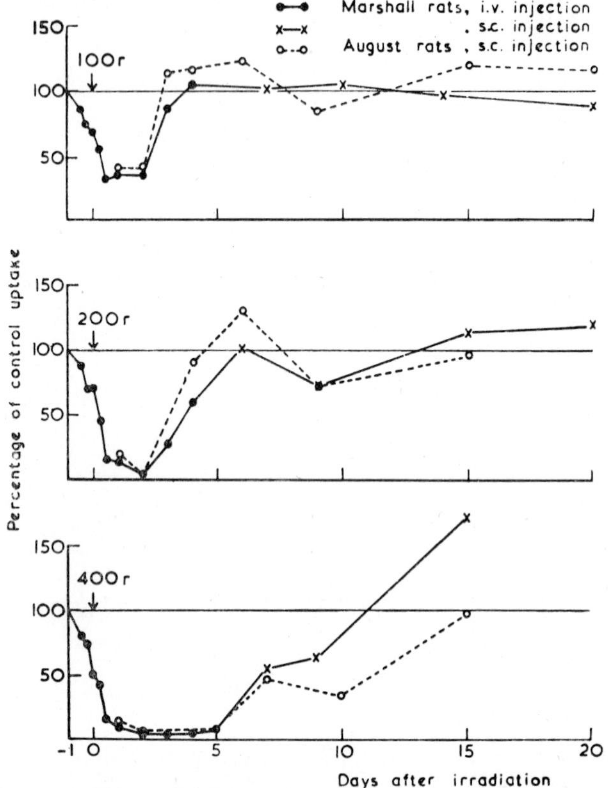

FIG. 6. 24-hour uptake in blood of ^{59}Fe injected at different times after whole-body irradiation.

The X-ray results are summarised in Figure 6, which shows the variation in blood uptake one day after ^{59}Fe injection (expressed as percentage of control values) with interval between irradiation and injection, for the three dosage levels used. All points represent a mean value from five animals and curves are presented for both the August and Marshall strains of rat. The general agreement between the results for the two strains is good. These curves show that after 200r and 400r recovery is not continuous. This is an effect also observed in our studies of tissue turnover in irradiated rats, as reported at last year's Conference (Belcher *et al.*, 1955).

Studies with drugs

It is of interest to study the effects on iron metabolism of agents other than radiation which affect bone-marrow function. In another paper at this Conference Dr. Elson will be comparing the patterns of blood change after irradiation with those produced by the administration of certain anti-leukæmic drugs, and, in particular, the nitrogen mustard CB1348, and

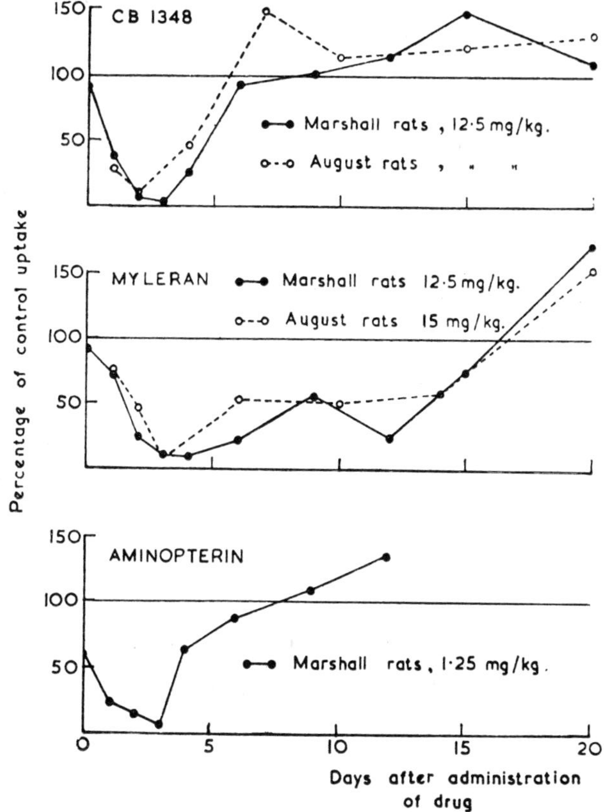

FIG. 7. 24-hour uptake in blood of ^{59}Fe injected at different times after administration of various drugs. Marshall rats intravenous injection of ^{59}Fe. August rats subcutaneous injection of ^{59}Fe.

Myleran.* We have carried out ^{59}Fe experiments with both these drugs, and the curves for the first-day uptake in the blood as shown in Figure 7, for the two strains of rat. Although the general shape of each curve is the same for the two strains there are some differences, which may however be the result of variation in preparation of the particular sample of drug used and, also, in the case of Myleran, to a difference in the dosage of drug and route of administration.

* Elson: This volume, p. 372.

The curves for 12·5 mg./kg. CB1348 and 200r resemble each other closely up to six days after administration of the drug or irradiation, but for later intervals there is little evidence of the interruption in recovery observed with 200r. In the case of Myleran, with the dosage given, the time of maximum depression is a day or so later than for 200r or CB1348 at 12·5 mg./kg., which may be due to a more prolonged action of this drug, but little is known at present concerning either its duration of action or its tissue distribution. Myleran does, however, show the interruption in recovery characteristic of the 200r and 400r curves.

Also in Figure 7 are shown the results for a more limited study using Aminopterin at a dosage of 1·25 mg./kg., given intraperitoneally.

Discussion

As a sensitive criterion of the effect of radiation on the erythropoietic system the depression of radioactive iron uptake in the blood provides a very attractive technique. The present studies show, however, the need for care in interpreting the experimental data. The first-day uptake in blood at various times after radiation will depend both on the rate of emergence of new red cells and the rate at which these cells take up iron at the time of ^{59}Fe injection. The depression observed in the first-day uptake will depend on the time of maturation of the red cell precursors, their relative capacities for iron uptake and the time relation between radiation damage and marrow regeneration, and thus it is likely to be very species-dependent.

It must also be remembered that the iron metabolism of the animal is less affected if the irradiation of the body is not uniform, as shown by studies of hind leg shielding (Baxter *et al.*, 1955; Belcher *et al.*, 1955) so that the rate of appearance of ^{59}Fe into the peripheral blood may be considerably changed if the radiation exposure, or drug used, does not affect all erythropoietic tissues uniformly.

From measurements of uptake in peripheral blood alone it is difficult to assess the iron-incorporating capacity of the erythropoietic tissues at the time of injection. However, this may be possible if an estimate can be made of the extent to which storage iron is utilised, or if such utilisation can be prevented.

Summary

A study has been made of the uptake into the blood of the rat of radioactive iron given at different times relative to X-radiation or to the administration of various drugs. The results are discussed in terms of the mechanisms by which these agents may depress erythropoietic function.

Acknowledgments

This work has been made possible only by the enthusiasm and expert help of Miss K. Adams and Miss M. Winsborough. We are indebted to a number of colleagues for their collaboration and for helpful discussions, and to Professor W. V. Mayneord, Director of the Physics Department, Institute of Cancer Research, for his constant support and encouragement.

The financial help of the Medical Research Council and the British Empire Cancer Campaign is gratefully acknowledged.

REFERENCES

BAXTER, C. F., BELCHER, E. H., HARRISS, E. B., and LAMERTON, L. F., 1955. *Brit. J. Hæmat.*, **1**, 86.

BELCHER, E. H., GILBERT, I. G. F., and LAMERTON, L. F., 1954. *Brit. J. Radiol.*, **27**, 387.

BELCHER, E. H., HARRISS, E. B., and LAMERTON, L. F., 1955. *Progress in Radiobiol.*, 1956, p. 303 (Edinburgh, Oliver and Boyd).

BONNICHSEN, R., and HEVESY, G., 1955. *Acta Chem. Scand.*, **9**, 509.

HENNESSY, T. G., and HUFF, R. L., 1950. *Proc. Soc. exp. Biol. N.Y.*, **73**, 436.

LAJTHA, L. G., and SUIT, H. D., 1955. *Brit. J. Hæmat.*, **1**, 55.

SUIT, H. D., LAJTHA, L. G., ELLIS, F., and OLIVER, R., 1955. *Progress in Radiobiol.*, 1956, p. 506 (Edinburgh, Oliver and Boyd).

DISCUSSION

Hevesy. I wish to be permitted to add a few words to Dr. Lamerton's illuminating address. As he mentioned, and as was shown previously by Hennessy and Huff (1950), exposure to radiation depresses the ^{59}Fe incorporation into hæmoglobin. We found the same result, and also that the incorporation of ^{59}Fe into myoglobin is interfered with as well, by an exposure of rats to 500r of X-rays. As contrasted to this the ^{59}Fe incorporation into cytochrome and catalase was not affected. Similar results are obtained when investigating the effect of radiation on guinea-pigs. The amount of myoglobin present in the rat is about thirty times less than that of hæmoglobin. In the formation of the former, 40% of the body weight is involved, while in the formation of the latter only about 2%. The life cycle of myoglobin, furthermore, is longer than that of hæmoglobin. Thus the existence of radiosensitive 'myoglobetic' cells, in minute amounts in the muscle tissue, may explain the radiosensitivity of myoglobin formation in a similar way to that in which the great radiosensitivity of marrow cells explains the interference of radiation with hæmoglobin formation. An alternative explanation is that in the formation of myoglobin the hæm of hæmoglobin is involved, as the blood plasma always contains traces of hæmin.

About two-thirds or more of the ^{59}Fe leaving the plasma finds its way to the bone marrow and one-third into other organs. In the exposed animal, the number of marrow cells in which hæmoglobin can be laid down is greatly reduced after about one day. Therefore, a larger ^{59}Fe fraction is bound to be taken up by the extramedullary organs of the irradiated rat than by those of the control. This fact, also brought out by Dr. Lamerton's data, explains, at least partly, the increased ^{59}Fe uptake by the iron-containing compounds, hæmins and non-hæmins of the liver. Another hæmin is also interfered with by irradiation, namely the non-watersoluble hæmoglobin in the nuclear fraction of the liver of rats, mice, rabbits, and guinea-pigs, Bonnichsen *et al.* (1955). Incorporation of ^{59}Fe into the nuclear hæmoglobin fraction of the liver is seemingly less depressed than the incorporation of ^{59}Fe into the circulating hæmoglobin. But the difference almost disappears if we take into account the higher activity level of the liver in the exposed animal in which this hæmoglobin is formed. The effect of radiation on ^{59}Fe incorporation into the nuclear hæmoglobin shows a similar time lag to that of the incorporation into the circulating hæmoglobin. It is hardly manifest when the rats are injected at once after irradiation and killed 5 hours later, but is very pronounced one day after exposure to radiation.

Lajtha. Has Professor Hevesy studied the time factor in the increase of iron uptake after irradiation? Does the increase take place only when ^{59}Fe is given immediately after irradiation or does he find an increase also when given one or more days after irradiation?

Hevesy. Increased ^{59}Fe incorporation into the iron fractions of the liver, due to exposure, is observed only several hours or more after irradiation.

Lajtha. One part of the evidence Dr. Lamerton brought forward for the retention theory of reticulocytes in the marrow was that two days after radiation the reticulocytes were so much saturated with hæmoglobin that they showed hardly any ^{59}Fe uptake. Now, reticulocytes which are near to the erythrocyte stage are very difficult to count. I wonder whether you have done the reverse type of experiment, i.e. to add ^{59}Fe immediately after irradiation? In that case, if the reticulocytes are stained they should show a higher grain count than in the non-irradiated controls.

Lamerton. No, we have not done that, but it is a suggestion which is well worth taking up.

van Bekkum. After irradiation with a slightly larger dose, e.g. 700r, rats develop a severe anæmia during the second week after irradiation. According to several authors this anæmia is partly the result of blood loss during the hæmorrhagic stage. Have you obtained any data to support this theory, using the ^{59}Fe method?

Lamerton. Our own observations on the relation between platelet fall and the development of radiation anæmia suggest that the anæmia is largely the result of the severe platelet fall. This, of course, is the conclusion drawn from the experiments of Cronkite and other workers in the United States on the effect of platelet transfusion following whole-body radiation. With regard to studies of radioactive iron turnover in the various tissues of the rat we have worked with a dose of 550r, and found that, as in the case of lower doses, the erythropoietic regeneration commences in the spleen and is followed after an interval by bone-marrow regeneration.

IN VIVO UPTAKE OF RADIOACTIVE IRON BY THE ERYTHROID CELLS OF RAT BONE MARROW

EILEEN B. HARRISS

Physics Department, Institute of Cancer Research,
Royal Cancer Hospital, London, S.W.3, England

The use of ^{59}Fe in the study of the effects of whole-body irradiation and chemical agents on bone-marrow function has been discussed during this Conference by Dr. Lamerton.* In certain of the studies reported by him, measurements were made of the amount of radioactive iron appearing in the circulating blood when the interval between irradiation and administration of active iron was varied. The interpretation of such experiments requires a knowledge of the stage, or stages, in the development of the erythrocyte in which iron is incorporated, and also of the time spent by the developing erythrocyte in the various stages in the bone marrow. Information concerning these aspects of the problem can be obtained using high resolution autoradiographic techniques, since in this way the active cells can be identified individually and an estimate can be made of the relative activities of cells in the various stages. Such investigations have been carried out by Lajtha and Suit (1955), by Suit (1956) and by Austoni (1955). Suit, using rabbits, made autoradiographs of bone-marrow smears obtained at various intervals after injection of ^{59}Fe. He found that the initial uptake of iron was predominantly in the early forms (pronormoblast and early normoblast) and thus was able to follow the passage of the activity through the stages of development. Lajtha and Suit have obtained similar results using human bone marrow *in vitro*. Austoni used rats and, in contrast to Suit's results, he found iron was incorporated at all stages of development, including polychromatic erythrocytes (reticulocytes). He administered ^{59}Fe intraperitoneally whereas, in the work reported by Lamerton, radioactive iron was given intravenously. As the appearance of activity in the circulating blood is delayed after intraperitoneal injection and in view of the species difference shown by Austoni and Suit, it was necessary to repeat Austoni's work using intravenous injection, and in particular, our own inbred strain of rats. Also, it is possible to obtain more accurate results by grain counts of active cells. Although the quantitative results are not yet complete, preliminary information from our experiments is of interest and is in agreement with Austoni's results.

* Lamerton: this volume, p. 321.

Adult male 'August' rats, of about 250 gm., were injected intravenously with 20 μc. ^{59}Fe in citrate buffer, and sacrificed at 1, 2, 4, 8, 12, 18, 24, 36 and 48 hours after injection. The animals were heparinised just before death and the hind limbs perfused with saline, via the abdominal aorta, in order to remove as much blood as possible from the marrow and thus avoid a high background on the autoradiographs when blood activity was high.

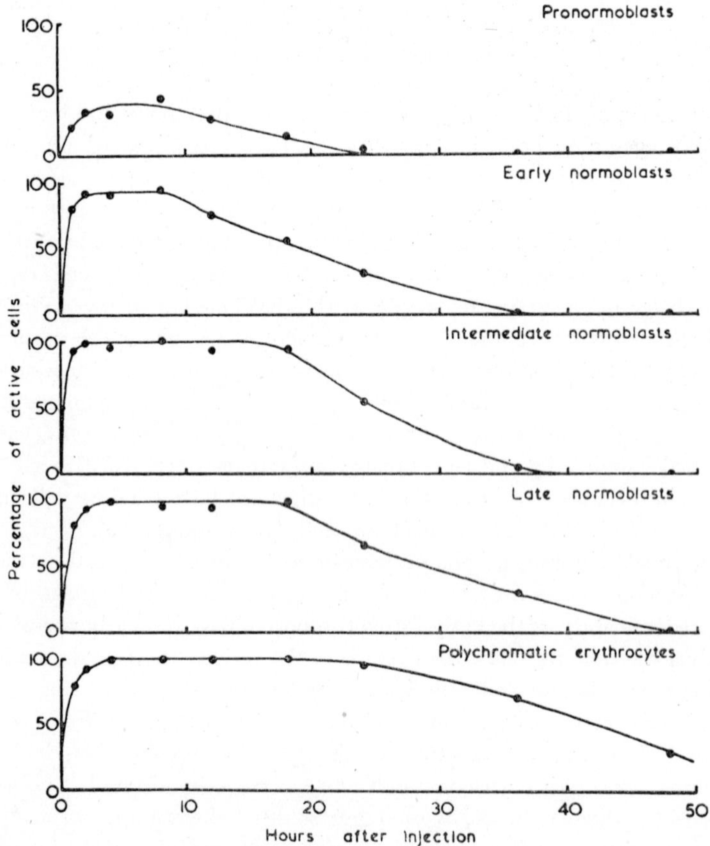

FIG. 1. Percentage of cells of each type found active at various intervals after injection.

Marrow from both femora and tibiæ was blown out into a specimen tube with about 3 ml. saline and shaken gently to break up the clumps of cells. The supernatant was removed from the débris and allowed to settle for about 5 minutes. After this time, the supernatant was again removed and the sediment was used for making smears. In this way, smears could be obtained in which the cells were sufficiently separated from one another to

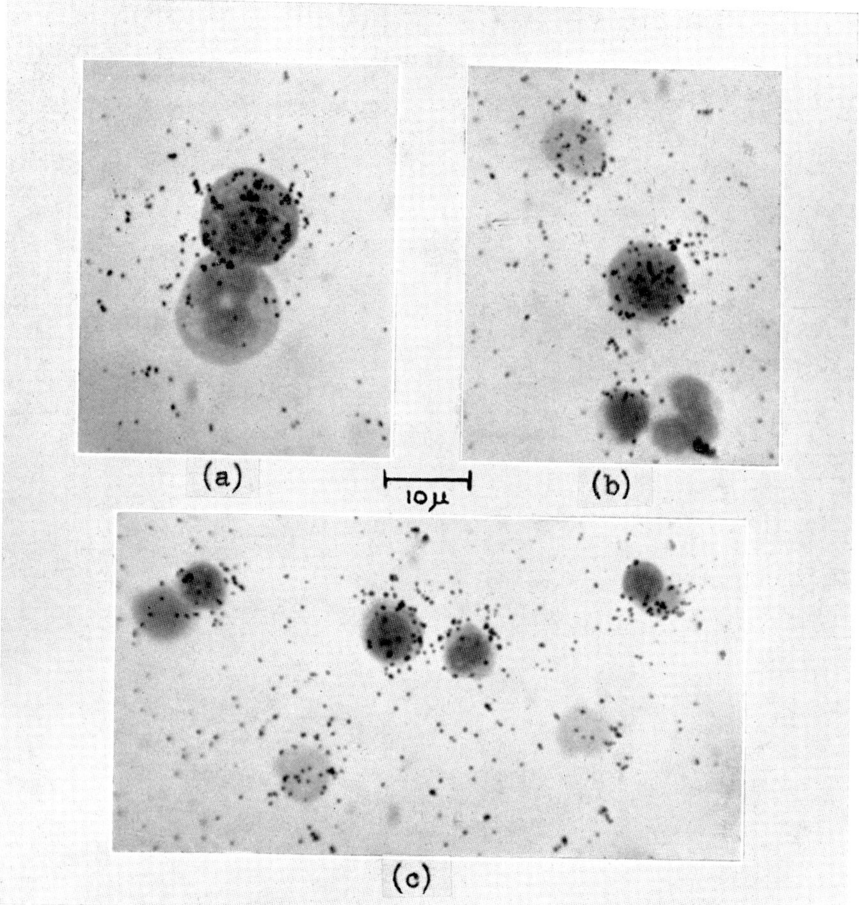

FIG. 2. Stripping film autoradiograph of bone marrow smear taken from rat killed 4 hours after intravenous injection of 20μc. ^{59}Fe. Exposure 14 days. Stained with May Grünwald-Giemsa. (*a*) early normoblast, (*b*) intermediate normoblast, (*c*) late normoblasts.

give good autoradiographic resolution, and the dilution with saline prevented a high background when the plasma activity was high. The smears were fixed in methyl alcohol and stripping film autoradiographs prepared using Kodak autoradiographic film. After two weeks' exposure the slides were developed for 5 minutes in D19b developer at 21° C., and fixed. They were washed thoroughly in running water, then washed in phosphate buffer at *p*H 6·8, and stained with May Grünwald-Giemsa stain. The final washing at *p*H 6·8 improved the differential staining. An example of a bone-marrow autoradiograph is shown in Figure 2.

TABLE

Stage	Active cells per 100	Mean grains per active cell	% of cells present	Active content of stage (arbitrary units)	Active content as percentage of total activity
Pronormoblast ...	32	25	2·5	0·22	0·7
Early normoblast ...	90	34	11·5	3·5	11·9
Intermediate normoblast	98	36	24	8·3	28·2
Late normoblast ...	93	30	31	8·7	29·6
Polychromatic erythrocyte	91	30	31	8·7	29·6

Cells were classified as pronormoblast, early, intermediate or late normoblast according to their nuclear structure; polychromatic erythrocytes could easily be distinguished from the mature erythrocytes remaining after perfusion. For each time of sacrifice, 100 cells of each type were examined and the numbers of active cells having a grain density above background level were counted. From these figures, the variation with time of the percentage of active cells could be plotted for each stage of development; these curves are shown in Figure 1. From these curves it can be seen that there is an immediate uptake into all types of cells, and that within two hours of injection practically all the normoblasts and the polychromatic erythrocytes are active. The polychromatic erythrocytes remain 100% active until 24 hours after the injection, this stage being filled by active cells from the previous stages. After this time, the percentage of active cells of this type begins to fall, and nearly all the activity has been released from the marrow into the blood stream by 48 hours. The time of development from early normoblast in the marrow to polychromatic erythrocyte in the blood is therefore about two days.

At the same time as the percentage of active cells is decreasing, the active content per cell is also decreasing. In order to obtain an estimate of the decrease in active content, counts were made of the mean number of grains over active cells of all stages at the various times of sacrifice after injection.

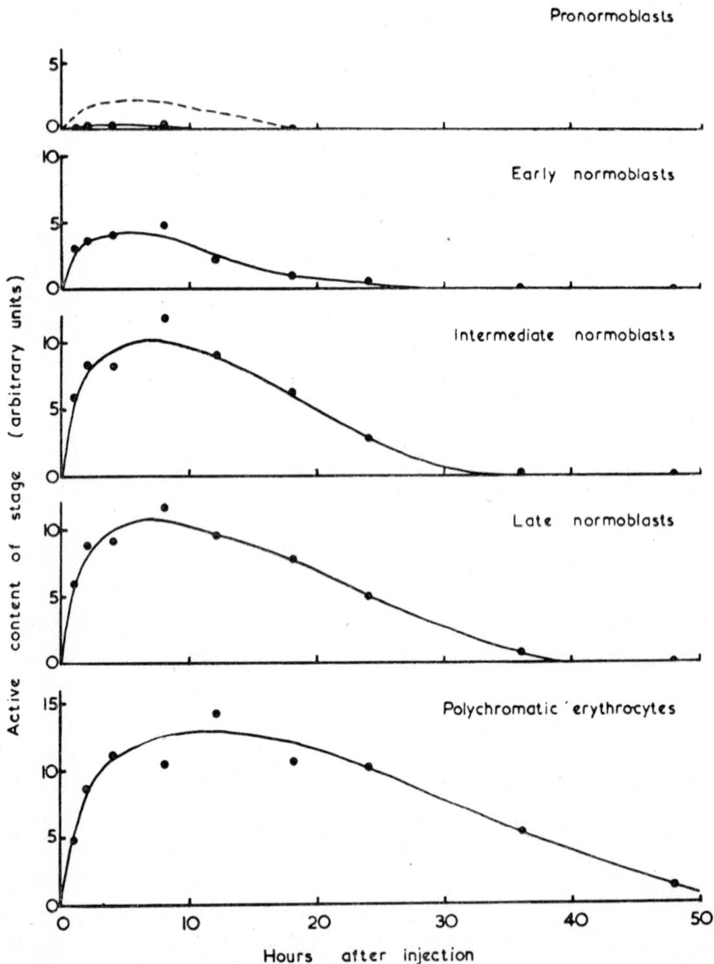

FIG. 3. Variation of active content of erythrocyte precursors with time after injection. Curve for pronormoblasts is also shown with ordinates $\times 10$ (dotted curve).

The product of number of grains per cell and percentage of active cells gives a measure of the active content of 100 cells of each type. If, in addition, the relative proportions of the various stages comprising the erythroid fraction of the marrow is known, an estimate can be made of the total active content of each stage of development. Table I shows figures obtained for a rat killed two hours after injection, when it may be expected that little maturation of cells has occurred.

It may be noted that although the early normoblasts are practically all active, the active content of the group is much lower because there are relatively few cells of this type present. It is also clear that there is very little activity in the pronormoblast stage. The bulk of the activity is in the

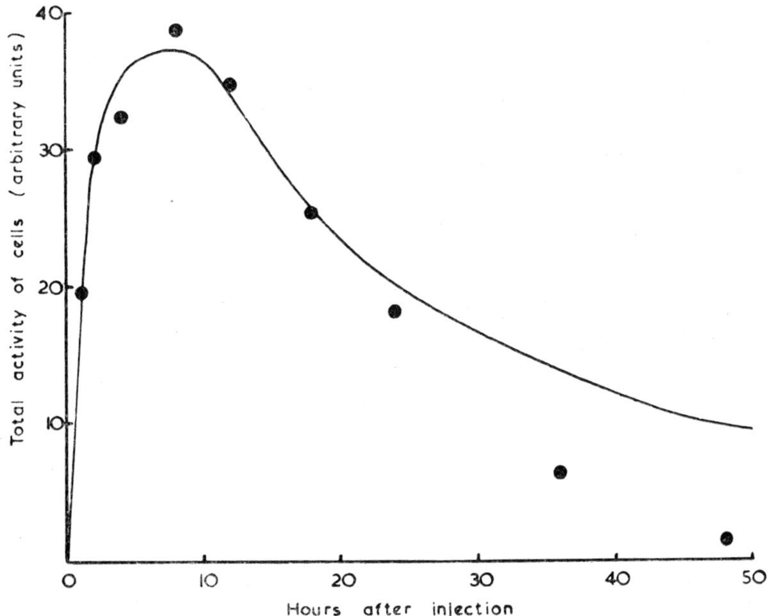

FIG. 4. Turnover of intravenously injected ^{59}Fe in rat bone marrow.
———: experimental curve obtained by scintillation counting of hind limbs.
●—●: turnover calculated from autoradiographic data.

intermediate and late normoblasts and little of the injected activity goes to the early forms, which is in agreement with the shape of the 'blood activity' curve of a normal animal. The active cells appearing in the circulating blood during the first day after injection are probably those which incorporated ^{59}Fe in the marrow in the late normoblast and polychromatic erythrocyte stage, whereas the active cells appearing during the second day after injection incorporated iron whilst in the early and intermediate normoblast stage. From measurements of blood activity after intravenous injection, we have found that of the activity appearing in the blood in 2 days, about 60% appears in the first day and 40% in the second day, which is in the same ratio as the percentage activities of the late and early forms (Table).

The variation of active content of each stage with time after injection is shown in Figure 3. Again it can be seen that the more mature stages lose their activity at increasing times after injection and that the time taken to clear the marrow activity is about 50 hours.

The sum of the active contents of the various stages represents the total activity present in the cells at any given time. Figure 4 shows the total activity present at various times after injection. This curve agrees very well with hind-limb turnover curves obtained by scintillation counting of the hind limbs of animals killed at various times after administration of ^{59}Fe.

Acknowledgments

This work was carried out in the Physics Department, Institute of Cancer Research, Royal Cancer Hospital, and the author would like to express her thanks to Professor W. V. Mayneord, Director of the Physics Department, for his support and encouragement, and to Dr. L. F. Lamerton for many helpful discussions.

REFERENCES

AUSTONI, M., and ZILIOTTO, D., 1955. *Acta Med. Patavina*, **15**, 1.
LAJTHA, L. G., and SUIT, H. D., 1955. *Brit. J. Hæmatol.*, **1**, 55.
SUIT, H. D., 1956. (Personal communication.)

DISCUSSION

Lajtha. Looking at the curve you have presented on the hind limb you get a halving of the ^{59}Fe content of the marrow by 24 hours. However, your grain counts over the individual cells do not appear to be half by the end of the first 24 hours. Does this mean that the cells leave the marrow faster than they divide—since each division after the first 2–3 hours following the ^{59}Fe injection should halve the grain counts? Probably the answer is that the high proportion of active reticulocytes leave the marrow quickly. Do you know how cell division keeps up with cell production?

Harriss. I think it is difficult to deduce anything about cell division from the results so far available.

Lajtha. You have shown that the pronormoblasts are never 100% positive. I think this may be due to the difficulty of defining a pronormoblast. Identical looking cells may take up ^{59}Fe or $^{35}SO_4$—which means that some of them are in fact erythroid, others myeloid—although they look the same. I think at that early stage it is very difficult to identify cells purely on morphological grounds.

Harriss. I am interested to hear from Dr. Lajtha of his difficulty in distinguishing pronormoblasts. I think this may account for the low percentage of active pronormoblasts recorded from the autoradiographs.

Hevesy. Miss Harriss found that almost all iron leaves the bone marrow within two days. Correspondingly, after that date the ^{59}Fe content of the circulating red corpuscles does not increase appreciably. In the human it takes several days until the maximal ^{59}Fe uptake into the red corpuscles is achieved. You have then, to suppose that ^{59}Fe makes an appreciably longer stay in the human bone marrow.

Harriss. Work with radioactive iron certainly indicates this difference in iron turnover between rat and man, and the sigmoid shape of the uptake curve of ^{59}Fe in human blood fits well with the assumption that more iron is incorporated into the earlier than into the later forms of erythrocyte precursors in the human bone marrow.

Hulse. I am very interested in your results, particularly those from which you have been able to deduce the time taken for the red cells to mature. The only previous results I have seen on this subject are those of Austoni. His results, however, do not show the logical sequence from one cell type to another which yours do. Do you know of any reason for the difference?

Harriss. One difference between Austoni's experiments and those I have described is the mode of injection. He injected iron intraperitoneally whereas I injected it intravenously. We know that the turnover rate in the bone marrow varies with the injection route, and this may account for some differences in our results.

Eldjarn. I would like to ask one question, possibly related to the small differences between your results and those of Austoni. What was the chemical form of the injected iron and what was the specific activity?

Harriss. Iron was injected as a citrate complex. ^{59}Fe as ferric chloride was buffered to a *p*H of about 5 with citrate buffer before intravenous injection. The specific activity of the iron was about 5 μc./μg. The total iron injected was well below serum iron levels, and was believed to be a true 'tracer' dose.

Taylor. The weight of iron which has been used by Miss Harriss has varied between about 4 and 10 micrograms.

Hevesy. Our rabbits have a plasma iron content amounting to about 100 μg., we increase this figure by injecting labelled iron to 101–102, thus the increase in our experiment in which we observe disappearance of a quite appreciable fraction of iron within one minute is almost negligible. This increase is much less than that observed in the course of the diurnal fluctuation of the iron level of the plasma. The binding capacity of the plasma for iron being about 300 μg.% we are very far from the saturation point of the plasma β_1-globulin for iron. It is, however, possible that prior to leaving the plasma, the iron was temporarily bound to proteins as even a quite appreciable fraction of injected protein-bound iron leaves the circulation of the rabbit within one minute or less. This fact vitiates the calculation of the plasma volume of the rabbit—which would be a very convenient procedure—from dilution figures of the injected protein-bound iron. In humans, however, Dal Santo working in our laboratory, by making use of this method, arrived at almost correct plasma volume figures.

Harriss. We have measured plasma clearance rates in our rats using solutions of citrated ^{59}Fe similar to those used in the autoradiography experiments and also using solutions of ^{59}Fe bound to rat plasma. We have found no appreciable difference in the clearance rates for the two solutions.

Hevesy. If you inject even small amounts of labelled FeCl into the circulation of the rat does not a fraction of ^{59}Fe leave the plasma prior to being bound to the β_1-globulin? In experiments with rabbits we obtained indications of such disappearance.

Harriss. In plasma clearance studies in which several blood samples were taken during the first minute after injection, we were unable to observe any rapid component of the clearance curve, whether iron was injected as a protein complex or as iron chloride in citrate buffer. However, the figures for plasma volume obtained by extrapolation of the clearance curve were about 20% greater than those which would be expected on the basis of observed red cell volumes and venous hæmatocrits. This, it is true, might imply an extremely rapid loss of iron from the circulation during the mixing period but might also be explained in terms of the difference between venous and body hæmatocrits. Further data on plasma volumes will help to elucidate this point.

Faber. The problem of the iron incorporation into human hæmoglobin is even more complex. If you separate the hæmoglobin into fœtal (i.e. alkaliresistant) and normal you find an iron incorporation in the fœtal at a rate 5 times that of the normal hæmoglobin. This suggests the presence of two types of precursors in the bone marrow.

Lajtha. Do you think that there is a special population of cells which make fœtal Hb, or do you think all cells have the enzyme system to manufacture both? It is more likely that one population of cells is concerned with fœtal Hb, but is there any experimental evidence for it?

ERYTHROPOIETIC ACTIVITY IN IRRADIATED RATS INJECTED WITH HOMOLOGOUS AND HETEROLOGOUS BONE MARROW STUDY WITH ^{59}Fe

H. Maisin, A. Dunjic, P. Maldague and J. Maisin

The Laboratory for the Protection of Civil Population, Department of Radiobiology, Cancer Institute, University of Louvain, Belgium

Bone marrow protection during irradiation, or transplantation of homologous normal bone marrow after irradiation, improves survival rate, reduces second weight drop and improves the hæmatological response of white inbred rats in our laboratory (Mandart *et al.*, 1952; Maisin J., *et al.*, 1954 and 1955).

In a previous experiment (Maisin, H. *et al.*, 1954 and Maisin J. *et al.*, 1955) taking iron-utilisation curves of the blood as a measure of erythropoiesis (Hennessy and Huff, 1950; Huff *et al.*, 1950, and Maisin, H. and Wolfe, 1953) we have observed a protective effect in irradiated rats with one hind leg shielded, and a therapeutic effect in rats receiving homologous bone marrow intraperitoneally after irradiation. In irradiated rats with one hind leg shielded, the time-concentration relationships of the tracer, in the shielded femur were almost identical with those noted in the femur of the non-irradiated controls, and, in the non-shielded femur with those noted in the femur of the irradiated controls. 0·33 μc. ^{59}Fe was injected subcutaneously on the 3rd post-irradiation day and rats were sacrified at various times after injection (6, 12, 24 hours, 3, 5, 9, 11, 13 and 18 days). All irradiated animals received an LD 50 dose from an X-ray source.

In a recent experiment, we have employed a technique described by Lamerton *et al.* (1955) (Baxter *et al.*, 1955). The rats were sacrificed at various times after irradiation (1, 2, 4, 7, 10, 13, 16 and 19 days). At a time of 24 hours before sacrifice, they were injected subcutaneously with 0·66 μc. ^{59}Fe as ferric chloride in citrate buffer. Blood, plasma and samples of various tissues after perfusion (liver, spleen, femur and omentum with pancreas), were assayed for radioactivity in a vial scintillation counter. Three groups of rats 4 months old and 145-155 gm. in weight were irradiated with 600r (LD 50). One group served as irradiated controls, a second had one hind leg shielded, a third received intraperitoneally a suspension of 120 mg. of homologous bone marrow in 1 ml. of saline immediately after irradiation. The injected cells are normally collected in the omentum (Congdon *et al.*, 1952; Maisin, J. *et al.*, 1955). Finally, a fourth group received ^{59}Fe 24 hours before sacrifice and served as non-irradiated controls.

The results are given in Figures 1 and 2, each point on the curves represents the mean obtained from 6 rats. The 24-hours uptake in the red cells in the

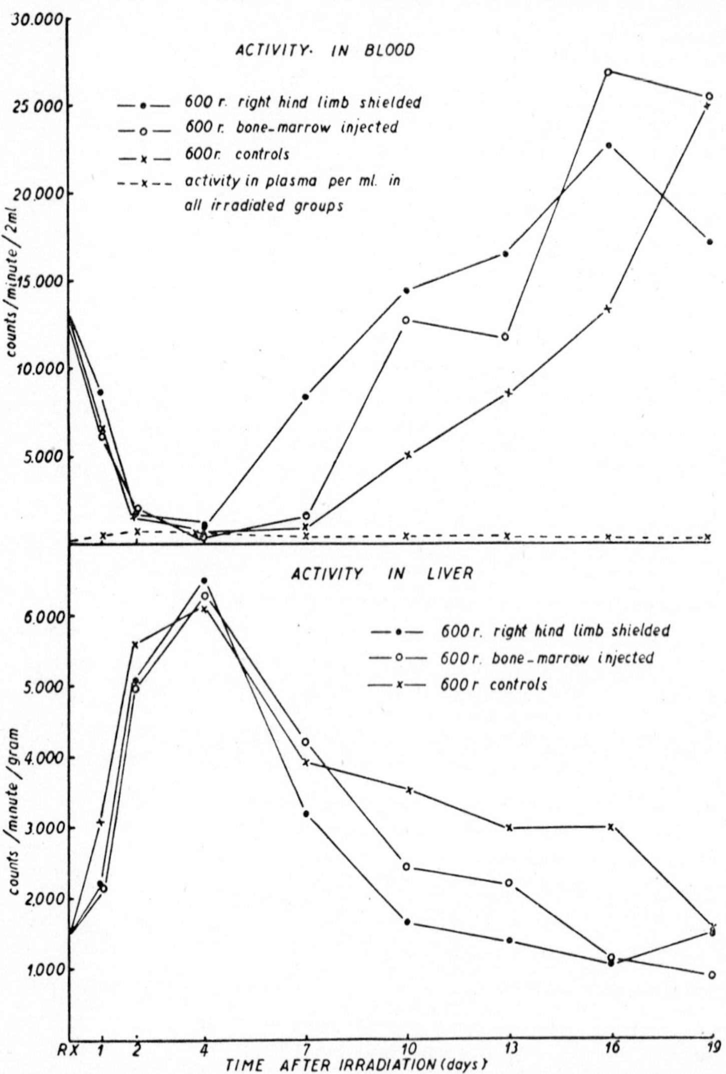

FIG. 1. Mean value of the 24-hours uptake of ^{59}Fe in circulating blood (2 ml). and in liver (1 g.). Rats injected at various times after 600r X-irradiation. 24-hours mean value uptake of non-irradiated control rats is at the starting point of the curves.

circulating blood is similar in the three irradiated groups during the first days after irradiation. The depression of uptake is seen to be maximal between 2 and 7 days. In irradiated controls, after the seventh day, the

24-hours uptake rises and reaches a normal level after 16 days. In groups with one hind leg shielded or receiving bone marrow after irradiation, the

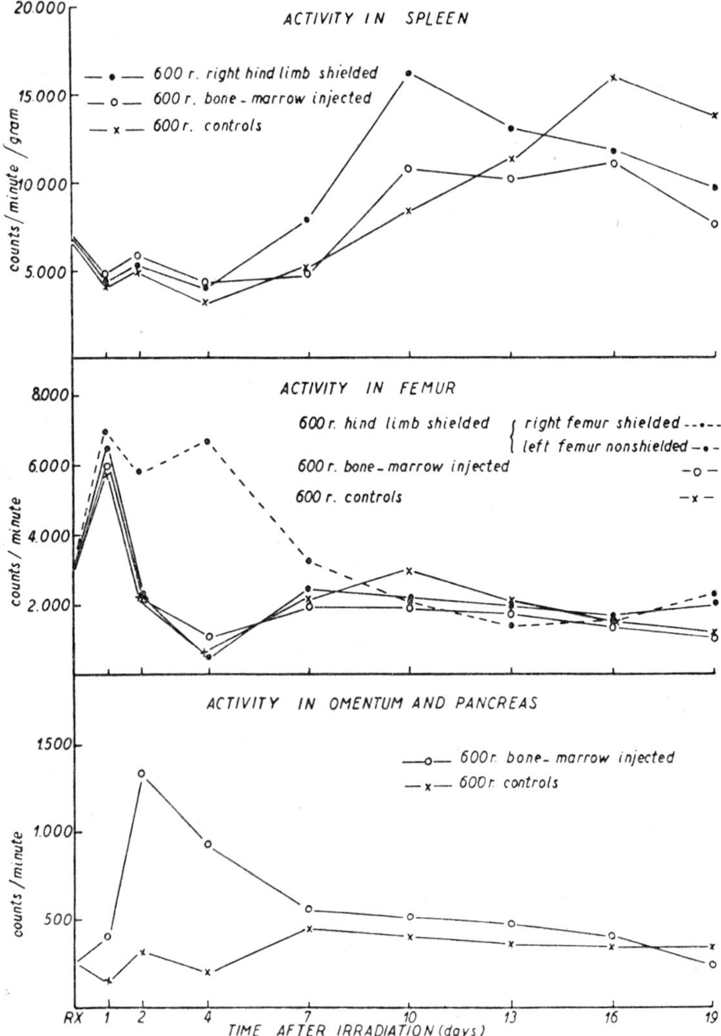

FIG. 2. Mean value of the 24-hours uptake of ^{59}Fe in spleen, femur and omentum + pancreas (intraperitoneally injected marrow cells are normally collected in the omentum). Rats injected at various times after 600r X-irradiation. 24-hours mean value uptake of non-irradiated rats is at the starting point of the curves.

24-hours uptake begins to rise earlier. In all irradiated animals (injected with ^{59}Fe immediately after irradiation) the first-day uptake in femur is above normal, this hyperactivity is maintained in shielded femurs up to the

7th day while uptake in all unshielded femurs is depressed between the 2nd and 7th day. In the shielded and bone-marrow injected animals recovery of irradiated bone marrow is no faster than in irradiated controls. Later, the activity of shielded and unshielded femurs remains similar. The shape of the curve of the iron uptake in the graft is not so high but is similar to that of a shielded marrow. From our histological data it appears that this graft is functional. In all irradiated groups erythropoiesis appears in spleen during recovery. The spleen of our non-irradiated rats is not erythropoietic. This is confirmed by histological examination. The splenic erythropoiesis begins earlier in the shielded and bone-marrow injected animals, than in irradiated controls. The liver uptake is far above normal in all irradiated groups between 2 and 7 days. Later, for each group it is the opposite of the activity in red cells; this shows that during the inactivity of the marrow the iron is stored in the liver.

We confirm the findings of Lamerton *et al.* in irradiated controls and in leg-shielded rats during irradiation (1955, Baxter *et al.*, 1955). In addition we study the behaviour of bone marrow the first day after irradiation, and during the recovery period and we analyse the role of storage by the liver. We complete and extend this work with animals receiving homologous bone-marrow suspensions after irradiation. We failed to observe, employing this technique, a faster recovery of erythropoiesis in the irradiated bone marrow in the shielded animals and in the bone-marrow injected animals. We conclude from this experiment that it is the shielded bone marrow or the graft which carries on erythropoiesis during the inactivity of the irradiated bone marrow and tides the animal over this critical period in the LD 50 dose range. Naturally it could be that after a very high dose of irradiation the injected bone marrow can entirely replace the host's own marrow as Ford *et al.* suggested recently (1956).

This experiment had given indirect evidence of the erythropoietic activity of the graft. We thought that the study of the fate of ^{59}Fe-labelled bone marrow, injected intraperitoneally or intravenously, would be a more direct proof and at the same time a new approach to a problem already studied histologically by Congdon *et al.* (1952).

Young rats 10 weeks old, 130 gm. in weight and young guinea-pigs 4 weeks old, 150 gm. in weight, were injected subcutaneously with 0·66 μc. ^{59}Fe as ferric chloride in citrate buffer. Eighteen hours after injection, at a time when radioactive iron uptake in bone marrow is nearing its maximum and when plasma contains insignificant amounts of radio-iron, the animals were sacrificed. All bone marrow from femora, tibiæ and humeri were collected and the activity assayed. Activity in bone marrow from three rats or from two guinea-pigs was about 15,000 counts per minute. Such an amount of ^{59}Fe-labelled rat or guinea-pig bone marrow suspended in 0·5 to 1 ml. of saline was injected intraperitoneally or intravenously (femoral vein) into rats 24 hours after total-body irradiation of 1,000r at a time when the erythropoiesis of the host is entirely depressed. The dose of X-rays used is LD 100 between 4 and 7 days in our laboratory. Two groups of irradiated rats which served as controls received the same dose (15,000 counts per minute) of

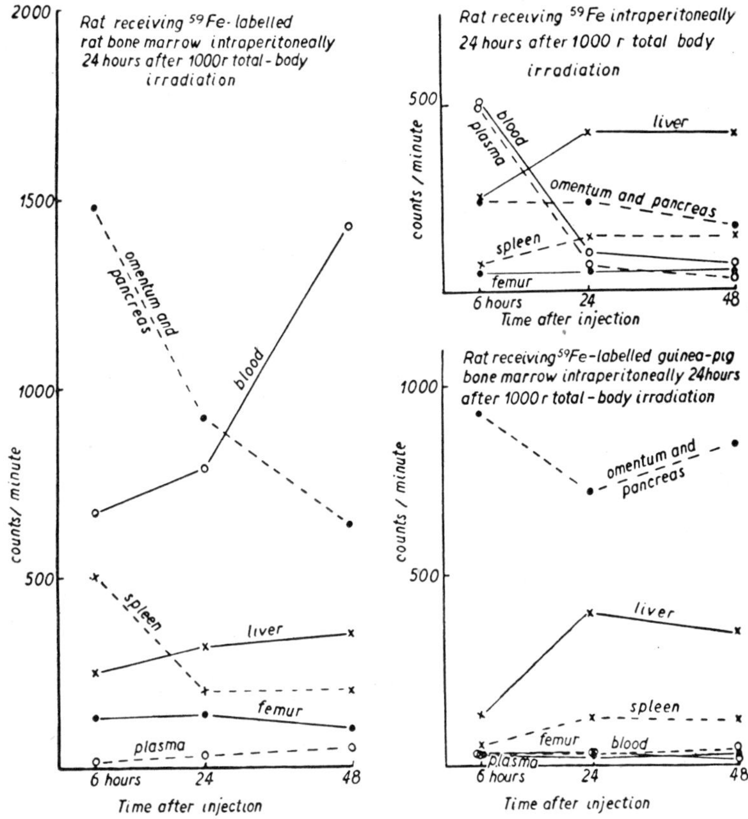

FIG. 3. Iron turnover in various tissues, in rats receiving 15,000 counts per minute, ^{59}Fe-labelled rat or guinea-pig bone marrow or 15,000 counts per minute ^{59}Fe intraperitoneally 24 hours after 1,000r total-body irradiation. Rats sacrificed at various times after injection.

^{59}Fe as ferric chloride in citrate buffer. All the irradiated rats were 4 months old and 140–150 gm. in weight. The rats were sacrificed at various times after injection (6, 24 and 48 hours). In all groups 2 ml. of blood, 1 ml. of plasma and samples of various tissues after perfusion (1 gm. of liver, spleen and femur), were assayed for radioactivity. Furthermore, in irradiated controls and in intraperitoneally bone-marrow injected rats, we examine the activity of the omentum and the pancreas, and in intravenously bone-marrow injected rats the activity of the lungs.

The results are given in Figures 3 and 4. Each point on the curves represents the mean obtained from 3 to 5 rats.

In the rats receiving ^{59}Fe-labelled rat bone marrow intraperitoneally, activity in graft (omentum) is decreasing while uptake in blood increases between 6 and 48 hours after injection. The two curves cross each other.

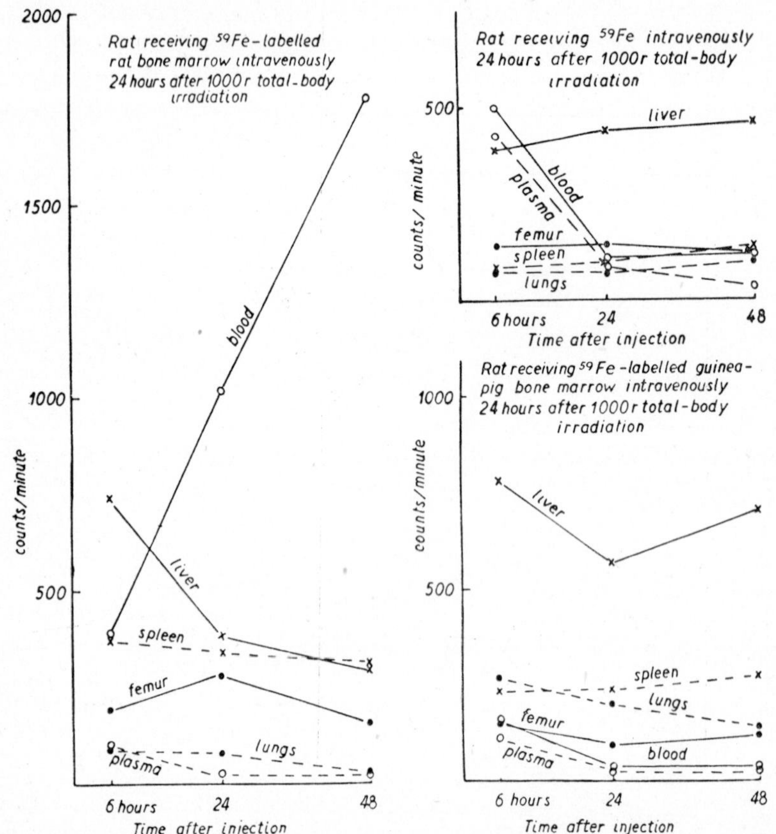

FIG. 4. Iron turnover in various tissues in rats receiving 15,000 counts per minute, ^{59}Fe-labelled rat or guinea-pig bone marrow or 15,000 counts per minute ^{59}Fe intravenously 24 hours after 1,000r total-body irradiation. Rats sacrificed at various times after injection.

Insignificant activity in the plasma proves that ^{59}Fe is incorporated in red cells. 1,000r total-body irradiation excludes formation of labelled red cells from host bone marrow. Furthermore, the irradiated controls do not show any activity in the blood 48 hours after intraperitoneal injection of ^{59}Fe. We conclude that the homologous graft pours red cells into the circulating blood very early after its implantation and is functional. In the rats receiving ^{59}Fe-labelled guinea-pig bone marrow intraperitoneally, the graft retains the largest quantity of ^{59}Fe, and activity in the blood does not appear. Thus the heterologous graft is not functional. Our histological data reveals that these grafts are necrotic. The iron uptake in liver increases in the three groups and is higher in irradiated controls and less important in homologous bone-marrow injected animals.

The largest part of the activity 6 hours after intravenous injection of ^{59}Fe rat or guinea-pig bone marrow is in the liver. Spleen, femur and lungs are much less active. Later the liver activity of the animals injected with homologous bone marrow falls while their red cell uptake rises rapidly like animals which were injected intraperitoneally. On the contrary, the activity of the liver of the rats which received intravenous guinea-pig bone marrow, remains high, and labelled red cells in their blood do not appear. The irradiated controls injected intravenously with ^{59}Fe do not, of course, show any activity in their blood and the uptake in liver rises.

Finally we can add that the fate of ^{59}Fe-labelled rat or guinea-pig bone marrow injected intraperitoneally or intravenously is the same in non-irradiated rats.

We conclude that in irradiated rats homologous bone marrow is functional and pours erythrocytes into the circulating blood very soon after its intraperitoneal or intravenous injection. Guinea-pig marrow is unable to take over erythropoiesis in irradiated rats.

These results are not in favour of the humoral theory.

REFERENCES

BAXTER, C. F., BELCHER, E. H., HARRISS, E. B., and LAMERTON, L. F., 1955. *Brit. J. Hæmatol.*, **1**, 86.

CONGDON, C. C., UPHOFF, D., and LORENZ, E., 1952. *J. Nat. Cancer Inst.*, **13**, 73.

FORD, C. E., HAMERTON, J. L., BARNES, D. W. H., and LOUTIT, J. F., 1956. *Nature, Lond.*, **177**, 452.

HENNESSY, T. G., and HUFF, R. L., 1950. *Proc. Soc. exp. Biol. N.Y.*, **73**, 436.

HUFF, R. L., BETHARD, W. F., GARCIA, J. F., ROBERTS, B. M., JACOBSON, L. O., and LAWRENCE, J. H., 1950. *J. Lab. clin. Med.*, **36**, 40.

LAMERTON, L. F., and BELCHER, E. H., 1955. *Radiobiol. Symp.* 1954, p. 136 (London, Butterworth).

MAISIN, H., DUNJIC, A., MALDAGUE, P., 1954. *Symp. Radiosotopes Institut Interuniversitaire Sciences nucleaires. Bruxelles* (1954) *Rapport Annuel*, p. 260.

MAISIN, H., and WOLFE, R., 1953. *Observation, U.C.R.L.* 3328 *U.S. Atomic Energy Com. Report* (1956).

MAISIN, J., MAISIN, H., and DUNJIC, A., 1954. *C.R. Soc. Biol. Paris*, **148**, 1293.

Idem, 1955. *Radiobiol. Symp.* 1954, p. 154 (London, Butterworth).

MAISIN, J., MAISIN, H., DUNJIC, A., MALDAGUE, P., 1955. *J. Belge. Radiologie.*, **38**, 394.

MANDART, M., LAMBERT, G., and MAISIN, J., 1952. *C.R. Soc. Biol.*, **146**, 1392.

Discussion

Lamerton. I would like to ask Dr. Dunjic whether he is certain that there is no erythropoiesis in the spleen of his control rats. Also, is it the case that the erythropoietic regeneration occurs first in the spleen when bone marrow is given to his irradiated rats?

Dunjic. On microscopic examination we do not find any sites of erythropoiesis in the spleen of our normal 4-months-old rats. After injection of ^{59}Fe we find no sharp maximum in the activity curve in the spleen during the first 24 hours in these normal rats as contrasted to the animals in which erythropoiesis occurs (Maisin and Wolfe, 1953). Our irradiated control animals show erythropoiesis in the spleen but much later than in the protected or bone-marrow injected animals, although activity is less important in the latter. The appearance of erythropoiesis in all irradiated animals is difficult to explain. Anyway, the spleen does not seem to be necessary to regeneration as you have yourself found in splenectomised limb protected animals (Baxter *et al.*, 1955).

REFERENCES

Baxter, C. F., Belcher, E. H., Harriss, E. B., and Lamerton, L. F., 1955. *Brit. J. Hæmatol.*, **1**, 86.

Maisin, H., and Wolfe, R., 1953. *U.C.R.L.* 3328 *U.S. Atomic Energy Com. Report* (1956).

QUANTITATIVE CHANGES IN THE ERYTHROPOIETIC CELLS OF THE RAT BONE MARROW DURING THE FIRST FORTY-EIGHT HOURS AFTER WHOLE-BODY X-IRRADIATION

E. V. HULSE

Medical Research Council, Radiobiological Research Unit,
Atomic Energy Research Establishment,
Harwell, Berkshire, England

Damage to the bone marrow is acknowledged to be one of the important consequences of irradiation but no truly quantitative assessment seems yet to have been made. This has now been attempted using a technique which is capable of giving the absolute number per cu. mm. of each cell type present. Data on the red cell precursors for the first 48 hours after irradiations are given here.

Method

Male rats of an inbred albino strain aged between 4 and 7 months were used in groups of litter mates. At least one animal from each litter was used as a non-irradiated control. The animals were irradiated either singly or in pairs in aluminium boxes. The radiation factors were 240 kV, 14 mA, 0·25 mm. Cu + 0·75 mm. Al. half value layer 1·2 mm. Cu. The dose rate was either 68r/min. at a distance of 60 cm. or 43r/min. at 70 cm.

The procedure used is based on that of Yoffey (1954). A representative sample of marrow is dispersed in a suitable fluid by mechanical shaking and cell counts are performed on the resulting suspension. These figures in conjunction with differential counts of a marrow smear give the absolute number of each cell type.

Yoffey's method has been modified and a more detailed account of the technique will be published elsewhere. Perhaps the most important modification has been to replace serum as the dispersing medium by a fluid made from dipropylene glycol, one part: 4% sodium citrate, three parts. Marrow was used from one femur for the cell counts and from the other for smears. As smears made from marrow suspensions give a large number of damaged cells small portions of marrow were chopped up in rat serum on a glass slide and then spread in the usual manner for blood films. Normally a differential count was made on 1,000 cells but in some smears 10,000 cells were counted. The cells of the erythroid series were divided into four classes: late normoblasts, intermediate normoblasts, early normoblasts and pronormoblasts. The method gives the number of surviving cells, not the number killed.

Results

Two series of experiments were performed. In one the effect of various doses of X-rays ranging from 25r to 5,000r were compared at a time interval of 24 hours after whole-body irradiation and in the other the effects of three doses of irradiation, namely 100r, 600r and 5,000r, were followed at various time intervals from 2–48 hours.

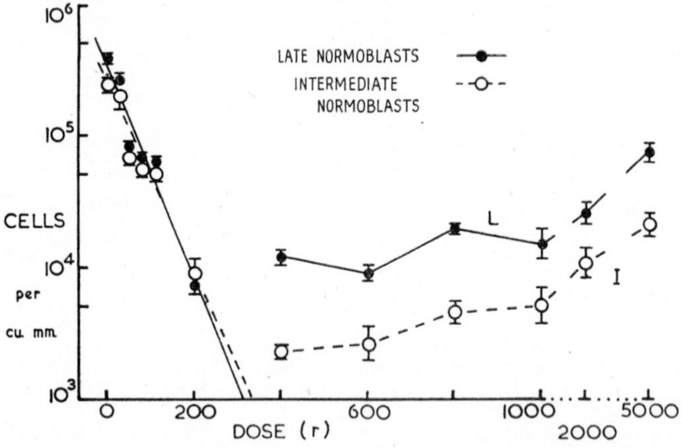

FIG. 1. Numbers of late normoblasts (L) and intermediate normoblasts (I) per cu. mm. of marrow (logarithmic scale) and dose of whole-body irradiation (roentgens). Straight lines fitted by method of least squares to data from 25–200r.

Late and intermediate normoblasts

Both these categories respond very similarly. Figure 1 shows the numbers remaining in the marrow 24 hours after various doses of irradiation. There was a rapid fall with increasing dose; after 50r both types were reduced to less than half and after 200r to between 2% and 4% of their normal numbers. They are therefore amongst the most radiosensitive mammalian cells. At doses above 200r there was an abrupt change in the response, and the number of surviving cells began to increase with increasing dose so that with a dose of 5,000r these cells were ten times more numerous than they were after 200r.

The change in the number of late and intermediate normoblasts at various time intervals after irradiation is shown in Figures 2 and 3. The late normoblasts do not divide and the intermediate normoblasts do so only rarely. There is not, however, any evidence of a time-lag in their response to irradiation, thus indicating that irradiation has a direct effect on them and that their decrease in numbers is not merely due to a hold-up in mitosis amongst more primitive cells. A dose of 600r produced a progressive decrease in the number of cells, the minimum being reached in 24 hours. The effect of 5,000r was at first far less than 600r and in fact very similar

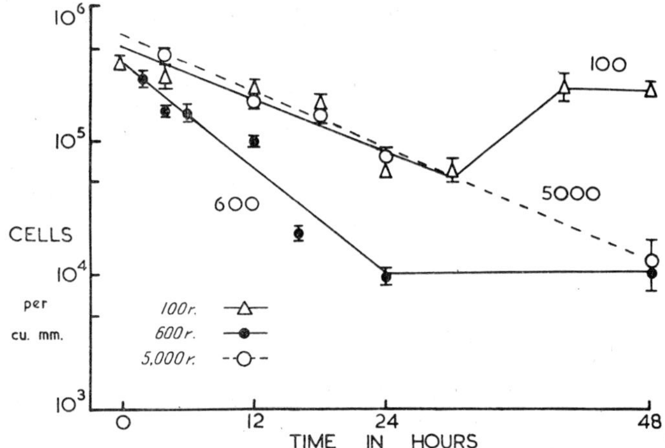

FIG. 2. Numbers of late normoblasts (logarithmic scale) and time after irradiation for three doses (100r, 600r and 5,000r). Straight lines fitted by method of least squares to data from 4 to 24 hours for 100r results, from 4 to 48 hours for 5,000r results.

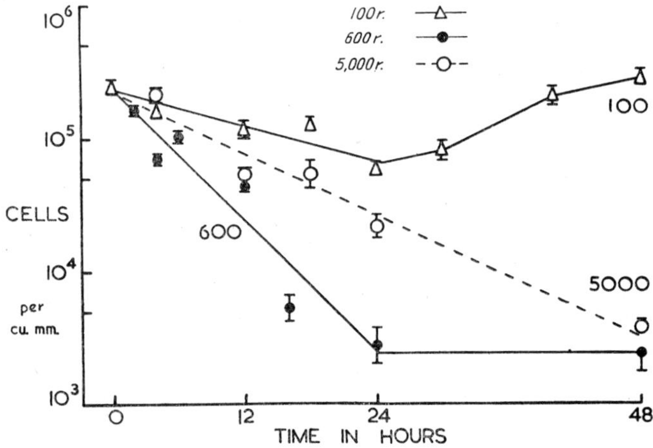

FIG. 3.—Numbers of intermediate normoblasts (logarithmic scale) and time after irradiation for three doses (100r, 600r, and 5,000r). Straight line fitted by method of least squares to data from 4 to 24 hours for 100r, from 2 to 24 hours for 600r and from 4 to 48 hours for 5,000r.

to that of 100r. With 100r, however, recovery occurred after 24 hours whilst with 5,000r the cells continued to decrease in numbers but a further 24 hours had to elapse before they had descended to the 600r level.

Early normoblasts and pronormoblasts

These cells are fewer in number than the more mature cells and the mean values therefore are less exact. For the larger doses the precision was improved for the later time intervals by differential smear counts of 10,000 cells. The dose response curves for the early normoblasts and pronormoblasts (Fig. 4) show an increasing effect with increasing dose up to 400r but higher doses show scarcely any further change. Both curves are of the exponential type but are much less steep than for the more mature cells.

The time curves (Fig. 5) show that 100r produced relatively minor fluctuations in both cell types. With 5,000r there was a progressive decrease until at 48 hours not one cell of either type was seen amongst 40,000 cells. After 600r there was a progressive fall in numbers for 24 hours but no further decrease thereafter.

Discussion

This investigation has shown that the red cell precursors disappear from the marrow in the classical exponential manner, the logarithm of the number of surviving cells when plotted against the dose giving a straight line. There are, however, two rather surprising anomalies. The first is that, contrary to what would be expected on general grounds, the more mature cells proved more radiosensitive than the immature. The second is that the damage (measured at 24 hours after irradiation) does not increase indefinitely as the dose is increased. This is most pronounced for the late and intermediate normoblasts, the curves of which completely reverse their direction between 200r and 400r. This observation is explained by the time response curves which show that an increase in dose from 600r to 5,000r slows down the rate of cell loss.

The explanation of these anomalies may lie in the quite exceptional characteristics of the erythrogenic cells. Even though in 3 of the 4 categories the nucleus is capable of mitosis it is, in normal animals, continually progressing towards a state of pyknosis. This degeneration, which is particularly prominent in the intermediate and late normoblasts, is completed by the disappearance of the nucleus by extrusion or by lysis leaving the non-nucleated red cell. Viewed as the mode of red cell production this is a maturation but from a strictly cytological point of view it is a degeneration. We are therefore irradiating cells which are already undergoing degeneration albeit of a rather specialised type in that whilst the nucleus becomes pyknotic the cell prepares for its removal in such a way that a functioning entity, the red cell, remains.

One class of non-dividing cell, namely lymphocytes, can be very radiosensitive and the extreme sensitivity of the already degenerating late and intermediate normoblasts provides another such example. The process of

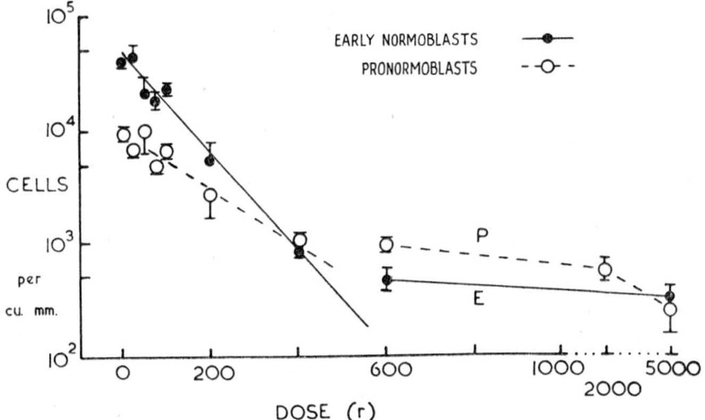

FIG. 4. Numbers of early normoblasts (E) and pronormoblasts (P) per cu. mm. of marrow (logarithmic scale) and dose of whole-body irradiation (roentgens). Straight lines fitted by method of least squares to data from 25r–400r.

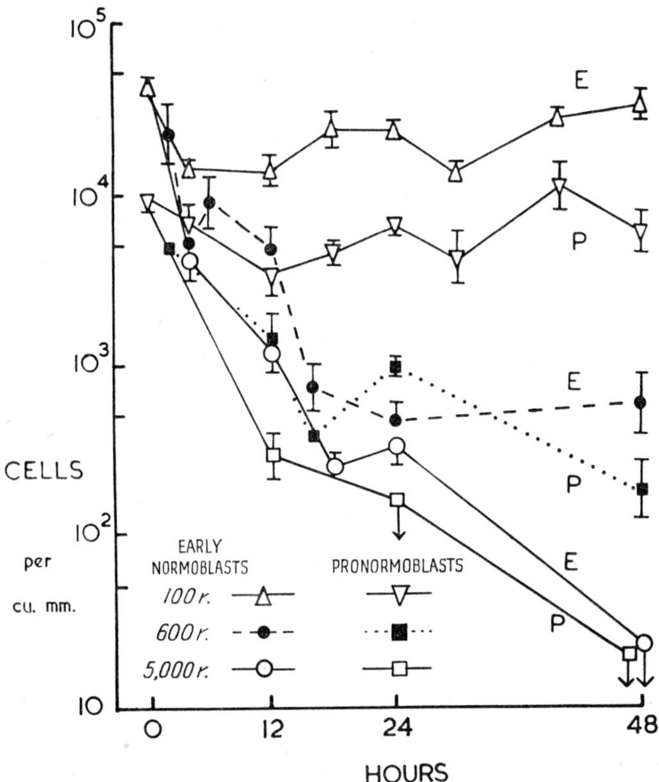

FIG. 5. Numbers of early normoblasts (E) and pronormoblasts (P) per cu. mm. of marrow (logarithmic scale) and time after irradiation for three doses (100r, 600r and 5,000r).

degeneration seems to be potentiated by irradiation and the earlier normoblasts which are much further away from the pyknotic state are consequently much less sensitive than the later groups.

The concomitant process of nuclear removal which is accomplished by the maturing cytoplasm appears to be retarded by higher doses of irradiation so that the pyknotic nucleus is preserved in the cell longer than it would otherwise be. In this way, with increasing doses of irradiation, increasing numbers of late normoblasts are retained in the marrow, and similarly, though to a lesser extent, intermediate normoblasts are retained. This process, which may be regarded as an interference with maturation, may have something in common with the curious behaviour of the rat thymus which 24 hours after irradiation contains more cellular elements (usually pyknotic) after very large doses than after much lower doses (Vogel and Ballin, 1955).

Summary

A method for the quantitative assay of the numbers per cu. mm. of the bone marrow of each variety of marrow cell has been applied to rats given whole-body irradiation. The results show that the late and intermediate normoblasts are extremely radiosensitive, e.g. 24 hours after a dose of 50r they are reduced to less than half their normal numbers.

There were two unexpected findings: (1) the sensitivity of the erythroid series was inversely proportional to the degree of maturation, (2) as the dose of radiation was increased the rate of development of radiation damage was reduced.

An explanation is offered based on the degenerative aspects of erythroid maturation.

Acknowledgments

I am indebted to Dr. R. H. Mole for helpful criticism and to Miss B. C. Dempsey for technical assistance.

REFERENCES

VOGEL, F. S., and BALLIN, J. C., 1955. *Proc. Soc. exp. Biol. N.Y.*, **90**, 419.

YOFFEY, J. M., ANCILL, R. J., HOLT, J. A. G., OWEN-SMITH, B., and HERDAN, G., 1954. *J. Anat.* Lond., **88**, 115-130.

DISCUSSION

Elson. Do you see any evidence of cell destruction, such as cell fragments, etc. or are you inferring destruction from the fact that the cells are not there?

Hulse. I have seen very few remnants of degenerate cells. The occasional ones I have seen could usually be identified as coming from the myeoloid series.

Lamerton. Is it not possible that the cellular depletion in the bone marrow of the femurs is not the result of destruction by radiation, but the result of loss from the bone marrow as a result of change in capillary permeability or some similar factor?

Hulse. I think this question raises a difficulty in nomenclature. The normal process of maturation is a degeneration, so whether the decrease in the number of late and intermediate normoblasts means that they become non-nucleated red cells, or that the cells are disrupted, the process is still a degeneration. With the higher doses of irradiation I have frequently noticed late normoblasts with nuclei that are more pyknotic than usual, and that have cytoplasm which is rather more red than usual. This of course fits the theory that these cells are retained longer than usual in the marrow by the higher doses.

Bacq. I would like to mention that Nizet, Herve and myself have found the maturation of reticulocytes *in vitro* to be markedly slowed down by high dose irradiation *in vitro*.

Fliedner. Our experiments on the effects of whole-body irradiation of rats give results similar to those found by Hulse. We irradiated rats weighing between 220 and 280 gm. with 800 rep fast electrons from the 15 MeV-Semens-Betatron, at Heidelberg. This is nearly the LD 50/30 dose. We used marrow from one femur to make smears and from the other for histological studies. We counted nucleated cells per square centimetre of smear using Sandkuhler's method. By this method we find the whole of the red cell precursors in the marrow to be reduced, after whole-body irradiation (W.B.I.), from about 150,000 to less than 10,000/cm.^{2} of smear within 24 hours, reaching a minimum after 72 hours.

At this time we find complete destruction of the marrow with alterations of the sinus-system, hæmorrhage, œdema and fat. When counting the different types of erythroblasts we used Weicker's method for measuring the diameter of the nucleus. We find the younger types (K_2 and K_1) have disappeared from the marrow within 9–12 hours, while the older types (K 1/4 and K 1/8) show an initial increase in number after 2–4 hours, and after that time a decrease, reaching a minimum 72 hours after TBI. The reticulocytes of the peripheral blood show a transient increase after 24 hours. After seventy-two hours nearly all reticulocytes have disappeared from the blood. This curve for the reticulocytes is very similar to that found for the neutrophil granulocytes, indicating that the increase after 24 hours depends on an output of mature cells from the marrow and that the following decrease is a result of the lack of formation of new blood-cells in the marrow. We find no significant changes in the number of erythrocytes of the peripheral blood in the first 5 days after irradiation, perhaps a slight decrease occurs within the first 3 days after irradiation.

In considering the mechanism of bone-marrow destruction we came to the conclusion from cytological bone-marrow examinations, that the disorganisation of the circulation in the marrow and the breakdown of the sinus-walls within 3 days of irradiation (with œdema and hæmorrhage of the marrow) is one of the most important factors, in addition to the direct damage done to single cells by ionising radiation. This destruction is not specifically due to ionising radiation, for we see the same type of disorganisation after application of TEM or benzene; it is, in fact, the specific reaction of the marrow to various drugs.

THE LIFE SPAN OF RED CELLS OF THE *RHESUS* MONKEY FOLLOWING WHOLE-BODY X-RADIATION

C. W. Gilbert, Edith Paterson and Mary V. Haigh
Christie Hospital and Holt Radium Institute, Manchester 20, England

Introduction

One of the effects of whole-body radiation at LD 50 levels is a marked disturbance in erythropoiesis. There is a sharp fall in the red cell count and a complete disappearance of reticulocytes. Onset of recovery is accompanied by a large rise in the reticulocyte count. The work reported here is part of an extensive study of the effects of whole-body radiation on *rhesus* monkeys, and presents some information about the life span of the red cells. ^{59}Fe (half-life 45·1 days) was used as a tracer in these experiments.

Methods

In these experiments male *rhesus* monkeys 4 to 5 years old, were used. Whole-body radiation was given with a Resomax X-ray tube operated at 300 kVp, and doses between 450 and 600r were given at a rate of 4·5r/min. at the centre of the animal.

The tracer isotope ^{59}Fe was injected intravenously as ferric chloride in normal saline containing 0·1% sodium citrate (Loeffler *et al.*, 1955). Plasma and red cell samples were assayed with a *Burndept* scintillation counter having an efficiency of 6% for a 5 ml. sample. Packed red cell volumes for radioactive assay were determined by centrifuging for 30 minutes at 700 g. in calibrated hæmatocrit tubes. Plasma volumes were determined by weighing.

Iron cycle

To help understand the results of these ^{59}Fe experiments a brief resumé is given of the relevant parts of the iron cycle in the monkey. A considerable part of the iron recirculates and it is convenient to begin with the protein-bound iron in the plasma as it is into this compartment that the tracer is added initially. The iron is cleared from the plasma exponentially with a mean time of about 200 minutes. A large fraction of this iron is incorporated into new red cells and appears in circulation with a mean delay of about 3 days. Thereafter the iron remains firmly in the cells until the end of the cell life, when the iron compounds are broken down and the iron again circulates in the plasma on its way to re-incorporation into another generation of cells.

Results

One monkey was given 8·6 μc. of ^{59}Fe, 14 days before 500r X-radiation. The concentration of the ^{59}Fe in the red cells was followed for 145 days, and the changes during the first 80 days are shown in Figure 1. It will be seen that incorporation of the iron tag into the red cells was complete after about 7 days and the concentration then remained constant until the radiation was given. During the 18 days following radiation there was a steady increase in the concentration of tag in the cells. This increase is due to the selective

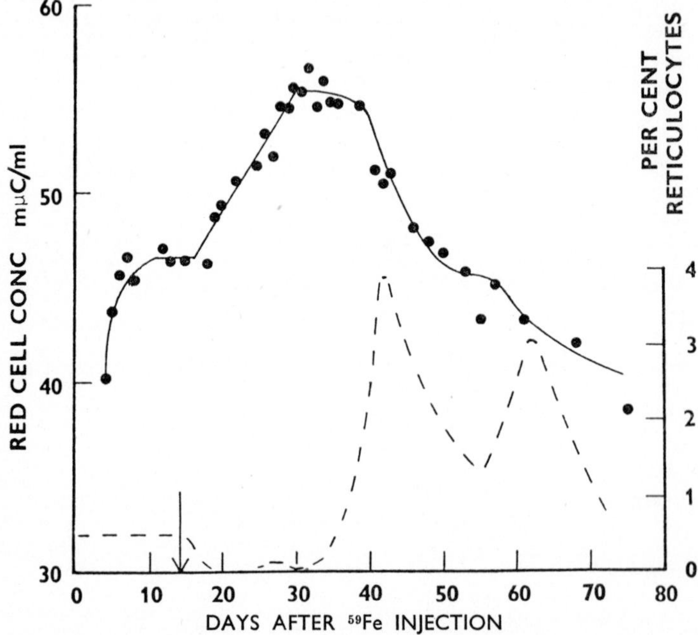

FIG. 1. Concentration of ^{59}Fe in red cells of Monkey (No. 151) given 500r X-radiation 14 days after 8·6 μc. ^{59}Fe. Full curve calculated for red cell life 86 days and reticulocyte life 1·05 days. Dashed curve reticulocyte percentage.

removal from circulation of old cells compared with the relatively young cells which carry the tag. These results give a clear indication that the red cells in the monkey have a finite life span, for if cells were lost from circulation in a random manner irrespective of their age, the ^{59}Fe concentration would have remained constant. At 18 days post-irradiation there is a sharp decrease in tag concentration, indicating the re-commencement of erythropoiesis which produces a large quantity of untagged red cells. This is confirmed by the high reticulocyte percentage at this time. The full line in Figure 1 shows the expected changes in ^{59}Fe concentration calculated from the assumption that the red cell life was 86 days and the reticulocyte life was 1·05 days.

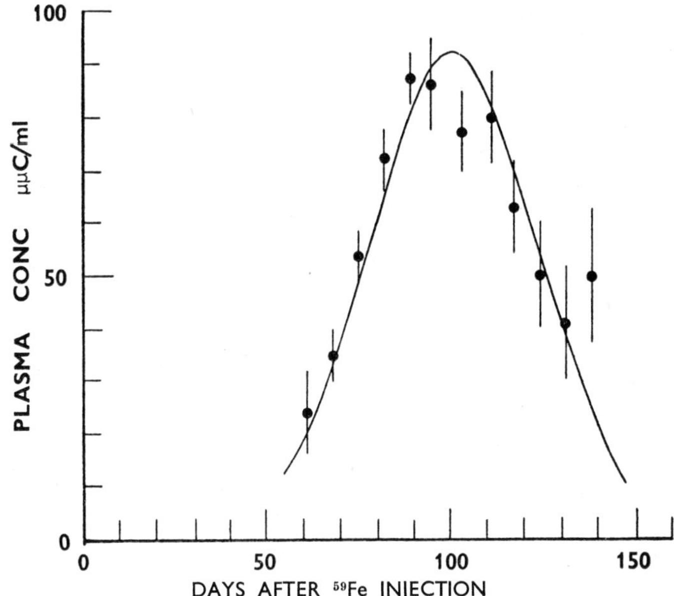

FIG. 2. Plasma concentration of Monkey (No. 151) given 500r X-radiation 14 days after 8·6 μc. ^{59}Fe. The fitted Gaussian curve has mean at 101 days and spread $\pm$ 23 days.

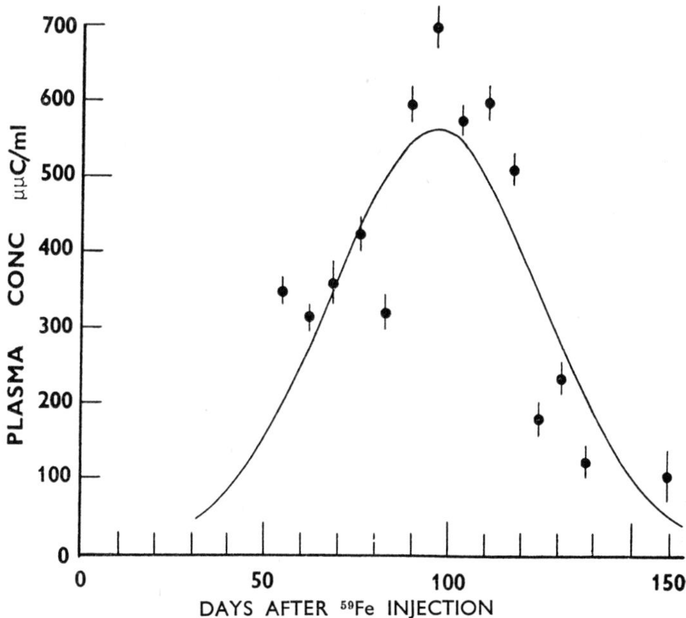

FIG. 3. Plasma concentration of unirradiated Monkey (No. 159) given 75 μc. ^{59}Fe. The fitted Gaussian curve has mean at 97 days and spread $\pm$29 days.

For this monkey and for another one given 75 μc. ^{59}Fe and not irradiated, the concentration of ^{59}Fe in the plasma is shown in Figures 2 and 3. The activities were low and the vertical lines indicate statistical errors due to counting. The maximum in plasma concentration at about 100 days, again indicates that red cells have a finite life span, and gives the time for the complete iron cycle from plasma back to plasma. The values of the life spans for the irradiated and unirradiated monkeys were 101 and 97 days. It therefore seems that the irradiation does not appreciably change the life span of red cells already in circulation.

Discussion

These experiments show that the red blood cells in the monkey have a finite life span of about 100 days, and that irradiation of the animals at LD 50 levels has no great effect on the life of the red cells in circulation, although this amount of radiation completely stops erythropoiesis. The drop in red cell count observed after irradiation may be partly accounted for by the cessation of erythropoiesis and the normal ageing loss of red cells, together with changes in plasma volumes. Further work is in progress to study this.

REFERENCE

LOEFFLER, R. K., RAPPOPORT, S., and COLLINS, V. P., 1955. *Proc. Soc. exp. Biol. N.Y.*, **88**, 441.

DISCUSSION

Fliedner. Have you any experience if the life span of red blood cells is reduced by the post-irradiation hæmorrhages, indicated by an increase of erythrocytes in the lymphatic duct as seen by Brecher?

Gilbert. Hæmorrhage, whether into the lymphatic duct or elsewhere, gives a loss of erythrocytes of all cell ages and our observations of the concentration of ^{59}Fe is shown in the paper to be insensitive to random loss of cells. Our other observations of the plasma radioactivity occur well after hæmorrhage ceases. Our experiments give no information as to whether erythrocytes that may get into the lymphatic duct are destroyed or whether they eventually reappear in the circulation again.

II. OTHER HÆMOPOIETIC FUNCTIONS

READ-OFF METHODS IN RADIO-HÆMATOLOGICAL CONTROL

Matts Helde
Institute of Radiophysics, Stockholm, Sweden

In the spectrum of opinions on the value of blood picture control in radiological work the two most extreme ones are, on the one hand, that hæmatological control is of no (or very little) value and ought to be dropped in favour of physical dose control, and on the other hand that it must be retained, because it gives useful information to the doctor in his care of the employee. But in one respect both views do agree: blood control as a diagnostic tool must be sharpened. This point is particularly significant when our traditional hæmatological methods of forming an estimate are applied to individuals. To a large extent the present scepticism in regard to the individual blood picture is due to a vague conception of the sharpness of the actual observations.

In a group of persons there may have been observed a relative frequency, ν, of persons with a certain type of blood picture. In an analogous group of persons but with an additional irradiation factor influencing this blood picture, the corresponding frequency of the picture in view may be $\alpha(> \nu)$. Then, under some simplifying assumptions an expression may be derived concerning the probability (p_{ir}) that the considered blood picture of the group is caused by radiation:

$$p_{ir} = c\left(1 - \frac{\nu}{\alpha}\right)$$

where c is a coefficient depending on the above-mentioned assumptions. In the following c is equal to 1.

We consider the following cases:

$$\alpha \neq 0, \text{ and } \nu = 0 \qquad \ldots\ldots(1)$$

i.e. a type of blood picture which does not exist in normal material but occurs in an irradiated group; then $\frac{\nu}{\alpha} \neq 0$, and $p_{ir} = 1$ (100%), and we are quite sure that the reaction is caused by radiation.

$$\alpha \gg \nu \qquad \ldots\ldots(2)$$

i.e. we consider a type of blood picture much more frequent in persons in radiological work than in similar but non-radiological work. $\frac{\nu}{a} \to 0$, and $p_{ir} \to 1$. Such a blood picture (and the reaction behind it) is a good indicator of irradiation.

$$a = 2\nu; \; p_{ir} = 0{\cdot}5 \; (50\%) \qquad \ldots\ldots(3)$$

i.e. this picture may, or may not, with the same degree of probability, be caused by radiation.

$$a = \nu; \; p_{ir} = 0 \qquad \ldots\ldots(4)$$

and there is no indicator reaction present. This type of blood picture is useless as an indicator.

As an example of (2) we may choose Dickie and Hempelmann's (1947) material on the increasing number of refractive neutral red bodies in the cytoplasm of the circulating lymphocytes. With their arbitrarily chosen limits for abnormality the frequency, ν, is 4% (in their normal group AI), and the frequency, a, in four groups with increasing gamma-ray exposure is respectively 48, 83, 100, and 100%. The probability (p_{ir}) calculated from the formula above is, for each of these four irradiated groups 92, 95, 96, and 96% respectively, a very high degree of probability that the observed blood picture is caused by radiation.

Now, each member of a group may be looked upon as a carrier of the mean probability of the group. For that reason a p_{ir}-value calculated from the group studied is also valid for individuals of the group, the more so the higher the value of p_{ir}. When the incidence probability is as high as in Dickie and Hempelmann's cases (roughly 95%), the reaction (and read-off method) must be regarded as a good individual test of irradiation (next to specific = case (1)).

Usually a much lower degree of confidence is attained when the conventional 'normal-limit-method' is used to judge the significance of an actual count. Instead of probabilities of ~90%, one obtains 10–50% only, or perhaps if groups which have been exposed to very high radiation levels are considered, probabilities of ~85% may be expected.

In medico-radiological work in Sweden, the irradiation circumstances for a large group of the personnel are very often such that the probability (p_{ir}) of a radiation-induced leucopenia (limit 4,000) is about 50% ($a = 2\nu$; case (3)). In other words, this particular type of blood picture equally may or may not be caused by radiation. Thus, use of fixed blood value limits in cases like this one is of no or little help; and sometimes this method will systematically mislead. We know that some persons constitutionally have low counts. Such a person may often or always be on the 'wrong' side of a warning or reject limit and therefore be considered as if he were affected, though he is perhaps fitter for radiological work than others with normal high counts. For instance, Turner (1953) has shown that persons with low pre-employment white counts have better health than those with higher (normal) counts. Among Swedish nurses in radiological work Helde and Hessler (1955) have observed that those with low white counts have had shorter sickness-absence than those with higher counts.

Of course it is possible to sharpen the conventional test methods by ensuring that the blood counts must be carried out in some relation to the meals, time of the day, or of the year, by counting more and more cells on the slide, etc. By these means the physiological and statistical variations are reduced, but the individual is still judged by comparison with the 'standard man' and not as an individual. The greater technical and physiological precision consumes much time and money. I think it would be better if this labour and expense were used to make more observations on fewer cases and thereby to obviate the necessity for strict adherence to the time schedule of blood sampling. The width and type of an individual's reactions are indicated better by multiple observations than by a smaller number of observations, however carefully the smaller number are made. Life consists of a series of reactions and without knowing an individual's normal reactions it is impossible to know when he is reacting abnormally.

In order to limit the discussion, we shall consider the total white count. The count may be in general considered to be the result of a complex of factors of both essential nature, let us call them 'causes', and of a more accidental nature, such as sickness, irradiation, etc., let us call them 'conditions'. It is likely that an individual's blood picture is normally governed by a relatively well defined complex of causes and conditions. If so, this complex will give rise to a typical mean count with the different observations building up a normal distribution. In order to test such distributions probability paper is useful.

Test method

If normally distributed and plotted on ordinary arithmetic grid paper against their frequencies the observations give a bell-shaped Gaussian curve, which when integrated gives the sigmoid, cumulative normal frequency curve. On probability paper, however, the scale of the cumulative frequency axes is divided into such suitable, unequal scale-divisions that the sigmoid curve is drawn out to a straight line. Thus, a probability paper diagram is nothing but a cumulative probability diagram, where an ideal normal distribution is characterised by a straight line. A straight line is more suitable for control of normality than the sigmoid curve on an arithmetic grid.

To calculate the cumulative probability ($p\%$) of each of the n observations of the series of observations the simple formula: $p = (i - 0{\cdot}5){\cdot}100/n$ is used. Here i is the number of any observation in the series one obtains when the observations are arranged according to magnitude. If more observations (e.g. number 6 and 7 in a series) are equal they will, as a rule, be ascribed a joint average number (in the example $i = 6{\cdot}5$).

With this simple method of calculating p the importance of the first and the last point of the test series must not be overestimated.

Observations in man

In Figure 1, left, are plotted two groups of total white counts from a nurse in non-radiological work made during 5 days. In each group there are an equal number of morning counts and afternoon counts (performed on an empty stomach for the morning counts). The straight line through the inter-menses group of observations shows that a normal distribution probably exists during this time. None of the observations in this group differ by more than 2% from the ideal straight line, characterising her (5 day) normal blood picture; but it must be pointed out that this series fits an ideal normal curve closer than usual. The impossibility of a straight line occurring during the mens-group of observations is obvious and may be interpreted to indicate that the normal complex of causes and conditions, determining the blood picture is disturbed.

Figure 1, right, shows a several years' series of total white counts from a nurse in radiological work. Lines of this broken type are quite common (both in long and short time series) among persons exposed to some risk in radiological work. The broken line indicates in this case that two complexes of causes and conditions are present. Generally, this nurse had received relatively small weekly doses, but because of headache, lassitude, and sometimes very low counts together with sporadic relatively high radiation doses, radiation injury was suspected. It is unfortunate that there is only one count before her employment in radiological work to be compared with her post-employment characteristics.

In an analysis of mammalian radiation injury and lethality Brues and Sacher (1950) in a diagram have published a sample from Jacobson's (1950) material on a two-year series of blood tests drawn every day from a male test person. Their comment that 'it is by the analysis of time series such as this that the parameters of physiological random processes must be determined', is very important. Plotting groups of, e.g., 25 successive observations of this series on probability paper diagrams we find that, as a rule, they form straight lines, but that in some groups the characteristic straight line is obviously disturbed. Something has happened during such a period. Infections, or what? In some of my own material I have found that infections make themselves known in an early stage with obvious deviations from normally straight characteristics. Studies of individual reactions will much sooner give us a deeper understanding of the phenomena considered. Group tests conceal individual and fine structure reactions.

The three straight lines (EX, GX, HX) (Fig. 2, left) illustrate the total white count characteristics of the normal cases of Goodfellow's material

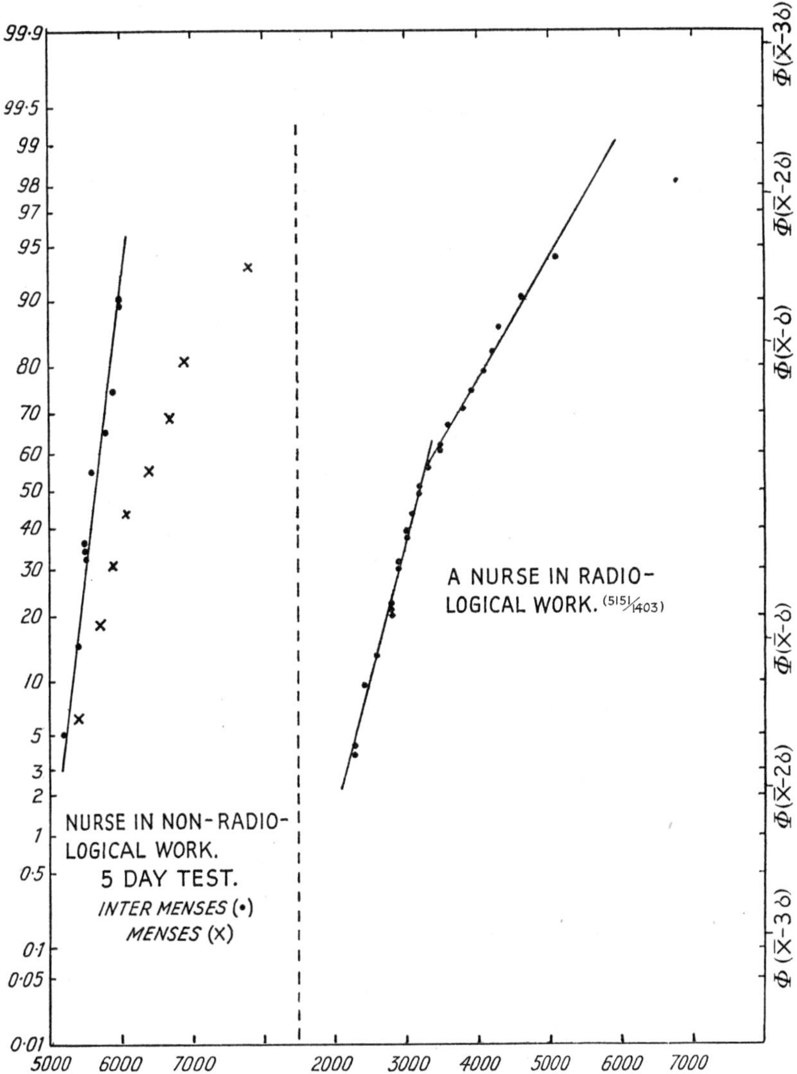

FIG. 1. Comparison of blood counts in nurse engaged in non-radiological work and nurse in radiological work.

(1935) with more than seven test values. The counts are drawn roughly once a month during periods of 24, 8 and 12 months respectively. The greatest differences between the observations and the ideal straight lines are found in EX (200, 300, and 400 cells); two of these differences are, moreover, the first and the last points, which should not be overestimated. HX deserves particular attention as she is an excellent example of the

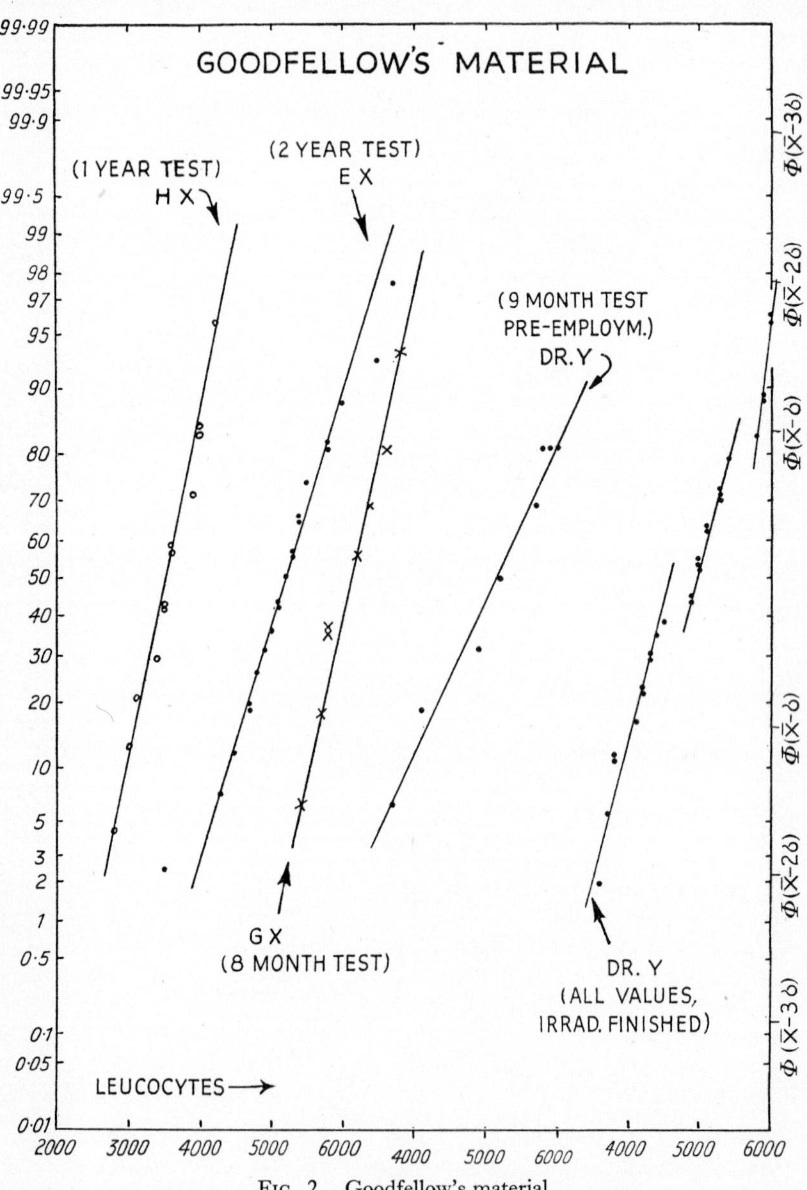

FIG. 2. Goodfellow's material.

type we call 'normal very low', with only one count from twelve higher than 4,000 cells. She would have been seriously suspect if occupied

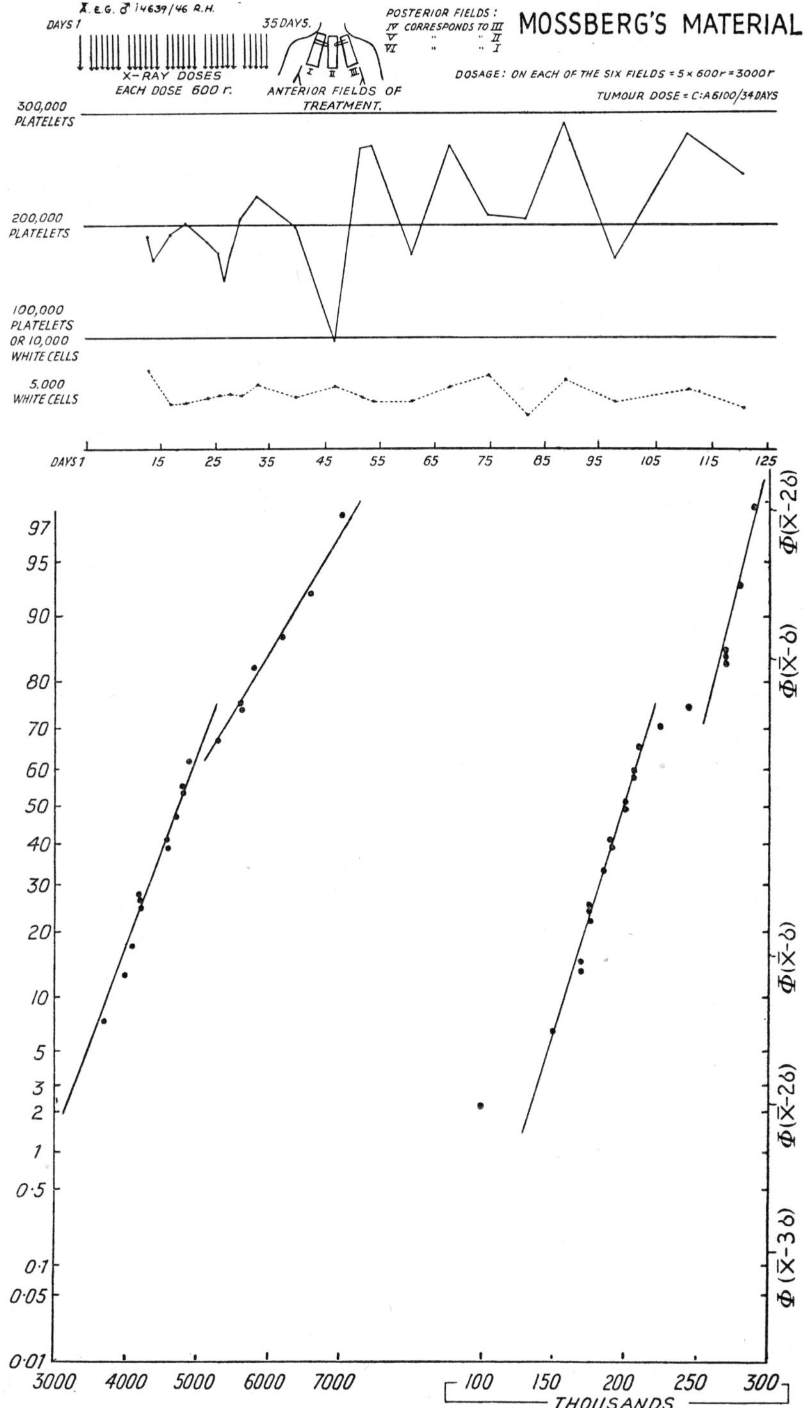

FIG. 3. Mossberg's material.

in radiological work and judged by conventional hæmatological 'limit principles'.

An interesting instance of the detection of an irradiation reaction is given in Goodfellow's case 'Dr Y'. Here eight tests were drawn before starting on radium work (Fig. 2, no. 4 from left). The fact that we have one straight line shows that there is one complex of causes and conditions behind the distribution of the observations before his radium work. All Dr Y's blood values, including those drawn before and after his occupation with the radium work, are plotted in Figure 2, right. The broken line refers to two (three) complexes of causes and conditions; his normal complex is disturbed by radiation.

In Figure 3 is an X-ray therapy patient from Mossberg's material (1947) on thrombocytopenia. The left broken line shows the white cell count during and after treatment, the right broken line is the corresponding platelet count. The total therapy material of this author is of the broken line type, as contrasted to his normal material which gives straight lines. In the top of Figure 3 the corresponding white cell and platelet counts are plotted on an arithmetic scale.

Observations in animals

Langendorff (1953) irradiated rats during a period of one year with 2·5r per day, and studied the white cell count, especially the lymphocyte count. When the results were treated by the usual methods, no changes were found in the blood picture. Histological changes in testes and spleen were, however, observed. It may be of interest to plot his material by the present method. In the top of Figure 4 the means of the relative lymphocyte count of the control group, B (10 animals), and of the irradiated group, A (20 animals) are plotted on arithmetic grid. Below in Figure 4 the values read from the above figure are plotted on probability diagram as before. The normal group gives a straight line and the irradiated group a broken line. As in preceding cases the observations on the irradiated material consist of different subgroups, each of them normally distributed. It is of interest to demonstrate their normality. For that reason the observations of these subgroups are plotted separately on probability diagram (A_1 and A_2), Figure 4, right. Distribution A_1, with the lower counts, consists of 40 values, and distribution A_2 the remaining 10 values. The difference between the two distributions is significant as seen in the Figure.

In the top diagram of Figure 5 four of Langendorff's cases are plotted. The upper ones show observations from two rats, one irradiated, C, and the control, D, which both died in the 53rd week. The values plotted on the probability diagram of Figure 5 show a straight line for the unirradiated as contrasted to the broken line of the irradiated one. In Figure 5, lower

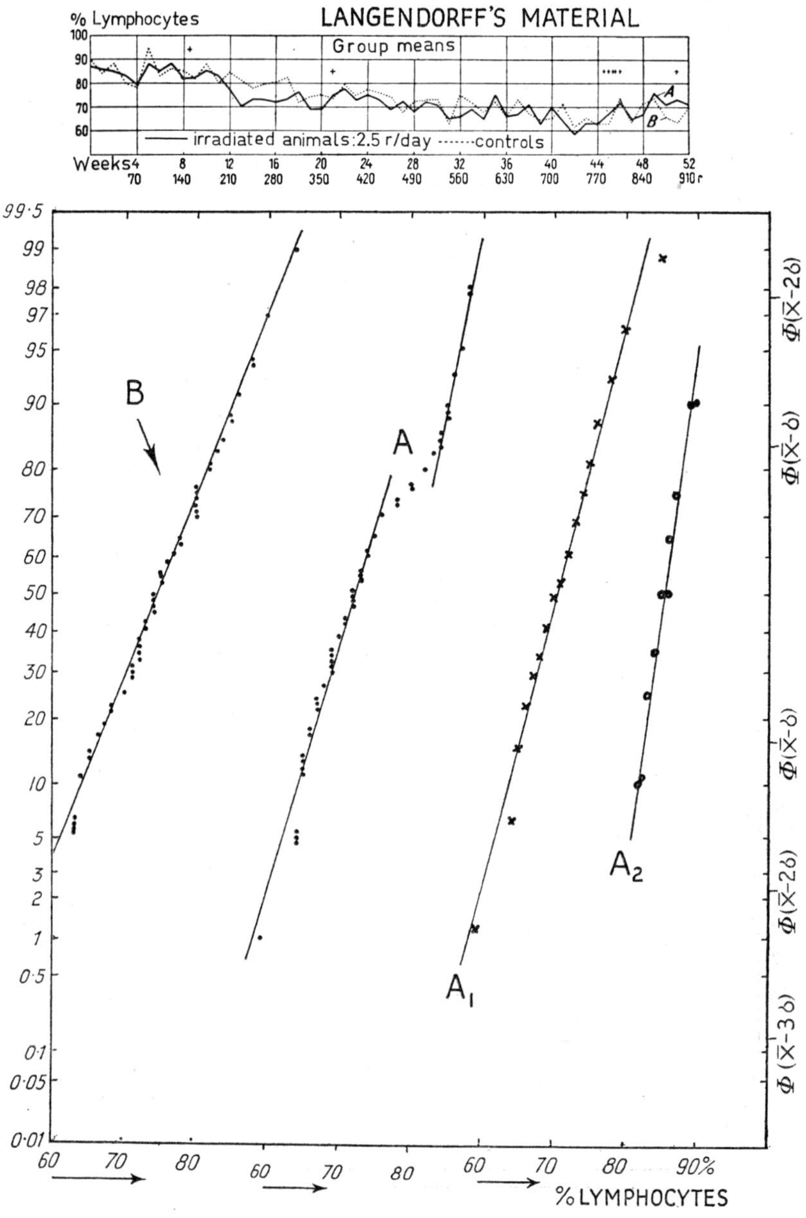

FIG. 4. Langendorff's material (i).

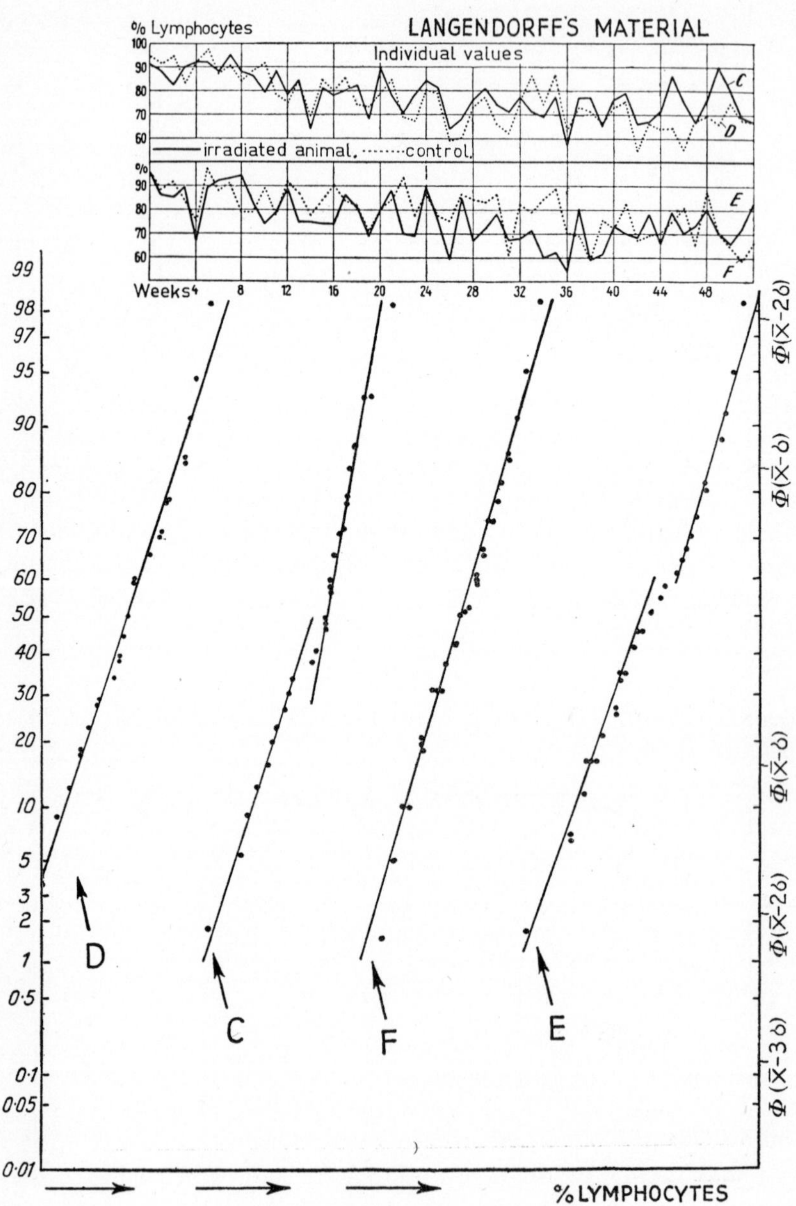

FIG. 5. Langendorff's material (ii).

arithmetic grid, Langendorff's data are reproduced for two more animals (E and F) which both survived for 100 weeks. In probability diagram, Figure 5, right, they show the usual characteristics, i.e. one defined distribution in the normal animal and double (or multiple) distributions in the irradiated.

* * *

When animals are observed during a longer pre-test period it is possible to separate animals with different normal pictures: with slow or steep slopes of the characteristic line or with multiple distributions. Mixed in groups they decrease the sharpness of the conventional group observations. Applying the read-off-method discussed here during pre-test periods it is also possible to ascertain if animals are reliable as a test object, so that correct conclusions can be drawn. It is also possible to decide when the very process of blood sampling no longer disturbs the blood picture, and if the animal is 'stable' as a test object. If investigations are started on animals with multiple distributions false conclusions as to the agent causing the 'effects' are probably unavoidable.

With the application of these read-off methods to individuals it is possible to detect irradiation effects of much smaller radiation doses than before. Also other routinely studied blood variables can be observed separately and the reactions of each of them may be followed.

REFERENCES

BRUES, A. M., and SACHER, G. A., 1950. *Symposium on Radiobiol.*, p. 441 (John Wiley & Sons).

DICKIE, A., and HEMPELMANN, L. H., 1947. *J. Lab. clin. Med.*, **32**, 1045.

GOODFELLOW, D. R., 1935. *Brit. J. Radiol.*, **8**, 669 and 752.

HELDE, M., and HESSLER, T., 1955. Paper read at 20th Congr. Scand. Assoc. Radiol., Gothenburg. Some material in: *Sveriges Offentliga Utredningar*, 1954, Nos. 22 and 23.

JACOBSON, L. O., Quoted in 1950. *Symposium on Radiobiol.* (John Wiley & Sons).

LANGENDORFF, H., 1953. *Strahlentherapie*, **90**, 408.

MOSSBERG, H., 1947. *Acta Radiol.*, Suppl. LXVII, Stockholm.

TURNER, F. M., 1953. *Brit. J. Radiol.*, **26**, 417.

DISCUSSION

Hollaender. I think you have demonstrated that this method is extremely useful in certain cases. Could you tell us how much radiation that nurse which you gave in one of the curves had received?

Helde. I think she had normally 0·2r/week and occasionally perhaps as much as 0·8r/week.

COMPARISON OF THE PHYSIOLOGICAL RESPONSE TO RADIATION AND RADIOMIMETIC CHEMICALS. PATTERNS OF BLOOD RESPONSE

L. A. ELSON

The Chester Beatty Research Institute, Institute of Cancer Research, The Royal Cancer Hospital, London, S.W.3, England

In a study of the blood response patterns in the normal rat, of chemicals used in treatment of leukæmia, etc., compared with the response pattern to X-radiation, characteristic patterns for the different types of chemicals have been obtained. The 'nitrogen mustards', used in treatment of lymphoid leukæmia show predominantly the lymphoid effects of X-radiation, whilst, with the 'Myleran' series of compounds (used in treatment of chronic myeloid leukæmia) the myeloid effects predominate.

Combined treatment with the two types of chemical is necessary to reproduce the complete X-radiation response pattern (Elson, 1955*a*, *b*).

Analysis of the effects of the 'nitrogen mustards'—represented by CB1348

$$(Cl\,CH_2CH_2)_2N{-}C_6H_4{-}(CH_2)_3.COOH$$

and of Myleran

$$CH_3.SO_2.O.(CH_2)_4.O.SO_2.CH_3$$

on the bone marrow (Elson, Galton, Lamerton and Till, 1956) has now revealed that some of the effects in animals undergoing combined treatment with the two drugs are not as intensive as in rats treated with the same dose of Myleran alone. CB1348 does in fact exert a protective action against the Myleran effect. Out of 9 rats treated with a toxic dose of Myleran (20mg./kg.) plus CB1348 (12·5 mg./kg.) only 1 death occurred, whereas 7 deaths occurred out of 10 rats treated with Myleran (20 mg./kg.) alone. This protection is almost certainly due to the intense bone-marrow activity which follows the initial depressive action of CB1348. One of the most obvious results of this stimulation of bone-marrow activity is the marked neutrophilia causing a 'hump' in the neutrophil response curve at about ten days after treatment with a nitrogen mustard. The hump is not confined to the neutrophils but is seen, although to a lesser degree, in the other blood elements, particularly the platelets (see Fig. 1).

This hump is also seen in a less intense form in the blood response curve to whole-body X-irradiation. Figure 2 shows the response to doses of whole-body X-irradiation of 100r, 200r, 400r and 550r.

CB. 1348.

$ClCH_2CH_2 \backslash N \langle\rangle CH_2CH_2CH_2COOH$ / $ClCH_2CH_2/$

12·5mg./Kg.

CB. 3025.

$ClCH_2CH_2 \backslash N \langle\rangle CH_2CHCOOH$ / $ClCH_2CH_2/$ NH_2

5mg./Kg.

CB. 3039.

$ClCH_2CH_2 \backslash N \langle\rangle CH_2CH_2N \langle CH_3, CH_3$ / $ClCH_2CH_2/$

15mg./Kg.

PER CENT OF NORMAL

300 280 260 240 220 200 180 160 140 120 100 80 60 40 20

4 8 12 16 20 4 8 12 16 20 4 8 12 16 20

DAYS

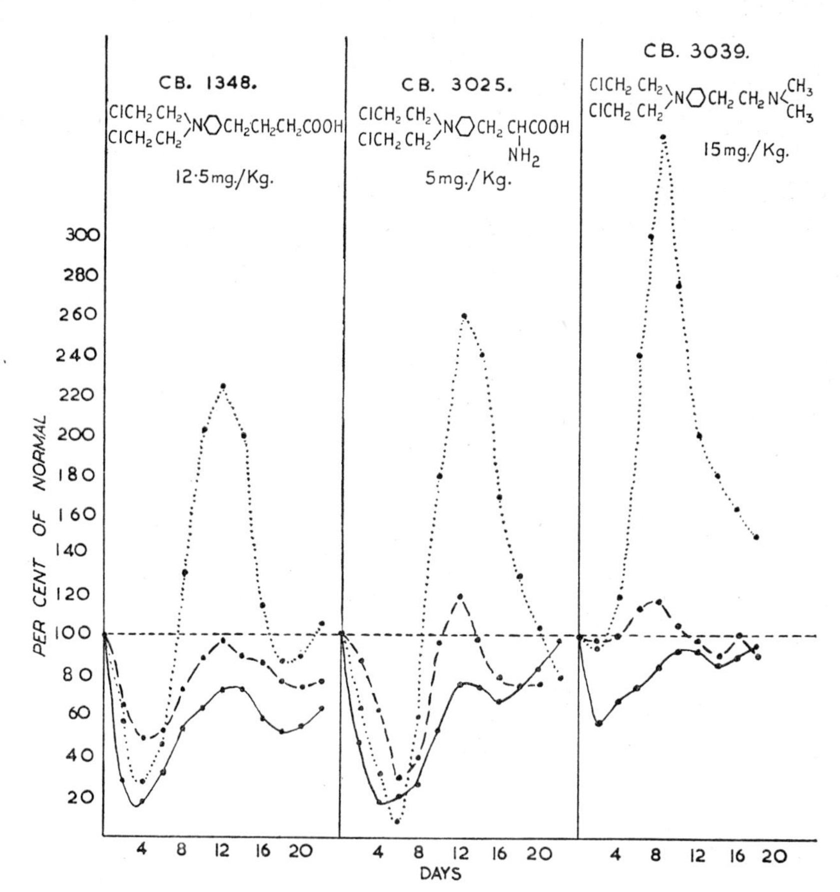

FIG. 1. Blood response patterns to the 'aromatic nitrogen mustard' derivatives CB1348, CB3025 and CB3039, showing the 'hump', most marked in the neutrophil curves, at about 10 days after a single treatment with the drug.
——— Lymphocytes ...... Neutrophils – – – Platelets

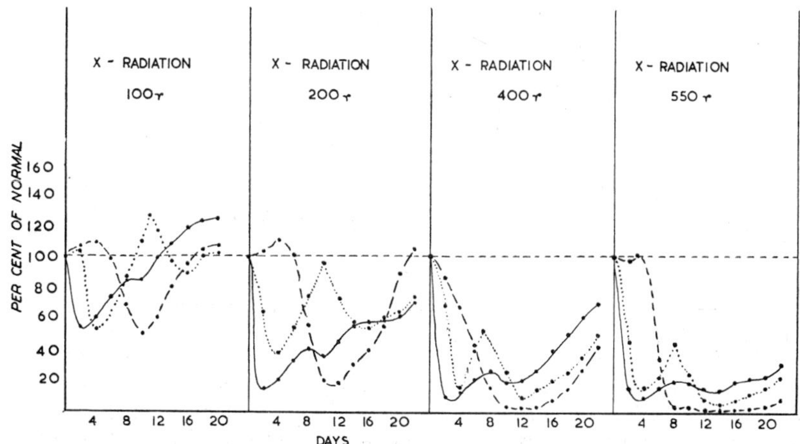

FIG. 2. Blood response patterns to whole-body X-radiation, 100r, 200r, 400r and 550r, showing the modified 'hump' most noticeable in the neutrophil curve.
——— Lymphocytes ...... Neutrophils – – – – Platelets

It is therefore most probably this bone-marrow stimulation, particularly of platelet formation, occurring just before the maximum depressive action of Myleran on platelets, and thus counteracting the intense thrombocytopenia, the main cause of the toxicity of Myleran, which is responsible for the

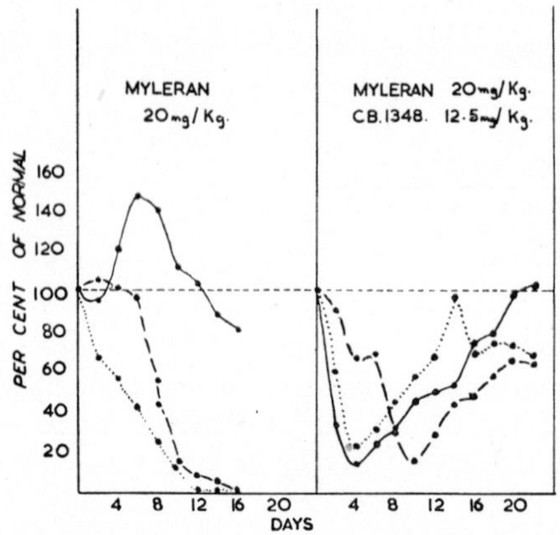

FIG. 3. Blood response pattern to Myleran and to Myleran 'protected' with CB 1348.
——— Lymphocytes ...... Neutrophils ---- Platelets

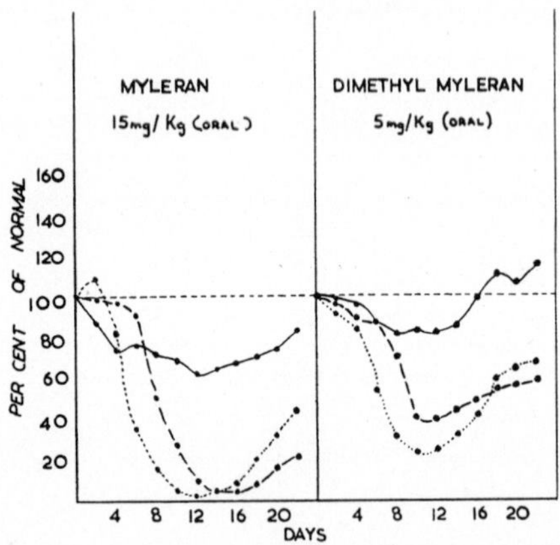

FIG. 4. Blood response patterns to Myleran $CH_3 . SO_2 . O . (CH_2)_4 . O . SO_2CH_3$ and to Dimethyl Myleran $CH_3 . SO_2 . O\, CH(CH_3) . CH_2CH_2CH(CH_3) . O . SO_2 . CH_3$
——— Lymphocytes ...... Neutrophils ---- Platelets

protective action of CB1348. Figure 3 shows the blood response curve of 'CB1348 protected' and unprotected rats to a toxic dose of Myleran.

The rapid, short-lived regenerative activity of the bone marrow resulting in the 'hump' in the peripheral blood curve is obviously of great importance in protection against the delayed effects of X-radiation. Indeed the similarity of the 'protected' curve (Fig. 3) to the blood response curve to 200r whole-body X-radiation is most remarkable.

With regard to the nature of the bone-marrow stimulation leading to the 'hump' it would appear to be a response to the acute but relatively short-lived damaging action of the nitrogen mustard type of chemical, or of the initial damaging effects of radiation on the bone marrow, as no hump is observed in response to the less acute but more prolonged action of compounds of the Myleran type (see Fig. 4).

Acknowledgments

This investigation has been supported by grants to the Royal Cancer Hospital and Chester Beatty Research Institute from the British Empire Cancer Campaign, the Jane Coffin Childs Memorial Fund for Medical Research, the Anna Fuller Fund and the National Cancer Institute of the National Institutes of Health, U.S. Public Health Service.

REFERENCES

ELSON, L. A., 1955*a*. *Radiobiol. Symp.*, p. 235 (London, Butterworth).

Idem, 1955*b*. *Brit. J. Hæmatol.*, **1**, 104.

ELSON, L. A., GALTON, D. A. G., LAMERTON, L. F., and TILL, M., 1955. *Progress in Radiobiol.*, 1956, p. 285 (Edinburgh, Oliver and Boyd).

THE EFFECT OF WHOLE-BODY X-IRRADIATION ON THE MEGAKARYOCYTIC SYSTEM IN RAT FEMUR

B. LINDELL AND J. ZAJICEK

The Institute of Radiophysics and the Institute of Radiopathology, Karolinska sjukhuset, Stockholm, Sweden

During maturation the megakaryocytes, precursors of the blood platelets, divide by endomitosis, i.e. the nuclear divisions are not followed by corresponding separation of the cytoplasmic mass. Consequently, the megakaryocytes continuously increase in size throughout maturation. In unstained preparations of bone marrow those cells of the megakaryocytic system which have reached a diameter of about 24 to 60 microns can readily be distinguished from the other cell systems present.

When working with rat megakaryocytes (Zajicek, 1956*a*) it was noted that these cells are highly resistant to external influences. Even after shaking, washing and repeated centrifugation, the megakaryocytes in a saline suspension of bone-marrow cells showed practically no disintegration. It was also found that from the femur of young albino rats, which contains practically no fatty marrow, a homogeneous suspension of hæmopoietic tissue in 0·7% saline could be prepared. As megakaryocytes can be seen in the counting chamber, the total femoral content can thus be determined.

Method

The rats were killed by means of ether inhalation. Their femurs were dissected out, freed from muscle tissue and bisected. Using a braking pipette, the blood-forming tissue was carefully washed out of the femur and was suspended in 1 ml. of a solution of 5% sodium citrate (1 part) and 0·7% sodium chloride (4 parts). By shaking and mixing, a homogeneous suspension of bone marrow cells was obtained. The megakaryocytes were counted in a Bürker chamber. Each femur was examined separately and the number of megakaryocytes counted was in the order of 200 to 300. It has been established that the variations in the megakaryocyte counts between the left and the right femur of the same animal demonstrates the statistical spread and that by counting 200 to 300 cells a relative error of about 10% is to be expected.

Results

Figure 1 shows the femoral megakaryocyte content in the left and in the right femur of 10 albino rats from one litter. Each femur contained an average of 25,000 megakaryocytes and it is seen from Figure 1 that the variations between the rats in this respect were very small.

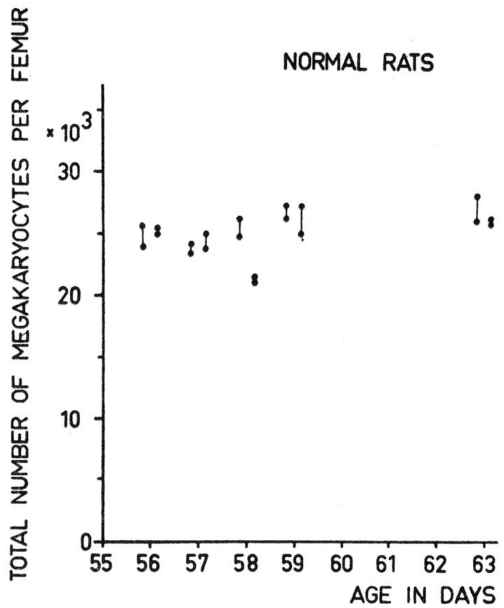

FIG. 1. The femoral content of megakaryocytes in the left and in the right femur of 10 albino rats from one litter. Each femur contained about 25,000 megakaryocytes.

In the irradiation experiments 2 or 3 rats from each litter were used as controls. The mean femoral count of megakaryocytes in these rats was

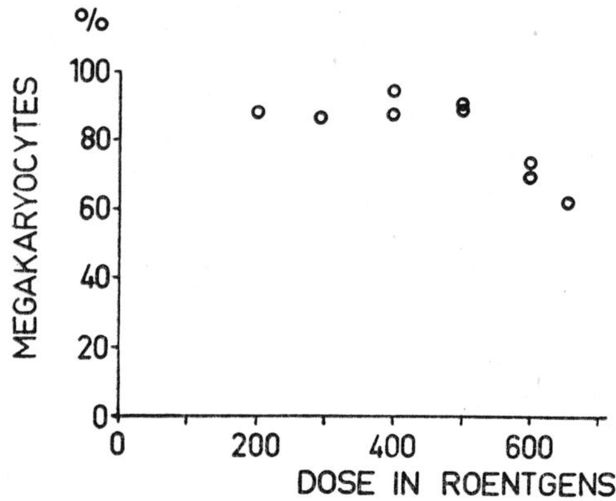

FIG. 2. Femoral megakaryocyte counts 24 hours after whole-body irradiation. Only the rats given 600r or more showed reduced counts.

plotted as 100%. The other rats of the litters were irradiated. Irradiation was performed with a General Electric 'Maximar' unit, with radiation quality corresponding to 370 kVp (half-value layer 1·8 mm. Cu). The irradiation time was determined from a dose-rate of about 50r per minute.

Figure 2 shows that 24 hours after irradiation only those rats which have received 600r or more had markedly reduced megakaryocyte counts. On the following day this reduction was also noted in the rats given 200r or

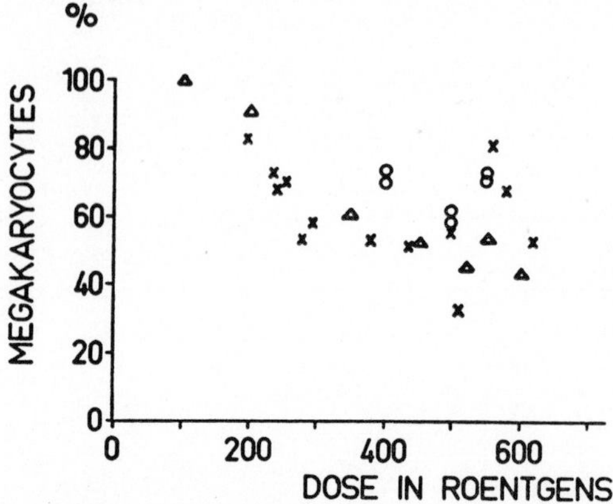

FIG. 3. Femoral megakaryocyte counts 48 hours after whole-body irradiation. Four litters were studied. (Crosses indicate two litters in which the non-irradiated controls showed practically identical megakaryocyte counts.)

more (Fig. 3). The reaction registered 48 hours after irradiation varied greatly between rats of different litters as well as between rats of the same litter. On the third and fourth post-irradiation days the number of mega-

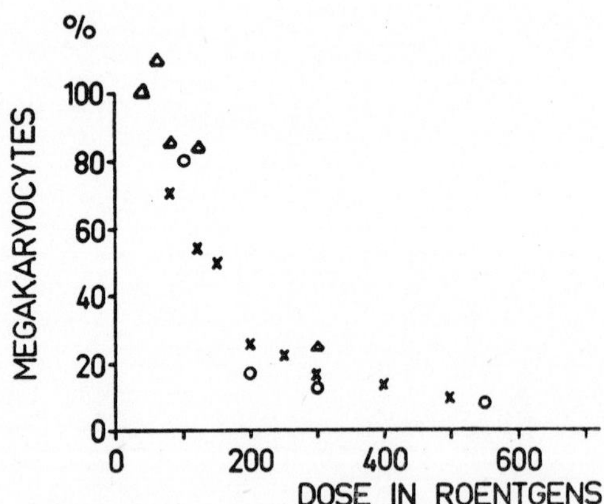

FIG. 4. Femoral megakaryocyte counts 5 days (crosses and triangles) and 7 days (open circles) after whole-body irradiation.

karyocytes continued to diminish, and by the fifth day a clear picture of the effect of whole-body irradiation on the megakaryocytic system was obtained (Fig. 4 crosses and triangles). A dose of only about 100r markedly reduced the femoral megakaryocyte count. About 250r effected a reduction of roughly 80%. Essentially similar results were recorded on the seventh day after irradiation (open circles in Fig. 4).

In a second series of the experiments only the left femur was irradiated whereas the rest of the body was shielded. The megakaryocyte counts in

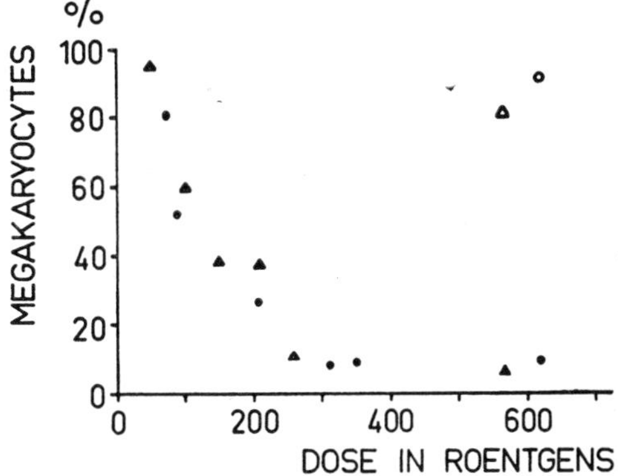

FIG. 5. The megakaryocyte counts in the left femur 6 days after irradiation. Only the left femur was irradiated, the rest of the body was shielded. The open triangle and the open circle indicate the megakaryocyte count in the right (shielded) femur.

the irradiated femur on the 6th post-irradiation day are shown in Figure 5. It is seen that the reduction of the megakaryocytes registered after the irradiation of the femur (Fig. 5) is essentially similar to that one recorded after whole-body irradiation (Fig. 4).

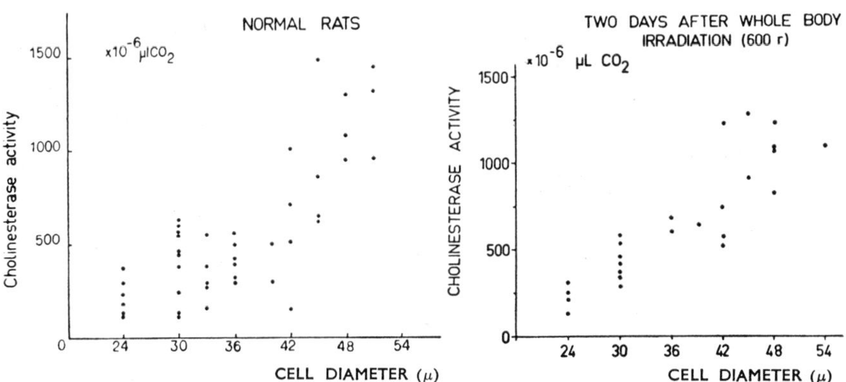

FIG. 6. Cholinesterase activity of individual megakaryocytes at various stages of development, isolated from bone marrow, of normal rats and from rats killed 2 days after whole-body irradiation with 600r.

It is reasonable to assume that the reduction of megakaryocytes after X-irradiation resulted from the block of the cell divisions. Due to the peculiar ability of the megakaryocytic cell system to develop during maturation by endomitosis, it can be postulated that the observed disappearance of megakaryocytes after a dose of about 250r, is due to the block of the mitotic divisions during the differentiation of the megakaryocytic system, i.e. before the actual maturation process, megakaryoblast-promegakaryocyte-megakaryocyte, begins.

The next problem we attempted to investigate was the effect of whole-body X-irradiation on the maturation process of megakaryocytes. In the past year we have evolved a micro diver technique, with which it is possible to make quantitative determination of the cholinesterase (ChE) activity in single cells (Zajicek and Zeuthen, 1956). The increase of this enzyme activity during maturation of megakaryocytes in normal rats (Zajicek, 1956*b*) and in rats irradiated with a dose of 600r has been studied. Figure 6 shows that on the second post-irradiation day the megakaryocytic cell system exhibited as high ChE activity as the megakaryocytes of the normal animals. The results thus suggest that the synthesis of the ChE enzyme is not impaired by a dose of 600r in the cells, which develop in the first 48 hours after irradiation. Further research in this direction is in progress.

Summary

The hæmopoietic tissue from the femur of irradiated rats was brought into suspension and the cells of the megakaryocytic system with a diameter exceeding 24 microns were counted in a Bürker chamber. The effect of X-irradiation on this system was studied by determining the femoral content of megakaryocytes on the first 7 days after irradiation. A dose of 100r markedly reduced the femoral megakaryocyte count. About 250r effected a reduction of roughly 80%.

Using a micro diver method for quantitative determination of cholinesterase activity in single cells it was shown that the synthesis of this enzyme is not impaired in the cells which develop in the first 48 hours after whole-body irradiation with 600r.

REFERENCES

ZAJICEK, J., 1956*a*. *Acta hæmatol.*, **15**, 298.
Idem, 1956*b*. *Experientia*, **12**, 226.
ZAJICEK, J., and ZEUTHEN, E., 1956. *Exper. Cell. Res.* **11**, 568.

DISCUSSION

Cohen. What type of cholinesterase was involved in your study?

Zajicek. The enzyme studied was an acetylcholinesterase (AChE). In man the enzyme is concentrated in the erythrocyte-erythropoietic system, platelets-megakaryocytes having no activity at all. A diminishing erythrocyte AChE content in the series from man, cow, guinea-pig, rat, rabbit to cat is accompanied by increased AChE activity of their respective platelets-megakaryocytes (Zajicek, 1956). In rat (the animal involved in the present study) the erythropoietic system contains little AChE and the enzyme is therefore synthesised largely in the megakaryocytes.

Cohen. One should be careful to interpret changes found in cholinesterase activity in this type of experiment in terms of changes in enzyme synthesis. Changes in the cell, like mineral content, physico-chemical changes, e.g., permeability, as a result of the irradiation may produce similar effects on enzyme activity.

Zajicek. In general it can be said that the AChE is one of the enzymes which is best suited for studying the problem of enzyme synthesis during the maturation process of somatic cells. This enzyme is very stable. A 42-year-old sample of horse blood (hæmolysed) collected aseptically and preserved in the dark, at room temperature, contained up to 85% of its original cholinesterase activity (Keilin and Wang, 1947). AChE is further firmly bound to the cell stroma. The difficulties met by earlier workers in attempting to elute this enzyme from the erythrocyte membrane are well known. In megakaryocytes the situation is in fact more complicated since the enzyme is located in the cell cytoplasm and thus it can be theoretically assumed that, under certain conditions, the cell membrane becomes impermeable to the substrate or to the enzyme-activating Ca ions. However, by comparing the activity of fresh, intact cells with that obtained after disruption of the cell membrane (e.g. by a short exposure of the cell to air) the influence of the cell membrane on the activity measurements can be controlled.

Hollaender. Have you observed the cells when irradiating them in the diver; it should be possible with the beautiful technique you have, to follow the activity of enzymes?

Zajicek. No, we have not. We were mostly interested in the effect of X-irradiation on the rates of AChE synthesis during the maturation of megakaryocytes. The maturation probably stops after the cell is introduced into the diver and is surrounded by the medium used for AChE determinations. We had therefore to irradiate the animals and pick out the cells from the femur at various intervals after irradiation. Data very recently obtained suggest that the synthesis of AChE was disturbed in cells investigated on the third day after whole-body irradiation with a dose of about 600r.

REFERENCES

Zajicek, J., 1956. *Acta hæmatol.*, **15**, 296.

Keilin, D., and Wang, Y. L., 1947. *Biochem. J.*, **41**, 491.

THE IMMEDIATE EFFECTS OF LARGE DOSES OF RADIOACTIVE PHOSPHORUS ON THE PERIPHERAL BLOOD COMPARED WITH THOSE OF EXTERNAL IRRADIATION IN PATIENTS WITH MALIGNANT DISEASE

E. M. LEDLIE
Radiotherapy Department, Royal Marsden Hospital, London, England

Since it is obviously not possible to ascertain the effect of a large dose of radioactive phosphorus on a normal person, observations have been made on thirty-six patients with advanced malignant disease who have been treated with a single large dose of ^{32}P, ranging from 10–15 mc. The hæmatological response has been observed and briefly compared with the effect of external whole-body irradiation.

The patients treated suffered from a variety of conditions in some of which, such as the reticuloses and carcinomata with widespread metastases, there may be interference with the normal production and destruction of blood cells. Furthermore, previous radiotherapy and or, chemotherapy may have diminished the recuperative power of the marrow.

It is now evident that a constant pattern of response occurs in the peripheral blood, and that this pattern differs from that observed after external whole-body irradiation.

Twenty-three patients with carcinoma and multiple metastases were treated and in some the reduction in circulating cells was so slight as to escape detection without frequent blood counts, whilst in others leukopenia and thrombocytopenia caused symptoms and signs. Figure 1 (*a*) shows the hæmatological response to 15 mc. ^{32}P in a man aged 29 years, suffering from a testicular tumour with multiple metastases, who had been previously treated by surgery and X-irradiation. The response in this case should be compared with that shown in Figure 1 (*b*) of a man aged 26 years, suffering from a similar condition, who had also been previously treated by surgery and irradiation, but who, on this occasion, was given wide-field abdominal X-irradiation (2 MeV, 4,200r in 30 days). The immediate fall in lymphocytes in the latter case is striking and a slight fall in the total number of white cells and platelets also occurred during treatment, but all elements had returned to normal 3 weeks later.

It was found that the response to ^{32}P was similar in patients with lymphoid tumours to that in patients with carcinomata, though the fall of platelets might be more marked. A patient with a reticulum celled sarcoma who

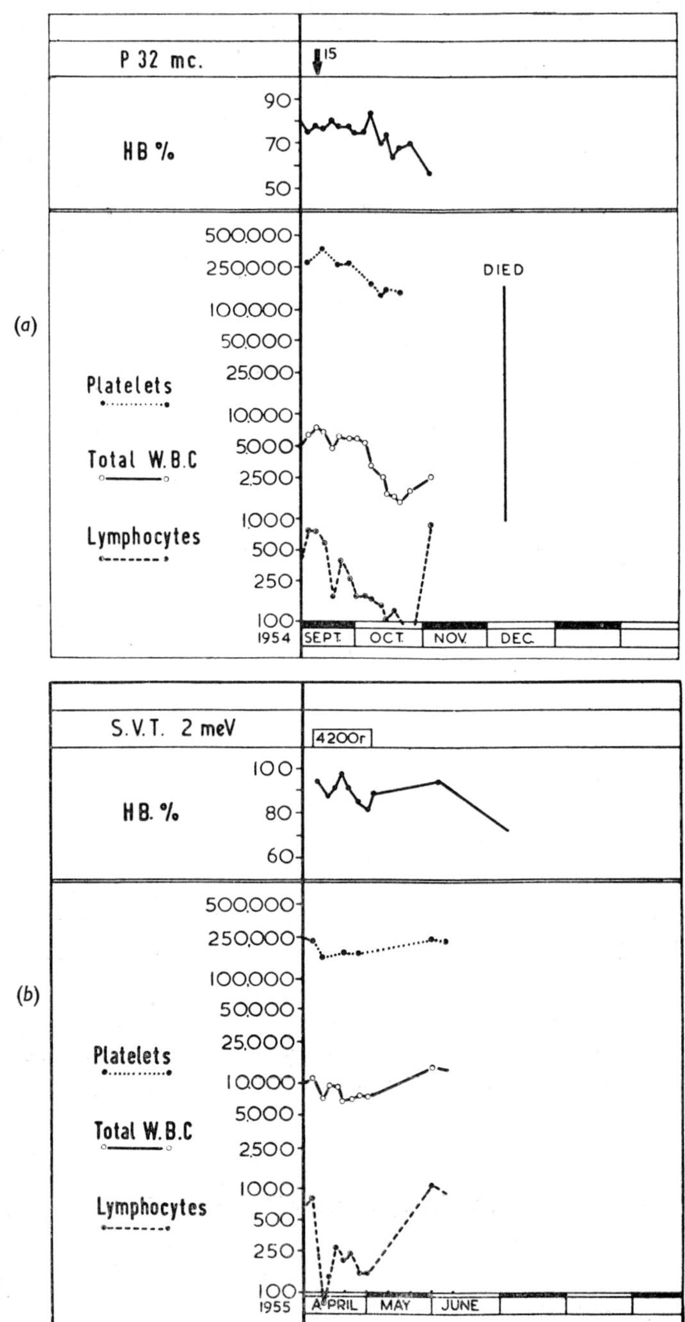

FIG. 1. Hæmatological response of two male patients with testicular tumours to:—
(*a*) 15·0 mc. ^{32}P. (*b*) 4,200r 2 MeV X-rays to abdomen.

received 15·0 mc. ^{32}P showed a platelet fall beginning during the third week and a sharp fall in white cells during the fourth week. This patient developed such a marked anæmia that a blood transfusion was required. Death occurred 14 weeks after ^{32}P administration (Fig. 2).

The most marked changes have been noted in patients suffering from lymphosarcomata. Figures 3 and 4 show the hæmatological response to 10·0 mc. ^{32}P and to a single dose of 3·9 megagramme roentgens whole-body external irradiation in two patients with lymphosarcomata. It will be seen

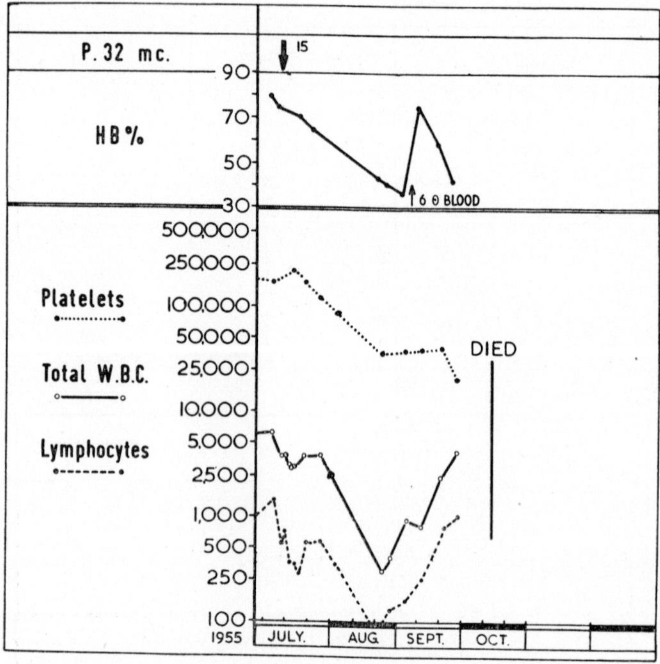

FIG. 2. Hæmatological response of a male patient aged 55 with a reticulosarcoma to 15·0 mc. ^{32}P.

that in the case of the external irradiation there was an immediate lymphocyte fall and rapid recovery in contrast to the ^{32}P-treated patient where a delayed pancytopenia occurred with petechial hæmorrhages in the mouth and skin.

Summary and Discussion

In the 36 patients treated with single doses of 10–15 mc. ^{32}P the lymphocyte count showed some diminution after 2–3 days; no marked change in the total white count was seen until 3–4 weeks after treatment. The minimum levels occurred 5–7 weeks after treatment, and the curve was characterised by a sharp fall and rapid recovery.

The hæmoglobin concentration, if affected, also reached its minimum level 5–7 weeks after treatment. A lowering of the platelet count was seen after about 3 weeks, and the low values sometimes persisted for about 6 weeks before rising again.

Experimental work on the effect of radioactive phosphorus on the blood counts of animals (chiefly rats) has been carried out with a far higher dose per kilogram of body weight than can be used clinically, and some of the results obtained are difficult to correlate with clinical work (Latta and

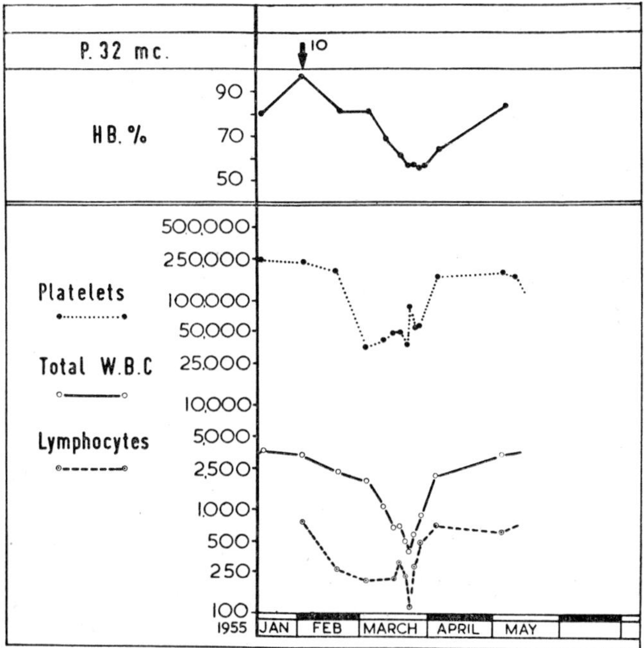

FIG. 3. Hæmatological response to 10·0 mc. ^{32}P in a female patient aged 67 with lymphosarcoma.

Waggener, 1954). The sharp fall and equally rapid recovery in platelet and hæmoglobin values show an essentially similar pattern, though the time interval is different (Lamerton and Baxter, 1955).

It would appear that the marrow is able to maintain a state of sufficiency for a few weeks after this protracted irradiation by ^{32}P, and signs of damage then appear rather suddenly (Brues, 1956).

A lymphocyte fall always occurs during the first few days, but the fall in total white cells may not be apparent for about 3 weeks. Recent work on the life span of granulocytes and lymphocytes may be of interest in this connection (McCombs and Lawrence, 1955).

It is well known that the platelet fall at about 3 weeks (maintained for perhaps 7 weeks) can be induced by much smaller doses of phosphorus than 10–15 mc. (Abbatt, 1953). So sensitive is this platelet response that

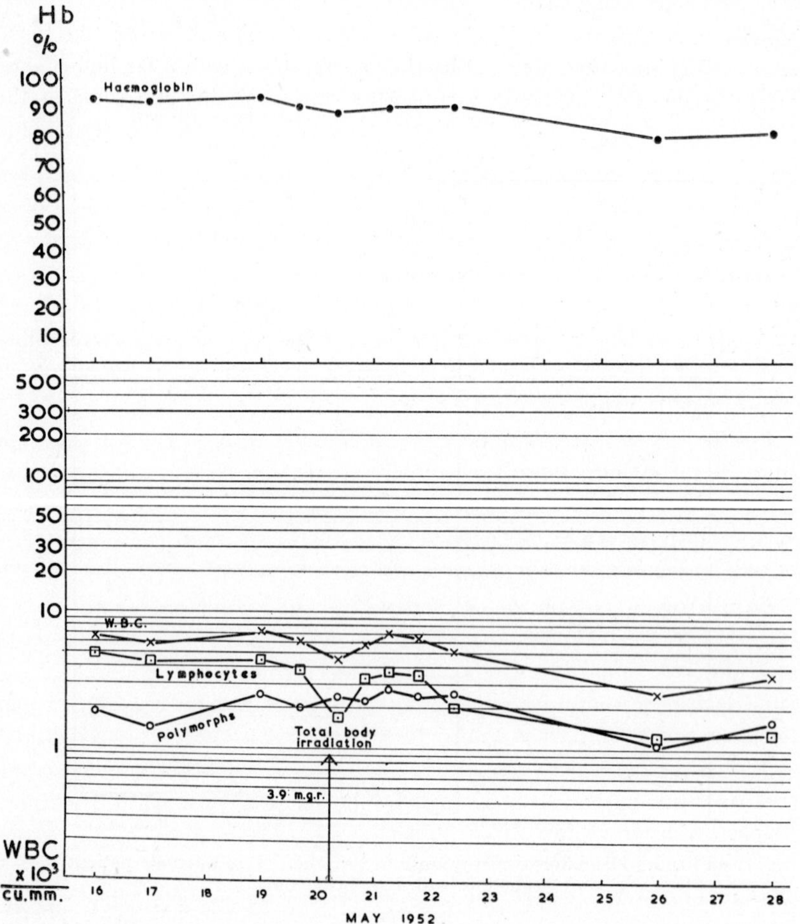

FIG. 4. Hæmatological response to a single whole-body dose of 3·9 mgm. r. external radiation in a patient with a lymphosarcoma.

it may be useful as an indication of the patient's response to treatment by ^{32}P or external irradiation if counts are performed sufficiently frequently (Brown, 1949).

Acknowledgments

I should like to thank Professor D. W. Smithers for his advice and encouragement, and to acknowledge the help of other members of the

Radiotherapy Department and Staff of the Royal Marsden Hospital. I am also indebted to the Staff of the Pathological Department and to Miss M. P. Leach.

REFERENCES

Abbatt, J. D., 1953. *J. Fac. Radiol. Lond.*, **5**, 141.
Brown, W. M. C., 1949. *Acta radiol., Stockh.*, **32**, 307.
Brues, A. M., 1956. *Proc. int. conf. Peaceful Uses of Atomic Energy*, **11**, 71.
Lamerton, L. F., and Baxter, C. F., 1955. *Brit. J. Radiol.*, **28**, 87.
Latta, J. S., and Waggener, R. E., 1954. *Anat. Rec.*, **119**, 357.
McCombs, R. K., and Lawrence, J. H., 1955. *Univ. Calif. Radiation Lab.*, p. 2942.

Discussion

Loutit. Have you any explanations for these odd results? Have you any views about the sparing effect of partial body irradiation? Presumably in your cases the ^{32}P is concentrated in the marrow and actively dividing tissues and much of muscle and soft tissue generally is relatively free of the radioactive material.

Ledlie. I can only assume that the marrow withstands the radiation from ^{32}P for a period, but that eventually a point is reached at which recovery does not keep pace with destruction. It is remarkable that this point occurs so constantly for each cellular element of the blood, and that the number of cells in the peripheral blood falls and rises so steeply.

Lamerton. The dose of ^{32}P used by Dr. Ledlie in her clinical studies corresponds to about 0·2 μC per gm. body weight. We have employed such doses in our own experiments and do not have any evidence of the late effects observed in the clinical work. If one makes the assumption that the bone uptake in man is similar to that in the rat, then an injected dose of 0·2 μC per gm. body weight will lead to concentration in bone in the region of 1 μC/gm. This will give radiation doses of the order of 40 rad/day to marrow in the vicinity of bone, but it is clear that there is a large part of the bone marrow which will receive very little radiation. The clinical observations of Dr. Ledlie suggest that one is dealing with bone marrow having a very abnormal type of radiation response.

Ledlie. I think that the fact that the effect on the peripheral blood is similar in patients with carcinoma and lymphosarcoma suggests that this is the usual response in humans and not an abnormal type of radiation response.

III. RADIATION LEUKÆMIA

INDUCTION OF LEUKÆMIA BY RADIATION

An Examination of Kaplan's Hypothesis

J. F. Loutit
Medical Research Council, Radiobiological Research Unit, A.E.R.E., Harwell, Berkshire, England

Kaplan (1954), in a most impressive review on the ætiology and pathogenesis of the leukæmias, drew largely on his own experience and experimental work to formulate a hypothesis of the causation of mouse lymphoma.

In summary, mice of certain strains such as C57BL are susceptible to the induction of lymphoma. In this particular strain and some others, the site of origin is frequently the thymus. Normally the thymus of the mouse involutes slowly through the first year of life; but, following acute involutional injury (following X-rays, methyl cholanthrene, œstrogen, transplantation and perhaps viruses) and repair, the incidence of tumours is raised often to nearly 100%. Kaplan's hypothesis invokes the 'bone-marrow factor'. If this is available, the thymus can regenerate normally under the influence of the usual stimuli for repair; thus the resulting incidence of lymphoma is low. If the 'bone-marrow factor' is destroyed, regeneration of the thymus is blocked and, under the influence of stimuli to repair, tumours are likely to develop.

The 'bone-marrow factor' is postulated as the outcome of a considerable number of experiments. Lymphomas are induced in C57BL mice by X-irradiation of the whole body, but not by local irradiation over the thymic region (Kaplan, 1949). Shielding of some bone marrow—such as that in one hind leg—during irradiation of the whole body, reduces markedly the incidence of tumours (Kaplan and Brown, 1951).

Furthermore, after irradiation of the whole body without shielding, the formation of tumours can be inhibited by the subsequent early injection of normal bone marrow (Kaplan *et al.*, 1953). This form of treatment is not only effective in preventing the formation of lymphomas from a large and near-lethal accumulated dose of X-rays, given in a number of fractions as is Kaplan's practice, but also permits recovery of mice from a single large dose of X-rays, which would otherwise be lethal (Lorenz *et al.*, 1952). Spleen which in the mouse has considerable myeloid activity is similarly effective in saving life (Barnes and Loutit, 1956; Cole *et al.*, 1953; Jacobson *et al.*, 1951) and in preventing leukæmia (Lorenz *et al.*, 1953).

At the time of the publication of this hypothesis of Kaplan it was widely believed that the activity of bone marrow and spleen in saving lives of irradiated mice was due to a humoral factor, present in these normal tissues, which accelerated the regeneration of the hæmopoietic tissues damaged or destroyed by the radiation (Cole *et al.*, 1953; Jacobson, 1952; Lorenz *et al.*, 1952). The alternative explanation was that the injected normal cells re-colonised the damaged tissues which regenerated as a graft from the donated cells. However, this latter explanation was discredited—chiefly from the experimental evidence that hetero-specific tissue from guinea-pigs (Lorenz *et al.*, 1952) and rats (Lorenz and Congdon, 1954) could on occasion effect the survival of the irradiated mouse. This observation conflicts with the normal laws of tissue-grafting, namely that a mammal will take a graft only from within its own species and only from individuals with an identical or at least compatible set of antigens.

A series of recent studies now demonstrates that, following a single large dose of X-rays in the lethal range, grafts of cells, which normally would be incompatible, do in fact 'take' under these abnormal conditions. Main and Prehn (1955) were able to show the survival of homografts of skin. Lindsley *et al.* (in rats, 1955) noted the production and persistence of foreign red blood corpuscles following injection of homologous bone marrow. Mitchison (1956), in this laboratory has demonstrated the persistence and increase in foreign antigens in the spleen and lymph glands of mice given homologous tissue. Nowell *et al.* (1956) illustrate that mice injected with bone marrow from rats produce marrow-cells and circulating granulocytes which are positive histochemically for alkaline phosphatase—indicating their origin from the rat. These findings all show that the tissues of the host either are colonised by the cells of the donor, or have incorporated its antigens perhaps by a process akin to transformation in bacteria. It is now unnecessary to invoke the, rather unlikely, latter explanation. Ford *et al.* (1956) have proved cytologically that, both in homologous transfer from mouse to mouse and in heterologous transfer from rat to mouse, marked chromosomes of the donor appear in the dividing cells of the recipient's hæmopoietic tissues—bone marrow, spleen, lymph glands and thymus. In most of these animals the markers have been found in virtually all the cells in mitosis. In mouse to mouse transfer these dividing cells have the normal murine content of 40 chromosomes: in rat to mouse they have the normal rat-complement of 42. There is thus no suggestion that the host has incorporated marked chromosomes (as well as antigens) from the donor. The conclusion is that it has regrown its damaged tissues from the cells of the donor. It must be inferred, therefore, that under these abnormal conditions of very severe irradiation the normal laws of histocompatibility no longer apply and that hæmopoietic tissues destroyed by irradiation can regenerate from suitable grafts—heterologous, homologous or autologous.

Two additional features are notable. Firstly, for survival of the heavily irradiated mouse, myeloid tissue seems to be essential. Thus bone marrow from the rat can effect the phenomenon, but not spleen—the spleen of the rat rarely contains myeloid tissue. In the mouse the spleen being normally partially myeloid, especially in early life, is as effective as bone marrow from the infant mouse, but less active in the adult. Liver from the newborn and adult mouse is without action but liver hæmopoietically active from the embryo-mouse is potent. Thymus and lymph glands from the mouse have little if any effect. Secondly, the injected material not only colonises the myeloid tissues, bone marrow and spleen, but also the lymphatic tissues —lymph glands and thymus. These facts suggest that the stem cells of the marrow, presumably reticulum cells, have greater potentialities than morphologically similar stem cells in lymph tissue; or that certain cells of myeloid tissue can de-differentiate and produce lymphatic tissue in place of their normally successive cells; or that the normal but minor lymphatic component of myeloid tissue is a facultative or perhaps essential source of regenerating lymph tissue.

In the light of these recent observations, the findings of Kaplan's group and his hypothesis of leukæmogenesis in the mouse can be reviewed. In earlier work Kaplan and Brown (1952) interpreted some of their results on induction of leukæmia not only according to the humoral theory of the bone marrow's action in recovery from radiation but also on the cellular theory. When part of the bone marrow was shielded during the irradiation, they commented that unirradiated marrow could act as 'a reservoir of healthy hæmatopoietic cells, which, by hæmatogenous dissemination, subsequently recolonised the thymus, spleen and other irradiated tissues'. In the light of the recent work quoted above I believe that this is more than a possibility; it is the probable explanation. It is meet, therefore, to examine other experimental results of Kaplan with this in mind.

Irradiation to the whole body of C57BL mice induces leukæmia in a high percentage and the thymus is the favoured site of origin. With the doses shown by Kaplan and Brown (1952), to be leukæmogenic the whole of the hæmopoietic tissue and lymphatic system must be severely damaged. My impression from histological evidence (Barnes and Loutit, 1956) is that bone marrow and spleen recover from damage by radiation before lymph glands and thymus. It is quite probable that young cells from this regenerating myeloid tissue enter the circulation and colonise the thymus and lymph glands. A subsequent lymphoma arising in the thymus could, therefore, trace its cellular ancestry from marrow or spleen equally as well as from an intrinsic thymic cell. To explain the site of election in thymus, one can invoke the conception of Berenblum (1954), namely that for carcinogenesis there is an *initiating action* (in this case the agent is radiation) and a *promoting action* in this case provided by the thymus.

Irradiation of the thymus alone produces no increase in thymic lymphoma in C57BL mice (Kaplan, 1949). On the thesis of colonisation of the damaged thymus with unirradiated cells from myeloid tissue this would be the expected result. The regenerated tissue would not have been directly exposed to the 'initiating' agent—the radiation.

Shielding of even part of the myeloid tissue—bone marrow or spleen, (which, it has already been noted, protects against the development of lymphoma)—would provide a similar focus of normal cells from which all the damaged hæmopoietic and lymphatic tissues could be recolonised. Once again the regenerated tissues would not have been directly exposed to the radiation.

Irradiation of the whole body followed by injection (intravenous or intra-peritoneal) of normal bone marrow or spleen gives virtually the same conditions. The damaged tissues are recolonised not from the irradiated host but from the unirradiated donor. Thus Kaplan (1953) finds a low incidence of leukæmia in C57BL mice irradiated and treated in this way. Similarly in this laboratory CBA mice, given a single dose of 950r and treated with intravenous injections of isologous spleen from CBA donors, also show leukæmia-rates so low as to be indistinguishable from the normally low rate for the stock.

Thymectomy before irradiation of the whole body, given in a way which induces leukæmia in intact C57BL mice, inhibits leukæmogenesis (Kaplan, 1950). In this case the regenerated tissues must have been exposed to the 'initiating' agent, but the 'promoting' agent of the thymus is absent. Thymectomy after irradiation of the whole body also inhibits leukæmogenesis if performed early (Kaplan *et al.*, 1953). Thus if the thymic 'promoting' agent is real, its action is confined to the thymus, which on other evidence is unlikely—or it is radio-sensitive and only restored to potency after some weeks.

Reimplantation of a thymus subcutaneously to thymectomised mice after their course of irradiation restored at least in part the susceptibility of C57BL mice to leukæmia (Kaplan *et al.*, 1953; Kaplan and Brown, 1954). In the first series (Kaplan *et al.*, 1953) the implant appeared to be involved in the leukæmic process inconstantly and 'the fact that some thymic implants were plainly unaffected by tumor in animals with disseminated lymphomas confirms the observation of Law and Miller that the tumors do not originate in the implants'. Lymphomas in most instances involved the mediastinal structures. In the second series (Kaplan and Brown, 1954) the implant was reported to be the site of origin in a selected number of instances. It is not clear from the two reports how these apparently differing results were reconciled. On the basis of a hypothesis of cellular transfer it is possible to do this and also seriously to doubt the claim in the second paper: 'the site of origin of these lymphoid tumours was in the thymic implants which received no radiation at any time . . . this represents the first definitive instance of the induction of a malignant tumour in a tissue that has not been exposed to the carcinogenic agent responsible for the neoplastic change'. The implanted thymus must first involute until its blood supply is established. Even in unirradiated mice such implants are colonised in part at least by the host—Law (1952). In irradiated mice conditions are more complex. Re-establishment of the blood supply at a favourable time, when cells from regenerating myeloid tissue are circulating and colonising damaged tissue, would lead to seeding with cells which in their stem line have been exposed to the initiating action of radiation and in the suitable thymic environment could produce lymphoma. If the blood supply is re-established at a time when the exodus from myeloid tissue is minimal, the graft may preferentially reform from its own surviving cells in which case the cells of the implant would not have been exposed to the initiating action of radiation. Nevertheless in the latter case there should have been at some stage a hæmatogenous dissemination of myeloid cells which helped to recolonise the host's own lymphatic tissue with descendants of irradiated marrow cells. The experimental findings were that in those thymectomised animals without a thymic re-implant leukæmia did not develop, but in those with such an implant it did to a significant extent. If, as postulated, the thymus was not connected to the circulation at the relevant time—and is not a focus of 'initiated' cells, it could still provide on its re-establishment with the circulation some humoral promoting agent which could act at a distance in the seeded lymphatic tissue. Its remote action might account for the incomplete restitution of susceptibility to leukæmia. This could be in accord with Kaplan's expressed view (1954): 'it would appear that the thymus plays a dual role in relation to lymphosarcoma development; in some strains, it appears to be the site of origin of the disease and in many and possibly all strains, it appears to furnish a factor which contributes to the induction process even when the disease arises elsewhere'.

At first sight the concept seems to be at variance with another observation of Kaplan and Brown (1951). C57BL mice, given four doses of 150r to the whole body at two-weekly intervals, developed a high incidence of lymphoma. Another batch, similarly irradiated but with the mediastinum (thymus) shielded, failed to develop leukæmia. In this case one might have expected 'initiated' cells to have been widely dispersed with an intact thymus for 'promotion'. However, these mice must have differed from those considered above. A shielded thymus is bound to differ from a re-implanted thymus and, more important, is the fact that in shielding the thymus and mediastinum some bone marrow (e.g. sternum) must also have been shielded. The animals are thus potentially similar to those whose hind limbs with contained marrow were shielded. Like them, their hæmopoietic tissue and lymphatic tissues (other than thymus) could have been repopulated with unirradiated myeloid cells from the shielded sternum.

The revised hypothesis outlined above is based on two premises. The first is that lymphatic tissue can be renewed from stem cells arising in bone marrow or myeloid tissue. This has experimental backing under the artificial conditions obtaining after severe irradiation. It has yet to be shown to occur in the physiological states and thus to be relevant to natural lymphoma. The second is that Berenblum's hypothesis of carcinogenesis, involving a primary initiating action and a secondary promoting action is applicable. In this respect it is opposed to another statement of Kaplan (1954): 'the fact that the thymus need not be directly exposed to the carcinogenic agent is of course difficult to reconcile with the recently revived somatic mutation theory of carcinogenesis.' In point of fact it could be claimed that the present concept is based on the theory of somatic mutation.

Summary

In so far as the 'bone-marrow factor', which permits recovery of mice from otherwise lethal doses of X-radiation, seems now not to be a hormone but living normal cells, it is probable that the similar 'bone-marrow factor' which inhibits the development of mouse leukæmia after X-radiation is also a description of living normal cells. It is, therefore, necessary to modify Kaplan's hypothesis concerning the induction of mouse lymphoma. It is here postulated (1) that X-irradiation given to the whole body 'initiates' a change in myeloid cells which, in the presence of a 'promoting' agent produced by the thymus, leads to neoplastic growth: (2) that lymphatic tissue damaged by the radiation is repopulated by cells from myeloid tissue from which the lymphatic tissue is regenerated and from which, if they have been subjected to 'initiation' and 'promotion', lymphoma arises.

REFERENCES

BARNES, D. W. H., and LOUTIT, J. F., 1955. *Progress in Radiobiol.*, 1956, p. 291 (Edinburgh, Oliver and Boyd).

BERENBLUM, I., 1954. *Cancer Res.*, **14**, 471.

COLE, L. J., FISHLER, M. C., and BOND, V. P., 1953. *Proc. nat. Acad. Sci. Wash.*, **39**, 759.

FORD, C. E., HAMERTON, J. L., BARNES, D. W. H., and LOUTIT, J. F., 1956. *Nature, Lond.*, **177**, 452.

JACOBSON, L. O., 1952. *Cancer Res.*, **12**, 315.

JACOBSON, L. O., SIMMONS, E. L., MARKS, E. K., and ELDREDGE, J. H., 1951. *Science*, **113**, 510.

KAPLAN, H. S., 1949. *J. nat. Cancer Inst.*, **10**, 267.

Idem, 1950. *Ibid.*, **11**, 83.

Idem, 1954. *Cancer Res.*, **14**, 535.

KAPLAN, H. S., and BROWN, M. B., 1951. *J. nat. Cancer Inst.*, **12**, 427.

Idem, 1952. *Cancer Res.*, **12**, 441.

Idem, 1952. *J. nat. Cancer Inst.*, **13**, 185.

Idem, 1954. *Science*, **119**, 439.

KAPLAN, H. S., BROWN, M. B., and PAULL, J., 1953. *J. nat. Cancer Inst.*, **14**, 303.

Idem, 1953. *Cancer Res.*, **13**, 677.

LAW, L. W., 1952. *J. nat. Cancer Inst.*, **12**, 789.

LINDSLEY, D. L., ODELL, T. T., and TAUSCHE, F. G., 1955. *Proc. Soc. exp. Biol. N.Y.*, **90**, 512.

LORENZ, E., and CONGDON, C. C., 1954. *J. nat. Cancer Inst.*, **14**, 955.

LORENZ, E., CONGDON, C. C., and UPHOFF, D., 1952. *Radiology*, **58**, 863.

Idem, 1953. *J. nat. Cancer Inst.*, **14**, 291.

MAIN, J. M., and PREHN, R. T., 1955. *Ibid.*, **15**, 1023.

MITCHISON, N. A., 1956. *Brit. J. exp. Path.*, **38**, 239.

NOWELL, P. C., COLE, L. J., HABERMEYER, J. G., and ROAN, P. L., 1956. *Cancer* Res., **16**, 258.

Addendum

Since this paper was submitted two further publications have appeared which are relevant.

(*a*) Kaplan *et al.* (1956) have taken C57BL × C3H(b)F_1 hybrid, irradiated, thymectomised mice and implanted C57BL thymuses. The subsequent tumours developing in the implant have been tested by passage to C57BL, C3H(b) and C57BL × C3H(b) mice. This shows the tumour to be genetically C57BL.

(*b*) Law and Potter (1956) have made similar experiments in hybrids of C57BL and A mice. The tumours in the thymic implants in their hands are, according to tests by histocompatibility, either C57BL or C57BL × A.

These results show that the situation is somewhat more complex than as stated originally by Kaplan or by myself and warrants further thought and investigation. This was my object in raising the problem in the first instance.

As a supplementary communication on leukæmia of the mouse, data were given on therapy by irradiation of the whole body. CBA mice to which leukæmia had been passed 1 week previously received 1,500r X-rays in 25 hours followed by intravenous injections of isologous or homologous bone marrow. Survivors after two or more months were 9/10, 8/10 and 8/15 in 3 experiments (in press).

REFERENCES FOR ADDENDUM

KAPLAN, H. S., BROWN, M. B., HIRSCH, B. B., and CARNES, W. H., 1956. *Cancer Res.*, **16**, 434.

LAW, L. W., and POTTER, M., 1956. *Proc. nat. Acad. Sci.*, **42**, 160.

BARNES, D. W. H., CORP, M. J., LOUTIT, J. F., and NEAL, F. E., 1956. *Brit. med. J.*, ii, 626.

DISCUSSION

Hollaender. I would like to mention that we have similar experiments in progress in our laboratory and as Dr. Loutit stated, similar results in regard to leukæmia were also obtained several years ago by Dr. Lorenz at the National Cancer Institute. There is another point I would like to bring out: leukæmia has been shown to be transmittable by a filterable agent to susceptible mice. Would this have any effect on the findings which Dr. Loutit reports?

Loutit. In our experiments the leukæmia was transmitted to the experimental mice by means of innocula of cells. The cells have undergone malignant transformation. It is to me immaterial whether, historically, this was induced by a virus or by other means. The interest in viruses lies with the original induction of the condition.

Cohen. If a therapy as described by Dr. Loutit was to be useful in tumours the sensitivity of human leukæmias should be reasonably high. Could Dr. Loutit give an idea how sensitive human leukæmias are as compared to the specific strain of the mouse leukæmia described by Lion? The maximal possible dose of X-rays will, of course, be limited by the level of the dose which causes death by other causes than bone-marrow destruction.

Loutit. I have no personal data on the radiosensitivity of human leukæmic cells. There must be evidence in the hands of radiotherapists, but from a rather different angle. We have been concerned with the LD 100 for the leukæmic cells of the mouse. Radiotherapists have so far been concerned only with palliative therapy. It is depressing that this LD 100 is so high and so close to the lethal dose for other tissues as well.

Lajtha. You said, Dr. Loutit, that 1,500r gave a better survival of the leukæmic mice than 950r. Do you think it is due to 1,500r killing more leukæmic cells, or do you think that the prolongation of the radiation to 25 hours versus 15 minutes, is important in this respect?

Loutit. Anything I say would be a guess. At the moment we have only the empirical observation that 1,500r in 25 hours may be, on occasion, the LD 100. We are unable to get recovery of the mice with bone-marrow treatment following 1,500r in 15 minutes; 950r given over 25 hours has been inadequate to kill all the leukæmic cells.

Simmons. In experiments which complement these results we have found that if one irradiates the CF No. 1 mouse (which is normally resistant to lymphatic leukæmia P—1534 of the DBA/2 mouse) and gives a supportive injection of cells from DBA/2 mice, the surviving mice are then sensitive to an inoculation of DBA/2 leukæmia cells.

Loutit. We have observations which are similar in principle but naturally differ in the details of the mice and tumours used.

RADIATION-INDUCED LEUKÆMIA IN DENMARK

MOGENS FABER
The Finsenlaboratory, Copenhagen, Denmark

During recent years there has been a shift of the point of interest in the effect of X-ray leukæmogenesis in man. When X-rays first became available the personnel using the apparatus was in danger. The result of adequate protection has been that leukæmia is rare or non-existent in the radiologist, although still present in other X-ray personnel, as is evident from Danish experiences.

The problem now is to evaluate the risk to the patient. During the last few years a series of reports have appeared on the occurrence of leukæmia in X-ray treated persons. Last year we heard of British experiences with leukæmia in patients treated for ankylosing spondylitis (Brown and Abbatt, 1955), and there have been reports on leukæmia after X-ray treatment of the thymus in the newborn (Simpson *et al.*, 1955). Studies on the increased incidence of leukæmia in treated cases of cancer of the cervix (Engelbreth-Holm, 1941) suggest similar effects.

The British report on radiation hazards (1956) suggests that the incidence of leukæmia is dependent on radiation dose to the bone marrow. If this is the case it should be possible also to demonstrate an increase in leukæmias with very low doses, i.e. in diagnostic irradiation.

The Danish health insurance system includes free hospitalisation for all members and should permit such an evaluation; since a major part of the diagnostic irradiation is given in hospitals where the records are available.

The cases of leukæmia notified to the Danish Cancer Registry during the years 1950–53 were selected for a preliminary study and so far 828 case histories have been studied. This preliminary report will concern only these cases. The case histories of these patients have been studied for re-evaluation and only cases with a clearcut diagnosis of leukæmia have been accepted. At the same time, the first symptom of the leukæmia has been determined when possible. From the case histories of the final disease and from all earlier case histories of these patients going back as far as possible, a record has been made of the amount and type of irradiation given. Due to difficulties in tracing hospital admissions more than 20 years old the incidence of remote irradiation represents a minimum figure only. Furthermore, no attempt has so far been made to study the incidence of irradiation given at the tuberculosis dispensaries. This also gives a trend towards minimum figures.

The distribution among the major types of leukæmias and the age distribution can be seen in the figures. So far, the material consists of

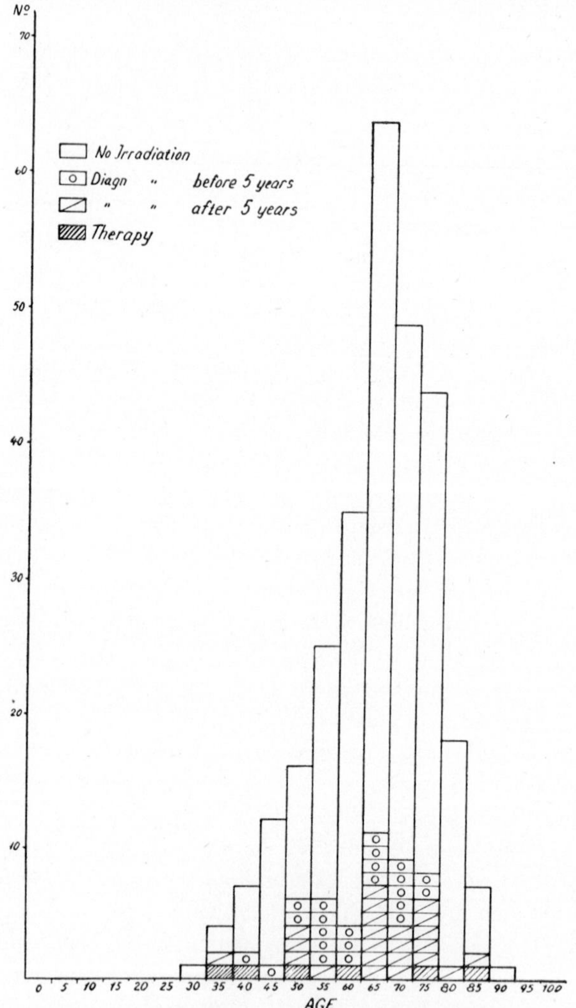

FIG. 1. The incidence of premorbid irradiation in lymphatic leukæmia.

283 cases of chronic lymphatic leukæmia, 150 cases of chronic myeloid leukæmia and 395 cases of acute leukæmia. This last group comprises both the acute myeloid, lymphatic and stem cell leukæmias and the few cases of monocytic leukæmias. The problem of control material has so far been

circumvented by using the chronic lymphatic leukæmias as such. The results of Moloney (1955) from Japan permit us to assume that chronic

TABLE I

The Incidence of Therapeutic and Diagnostic Irradiation.

	No.		Therapy		Diagnostic irradiation over 40 (Therapy cases deducted)					Total irradiation %
	Total	above 40	No.	%	before 5 yrs.		after 5 yrs.		Total	
					No.	%	No.	%	%	
Chronic lymphatic leukæmia	283	278	5	1·8	23	8·4	22	8·1	16·5	18·0
Chronic myeloid leukæmia	150	129	10	7·8	15	12·6	14	11·8	24·4	30·2
Acute leukæmia ...	395	163	13	8·0	13	8·7	26	17·3	26·0	31·9

lymphatic leukæmia arises independently of irradiation. The absence of lymphatic leukæmia in the lower age groups is not of significance in this study. The only large group of leukæmias in younger people, are the acute

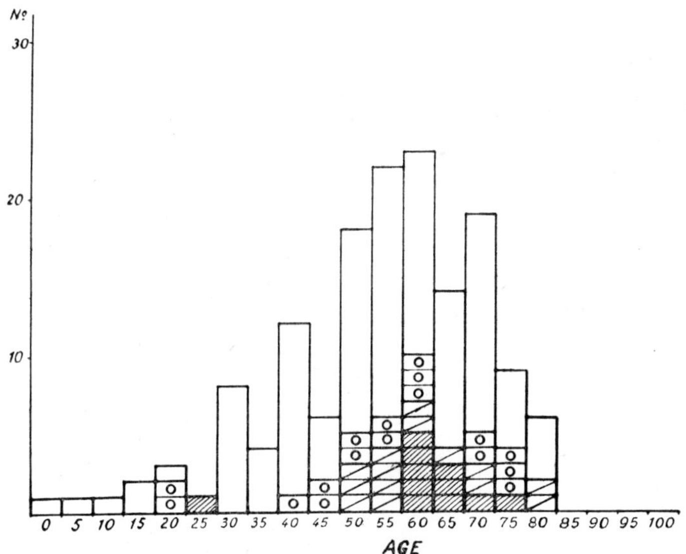

FIG. 2. The incidence of premorbid irradiation in chronic myeloid leukæmia.

forms and it can be seen from the figures that previous irradiation is of no significance for these patients. The study will thus be limited to the leukæmias occurring in patients dying above the age of 40 years.

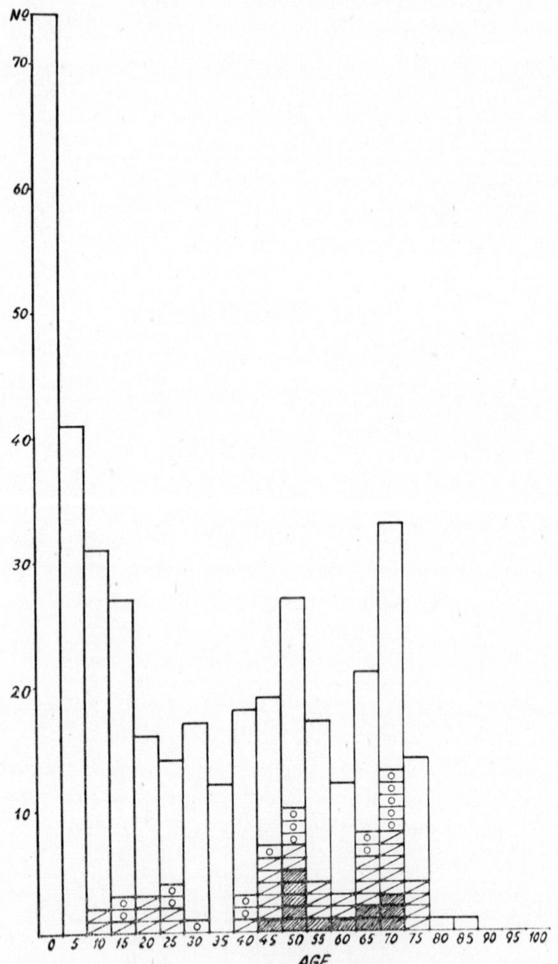

FIG. 3. The incidence of premorbid irradiation in acute leukæmia.

The incidence of exposure to X-rays, and in a few cases also to other types of ionising radiation, is seen in Table I. The irradiation has been divided into therapeutic or long-time (Thorium) irradiation and diagnostic irradiation. This last group has furthermore been divided into cases where irradiation was given more than 5 years before the first symptom of leukæmia or where the exposure occurred during the last 5 years.

In the case of the chronic lymphatic leukæmia we find an incidence of therapeutic irradiation of 1·8% and an incidence of diagnostic irradiation of 16·5% with 8·4% 'old' and 8·1% recent irradiation. For the group as a whole the incidence of radiation exposure is 18·0%.

In the case of the chronic myeloid leukæmia the figure for therapeutic irradiation rises to 7·8% and the incidence of diagnostic irradiation to 24·4% with 12·6% 'old' and 11·8% recent exposure. The total incidence was 30·2%.

The acute leukæmias show a slightly divergent picture. The incidence of therapeutic irradiation is the same as in the case of the myeloid form, i.e. 8·0% and the total incidence of diagnostic exposure equals the myeloids with a total incidence of radiation of 31·9%. The distribution of the

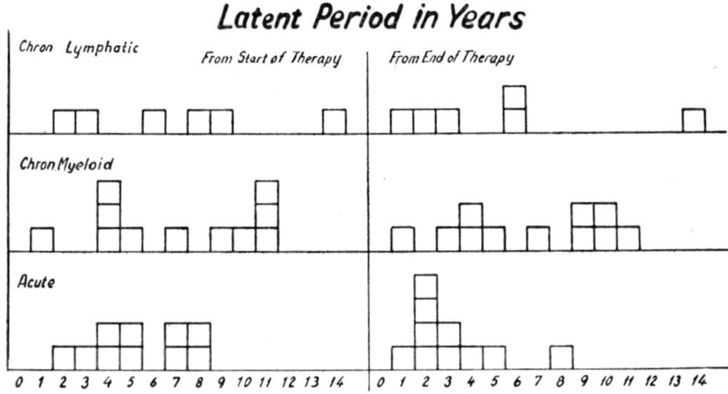

FIG. 4. The latent period between therapeutic irradiation and appearance of first symptom of terminal leukæmia.

diagnostic irradiation between 'old' and recent is, however, quite different. Irradiation more than 5 years before the appearance of leukæmia is found in 8·7% and equals the lymphatic leukæmias. During the last 5 years, however, we find an incidence of 17·3%.

These results indicate that both chronic myeloid and acute leukæmia can be induced by irradiation, and that this is the case in some 13% of the patients studied. This increase can not be explained only on the basis of the 8% of therapeutic irradiation but must partly be due to diagnostic irradiation as well.

Furthermore, it seems as if the latent period of myeloid leukæmia from the time of irradiation till the symptoms are recognisable is somewhat longer than is the case of the acute leukæmias where a latent period of less than 5 years must be accepted. In therapeutic irradiation the same emerges (Fig. 4) especially when the latent period is measured from the end of the treatment when this has been protracted.

TABLE II

The Localisation of Diagnostic Irradiation

	Chronic lymphatic leukæmia				Chronic myeloid leukæmia				Acute leukæmia			
	before 5 yrs.		after 5 yrs.		before 5 yrs.		after 5 yrs.		before 5 yrs.		after 5 yrs.	
	No.	% of all patients	No.	% of all patients	No.	% of all patients	No.	% of all patients	No.	% of all patients	No.	% of all patients
Caput	0	0	0	0	1	0·8	1	0·8	4	2·7	1	0·7
Sup. extr.	5	1·8	1	0·4	1	0·8	0	0	0	0	1	0·7
Inf. extr.	4	1·4	3	1·1	1	0·8	2	1·6	4	2·7	1	0·7
Spine and Pelvis ...	2	0·7	5	1·8	1	0·8	2	1·6	2	1·3	4	2·7
Thorax	8	2·9	8	2·9	7	5·9	5	4·2	8	5·3	7	4·7
Abdomen	14	5·1	12	4·4	12	10·0	6	5·0	10	6·7	18	12·0
Total	33		29		23		16		28		32	

The dose necessary to induce leukæmia is not easily calculated from the figures available so far, but some information can be obtained by studying the type of diagnostic procedures used in the patients. In Table II is listed the incidence of irradiation in different regions in the three groups studied. It is evident that the types of diagnostic procedures which predominate in the myeloid and acute groups are centred on the thoracic cavity and the abdomen.

When it is taken into consideration that these groups contain cases of pulmonary tuberculosis it must be accepted that the major risk is confined to those diagnostic procedures where the largest doses are used and where the largest dose will be delivered to the bone marrow. Diagnostic procedures confined to the extremities, however, do not appear to carry any risk of an increase in the incidence of leukæmia.

REFERENCES

BROWN, W. M. C., and ABBATT, J. D., 1955. *Lancet*, **2**, 1283.
ENGELBRETH-HOLM, J., 1941. *Nord Med.*, **9**, 791.
Medical Research Council's Report, 1956. Cmd. 9780 (London, H.M.S.O.).
MOLONEY, W. C., 1955. *New England J. Med.*, **253**, 88.
SIMPSON, C. L., HEMPELMANN, L. H., and FULLER, L. M., 1955. *Radiology*, **64**, 840.

DISCUSSION

Loutit. Dr. Faber, you have excluded on purpose those cases of acute leukæmia beginning in childhood. Perhaps you know that in England such cases are being investigated and the results are being quoted in conversation. It would appear that irradiation of mothers for diagnostic purposes during pregnancy is relevant.

Christensen. In connection with the discussion on radiation-induced leukæmia I should like to call attention to some points in connection with polycythæmia vera and its treatment with ionising radiation, especially ^{32}P.

It is a well-known fact that many patients with polycythæmia vera end with myeloid leukæmia without having had any treatment with ionising radiation.

With ^{32}P the patients receive quite a heavy dose to the bone marrow—hundreds of rads. This treatment was introduced in 1938–39 by John H. Lawrence, who in a very recent follow-up of about 250 patients points out that he saw no more leukæmias now, than before this treatment was introduced. We—at the Finsen Memorial Hospital—started similar treatment in 1948, and have now treated 86 patients, with an average observation time of a little more than 5 years. In a survey just finished we found 5 cases of myeloid leukæmia, all of whom showed a high white count with a shift to the left before treatment.

The above-mentioned dose to the bone marrow is delivered in approximately 3 months. I wonder whether the dose rate is of significance in connection with the induction of leukæmia or whether the observation period in Lawrence's series and mine are not long enough ?

Faber. The incidence of radiation-induced leukæmia in children is being studied by our group, but no results are so far available.

Concerning the problem of leukæmia in patients with polycythæmia I feel that our present knowledge of this disease is insufficient. I prefer to register leukæmia in this disease, appearing some time after radiation by X-rays or ^{32}P, as probably radiation-induced. So far only two such cases have been included.

SECTION IX

RADIATION GENETICS

PRINCIPLES OF BACK MUTATION AS OBSERVED IN *DROSOPHILA* AND OTHER ORGANISMS

H. J. Muller and I. I. Oster
Zoology Department, Indiana University, Bloomington, U.S.A.

Drosophila work of approximately the first decade of X-ray mutagenesis, by Muller (1928*b*), Patterson and Muller (1930), Timoféeff-Ressovsky (1930-39), and Johnston and Winchester (1934), clearly established, for a number of loci, the production of back as well as 'direct' point mutations, to diverse alleles, when X-rays were applied either to spermatozoa or to other stages. The interpretations of contamination, of suppression, and of spontaneous origin of the reversions found, were all ruled out in these studies. For most loci and alleles direct mutations could be induced by X-rays with a frequency several fold (commonly about an order of magnitude) greater than that of back mutations, but in the case of forked the frequencies were more nearly equal.

Neither the back nor the direct point mutations of *Drosophila* induced by X-rays were distinguishable as a class from those already known to arise spontaneously. That an individual mutant gene may, however, be different in structure from a superficially identical allele of independent origin was shown by Timoféeff-Ressovsky on examination of the pleiotropic effects of independently arisen white-eye mutants (1933*a*) and of the X-ray-induced mutation frequency of normal alleles of white eye derived from different sources (1932).

All the above principles have more recently been abundantly verified and much extended in *Neurospora* by Giles *et al.* (1948-56) and by Westergaard, Kølmark, Jensen *et al.* (1949, 1952). They have found that the frequency distribution of mutations to other alleles differs both from allele to allele of a given locus, from locus to locus, and from one mutagen to another (or to non-treatment). And a special study carried out by de Serres with Giles and others (1955) showed that alleles which had arisen as X-ray-induced point mutations were, like alleles of other origins, able to undergo back mutation, and that these back mutations could arise as a result of exposure to X-rays as well as to other mutagens and spontaneously.

Since the publication of the above-mentioned work on *Drosophila*, a series of authors (Demerec, 1938; Kaufmann, 1942; Hinton and Dibble, 1947; Yu, 1949; Lefevre *et al.*, 1950-55; Spofford, 1955) have reported experiments in which they failed to obtain back mutations by X-ray treatment of *Drosophila* spermatozoa, and some of these authors have publicly questioned the possibility of their being obtained. Lefevre in 1950 suggested that the cases recorded by the earlier workers might have arisen by contamination of the cultures, but in 1955 he and his co-workers stated, although without presentation of actual data, that forked mutated back so frequently in their control material as to explain the reversions of this mutant that had seemingly been induced by X-rays, on the basis of their having been spontaneous occurrences.

Although in our opinion these criticisms of the earlier work are invalid, and we believe the scale of the later experiments to have been too limited to afford ground for expecting significantly positive results from them, nevertheless, in view of these challenges, a careful repetition of some crucial portion of the earlier work has seemed to us to be desirable. Oster has therefore been engaged this year in conducting a series of experiments concerned especially with the most contested locus, that of forked. Although these experiments are as yet incomplete, especially in that tests of the suppressor possibility have not all been carried through, they are now at a stage justifying a preliminary report.

TABLE I

Sex-linked Markers in Parents

Cross	Fathers	Mothers
a	$y\ ct^{6}\ t^{2}\ v\ f^{1}\ car$	$y\ f^{1}$
b	$y\ f^{1}$, X^{c2} (ring)	$y\ f^{1}$
c	$w^{i}\ f^{3}\ bb$	$y\ w\ f^{1}$
d	$y\ v\ f^{x}\ car$	$y\ f^{1}$

All mothers have their X-chromosomes attached to form 'double X' and have a Y; expected daughters are therefore like mothers and sons like fathers.

As Table I shows, all X-chromosomes of both males and females carried some forked allele. They also contained other markers, chiefly as a check on contamination but also to supply additional loci in which back mutations might be observed. However, on the basis of the earlier work, there was little probability of finding back mutations in these other loci in experiments done on the present scale, since most of the other mutant genes had given evidence of much lower frequencies than forked. For gene symbols see Bridges and Brehme (1944).

Irradiation of 3,200r X-rays was applied to the inseminated females, which were bred for 6 days thereafter. Thus, germ cells of each sex were at irradiation at a stage especially susceptible to mutagenesis by radiation.

Since the females had ('double') attached X-chromosomes, the daughters received both of their mothers' X's and disclosed reversions arising in either one of them, while the sons received their father's X and disclosed reversions in it. The inseminated females were bred for 3 days before irradiation to furnish control material. The attempt was made to obtain this on such a large scale as to provide a number of spontaneous back mutants comparable with that resulting from irradiation, and thus to allow an estimate of the ratio of induced to spontaneous rates to be made.

While examining the cultures for mutants, the observer kept himself unaware of whether he was looking at treated or control material, in order to avoid bias. Doubtful cases of reversions were bred, for the purpose of detecting as many partial reversions as possible. Since, however, many of these cases proved to be mere phenotypic variants, and a few probably represented dominant autosomal suppressors, the verified cases, shown in Table II, are a culled-out remainder that proved to be inherited and sex-linked.

As previously stated, tests have not yet been completed to determine whether every reversion involves a change in the forked-containing region or represents a suppressor located elsewhere. All our tests so far, however, including tests of all reversions of forked 1 from treated males, agree with the back mutation interpretation. It should also be noted that whereas all these cases of reversion have proved to be sex-linked, only one sex-linked suppressor, *su-f* of Whittinghill (1938), has ever been recorded, despite the finding of numerous back mutations and of some dominant autosomal suppressors. Moreover, whereas *su-f* is recessive in its suppressing action, all our cases of induced reversions have been found to act as complete or partial dominants over f^1, when this is present in the homologous X of a female, as was to be expected of alleles or pseudo-alleles of the forked region.

Three different alleles of forked were used (see Table I). In the females forked 1, the allele used in all the earlier work on mutant-type forked, was present in all cases. In the males forked 1 was present in crosses (*a*) and (*b*), forked 3 in cross (*c*) and forked *x* in cross (*d*). The X-chromosome of the males in cross (*b*), unlike that in the other crosses, was ring shaped (denoted as X^{c2}). The purpose of using a ring was to ascertain whether this conformation, by causing the elimination, through their dicentricity, of chromosomes resulting from single exchanges between homologous or sister chromatids, would noticeably reduce the frequency of back mutations, as would be the case if most of them represented duplications. The more extreme mutant, forked 3, not suppressed by *su-f* according to Green (1955), was used because it is a pseudo-allele of forked 1 (Green, *ibid.*). The moderate but very stably expressed forked *x*, induced by X-rays, was used because it had been inferred (Muller, 1946 *et seq.*) to represent a deficiency of a part of a multiple locus and, as such, it might be found to be reversible with exceptional difficulty or not at all.

The results are shown in Table II. Considering first the allele forked 1 in the female germ cells, it is seen that the untreated females gave only 3.

TABLE II

Data on Reversions of Forked

Numbers in parenthesis indicate mutants occurring in a 'cluster' from same culture and therefore probably of common spontaneous origin; * cluster is omitted from reckoning of frequency after irradiation. Partial reversions are denoted as p, seemingly complete ones as $+$

Sex	Female	Male	Male	Male	Male	Male
Allele ...	forked 1	forked 1	forked 1	forked 3	Sum f^1, f^3	forked x
Cross ...	a, b, c, d	a	b	c	$a\ b, c$	d
Type of X	attached	rod	ring	rod	rod and ring	rod
Control count	2p, 1 + / 975,108	1p (2 +) / 119,075	0 / 36,010	1p / 160,017	2p (2 +) / 315,102	0 / 134,007
Control frequency	3×10^{-6}	$2{\cdot}5 \times 10^{-5}$	0	$0{\cdot}6 \times 10^{-5}$	$1{\cdot}3 \times 10^{-5}$	0
Irradiation count	4p, (2p) / 118,642	1p, 2 + / 19,797	2p, 1 + / 28,918	2p / 26,395	5p, 3 + / 75,110	0 / 30,434
Irradiation frequency	$34^* \times 10^{-6}$	15×10^{-5}	10×10^{-5}	$7{\cdot}6 \times 10^{-5}$	11×10^{-5}	0
Ind. control ratios*	10X	5X	—	11X	7X	—

*Induced: control ratio = (irrad. freq.-control freq.)/control freq.

reversions among nearly a million tested chromosomes, a rate of $3{\cdot}1 \times 10^{-6}$, whereas the irradiated oöcytes gave a rate of 34×10^{-6} (when two cases of common origin and therefore probably spontaneous are omitted from the count). In the males forked 1 gave in the untreated material 3 reversions (2 having probably been derived from the same mutation) among about 155,000 chromosomes (including about 36,000 of ring type, without reversions), a rate of $1{\cdot}9 \times 10^{-5}$, while among the irradiated chromosomes, including rings, there were 6 in about 49,000, or $12{\cdot}3 \times 10^{-5}$. It is notable that the forked 1 data on treated rings and rods are in good agreement. Forked 3 in untreated and treated males is seen (Table II) to give somewhat but not yet significantly lower rates than forked 1.

In contrast, forked x has given no reversions in either the untreated or treated series. This more or less expected result supports, together with the evidence from the ring-X, the interpretation of the forked reversions as arising by a qualitative change in the gene rather than by duplication. For the hypomorphic nature of the forked 1 mutant (Muller, 1933, 1950) should have allowed a duplication of forked 1 and presumably even one of forked *x* to manifest itself in the guise of a more nearly normal allele. Thus the absence of these reversions here testifies to the non-duplicational, qualitative nature of most of the reversions of forked 1 and forked 3.

Despite the small absolute numbers of mutations detected, the enormous differences in the totals among which they occurred in the untreated and treated series make the differences in the frequencies for these series highly significant, and allow a comparison between induced and spontaneous mutation rates. Since the frequencies for the chromosomes of paternal origin were not significantly different in crosses (*a*), (*b*), and (*c*), it is legitimate to add the data together for the treated males of all these crosses, and also for the untreated ones, in order to find the ratio of induced to spontaneous frequencies in this material as a whole. This is calculated by first subtracting the control mutation frequency ($1{\cdot}3 \times 10^{-5}$) for all these crosses together from the frequency observed after irradiation ($10{\cdot}7 \times 10^{-5}$). This gives us the frequency of induced mutations ($9{\cdot}4 \times 10^{-5}$). The quotient obtained when this 'induced frequency' is divided by the control frequency is 7. That is, the induced mutation rate in male germ cells is 7 times the control rate in them. In the females of all crosses combined, the corresponding ratio of induced to control rates is 10, a figure not significantly different from that for the male rate. Taking the average ratio as 8 for this dose of 3,200r, we find that 3,200r/8 is a dose giving about as many mutations as occur spontaneously. In other words, 400r is the 'doubling dose'. Although this is a good deal higher than the doubling dose (about 60r) previously found for recessive lethals after irradiation of spermatozoa in the male or of oöcytes, it is not significantly higher than the doubling dose reckoned from the data obtained by the Indiana group on point mutations induced in these stages at specific loci.

The 3- to 6-fold higher frequency observed among male than among female germ cells in both the treated and untreated material represents a significant difference in both cases. However, the reason for it depends in part, and perhaps in very large measure, upon partial reversions (which comprised most of our mutations) being more readily distinguishable from the forked from which they had originated when in hemizygous condition, as they were in the male, than when their expression was partially masked by the presence of the original forked in the homologous X-chromosome, as was the case in the female. Other work of Muller (1928*a*, *b*, *et seq.*) and of co-workers has indicated a much smaller difference than this between male and female germ cells in regard to both spontaneous and induced rates of lethals and of direct point mutations involving specific loci.

Comparison of the rate of reversion of forked here obtained in the treated male germ cells with that of direct mutation of non-forked to a forked allele, as found in work of our own and other groups, shows these two rates to be not very different from one another, unlike what is true for most loci. The spontaneous rates of back and direct point mutation of forked also seem to be very similar, so far as can be judged. However, we are here using only the very meagre figure of 2 spontaneous direct point mutations to forked among some 529,000 chromosomes derived from male germ cells, as found in data of Bonnier and Lüning and of our colleagues Frye and Schalet (unpublished), and this rate is therefore subject to large error.

As for the other loci investigated in our material, only one case of a reversion that proved fertile and was inherited was found. This was a spontaneous change of ivory eye (w^i) to normal red in cross (*c*). There were also two changes that appeared to be of the same kind in the treated material of this cross, but both of these mutants were sterile. Since, on the other hand, most of the loci here dealt with other than forked have been shown in earlier work of our group (Valencia and Muller, 1949) and in that of other authors to have direct rates of point mutation, after application of X-rays, comparable with that of forked, it is evident that forked has a frequency of induced reversions much more nearly equal to that of its direct mutations than most loci do.

All in all, it may be concluded that in *Drosophila*, as in *Neurospora*, mutations of the same general types, including back mutations, can be produced by ionising radiation as arise otherwise, and that these mutations include cases of chemical change, of the same categories as those that constitute the material for evolution.

Acknowledgment

New observations here reported were made with the aid of a grant from the American Cancer Society, and of a post-doctoral fellowship to I. I. Oster from the National Science Foundation of the U.S.A.

REFERENCES

Bonnier, G. and Lüning, K. G., 1949. *Hereditas*, **35**, 163.

Bridges, C. B., and Brehme, K. S., 1944. *Publ. Carneg. Instn.*, **552**.

Demerec, M., 1938. *Publ. Carneg. Instn.*, **501**, 295.

Giles, N. H., Jr., 1952. *Cold Spr. Harb. Sym. quant. Biol.*, **16**, 283.

Idem, 1956. *Brookhaven Sym. Biol.*, **8**, 103.

Giles, N. H., Jr., de Serres, F. J., and Partridge, C. W. H., 1955. *Ann. N.Y. Acad. Sci.*, **59**, 536.

Giles, N. H., Jr., Flynn, A. E., and Giles, D. L., 1948. *Rec. Genet. Soc. Amer.*, **17**, 38.

Giles, N. H., Jr., and Lederberg, E. A., 1948. *Amer. J. Bot.*, **35**, 150.

GREEN, M. M., 1955. *Proc. nat. Acad. Sci., Wash.*, **41**, 375.
HINTON, T., and DIBBLE, F., 1947. *Drosophila Inform. Serv.*, No. 21, 86.
JENSEN, K. A., KIRK, I., KØLMARK, G., and WESTERGAARD, M., 1952. *Cold Spr. Harb. Sym. quant. Biol.*, **16**, 245.
JOHNSTON, O., and WINCHESTER, A. M., 1934. *Amer. Nat.*, **68**, 351.
KAUFMANN, B. P., 1942. *Genetics*, **27**, 537.
KØLMARK, G., and WESTERGAARD, M., 1949. *Hereditas*, **35**, 490.
LEFEVRE, G., 1950. *Amer. Nat.*, **84**, 351.
LEFEVRE, G., Jr., and FARNSWORTH, P. C., 1953. *Drosophila Inform. Serv.*, No. 27, 97.
Idem., 1954. *Ibid.*, No. 28, 129.
LEFEVRE, G., Jr., GREEN, M. M., and FARNSWORTH, P. C., 1955. *Genetics* (Abstr.), **40**, 581.
MULLER, H. J., 1928*a*. *Ibid.*, **13**, 279.
Idem, 1928*b*. *Proc. nat. Acad. Sci., Wash.*, **14**, 714.
Idem, 1933. *Proc. 6th Int. Congr. Genet.*, **1**, 213.
Idem, 1946. *Drosophila Inform. Serv.*, No. 20, 88.
Idem, 1947. *Ibid.*, No. 21, 71.
Idem, 1950. 1947/48 *Harvey Lectures*, Series XLIII, p. 165.
PATTERSON, J. T., and MULLER, H. J., 1930. *Genetics*, **15**, 495.
SPOFFORD, J. B., 1955. *Drosophila Inform. Serv.*, No. 29, p. 165.
TIMOFÉEFF-RESSOVSKY, N. W., 1930. *Naturwissenschaften*, **18**, 434.
Idem, 1932. *Biol. Zbl.*, **52**, 468.
Idem, 1933*a*. *Proc. 6th Int. Cong. Genet.*, **1**, 308.
Idem, 1933*b*. *Z. indukt. Abstamm. u. VererbLehre*, **64**, 173.
Idem. 1937. *Experimentelle Mutationsforschung in der Vererbungslehre*, p. 83 (Dresden u. Leipzig, Theodor Steinkopff).
Idem, 1939. *Proc. 7th Int. Cong. Genet.*, p. 281.
VALENCIA, J. I., and MULLER, H. J., 1949. *Hereditas* (Abstr.), suppl. vol., p. 681.
WHITTINGHILL, M., 1938. *Genetics*, **23**, 300.
YU, S-C., 1949. *Drosophila Inform. Serv.*, No. 23, p. 104.

DISCUSSION

Auerbach. As regards Stadler and Roman's data, I feel that one of their pieces of evidence, i.e. the lack of induced back mutations at the alpha-locus, has become doubtful after McClintock's work, which showed that the so-called spontaneous back mutations in *Dotted* plants are not from back mutations, but position effects of a very special kind. Apart from this, however, I find Stadler and Roman's analyses very convincing, and I should like to ask Professor Muller whether he thinks there is an essential difference between maize and *Drosophila* genes in their ability to back-mutate after X-radiation?

Muller. The most reasonable interpretation seems to me to be that chromosomes of maize are broken more easily by radiation, while possibly the genes of maize are at the same time less sensitive to the point mutation effect of radiation. I mean, there is no reason why these two kinds of events should run completely parallel even though they do have a certain amount of parallelism. Whether there *is* this difference in sensitivity with respect to the production of point mutations, we do not know, but we do know that the chromosomes of maize are altered more by a given dose.

von Wettstein. Some experiments with barley show in preliminary results that for *erectoides* a^{23} the mutation rate for reversions is of the order of $1{\cdot}7 \times 10^{-7}$ per spike and r-unit. The expected number of mutations from *wild* type to *erectoides* a^{23} can be calculated from the observed mutation rate of the other *erectoid* loci and appears to be of the order of 5×10^{-7} per spike and r-unit. Of special interest are two partial reversions. The extreme pleiotropic effect of *ert* a^{23} can be split up by reversion in one character, but not in the others. One partial reversion has an ear changed towards *wild* type, whereas the culm-structure displays an even more extreme condition than is present in *ert* a^{23}. The other partial reversion comprises a complete reversion in culm-structure with an unchanged ear-morphology.

Ford. I should like to ask whether in view of the great disparity in numbers of flies examined in the irradiated and control series the observational effect was in proportion to the numbers? This would have been achieved, of course, if the origin of the flies examined was not known to the observer at the time of examination.

Oster. In order to avoid bias the observer was given flies to examine for reversions, which had been collected and separated into males and females by the assistants and given code numbers in order that he would not be aware of whether he was looking at treated or control material.

Carter. Professor Muller mentioned the fact that a spontaneous back mutation rate tends to be lower than the corresponding forward rate. This would appear to be disadvantageous to the species. Would Professor Muller care to speculate on the evolutionary implications of this observation?

Muller. The quick answer to that will be that I have not so much faith in Providence!

Howard. What is known about the rates of back mutation of alleles having different origins, that is, spontaneous, radiation-induced or chemically induced?

Westergaard. To my knowledge only Giles and de Serres have made a systematic survey of mutations induced by X-rays. They found that most X-ray-induced mutations which had the standard gene arrangement were able to revert through back mutations; whereas a second group of X-ray-induced mutations which showed altered linkage relationship (indicating small chromosome deletions or rearrangements) were unable to revert. Other observations have shown that mutations induced by UV or by mustard gas are also able to back-mutate, but no systematic investigation has been made.

Kanazir. May I have your opinion about back mutation in bacteria? We observed that different gene-loci gave different frequencies.

Muller. Well, I can only say that this fits very well with what is known from other material. In most bacterial work so far, however, the analysis could not be pushed as far down as the gene level, since for that purpose, work has to be done on forms which are subject to recombination.

GENETIC IMPLICATIONS OF IRRADIATION IN MAN

T. C. Carter
Medical Research Council, Radiobiological Research Unit,
Harwell, Berkshire, England

Radiation genetics is a bifid science. It is concerned with, first, mutation and, second, the structure of populations in which mutation has occurred. Spontaneous mutation, that is to say mutation due to natural background radiation and all other natural causes, occurs in all populations. Radiation genetics is therefore essentially a comparative and quantitative science; we have to compare populations with more or less mutation, rather than those with and without mutation.

This bifid nature of radiation genetics has, for historical and technical reasons, had a curious and rather unfortunate effect. Experimental workers studying mutation have concentrated their efforts on induced mutation, since high mutation rates are relatively easy to measure; the low and technically difficult spontaneous rates have been either disregarded entirely or measured with low statistical precision. Workers studying populations, on the other hand, have concentrated their efforts on populations unexposed to mutagens, since these are easier to manage in the laboratory and in the wild they manage themselves. As a result, there have been exceptionally few reports of studies in which populations exposed to both mutagenic treatment and natural selection were observed over several generations. Yet studies of this type will be an essential prerequisite to a better understanding of the effects of chronic radiation exposure of human populations.

The study of mutation has received much more attention than that of populations. Its growth dates from the early 1920s. Its landmarks include the 1921 attempts, unsuccessful as we now know, of C. C. Little and H. J. Bagg to induce mutations in mice by means of X-rays; the first successful demonstration of X-ray-induced mutation in *Drosophila* by H. J. Muller, in 1927, and in maize shortly after by L. J. Stadler; the induction of chromosome mutation in mice with neutrons by G. D. Snell, soon after the first cyclotron was built; and the demonstration of chemical mutagenesis by Charlotte Auerbach and J. M. Robson, first published in 1944. The last ten years have seen an enormous growth of studies of induced mutation, and especially of chemical mutagenesis, following the development of genetical techniques for bacteria and other micro-organisms.

Further study of populations, and especially of populations exposed to mutagens, is now, I think, a much more pressing need than further extension of the study of mutagenesis. Considered from the narrow viewpoint of the genetic hazard of radiations to man, knowledge of the number of mutations induced in a human gamete by exposure to a given radiation dose would be of limited value; it would tell us little about the net effect on the population, which is what we need to know. Nevertheless, though I think attention should now be directed more and more to population studies, it is one aspect of the study of mutagenesis that I wish to consider in this paper. I refer to the fact that there are currently two different systems of measuring radiation-induced mutation; they may be called the 'absolute' and 'relative' systems, their essential difference being that in the absolute system the spontaneous mutation rate is disregarded, whereas in the relative system it is used as a yardstick for induced mutation. It is important to recognise the existence of the two systems, for there are occasions when one is appropriate and the other not. Failure to recognise this may lead—I think it has led—to erroneous conclusions being drawn from mutagenesis experiments.

Measures of Induced Mutation

Both systems of measurement rest on the assumption that induced germinal mutation is linearly related to the applied radiation dose. This is commonly believed to be true of gene but not of chromosome mutation, and therefore their use is limited to this class of change. If the assumption is valid, the relationship between the mutation rate of a gene, M, and the applied radiation dose, x may be expressed by the equation

$$M = s + bx \qquad \text{......(1)}$$

where s is the spontaneous rate and b a constant of proportionality.

The absolute system

Most early workers in radiation genetics used high radiation doses, corresponding with the more easily measured high mutation rates. Thus precise estimates were obtained of b but not of s. The practice therefore arose of quoting induced mutation rates in absolute terms, that is to say, in terms appropriate to the parameter b only, namely, mutations per roentgen per locus. The absolute system is well adapted to experiments in which interest is confined to the induced mutation, and not in its relationship to the spontaneous rate. It has the disadvantage that the number of loci scanned for mutation must be known; and for this reason it appears to be inappropriate for use in human radiation genetic studies, since in man it is impossible to make the genetic tests necessary to distinguish between allelic and mimic genes.

The relative system.

The relative system uses the spontaneous mutation rate as a yardstick for induced mutation. The philosophy behind its use is the supposition that mutation rates are amenable to natural selection, and that each species will have achieved a spontaneous mutation rate adapted to its needs: not so small that evolutionary adaptation to environmental change is impossible, nor yet so large that the species has to carry an impossibly large 'load of mutations', to use Muller's (1950) phrase. Given this condition, some species will have acquired a high and some a low spontaneous mutation rate; the effect on a population of a given absolute change in mutation rate per locus would depend on the magnitude of the spontaneous rate, being proportionately greater in species with low spontaneous rates. The relative

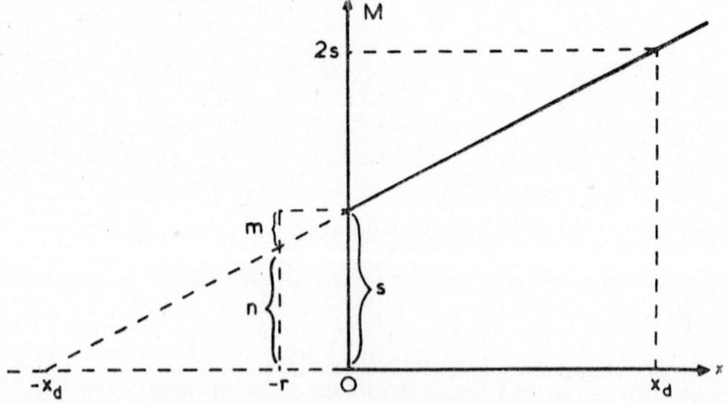

FIG. Mutation rate (M) plotted against applied radiation dose (x). s, spontaneous mutation rate; m and n, those parts of s due to natural background radiation dose (r) and to other causes; x_d, mutation-rate-doubling dose.

system is therefore more appropriate than the absolute system to comparisons between species (if, indeed, quantitative comparisons of this type are justifiable at all). It has the advantage that when considering a group of loci one need not know the number in the group; it is therefore appropriate for use in human radiation genetic studies.

The unit of measurement under the relative system is the 'doubling dose'; that is to say, it is the applied radiation dose such that the total mutation rate becomes equal to twice the spontaneous rate. Applied to a single gene, or to a whole group of genes taken together, the system is straightforward. However, one cannot suppose *a priori* that all genes in an organism will have the same doubling dose, and it it not immediately apparent how to calculate an appropriately weighted mean doubling dose from the individual doubling dose of the genes in a group. This is sometimes held to be a defect in the system. The difficulty can be resolved by considering the *relative mutability* of each gene in the group, defined as the proportion of spontaneous mutation which must be attributed to natural background radiation; it can then be shown that the spontaneous mutation rates are the appropriate weights for calculating a weighted mean.

This result is obtained in the following way. Consider Figure 1, in which the mutation rate, M, of a gene is plotted against the applied radiation dose, x. The spontaneous mutation rate is s and it consists of two parts: m, due to the accumulated natural background radiation dose, r, and n, due to all other causes. x_d is the doubling dose. Then the relative mutability, g, of the gene is by definition

$$g = m/s \qquad \text{......(1)}$$

The following relationships are consequent upon the geometry of the diagram

$$m/r = s/x_d \qquad \text{......(2)}$$

$$\begin{aligned} M &= s + xs/x_d \\ &= s + xm/r \\ &= s + (x/r)\, gs \end{aligned}$$

whence

$$\frac{M}{s} = 1 + \frac{x}{r} g. \qquad \text{......(3)}$$

For a whole group of genes,

$$\Sigma M = \Sigma s + (x/r)\, \Sigma gs$$

whence

$$\frac{\Sigma M}{\Sigma s} = 1 + \frac{x}{r} \frac{\Sigma gs}{\Sigma s} \qquad \text{......(4)}$$

Comparing the last terms of equations (3) and (4), it appears that the spontaneous mutation rates are the appropriate weights for calculating $\hat{g}$, the weighted mean value of g; and from equations (1) and (2) it follows that $\hat{x}_d$, the weighted mean value of x_d, is given by

$$\hat{x}_d = \frac{r}{\hat{g}}. \qquad \text{......(5)}$$

Implications for Human Radiation Genetics

The arguments put forward above lead to the conclusion that the relative system of measurement should be used in preference to the absolute system in at least two circumstances: first, when considering an organism, such as man, in which it is impossible to specify the number of genes under examination; and, second, when making comparisons between different organisms. This holds a number of implications.

Use of the relative system means measuring doubling doses. In order to do this accurately it is necessary to measure the spontaneous mutation rate with at least the same accuracy, proportionately, as the rate under irradiation. This means that for maximum accuracy with a given expenditure of effort the same number of new mutations must be observed in the control as in the irradiated series. The first implication, therefore, is that in planning any programme aimed at direct measurement of the doubling dose for human genes the number of individuals in the control series must be greater than that in the irradiated series, in inverse proportion to the probable mutation rates. Thus, for example, it has recently been reported (Court Brown and Doll, 1956) that there are case records of more than 11,000 males treated with X-rays in Great Britain for ankylosing spondylitis; and Mr. S. B. Osborn informs me that the gonad dose in this treatment may easily be as high as 90r. If we suppose *ex hypothesi* that the average doubling dose for human genes is likely to be about 30r, we should expect these spondylitics to show a mutation rate about four times the normal. Any plan for a genetic study of their progeny should therefore cover also an otherwise comparable population of control children at least four times as large. In general, the inference is that any human radiation genetics programme must consist preponderantly of measuring spontaneous mutation rates.

This leads to the second implication, which is that any measurement of human spontaneous rates, even if divorced entirely from radiation programmes, will help to provide the necessary substratum of information on which the planning of radiation genetic programmes can subsequently rest.

The third implication follows from the fact that the spontaneous mutation rates are the appropriate weights for use when calculating a weighted mean doubling dose for a group of genes. If they have widely differing spontaneous mutation rates, the weighted mean doubling dose will then closely reflect the doubling doses of the few genes with the highest spontaneous rates. Furthermore, there is evidence that human genes may vary widely in this respect; Haldane (1948) has pointed out that human spontaneous rates probably cover the range 10^{-4} to 10^{-10}. The problem of finding a representative doubling dose for human genes in general may therefore resolve itself into the simpler problem of finding which human genes are the most spontaneously mutable, and measuring their doubling doses.

The fourth implication is that we should re-examine the assertion, so often made in calculations of the genetic hazard of ionising radiations, that human genes are more sensitive to radiation-induced mutation than those of lower organisms; for it depends on a comparison of mouse and *Drosophila* mutation rates made by the absolute system. The philosophy behind this assertion is an assumption that human genes are at least as sensitive to radiation as those of the mouse, and an inference from laboratory experiments that mouse genes are fifteen times as sensitive as those of *Drosophila*. This inference derives from a comparison of mutation rates to recessive visible alleles at specified autosomal loci in spermatogonia of *Drosophila* (Alexander, 1954) and of the mouse (Russell, 1951, 1956). However, when this comparison is made by the relative system the data cease to support the inference of lower

Drosophila sensitivity; this is because no spontaneous mutations were observed in the *Drosophila* experiment, though they appeared among comparable numbers of control progeny in Russell's mouse experiment and also in further mouse experiments carried out in Great Britain (Carter, Lyon and Phillips, 1956 and new data in the Table). Taken together, these mouse data indicate a doubling dose of about 36r; they are thus in fair agreement with those of Charles (1948), which indicated a doubling dose of about 50r for another class of mutation in the mouse. The doubling

TABLE

Mutations at Seven Specified Loci in the Mouse

Author	Dose (r)	Number of animal examined	Number of mutations at locus							Total
			a	*b*	*c*	*d*	*p*	*s*	*se*	
Russell, 1951	600	48,007	0	11	3	6	8	25	0	53
Russell, 1951	0	37,868	0	0	0	1	0	1	0	2
Carter, Lyon and Phillips, 1956	37·5	10,024	0	0	0	0	0	1	0	1
Carter, Lyon and Phillips, 1956	0	18,355	0	0	0	1	0	1	0	2
Carter, Lyon and Phillips, new data	37·5	14,423	0	2*	0	0	0	0	0	2
Carter, Lyon and Phillips, new data	0	22,907	0	0	0	1†	0	0	1†	2

* Two clusters: one was of two mutants among more than thirty classified progeny; the other was of more than thirty mutants and showed that the father was heterozygous for *brown*.

† One animal was mutant for both *dilution* and *short-ear*; this has been counted as a single mutational event in the analysis.

dose for *Drosophila* spermatogonia is indeterminate from Alexander's data, owing to the absence of spontaneous mutations. For the induction of recessive sex-linked lethals in mature *Drosophila* sperm Spencer and Stern (1948) found a doubling dose of about 50r; both higher and lower values have been found for sex-linked lethals under various other conditions (Caspari and Stern, 1948; Muller, Herskowitz, Abrahamson and Oster, 1954). Thus at present I consider the statement that mouse genes in general are fifteen times as sensitive to radiation as those of *Drosophila* is not yet proven.

Finally, we may ask by how much human mutation rates may have been increased by modern uses of ionising radiations. An answer can be derived from the recent reports of both the British Medical Research Council and the United States National Academy of Sciences—National Research Council, though neither states it explicitly. Both consider 30 to 80r a reasonable guess of the range in which the representative doubling dose for human genes may lie. Taking the lower value, 30r, we may say that the total spontaneous mutation, from all causes, is equivalent to that induced by 30r. The British report estimates that present practices give an additional gonad dose, up to the mean age of reproduction, that is greater than 0·75r; how much greater is not known, owing to the absence of an estimate of the contribution due to radiotherapy and to uncertainty of that due to radiodiagnosis. Osborn and Smith (1956) think a realistic estimate of the latter might even be ten times the lower limit quoted in the report. This would imply an additional gonad dose of 6·75r, corresponding with a 22% rise in the mutation rate. Corresponding figures from the American report are 5r and 17%.

This possible increase of human mutation rates by as much as a fifth, due almost entirely to medical radiology, is something that appears to have escaped public notice, though it may well in the long run be a much greater hazard than that due to the much publicised accumulation of strontium 90 from nuclear tests in human skeletons.

Summary

Human radiation genetics has developed out of the descriptive study of human populations and the experimental study of mutagenesis. As a result of this hybrid ancestry the absolute system of measurement, which expresses induced mutation in terms of mutations per roentgen per locus, is sometimes used with reference to man, whereas the relative system, in which the unit is the mutation-rate-doubling dose, would be more appropriate. Misleading conclusions may be drawn if the inappropriate system is used. Implications for human radiation genetics of use of the relative system are discussed.

Acknowledgments

I am grateful to my colleagues Dr. Mary F. Lyon and Miss Rita J. S. Phillips for permission to quote the unpublished data in the Table; and to Mr. S. B. Osborn for allowing me to quote his unpublished measurements of the gonad dose in the radiotherapy of ankylosing spondylitis.

REFERENCES

Alexander, M. L., 1954. *Genetics*, **39**, 409.

Auerbach, C., and Robson, J. M., 1944. *Nature, Lond.*, **154**, 81.

Brown, W. M. C., and Doll, R., 1956. Appendix B, Medical Research Council's Report, Cmd. 9780 (H.M.S.O.).

Carter, T. C., Lyon, M. F., and Phillips, R. J. S., 1956. *Brit. J. Radiol.*, **29**, 106.

Caspari, E., and Stern, C., 1948. *Genetics*, **33**, 75.

Charles, D. R., 1948. *Radiology*, **55**, 579.

Haldane, J. B. S., 1948. *Proc. Roy. Soc.* (B), **135**, 147.

Little, C. C., and Bagg, H. J., 1923. *Amer. J. Roentgenol.*, **10**, 975.

Medical Research Council's Report, 1956, Cmd. 9780, p. 128 (H.M.S.O.).

MULLER, H. J., 1928. *Z.I.A.V. Suppl.* 1, 234.

Idem., 1950, *Amer. J. Human Genet.*, **2**, 111.

MULLER, H. J., HERSKOWITZ, I. H., ABRAHAMSON, S. and OSTER, I. I., 1954. *Genetics*, **39**, 741.

National Academy of Sciences—National Research Council, 1956. *The Biological Effects of Atomic Radiation*, p. 108 (Washington, D.C.).

OSBORN, S. B., and SMITH, E. E. 1956. *Lancet*, **1**, 949.

RUSSELL, W. L., 1951. *Cold Spring Harbor Symp. on Quantitative Biology*, **16**, 327.

Idem, 1956. *Amer. Nat.*, **90**, 67.

SNELL, G. D., 1939. *Proc. nat. Acad. Sci. Wash.*, **25**, 11.

SPENCER, W. P., and STERN, C., 1948. *Genetics*, **33**, 43.

STADLER, L. J., 1928. *Science*, **68**, 186.

DISCUSSION

Muller. Experiments done in our laboratory on the induction of mutations by X-rays in *Drosophila* oogonia, which we have good reason to believe react like spermatogonia in this respect, showed the doubling dose for point mutations to be at least 200r, so that mice would be at least six times as sensitive as *Drosophila*, as measured by their doubling dose. There is a source of error in judging the doubling dose from that of the genes having the highest spontaneous mutation rate. For the imperfectness of the correlation between spontaneous and induced mutability, as between different loci, would lead to the doubling dose being highest for the loci with highest spontaneous rate.

Carter. I based the comparison of mouse and *Drosophila* specific locus mutation rates on the spermatogonial mutation data of W. L. Russell and of M. L. Alexander, since they were obtained in experiments specially designed to test this point. Furthermore, they are the data commonly cited to support the statement that mouse loci are ten or fifteen times as sensitive as *Drosophila* loci. Alexander's data show an upper fiducial limit of 194r for the doubling dose, taking a 21/2% probability level with a single-tailed test. Thus, the spermatogonia in Alexander's experiments appear to have been more sensitive to induced mutation, judged by the doubling dose, than the oogonia in Professor Muller's experiments.

I agree with Professor Muller that of two loci with the same induced mutation rate, the one with the higher spontaneous rate will have the higher doubling dose. This is a matter of definition. But it does not appear to be a source of error.

Auerbach. I saw from your table that the *s*-locus mutated several times spontaneously. This was also the most mutable locus in the X-ray experiments. The other locus which mutated twice spontaneously was the *d* one. Was this also highly mutable in the X-ray experiments?

Carter. In Russell's experiments the *b*, *p* and *d* loci followed the *s*-locus in frequency of induced mutation. In our experiments the *d*-locus has mutated twice spontaneously and the *b*-locus twice after 37·5r, but we have not yet seen any mutations at the *p*-locus. This absence, however, is not statistically significant.

Tobias. It is clear from the data you presented and from other available information, that only a small part of the natural mutation rate is due to cosmic radiation and natural radioactivity. Would you care to discuss some of the other factors that cause the background, natural mutation rate in humans? The reason I mentioned this point is the need to consider additional factors beside radiation in any experiments or discussions involving radiation-caused mutations. Many environmental factors, possibly foodstuffs and the composition of the atmosphere we inhale, can cause mutation and can alter the natural mutation rate in the course of time.

Carter. Russell's mouse data and our own together indicate a doubling dose of about 36r. The gonads of a mouse receive about 0·03r per generation from natural background radiation. It follows that only about one thousandth of mouse 'spontaneous' mutation can be attributed to this cause. The origin of the remainder is a subject for speculation; chance imperfections of the gene replication process and chemical mutagens come to mind as two possibilities.

THE INDUCTION OF DETRIMENTAL MUTATIONS IN *DROSOPHILA* BY X-RAYS

K. G. Lüning and S. Jonsson
The Institute of Genetics, University of Stockholm,
Stockholm, Sweden

When living cells are exposed to ionising radiations a spectrum of mutations are induced among which only a fraction are detectable by conventional methods. By these methods one can pick up mutations with morphological effects, visibles or lethals—which cause the death of all homo- or hemizygous carriers. These two groups are not alternative but are independent. Visibles exist which also are lethals. Among the visibles are cases which are not lethals but are subvital, i.e. *detrimentals*. Hence subvitals may be expected which are not associated with morphological changes. In the pioneer works in this field Timoféeff-Ressovsky (1935) and Kerkis (1938) using *Drosophila melanogaster* showed that sex-linked detrimentals were induced and also that they appeared in considerable numbers, in fact 2–3 times more than the corresponding sex-linked recessive lethals. In these studies they used the proportion of males with the chromosome to be tested compared to either one class of sisters or to brothers with another marked chromosome. Mutations obtained by this method are those which in any stage prior to maturity kill all or parts of the males with this chromosome. By other means it might be possible to detect other types of mutations influencing the length of the adult life or time of development. The latter problem will be dealt with by Dr Bonnier in another paper to be presented to this Conference.* In addition to these mutations which interfere with the life of the individual it should be remembered that there exist mutations which affect fertility and fecundity. These latter mutations affecting viability and productivity are not alternative but may be present together or separately.

The problem of detrimentals can also be looked at from other points of view. On the one hand there are studies of the mutation rates and on the other there is occurrence of detrimentals in populations. The latter have been studied thoroughly by Dobzhansky and Wallace and their groups and will not be further discussed here.

We will concentrate our interest on the former problem of mutation rates. This field includes studies of subvitals, decreased developmental rate, decreased fertility and/or fecundity. All these types need special techniques

* This volume, page 433.

and some of them are strongly influenced by environmental factors and are for that reason difficult to establish. The easiest type to study are the subvitals which are affected prior to the adult stage. Studies along these lines and similar to those of Timoféeff-Ressovsky and Kerkis have been carried out by Hadorn (1950) and Käfer (1953). Hadorn studied the effect of a chemical mutagen, phenol, on ovaries *in vitro* and concluded that there, recessive lethals were induced as well as subvitals, in the second chromosome. Käfer made a comparison of X-ray-induced mutations in the X- and in the second chromosome in sperm, and found that in both types of chromosomes slightly higher rates of recessive lethals were induced than of 'strong' detrimentals.

From the detection of variations in the rates of chromosome-breaks in sperm irradiated at different stages of spermiogenesis (Lüning, 1952*a*), it seemed to be of importance to study not only recessive lethals but also other types of mutations affecting viability to a greater or lesser extent.

Before reporting this we will first draw attention to another problem. Up to now, all studies of this kind, except Hadorn's study with chemical mutagenes, have been confined to irradiation of sperm. It is, however, also of importance to study the induction of detrimental mutations in unfertilised eggs by ionising irradiation. This is specially important since the mutational processes in sperm and egg seem to differ, according to reports by Glass (1940), Bonnier, Lüning and Arnberg (1952) and Lüning (1954); in the last few months such an experiment has been carried out.

In these studies sex-linked mutations affecting viability have been looked for by means of the Muller-5 technique. In the studies of mutations induced in sperm, an X-chromosome was used which contained a marker-gene—yellow, called y^{16} which was obtained in an experiment some years ago. In the female irradiation series an X-chromosome was used which contained the marker-gene white.

In both experiments one (1) Muller-5 (= M5) male was mated to some females of the y^{16} and *w* stocks respectively. The P. M5 male was again mated to his heterozygous daughters which were then allowed to lay eggs singly. From one of them the stock to be used was then prepared. Hence the experimental stocks were started from one X-chromosome of each kind and differences which later appeared must have originated after this process.

In the male irradiation series y^{16} males were taken out when less than 24 hours old and were irradiated with a dose 2,160r. They were immediately mated for 24 hours. The females were discarded. The males were mated to new M5 females for another 48 hours. From this cross F_1 females were collected. The males were mated to a third group of females on the 4th–7th day. On the 7th day they were mated again to new females for another 2 days and were then discarded. From this last cross F_1 daughters were collected. The offspring from these two groups of F_1 daughters will be referred to below as the '2–3' and '7–8' series. There was also a control series started with untreated y^{16} males. The F_1 daughters were mated singly to M5 males from the stock. Heterozygous F_2 daughters were again mated to M5 males, 3–5 pairs per vial and were shifted to new vials every 3rd day. The F_3 offspring was examined as for the number of M5 and y^{16} males. About 400 males were counted for each chromosome tested. If there was not a sufficient number in F_3 another generation was produced in the same way.

In the female irradiation series *w* females were irradiated with 3,240r and were subsequently mated to M5 males. Eggs were collected for the first 4 days after the irradiation. The F_1 daughters were collected twice a day as virgins in order to avoid contamination with other irradiated *w*-chromosomes kept by the brothers. Only one case in F_2 was excluded because of non-virginity. The procedure was then the same as in the male irradiation series. There was also a control series in which one *w* male was repeatedly mated to M5 females. Hence this control series can give some information on spontaneous mutations. In these series about 300 males were counted for each chromosome tested.

It should be pointed out that this comparison between the two types of males, presupposes that the viability of the M5 males is the same in all cases. This is of course not always true as spontaneous mutations might appear, but by using several M5 males this is avoided as far as is possible.

In an experiment arranged in this way one would theoretically expect to get 50% M5 and 50% y^{16} respectively *w* males. The results in the control series show that there were proportionally more y^{16} and *w* males than M5 males. The mean values are 55·85% y^{16} and 56·52% *w* males. This means that the chromosomes to be tested for detrimentals were originally superior to the M5 chromosomes.

The results are presented in two sets of diagrams. The percentages of y^{16} and *w* males respectively are put on the *x*-axis with one interval for excessive lethals = 0% and the following intervals representing 5%. From 40% each interval corresponds to 1%. This arrangement gives us a more informative picture of the 'quasi-normal' region, i.e. that around the control values. On the *y*-axis are the frequencies in the different intervals.

In Figure 1 is shown the male irradiation series and the corresponding control series. The lower diagram presents the results of the control series. This consisted of only 86 tested chromosomes. The bimodal appearance of this curve need not reflect a heterogeneity in this series as it is not significantly different from a normal distribution.

The two upper diagrams represent the results of the irradiation series. It should again be emphasised that these two series come from the same irradiated males but from sperm ejaculated at different periods after irradiation. From earlier experiments it is known that about 4 times as many chromosome aberrations are induced in sperm corresponding to the present '7–8' series than in the '2–3' series. From the diagrams of these two series

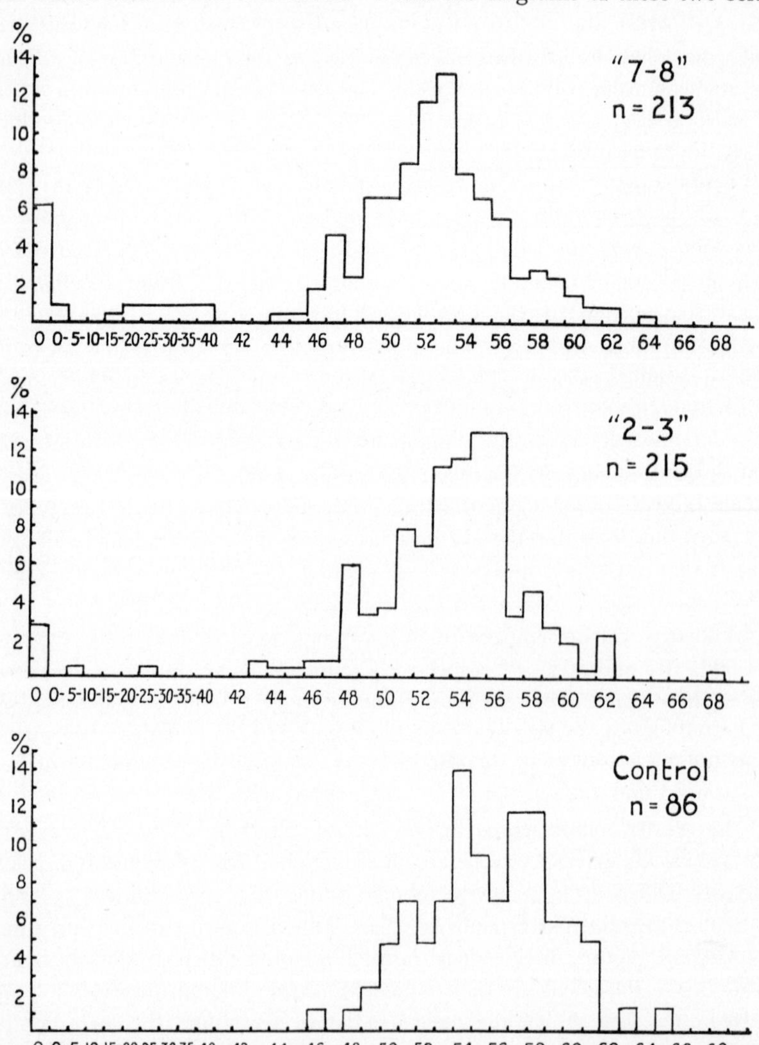

FIG. 1. Diagrams showing the frequencies of tests against different proportions of y^{16} males in the male irradiation series and the control. Ser. '2–3' represent offspring from sperm delivered the 2nd-3rd and the Ser. '7–8' those from the 7th-8th day after irradiation. On the x-axis are marked the percentages of y^{16} males over the total number of males per tested chromosome. Interval '0' represents recessive lethals. Intervals '0–' to '35–' represent 5% each. From 40% each interval is 1%. On the y-axis are marked the frequencies in each interval.

it is seen that there are more recessive lethals and considerably more detrimentals with 'strong' effect in the '7–8' than in the '2–3' series, i.e. in the range 0–45% y^{16} males. The rates of the 'strong' detrimentals seem to be about the same as that of the corresponding recessive lethals, which agrees fairly well with the results of Timoféeff-Ressovsky, Kerkis and Käfer. The majority of the tests showed in both series more than 45% y^{16} males. The mean of tests with more than 45% y^{16} males is in the control, 55·85 and in the '2–3' 54·16 and in the '7–8' series 52·83. It should be mentioned that at the same time as the rates increased of recessive lethals and 'strong' detrimentals, there was a shift in the 'quasi-normal' region towards a lower mean. This observation coincides with that of Kerkis that in the irradiation series there were more negative than positive deviations from the control mean, in contrast to the control, with about the same proportion on both sides of the mean. A similar predominance of negative deviations was also observed in one of Käfer's series but was not so marked in the rest.

These results indicate that a considerable number of mutations with detrimental effects were induced which are so weak that the individual cases were not detected; but they none the less show their effects in an experiment of this type. A calculation of the rates of these 'weak' detrimentals is very delicate as there are several variables. One can, however, infer from the present material that they are at least 5 times as frequent as the recessive lethals. This figure is slightly higher than that estimated by Kerkis but agrees with Muller's calculations (1950).

In Figure 2 the results are shown from the female irradiation series and their corresponding controls. As was mentioned above, this control series represents offspring from one single male and could hence give us some information about the spontaneous mutation rate. We obtained 2 recessive lethals but no 'strong' detrimentals. 'Weak' detrimentals, which fall within the 'quasi-normal' region, can be detected only if they appear in considerable numbers, so that they cause a skewing of the curve. This, however, does not seem to be the case. The two cases with extremely high rates of *w* males simulate supervital mutations. It is, however, possible that this is due to changes in the M5 chromosomes making the *m* more inferior to the *w* chromosomes. Because of lack of time this could not be tested.

The upper diagram in Figure 2 represents the results of irradiation of females. The rate of recessive lethals, although based on a very small number, agrees very well with that observed in other series (Lüning, unpublished). Two cases appeared with 'strong' detrimental effects. This is not significantly lower than that observed in the male irradiation series. In the 'quasi-normal' region there is a slight predominance of positive deviations which is different to the results in the male irradiation series. By a comparison between negative and positive deviations from the control mean, it is not possible to establish the occurrence of 'weak' detrimentals. On the other hand, it does not prove that 'weak' detrimentals do not exist in small numbers. That they really do occur is shown by one of the cultures with a low proportion of *w* males, which in the first test gave 45% and in a repeat gave 47·8%. That two such low values would occur due to random variations is highly improbable, and for that reason it is believed that this case is a real 'weak' detrimental mutation. It is possible also that some

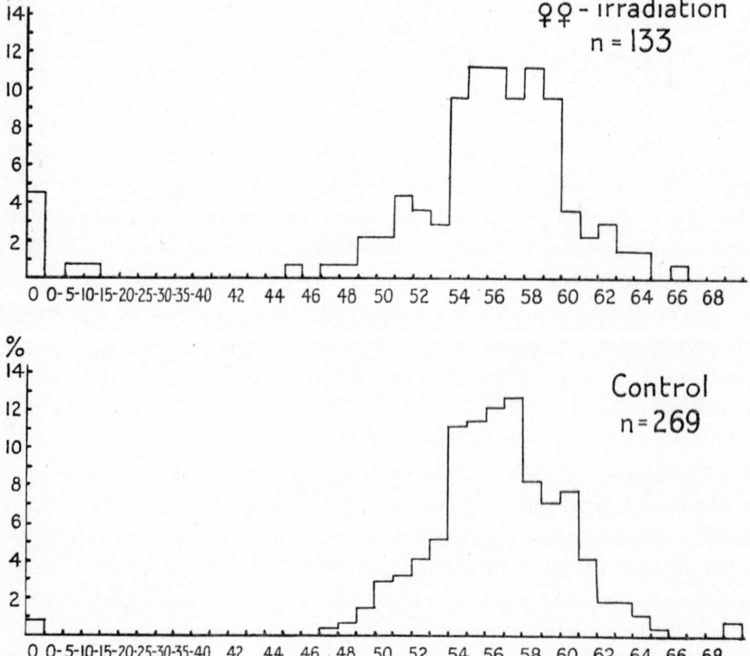

FIG. 2. Diagrams showing the frequencies of tests against different proportions of *w* males in the female irradiation series and the control. For further information, see the text to Fig. 1.

other cases with less than 50% *w* males were detrimentals, but this was not tested further. We can, in any case, conclude that there are induced, proportionally fewer 'weak' detrimentals as compared to recessive lethals, in the female irradiation series, than after irradiation of males.

Summarising the results we can state that the rates of detrimentals induced in sperm chromosomes vary with the stage treated in a manner similar to that shown for visibles and recessive lethals (Lüning, 1952*b*). Further the proportion of weak detrimentals appeared to be at least 5 times as high as the proportions of strong detrimentals or recessive lethals, which are similar to one another. In the female irradiation series the results have not contradicted the earlier observations that recessive lethals induced in unfertilised eggs appear at a slightly lower rate than in sperm, used the first 3 days after irradiation. Further, the relation between the rates of recessive lethals and 'strong' detrimentals need not be different from that observed in the male irradiation series. It is a remarkable fact that 'weak' detrimentals are so rare in the female irradiation series.

Beside studies of the mutation rates, there is also another part of this problem which should be emphasised and that is the degree of dominance of the detrimentals. In a recent study of vitality mutations Falk (1955) used 'hatchability' of eggs as a criterion. On his material he estimated that there were induced about 3·5 times more 'weak' detrimentals than recessive lethals in the third chromosome. He tested also the degree of dominance and found both dominant and partly dominant detrimental mutations with considerable frequency. Unfortunately he did no tests on the number of adults, using the conventional technique with marked test chromosomes, which has been of great value for a comparison between the two different methods. In a recent paper Dobzhansky, Pavlovsky, Spassky and Spassky (1955) tested some detrimentals isolated from wild populations of *Drosophila pseudoobscura* and concluded that they did not show any dominant detrimental effect. This difference between induced and spontaneous detrimentals may be due to the establishment of recessive detrimentals and/or, to the selection of modifiers buffering the effect of the detrimental gene in heterozygotes.

From what has been said above it is clear that for calculations of the genetic risks of ionising radiation, we must intensify our studies of these problems, despite the fact that they are extremely laborious.

Acknowledgment

The present study was supported by the Swedish Natural Science Research Council and the Nilsson-Ehle Fund.

REFERENCES

Bonnier, G., Lüning, K. G., and Arnberg, B., 1952. *Hereditas*, **38**, 109.
Dobzhansky, T., Pavlovsky, O., Spassky, B., and Spassky, N., 1955. *Genetics*, **40**, 797.
Falk, R., 1955. *Hereditas*, **41**, 259.
Glass, H. B., 1940. *Genetics*, **25**, 117.
Hadorn, E., 1950. *Pubbl. Staz. zool. Napoli. suppl.*, **22**, 32.
Käfer, E., 1953. *Z. ind. Abst. Vererb.*, **84**, 508.
Kerkis, I., 1938. *Izv. Akad. Nauk. U.S.S.R.* (Otd. mat. est. ser. biol.), p. 75.
Lüning, K. G., 1952*a*. *Hereditas*, **38**, 321.
Idem., 1952*b*. *Acta Zool.*, **38**, 193.
Idem, 1954. *Hereditas*, **40**, 295.

Muller, H. J., 1950. *J. cell. comp. Physiol.*, suppl. 1, **35**, 9.
Timoféeff-Ressovsky, N. W., 1935. *Nachr. Ges. Wiss. Goottingen, Math. Physik. kl. Biol. N.F.*, **1**, 163.

Discussion

Muller. A probable explanation of at least a part of Dr. Lüning's finding that irradiation of late oocytes gives a higher frequency of lethals in relation to detrimentals than does irradiation of a spermatozoa lies in the recent work of Herskowitz and Abrahamson. They found that a high proportion of the lethals induced by irradiation of late oocytes at the dosage-level used, represent unequal exchanges between homologous chromosomes. These vary faster than linearly with dose and show a marked time-dependence, as would be expected for multiple hit effects produced in cells in which union of broken ends can occur. On the other hand, the point mutations (which are more likely to result in detrimentals) are produced at a somewhat lower rate in oocytes than in spermatozoa. Thus, the ratio of lethals to detrimentals would be higher after irradiation of oocytes.

REFERENCE

Herskowitz, I. H., and Abrahamson, S., unpublished.

RATE OF DEVELOPMENT OF X-RAY-INDUCED DETRIMENTALS AND THE INFLUENCE OF SELECTION PRESSURE

G. Bonnier

Institute of Genetics, University of Stockholm,
Stockholm, Sweden

The results of the experiments of which I am going to speak are in a very preliminary state. I will therefore confine myself to questions which mostly have to do with methods and experimental design.

My starting point is the assumption that detrimental mutations, where effects are not too severe, will in the future constitute the most serious group of mutations within human populations. The reasons for this assumption are twofold. On the one hand, these mutations are probably the commonest types. On the other hand, persons suffering from such mutations, even in the future and in spite of medical and social progress, will probably really suffer, at least enough to decrease their happiness in living. We may, however, also assume that these same persons, because of medical and social progress, will be able to reproduce at about the same rate as non-mutants. This again will have the consequence that their detriment-causing genes will be easily spread in the population.

If we want to know anything about physiological effects of mutational changes in Man, then in genetical experiments, we must of course use a mammal. But in a mammal, for example in the mouse, we do not have access to the many marker genes which we have in *Drosophila*, and with the aid of which we often are able to isolate strains of detrimentals. How, then, will it be possible to make these observations in mice? I believe that the answer is that we may use growth rates as indices of viability.

To study this problem we are making some model experiments with *Drosophila melanogaster*. After X-ray irradiation of males a number of strains were produced, within each of which the second chromosome emanated from one single irradiated gamete, whereas different strains emanated from different gametes. In all strains the first and the third chromosomes were unirradiated. Use was also made of non-irradiated control strains which were synthesised in the same manner as the irradiated ones. Matings were made within each strain between females and males which were heterozygous with regard to the irradiated second chromosome and carried the dominant marker genes Cy L in the other second chromosome. With a limited number of such strains, females were permitted to lay eggs within intervals of 1 hour and, when the adult flies hatched, these were collected every 2 hours. Each line in Table I corresponds to one strain. There are 10 irradiated and 3 control strains.

The growth rate index is computed in the following way: For each of the strains I have taken the difference between the average time of development of the *wild* types, and the corresponding average of the *Curly Lobes* and expressed this difference in percent. of the latter average. The averages

TABLE I

Record number of strain	Non-pair matings		Pair matings		Growth rate index, see text for explanation
	Total offspring	% *wild* type	Total offspring	% *wild* type	
58	541	21·4	923	23·3	221·1
23	220	26·4	607	29·5	52·1
11	775	24·5	717	36·5	47·9
38	487	29·6	1114	31·1	13·3
81	562	28·5	981	32·8	10·6
96	866	29·6	644	28·0	8·4
14	535	37·0	1207	30·6	6·6
57	245	24·1	1143	30·6	5·2
37	507	29·8	816	33·7	−5·4
97	170	25·3	789	34·3	−25·3
Controls					
1002	342	32·7	1245	31·2	−21·2
1003	234	34·2	1415	31·7	−2·3
1004	532	33·6	1268	32·6	−2·5

themselves refer to the time from the eclosion of the first adult fly of the strain in question. When the *wild* types develop faster than the *Curly Lobes* the index will be negative as is the case with the controls.

In special tests, we have from the same kind of crosses also computed the percentage of *wild* types. When there is no selective advantage or disadvantage this percentage is, of course, expected to be 33·3. These tests have been made both by taking several females per vial and by making pair matings.

The two percentages of *wild*-type flies are in some cases rather different, to this I shall return presently. Here I wish to emphasise what the table clearly shows, namely, that several detrimentals, though perhaps not all, could have been picked out by their average rates of development alone. Now, with regard to the individual rate of development, it varies very much, its standard deviation being of the order of magnitude of 12 hours. In mice averages must always be based on a much smaller number of animals than in *Drosophila.* Nevertheless the preliminary results from *Drosophila,* presented here, seem to me to indicate strongly enough that it is worth while to try well-planned X-ray experiments with mice; and by using growth rate as a tool to find and to isolate detrimental strains.

Now, turning back to the segregation ratios of Table I, it is seen that in most of the irradiated strains, progenies from non-pair matings have a

TABLE II

	Cy L Pm	Cy L	Pm	+
Theoretical expectation ...	100	100	100	100
(*a*) 25 larvæ per vial Adult flies hatching	Cy L Pm	Cy L	Pm	+
within 10 days from egglaying	100	91	89	43
within 12 days from egglaying	104	100	90	63
more than 12 days from egg-laying	104	101	90	63
(*b*) 200 larvæ per vial Adult flies hatching	Cy L Pm	Cy L	Pm	+
within 10 days from egglaying	77	94	90	—
within 12 days from egglaying	91	100	97	35
more than 12 days from egg-laying	91	100	99	47

lower percentage of *wild*-type flies than progenies from pair matings. This among other things is the case for the five first strains, which, judged by their rates of development, could be assumed to be detrimentals. The reason probably is that there was greater competition between larvæ in vials where there have been several mothers than in vials with only one dam.

For some of the strains, for instance with strain 58, I have mated females and males heterozygous for the irradiated *wild*-type second chromosome. In this case, however, the females had Cy L but the males Pm in the other second chromosome. From this mating new hatched larvæ were collected in vials. In one comparison 400 larvæ were distributed between 16 vials with 25 in each, and simultaneously 400 more larvæ to 2 vials with 200 in each. The theoretical expectation then is 100 adults in both cases in all classes. Table II shows clearly the effect of differences in selection pressure :

the average viability is necessarily higher at a high selection pressure than at a low one.

With regard to experiments with *Drosophila*, I wish to conclude by saying that, if possible, different characteristics should be used when studying detrimentals, questions of viability and the rate of development. Furthermore, if we are to study X-ray influence on changes in viability, extrapolation to human populations can only be valid after close attention to selection pressure.

Discussion

Auerbach. I believe that Paula Hertwig found reduced growth rate in F_1 mice after irradiation of the fathers.

Muller. When Kerkis did his work on detrimentals he also examined his lines, derived from irradiation, for total body size, by weighing a given large number of individuals *en masse*, but found no significant differences. Evidently, then, final size is not nearly as sensitive an index of detrimental mutations as growth rate as studied by Dr. Bonnier.

COMPARISON OF CHEMICALLY- AND X-RAY-INDUCED MUTATIONS IN *DROSOPHILA MELANOGASTER*

O. G. FAHMY AND MYRTLE J. FAHMY
Chester Beatty Research Institute, Institute of Cancer Research, Royal Cancer Hospital, London, S.W.3, England

Introduction

The main discovery that emerges from the comparative study of the mechanism of mutagenesis under the effect of chemical agents as compared to radiation in *Drosophila*, is that the mutation process is selective, not random, and is somehow dependent on the nature of the mutagen.

Of course, there is nothing new in the concept of selective mutagenicity in micro-organisms. It has been demonstrated by Demerec (1955) for *Escherichia coli* and by Giles (1951), Kølmark and Westergaard (1953), and Kølmark (1953) for *Neurospora*. Also some of the geneticists of the Swedish school, under the guidance of Professor Gustafsson (see Gustafsson and Mackey (1948); Gustafsson and Nybom (1949)) have reported that the yield of mutations in cereals, especially barley, is dependent on the mutagen and the method of treatment.

In spite of the evidence from micro-organisms and plants, most geneticists were reluctant to accept the principle of differential genetic response to mutagens, mainly on the basis of the results with *Drosophila*. Muller's long experience on mutagenesis in this material, both spontaneously and after radiation, led him to restate quite recently (Muller, 1952 and 1955) the case for random mutability. Furthermore, the early results with chemical mutagens seemed to be not incompatible with the principle of random mutability. Thus Auerbach and Robson (1947) analysed a very small sample of mustard gas 'visibles' (viz. 10 sex-linked recessive mutations) and naturally failed to detect any specific effects on individual loci.

That was the situation when we started the utilisation of the alkylating compounds in mutagenesis work on *Drosophila*, some seven years or so ago. It was not long before the evidence for the differential genetic response to the various mutagens became apparent. This manifested itself not only in the different relative frequencies of the various types of mutations induced, but was often of a much finer nature, being a selectivity for certain gene loci. Some of this evidence will be presented to you now.

Material and technique

The mutagens used are the so-called alkylating agents, which from the chemical standpoint are chloroethylamines (or nitrogen mustards), dimesyloxyalkanes, epoxides and polyethyleneimines. All compounds were dissolved at the required concentration in isotonic saline, and administered intra-abdominally by injection into adult Oregon-K males. A list of some of the compounds used, and their comparative mutagenic efficiency in relation to each other and to X-radiation, has recently been published, Fahmy and Fahmy (1956*a*). This comparison has revealed that the alkylating compounds provide a tool for the induction of mutations as efficient in potency and as wide in range as any type of radiation.

Irradiation experiments were carried out on males comparable to those used in the chemical mutagenesis work. In all experiments irradiation was done by the same X-ray machine: a 'Picker' therapeutic machine, 250 kV—15 mA, with built-in filters and delivering a dose rate (under the conditions used) of 250–260 r/min.

The mutations analysed were the dominant lethals (both cytologically and by hatchability tests—see Fahmy and Fahmy (1954)), the viable chromosome breaks detected in the F_1 larval salivaries, as well as the sex-linked recessive lethals and visibles. Mutations at specific loci were detected by treating the wild-type males and mating them to females carrying the sex-linked recessive markers and scoring for these markers in the F_1 daughters.

Observations

Though the range of mutagenic efficiency of the alkylating agents is as good as that of radiation, their fundamental mode of action is different. This manifests itself in all the primary genetic reactions these compounds produce with the hereditary material: viz. chromosome breaks, deficiencies and point mutations whether leading to recessive lethals or recessive visibles. I propose to analyse these primary reactions one by one, and show how they vary with the mutagen.

1. *Chromosome breaks.* Table I gives the properties of three mutagens showing marked differences as regards chromosome breakage. It can be seen that at mutagenically equivalent doses, as regards the induction of recessive lethals, the imine is as effective in the induction of primary chromosome breaks as X-rays, but the amino-acid mustard is considerably weaker. The properties of the breaks themselves differ. The imine breaks tend to undergo more sister union and other inviable rearrangements than to exchange viably. This is the opposite tendency to that for the breaks induced by the amino-acid mustard and X-rays.

Not only do the frequency and properties of the chromosome breaks vary with the mutagen, but so also do their loci of induction. We have shown that the distribution of the F_1 viable breaks induced by the imine and localised along the salivary gland X-chromosome, was significantly different from that of the same breaks induced by X-rays, Fahmy and Fahmy (1956*a*).

TABLE I

Chromosome Breaks Induced by X-rays and 2 Alkylating Compounds: an Imine: 2 : 4 : 6-tri(ethyleneimino)1 : 3 : 5-triazine and a mustard: p-N-di(chloroethyl)aminophenylalanine at Mutagenically Equivalent Doses (as regards Sex-linked Recessive Lethals)

Genetic property	Mutagen and dose of equivalence		
	X-rays 1,000r	Imine $0{\cdot}6 \times 10^{-4}$ M	Mustard $0{\cdot}45 \times 10^{-2}$ M
Mean number of primary breaks per sperm a	0·78	0·77	0·10
Probability that a break exchanges eucentrically q	0·67	0·41	0·79
Probability that a break exchanges aneucentrically $p = (1 - q)$	0·33	0·59	0·21
Mean number of breaks exchanging eucentrically aq	0·52	0·32	0·08
Mean number of breaks exchanging aneucentrically ap	0·26	0·45	0·02

2. *Small deficiencies.* In Table II we have compared the efficiency of mutagenically equivalent doses of 3 alkylating agents as well as X-rays as regards the induction of chromosome breaks and small deficiencies. In view of the 'subjective element' in the detection of small deficiencies, it should be emphasised that these aberrations have been scored by the same observer, on coded lethal cultures. Some 350 lethals were analysed for the 4 mutagens over a period of 4 years and no less than 50 lethals were looked at for each agent.

The 3 chemical mutagens induced almost identical deficiency rates, which is roughly double that induced by X-rays. This is so in spite of the fact that the imine is a strong chromosome breaker, the phenylalanine mustard is a weak breaker and its phenoxy-derivative is practically ineffective in this respect. Thus different mutagens do not only induce different rates of deficiencies, but this rate is independent of the agent's efficiency in breaking the chromosome.

3. *Recessive lethals.* Differential induction of recessive lethals could only be proved through studies of distribution. Genetical localisation was undertaken of 567 sex-linked recessive lethals induced by 3 alkylating compounds, as well as 70 induced by X-rays. The same marker chromosome carrying *sc—ct—v—f—car* was utilised in all experiments. A statistical analysis was then undertaken to compare the distribution of the lethal loci induced by various agents. A certain amount of pooling had to be undertaken and it was thought that the most objective way of doing so was to group the loci placed between the markers (Table III). This of course has the great advantage of making the compared distributions independent of all intra-segmental errors in the localisation of the individual lethals.

TABLE II

Small Deficiencies Induced by X-radiation and Three Alkylating Compounds

Compound	Dose of equivalence	Primary breaks per 100 X-chromosomes	Deficiencies per 100 X-chromosomes
X-rays	1,000r	15·6	0·56
Imine	$0{\cdot}6 \times 10^{-4}$ M	15·4	1·02
Phenylalanine-mustard ...	$0{\cdot}45 \times 10^{-2}$ M	2·0	1·12
Phenoxy-derivative of above mustard	$1{\cdot}19 \times 10^{-2}$ M	—	1·00

Imine: 2 : 4 : 6-tri(ethyleneimino)-1 : 3 : 5-triazine.
Phenylalanine-mustard: p-N-di(chloroethyl)amino-phenylalanine.
Phenoxy-derivative of above mustard: p(p′-N-di(chloroethyl)aminophenoxy) phenylalanine.

It was found that the distributions of the lethal loci induced by the 3 alkylating compounds were not significantly different. We, therefore, pooled all the chemical data and compared the overall distribution of the chemically-induced lethals with that for X-rays. This comparison is presented as a χ^2 contingency table (Table IV), with one degree of freedom for each cell, and (cells-1) degrees for each row. Our data show significant differences in the distribution of the chemical and radiation lethals at the tip of the chromosome (between *sc—ct*), as well as at the base (proximal to *car*). The actual mutation frequency (Table III) shows that X-rays are more effective distally, whereas the chemicals are more so proximally.

We have also presented in Tables III and IV the distribution of some X-ray recessive mutations (mainly lethals) which have been localised by Spencer and Stern (1948). They used the same markers as in our chemical experiments, except for *ct* which was replaced by *cv*. This necessitated the pooling to be between the common markers in the 2 sets of data, viz. *sc—v—f—car*. It is interesting to note that our data for X-rays are not significantly different from those of Spencer and Stern, but our chemical data are. Also when we increased the sample of the X-ray mutations by pooling our data with those of Spencer and Stern, the significance of the difference in the distribution of the lethals induced by chemicals and X-rays has actually been increased.

TABLE III

The Distribution of the Sex-linked Recessive Lethals Induced by the Alkylating Compounds and X-radiation between the Localisation Markers

	Genetical map of the X-chromosome					
	sc	*ct*	*v*	*f*	*car*	centromere
Mutagens and authors	0–19·9	20·0–32·9	33·0–56·6	56·7–62·4	62·5 +	Total
Alkylating compounds (Fahmy and Fahmy, 1956)	172	92	182	42	79	567
X-radiation ... ... (Fahmy and Fahmy)	24	16	21	6	3	70
X-radiation ... ... (Spencer and Stern, 1948)	116		80	21	12	229
Alkylating compounds (regrouped)	264		182	42	79	567
Pooled X-radiation data (Fahmy and Fahmy, 1956), (Spencer and Stern, 1948)	156		101	27	15	299

4. *Visible mutations.* It is in the differential yield of visibles that selective mutagenicity under different mutagens is most apparent. Of course, it is well known that the aptitude in the detection of visibles varies markedly from observer to observer. That is why when the differential yield of visibles with different treatments passed the phase of impression and approached a state of reality, scoring for visibles was undertaken by one of us only (M. J. F.), on coded cultures.

Most of the alkylating agents yield 1 visible to 10 lethals in the same sample of treated X-chromosomes. X-radiation yields 2 visibles to 10 lethals. The

TABLE IV

The Significance of the Variations in Distribution (expressed in χ^2) of the Sex-linked Recessive Lethals Induced by the Alkylating Compounds and X-radiation between the Localisation Markers. (Based on data in Table III)

Comparison and authors	Genetical map of the X-chromosome: *sc* 0–19·9	*ct* 20·0–32·9	*v* 33·0–56·6	*f* 56·7–62·4	*car* 62·5 + centromere	Total χ_2	P
Alkylating compounds and X-radiation (Fahmy and Fahmy)	4·563	1·945	0·126	0·121	5·170	11·925	0·018
X-radiation (Fahmy and Fahmy) and (Spencer and Stern)	0·094		0·584	0·023	0·102	1·613	0·660
Alkylating compounds (Fahmy and Fahmy), and X-radiation (Spencer and Stern)	1·096		0·594	0·695	12·16	14·545	0·004
Alkylating compounds and pooled X-radiation data	2·470		0·251	0·703	16·10	19·524	<0·001

amino-acid mustards give 3 visibles to 10 lethals, Fahmy and Fahmy (1956*a*). A particular sulphonate 2-chloroethyl methane-sulphonate gives 4-5 visibles to 10 lethals and for the sensitive stage to the action of this compound (the very young male germ cells) the yield of visibles is equal to that of the lethals; the ratio being 1 : 1, Fahmy and Fahmy (1956*b*).

Many of the chemically-induced visibles are different in their phenotypic expression and genetic position from those induced by other agents. Figure 1 is a histogram giving the incidence and distribution of the visible loci mutated chemically (above, in black) as compared to those mutated by other means mainly X-rays (below, in white). The stippled parts represent the common sectors; that is, the number of loci which have been mutated by chemicals as well as by other means. It can be seen at a glance that the use of the chemical mutagens has resulted in the induction of a great many apparently 'new' visibles.

Figure 2 illustrates two superimposed histograms showing the percentage distribution of: (*a*) the chemical visibles (in black) and (*b*) the X-ray visibles (in white); where the 2 columns are superimposed is represented dashed.

It is of interest to note that the white summits predominate for the first half of the chromosome, whereas the black summits predominate for the second half. This result recalls the situation as to the distribution of the lethals, where it was found that X-rays are more effective at the distal part, whereas the chemicals are more effective at the proximal part of the chromosome.

5. *Mutability at specific 'visible' loci.* The indisputable proof for differential genetic response under the effect of the alkylating compounds and X-radiation, comes from the study of mutability at specific visible loci. The technique used is that usually adopted for such purposes. Females carrying the tested loci were mated to irradiated males carrying the normal allelomorphs and the F_1 daughters were scored for the genes tested; 30,000–80,000 chromosomes were scored for each locus.

A daughter could show one of the characters tested for, if the paternal X has been affected so as to carry:

(*a*) a 'visible' allelomorphic to the marker

or (*b*) a 'visible' as above together with a lethal somewhere else on the chromosome

or (*c*) a deficiency or deletion (in itself a lethal) covering the locus of the marker.

There are some classical techniques for the differentiation between these possibilities and they have been undertaken in our analysis.

Table V summarises the results. The left half of the table represents the mutability at the apparently 'new' chemically mutated visibles, whereas

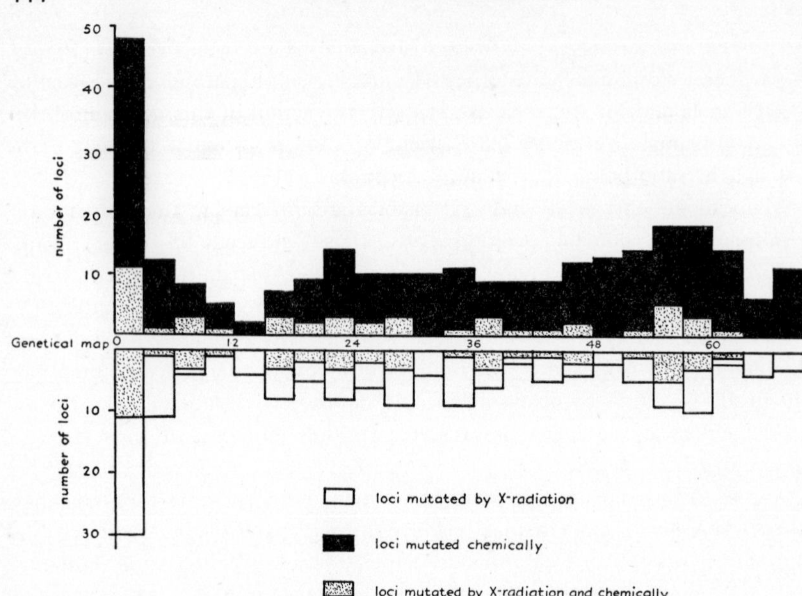

FIG. 1. The distribution of visibles along the genetical map of the X-chromosome (pooled in 3 unit segments).

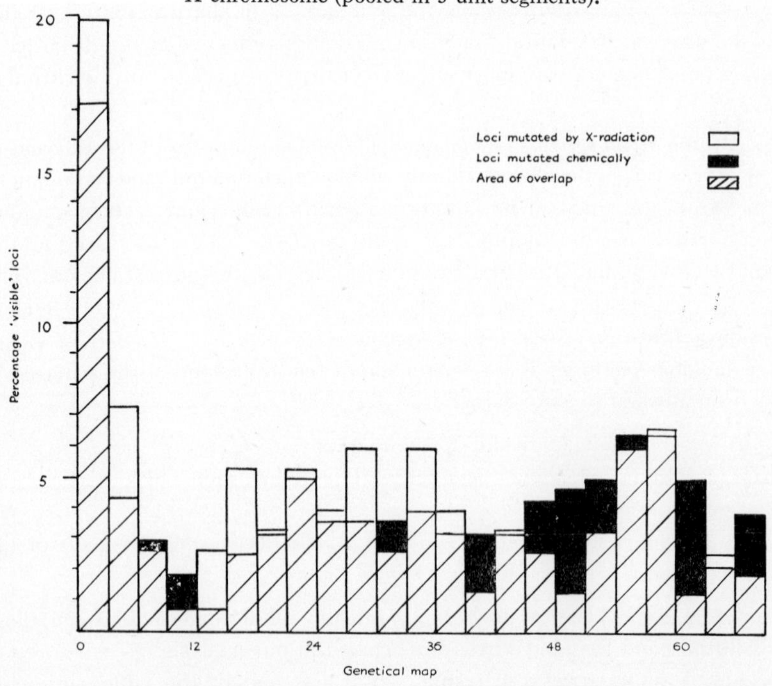

FIG. 2. Superimposed histograms for the percentage distribution of visible loci along the genetical map of the X-chromosome (pooled in 3 unit segments).

TABLE V

Mutability at Specific 'Visible' Loci under the Effect of X-radiation

'New' chemically mutated loci	Genetic position	Mutation rate per locus per r × 10^{-8}	
		Deficiencies and deletions	Point mutations
v 1230	0·0	—	—
v 638	2·7	—	—
v 1998	3·3	1·27	—
v 1276	4·1	1·24	—
v 1243	5·7	1·82	—
v 1255	14·8	1·82	—
v 1287	23·4	1·52	—
v 1726	29·9	—	—
v 1892	40·5	1·17	—
v 1523	43·1	—	—
v 2161	51·7	—	—
v 2100	53·6	—	—
v 2212	59·1	—	—
v 2073	61·6	1·05	8·42
v 1663	64·5	—	—
v 1225	66·9	8·26	—
v 1920	67·2	6·07	—

Known loci (control)	Genetic position	Mutation rate per locus per r × 10^{-8}	
		Deficiencies and deletions	Point mutations
scute	0·0	1·74	3·49
cut	20·0	5·23	0·87
vermilion	33·0	0·7	1·44
wavy	41·9	6·10	3·49
garnet	44·4	5·23	4·36
forked	56·7	0·87	2·62
carnation	62·5	3·06	—

Tested sample: 30,000–80,000 chromosomes per locus. Dose range: 2,500–4,250 roentgens.

the right half represents the response of some classical visibles for control purposes.

Of the 17 chemical visibles only one has been mutated intragenically by radiation and also eliminated within deletions; 8 others have been eliminated in deletions, 6 at a low frequency, and the 2 near the heterochromatin, at a much higher frequency but were not mutated intragenically. The remaining 8 were stable to radiation at the size samples utilised. That this size sample is reasonably adequate is shown by the fact that it was sufficient for the induction of intra-genic mutations, as well as deletions for the tested known loci, except *car* which was only eliminated.

The mutability of the very same 'chemical' and known visibles will be tested under the effect of the phenylalanine-mustard, and the full comparative results and their statistical analysis will be presented in due course. Nevertheless, it is perhaps worth mentioning that in a sample of 50,000 X-chromosomes treated with various alkylating compounds, all the 'new' visibles were detected at a minimum frequency of 2 and a maximum of 5. Most interesting is that *car*, which is mutated by radiation at a very low rate indeed (probably 1 : 60,000 chromosomes), occurred 3 times in the chemical experiments (i.e. roughly 3 in 50,000 chromosomes). Also *sc* and *ct* which are mutated fairly frequently with radiation, have never been encountered in the chemical experiments.

It is abundantly clear, therefore, even at this stage of our investigation, that the results on mutability at specific loci are in harmony with the other evidence outlined in this communication, and support the principle of selective mutagenicity.

Acknowledgments

This investigation has been supported by grants to the Royal Cancer Hospital and Chester Beatty Research Institute from the British Empire Cancer Campaign, the Jane Coffin Childs Memorial Fund for Medical Research, the Anna Fuller Fund and the National Cancer Institute, National Institutes of Health, U.S. Public Health Service.

REFERENCES

Auerbach, C., and Robson, J. M., 1947. *Proc. Roy. Soc. Edinb.*, **62**, 271.
Demerec, M., 1955. *Proc. 9th Int. Congress. Genet. Suppl. Caryologia*, **6**, 201.
Fahmy, O. G., and Fahmy, M. J., 1954. *J. Genet.*, **52**, 603.
Idem, 1956*a*. *Ibid.*, **54**, 146.
Idem, 1956*b*. *Nature, Lond.*, **177**, 996.
Giles, N. H., Jnr., 1951. *Cold Spring Harbor Symp. Quant. Biol.*, **16**, 283.
Gustafsson, A., and J. Mackey, 1948, *Hereditas*, **34**, 371.
Gustafsson, A., and Nybom, N., 1949. *Ibid.*, **35**, 280.
Kølmark, G., 1953. *Ibid.*, **39**, 271.

KØLMARK, G., and WESTERGAARD, M., 1953. *Ibid.*, **39**, 209.
MULLER, H. J., 1952. *Symp. on Radiobiol., Nat. Acad. Sci.*, **17**, 296, (New York, Wiley).
Idem, 1955. *Bull. Atomic Scientists*, **11**, 329.
SPENCER, W. P., and STERN, C., 1948. *Genetics*, **33**, 43.

DISCUSSION

Oster. I would like to know if these so-called 'new mutations', which you claim to have induced with your special chemicals, have been tested for allelism by being crossed to stocks bearing similar-looking sex-linked mutations of induced and/or spontaneous origin, or have you only used linkage data for determining whether or not some of these mutations occurred previously?

M. J. Fahmy. All the 'new' mutants have been tested for allelism whenever possible. Also, whenever the position of the new mutant was within 5 units of a previously described visible of a comparable phenotype, not available for testing, it was assumed that the visible obtained was not a repeat. As regards the occurrence of normal alleles on the Y to some of the new visibles in the proximal part of the X-chromosome, one 'new' mutant fell into this class. The expression in the homozygous male with the normal Y was not nearly as extreme as in the homozygous female. A large deficiency covering this visible was placed at 66–7, that is, beyond 'bobbed'.

Oster. In that case I would like to question the accuracy of these determinations for deciding whether or not a mutation had occurred previously at a particular locus. I ask this because with your genetic scheme which allowed for the detection of induced mutations in males which ordinarily carry a Y-chromosome, you reported finding a few recessive mutations which you subsequently placed on the basis of linkage data as being located to the right of the gene, 'bobbed'. Now I would like to know how you can reconcile the presumed exactness of these localisations, with the fact that it has been known for some time that the locus of 'bobbed' as well as deficiencies for the entire region of the chromosome to its right, are covered by the normal alleles in the pairing segment of the Y-chromosome?

O. G. Fahmy. The accuracy in placing the individual lethals in the proximal segment is irrelevant to the comparison of the distribution, since we have pooled all the loci to the right of *carnation*. In the cytological placings also, we pooled all the loci in the heterochromatic segment (segment 20 of the X-chromosome) to get over the difficulty in locating aberrations within sub-divisions in this confused heterochromatic part.

Westergaard. Do you find any differences in the mutations induced by different chemicals?

O. G. Fahmy. Differences occurred in the ratio of visibles to lethals induced by the different chemicals in the same sample of X-chromosomes. However, the distribution of the lethals induced by these related chemicals was not significantly different. That is why the chemical lethals have been pooled together and compared with those induced by X-rays.

Ford. Have you found repeats in the salivary gland chromosomes, and if so, do they include both direct and reverse repeats? In nitrogen-mustard treated *Vicia* root tips, chromatic intrachanges are very common. Many of them would yield direct repeats and some, reverse repeats.

Auerbach. I wanted to ask the same question as Dr. Ford. Dr. Slizgreska in our Institute has just analysed cytological changes which have been produced by formaldehyde feeding. She finds that compared with X-rays, a shortage of inversions and, especially, translocations is almost counter-balanced by a great excess of deficiencies and, particularly, of duplications. Most of the duplications were repeats and a proportion of them were reverse repeats. In some mosaics she found both a deficiency and its complementary duplication. Now, if I remember rightly, you too found mosaics for deficiencies in your material. Did you also find duplications, and if not, how do you explain their absence?

O. G. Fahmy. The frequency of duplications after treatment with the alkylating agents is very low; much lower than other major rearrangements such as inversions, deletions and translocations. Of the latter rearrangements the frequency of intra-chromosomal aberrations (deletions, inversions) is appreciably higher than inter-chromosomal aberrations (i.e. translocations).

THE POSSIBLE ROLE OF PEROXIDES IN RADIATION AND CHEMICAL MUTAGENESIS IN *DROSOPHILA*

F. H. Sobels
Institute of Genetics, The State University of Utrecht, Netherlands

Previous observations have shown an enhancement of the mutation rate induced by X-rays after pretreatment with cyanide and azide (Sobels, 1955). This finding suggested that an increased content of hydrogen peroxide, due to inhibition of the cytochrome and catalase enzyme systems, made the genetic material more susceptible to the effects of radiation. Recent experiments comparing the effect of cyanide pre-and post-treatment, on the rate of X-ray-induced translocations and lethals will be discussed here.

Because formaldehyde is also known to act as a catalase inhibitor (Stern, 1932) and presumably produces mutations via the formation of an organic peroxide (see below), the effect of formaldehyde pretreatment on the mutagenic action of X-rays was studied. For a further test of the assumption that the presence of increased amounts of a peroxide enhances the effects of irradiation, flies were pretreated with an organic peroxide, dihydroxydimethyl peroxide ($HO—CH_2—O—O—CH_2—OH$).

The hypothesis, put forward by Jensen *et al.* (1951) that in *Neurospora*, peroxide formation may be involved in the mutagenic action of formaldehyde was tested for *Drosophila* by means of catalase inhibition and by a study of the mutagenicity of dihydroxydimethyl peroxide, the organic peroxide formed by the reaction of formaldehyde and hydrogen peroxide.

Materials and methods

Except when otherwise stated, all mutation experiments were done with males from the Oregon-K strain which were tested for the incidence of sex-linked lethals by means of the Basc or Muller-5 method. Differences in sensitivity in the treated germ cells were studied by remating the males individually to three fresh females at specific time intervals; the details of exposure to cyanide gas are published elsewhere (Sobels and Simons, 1956). Mention should be made of the fact that cyanide has no mutagenic effects by itself (Sobels, 1955, and Sobels and Simons, 1956). With the exception of mustard gas all the chemicals were injected intra-abdominally ($\pm$ 0·28 mm.^{3}). X-radiation was delivered at 100 kVp with 1 mm. Al filtration at an intensity of 244r/min. or 590r/min.

Results and discussion

The results obtained after pretreatment with formaldehyde, using daily brood changes, will be considered first. Injection of 0·033 M formaldehyde prior to irradiation with 1,700r caused, from the second day onwards a significant enhancement of the mutation rates induced by irradiation. Fully mature sperm, however, did not respond any more to the potentiating effect of formaldehyde on X-ray mutagenesis (Sobels, 1956). These findings support the assumption that an increased production of peroxides exerts a potentiating effect on the mutagenic action of X-rays in immature germ cells.

The results obtained with dihydroxydimethyl peroxide will be considered next. Pretreatment with 0·0138 M of this organic peroxide also caused a pronounced enhancement of the mutagenic effect of X-rays (Sobels, 1956*a*). A comparison with the effects of cyanide and azide shows that all three substances exert their potentiating action in one particular stage of spermatogenesis, characterised by peak sensitivity to the mutagenic action of X-rays. The observed correlation between peak sensitivity to irradiation and preferential response to pretreatment with an organic peroxide, and substances which are thought to increase the content of peroxide in the cell, suggests that the formation of peroxides may account for at least part of the genetic effects of X-rays in *Drosophila* (Sobels, 1956*a*).

The interpretation of the previous results has been based on the assumption that increased amounts of peroxide sensitise the chromosomes to irradiation. The reverse interpretation that radiation sensitises the chromosomes to the action of peroxide would bring our results into line with those obtained by Alper (1954) on bacteriophage. Our material does not, however, allow us to decide which of these interpretations is correct.

It is not possible to say whether, in the case of cyanide, we are primarily dealing with an oxygen effect. The similar results obtained after pretreatment with the organic peroxide and formaldehyde do not seem to support such an interpretation.

The idea that hydrogen peroxide produced by the irradiation itself accounts for the genetic effects of X-rays has recently been contradicted by Kimball (1955) on the basis of his experiments with *Paramecium*. It should be remembered, however, that this source of hydrogen peroxide has not been considered here, but rather the hydrogen peroxide produced via the metabolic pathways which eventually may serve as the initial step in the formation of organic peroxides.

By using Muller's (1954) multipurpose stocks a comparison between the effects of cyanide pretreatment on the rate of translocations, chromosome breaks and lehtals induced by X-radiation in stages with different sensitivity could be made, Figure 1. The most striking effect is that after cyanide pretreatment the rate of translocations induced at the stage of peak sensitivity, shows a disproportionate increase, as compared to that of lethals and of chromosome loss. This result can be interpreted by assuming that cyanide,

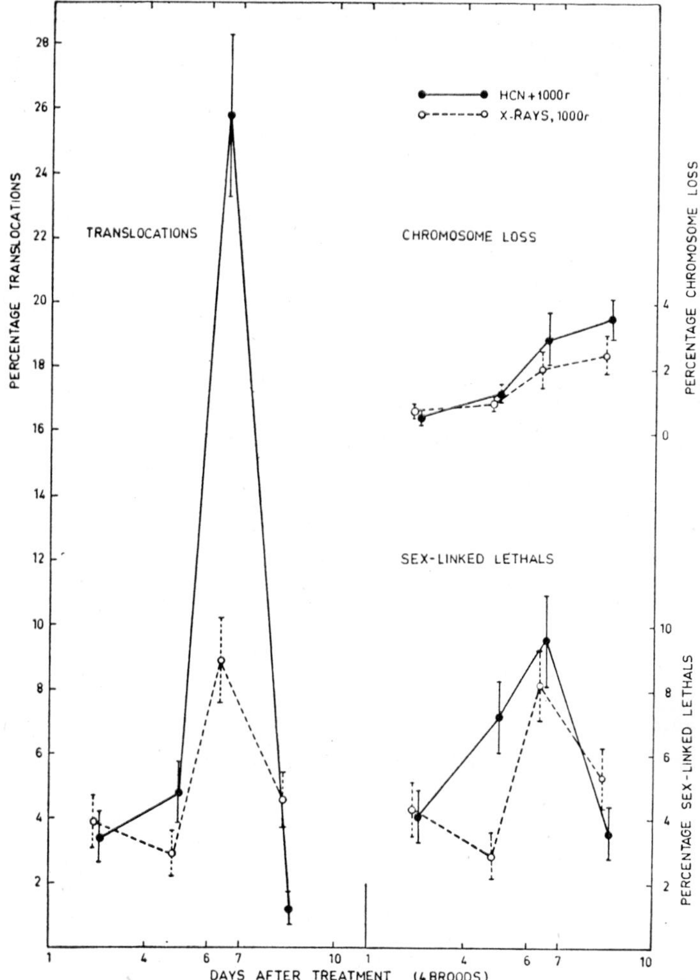

FIG. 1. The frequency of translocations (2–3, 2–Y, 3–Y, 4–Y), chromosome loss (X or Y, or partial loss of X or Y), and sex-linked lethals in four successive broods from *sc*^{8}.Y/*y* In49 B; *bw*^{D} males which had been exposed to cyanide gas prior to X-radiation with 1,000r, as compared with that from males which were irradiated only.

apart from its effects on chromosome breakage and gene mutation, by raising the peroxide level in the cell, also effects reunion of the broken chromosomes, so as to increase the chances that rearrangements are formed. Such an explanation would be in agreement with the observations on plant material by Wolff and Luippold (1955) that cyanide and other agents interfering with the synthesis of ATP, delay the restitution of radiation-produced chromosome breaks.

Further confirmation of this idea has been obtained in a recent experiment in which the flies were irradiated first and then post-treated with cyanide gas. The data are set out in Figure 2. They show that, compared to the control group which received only irradiation, the frequency of translocations is significantly increased in the sensitive stage of the post-treated group.

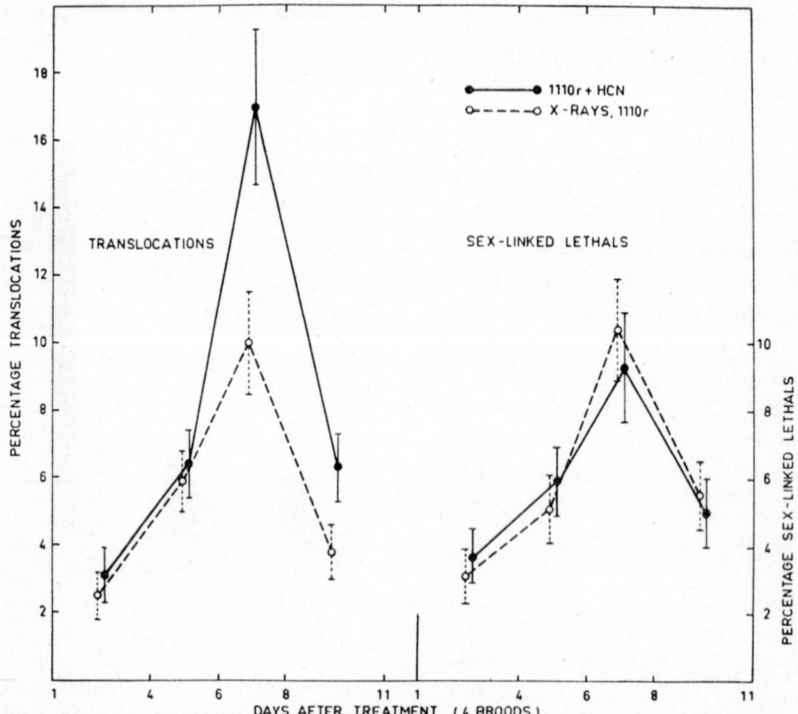

FIG. 2. The frequency of translocations (2–3, 2–Y, 3–Y, 4–Y) and sex-linked lethals in four successive broods from sc^8.Y/*y In*49 B; *bw*ᴰ males which had been exposed to cyanide gas after X-radiation with 1,110r. *Note in press.* The X-ray dose used for this experiment was 1190r, as compared with that from males which were irradiated only.

The rate of induced sex-linked lethals, however, is the same in four broods tested for both groups of flies. This result might indicate that, during the sensitive period, only a negligible proportion of gross rearrangements is involved in the production of sex-linked lethals by irradiation.

Catalase inhibition has also been used to test the idea that peroxide formation plays a role in the production of mutations by injected formaldehyde in *Drosophila.* The result obtained in collaboration with Simons favoured this assumption; a given dose of 0·031 M formaldehyde produced twice as many mutations in males which had been pretreated with cyanide gas (5·0% lethals in 2,030 chromosomes) as in flies which received formaldehyde only (2·5% lethals in 2282 chromosomes) (Sobels and Simons, 1956).

This hypothesis is further supported by the finding that dihydroxydimethyl peroxide acts as a fairly potent mutagen in *Drosophila* (Sobels, 1956). Moreover, there is a marked similarity in the sensitivity pattern to the mutagenic effects of formaldehyde and of dihydroxydimethyl peroxide, since both chemicals produce mutations in mature sperm and in a much earlier stage of spermatogenesis.

It was thought that the lack of mutagenic effects of formaldehyde in females, as observed by Auerbach (1952) and Herskowitz (1955), might possibly be explained by assuming a greater activity of enzymes responsible for a rapid breakdown of the active peroxide. Sex-linked lethal tests in females were carried out according to a technique kindly suggested to us by Dr I. I. Oster. It was observed that after cyanide pretreatment, formaldehyde definitely acts as a mutagen in females, Figure 3. Formaldehyde by itself, in strong concentrations which either kill or sterilise the great majority of the injected flies, also raises the spontaneous mutation rate in females (Sobels, 1956*c*).

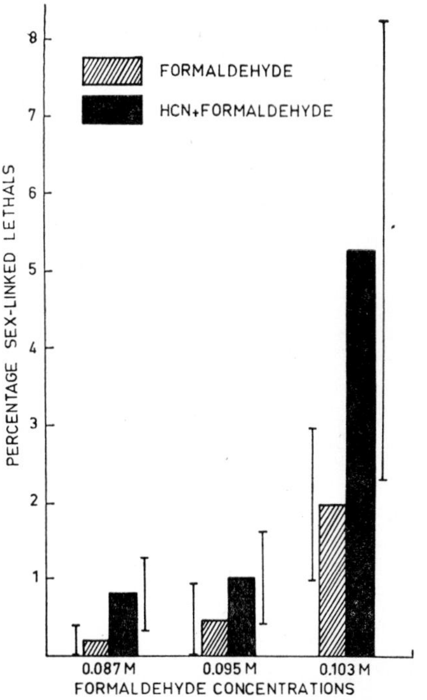

FIG. 3. The frequency of sex-linked lethals from *y w In*49 *f*/*y* sc^{s1} B *In* S females, during the first three days after treatment, which had been exposed to cyanide gas prior to injection with different concentrations of formaldehyde, as compared with that from females which received formaldehyde only.

In view of the great similarity between the genetic effects of X-rays and mustard gas the effect of cyanide pretreatment on the rate of mustard gas-induced mutations has been studied. Although the results are not as clear as those for X-rays, they suggest an enhancing effect of the pretreatment. In three out of four experiments the overall mutation rate was higher after cyanide pretreatment than in simultaneously treated mustard gas controls. These results do not exclude the possibility that comparable chemical products are involved in the production of the genetic effects of X-rays and mustard gas.

Experiments on chemical induction of crossing-over in *Drosophila* males, however, gave quite different results for mustard gas on the one hand, and formaldehyde and dihydroxydimethyl peroxide on the other (Sobels and van Steenis, 1957). Mustard gas, which is one of the most potent chemical mutagens, produces much fewer cross-overs relative to induced mutations, than the organic peroxide and formaldehyde. There is, however, a similarity between the frequency of crossing-over induced by formaldehyde and dihydroxydimethyl peroxide and by X-irradiation (Auerbach, 1954 and Whittinghill, 1955). In the light of the above discussion on the possible role of peroxides in radiation— and formaldehyde mutagenesis, it is possible that this correspondence between the effects of X-rays and of these two chemical mutagens is more than a mere coincidence.

Acknowledgment

This investigation was subsidised by the Radiobiological Group of the Health Research Council T.N.O.

REFERENCES

Alper, T., 1954. *Brit. J. Radiobiol.*, **27**, 50.
Auerbach, C., 1952. *Amer. Natural.*, **86**, 330.
Idem, 1954. *Z. ind. Abst. u. VererbLehre.*, **86**, 113.
Herskowitz, I. H., 1955. *Genetics*, **40**, 76.
Jensen, K. A., Kirk, I., Kølmark, G., and Westergaard, M., 1951. *Cold Spr. Harb. Sym. quant. Biol.*, **16**, 245.
Kimball, R. F., 1955. *Ann. N.Y. Acad. Sci.*, **59**, 638.
Muller, H. J., 1954. *Drosophila Inform. Serv.*, **28**, 144.
Sobels, F. H., 1955. *Z. ind. Abst. u. VererbLehre*, **86**, 399.
Idem, 1956 (*a*). *Nature, Lond.*, **177**, 979.
Idem, 1956 (*b*). *Experientia*, **12**, 318.
Idem, 1956 (*c*). *Z. ind. Abst. u. VererbLehre*, **87**, 743.
Sobels, F. H., and Simons, J. W. I. M., 1956. *Ibid.*, **87**, 735.
Sobels, F. H., and van Steenis, H., 1957. *Nature, Lond.*, **179**, 29.
Stern, K. G., 1932. *Hoppe-Seyl. Z.*, **209**, 176.
Whittinghill, M., 1955. *J. cell. comp. Physiol.*, Suppl. 2, **45**, 189.
Wolff, S., and Luippold, H. E., 1955. *Science*, **122**, 231.

Discussion

Auerbach. I should like to point out that there are two different ways of producing mutations by formaldehyde, namely injection and feeding, and that Dr. Sobels' experiments were done by the first method. It would be very interesting if cyanide pretreatment could be applied also before formaldehyde feeding, to which *Drosophila* females are entirely refractory.

Westergaard. Have you tried the effect of this organic peroxide on female *Drosophila*? Also it would be interesting to see if combined treatment with formaldehyde and X-rays enhanced the yield of mutations in female *Drosophila*, since, as Dr. Auerbach has pointed out, formaldehyde alone has no mutagenic effect on *Drosophila* females.

Sobels. We have not yet applied the organic peroxide to females or used formaldehyde as a pretreatment for X-rays in females. This is certainly very interesting and we hope to look at it in future experiments.

Wolff. In the experiments on *Vicia* we found that post-irradiation treatment with CW did not increase the translocation yield; as has been pointed out the time that the breaks stayed open did not influence the aberration yield in *Vicia*. There is no discrepancy between these *Vicia* experiments and your experiments performed on sperm from the earlier broods. I wonder if you would care to speculate as to whether or not the resting nuclei of *Vicia* and of the sperm from early broods are similar and if there can be a mechanical explanation (such as Oster attempted to disprove) of the discrepancy.

Auerbach. This is easily explained by the fact that the germ cells used for the first two broods had reached stages in which rearrangements no longer take place. Those used in the 3rd brood derive from treated spermatids in which, as Dr. Oster has shown, rearrangements do take place.

Alper. I wonder whether Dr. Sobels would care to comment on the possibility that some of his results may in fact be another expression of the oxygen effect, as Dr. Oster suggested for his own results. The potassium cyanide might be inhibiting cell respiration as well as catalase, so that more free oxygen may be available within the cells during irradiation, since it has to get in by diffusion. The fact that Kimball obtained different results may be due to the fact that his material is in free cell form, with a plentiful oxygen supply, whereas with organised tissue the oxygen supply in any cell is very dependent on the respiration rate of the cells between itself and the free supply of oxygen.

Sobels. To distinguish between the effect of hydrogen peroxide and of oxygen we have thought of an experiment in which the flies, prior to irradiation in nitrogen, are treated with nitrogen and cyanide gas. But under such conditions there is the possibility that no H_2O_2 is formed either, so that no decisive evidence for or against the effect of O_2 or H_2O_2 can be obtained.

The observation, however, that pretreatment with the organic peroxide has a similar effect to that of cyanide pretreatment is not, as the sole explanation, in favour of a heightened oxygen tension within the cell, for the observed enhancements in the pretreated groups.

Ehrenberg. The greater radiation effect after pretreatment with, e.g., formaldehyde can be understood on the basis that catalase is blocked and is unable to destroy hydrogen and other peroxides. Since several explanations are possible, it is desirable to eliminate some of them and obtain a further understanding of the basic mechanism. Is it possible to irradiate the flies (after the pretreatments used, in nitrogen and at such a low ion density with gamma-rays or X-rays harder than 100 kV) so that no H_2O_2 is formed in the absence of oxygen? The results of such an experiment would exclude some of the possible explanations.

Sobels. It seems plausible that formaldehyde enhances X-ray-induced mutation rates not only by its inhibiting effects on catalase, but also because formaldehyde, in higher concentrations, produces mutations via the formation of an organic peroxide. Your suggestion to study the effect of the pretreatment and irradiation at low ion density in the absence of oxygen might be very useful as a means of distinguishing between the various alternative interpretations of the observed results. It will take some time, however, to arrange suitable conditions for such an experiment.

INCORPORATION AND MUTAGENICITY OF ^{32}P IN *DROSOPHILA* SPERM

P. OFTEDAL AND J. C. MOSSIGE
Norsk Hydro's Institute for Cancer Research,
The Norwegian Radium Hospital, Oslo, Norway

The amount of ^{32}P incorporated during spermatogenesis has been found to vary with time; and a close correlation with the mutation curve with time was found.

Experimental

Newly eclosed *Drosophila* males are fed a sugared solution of ^{32}P containing about 1 mc./c.c. for 1 hour on day 0. The activities of the males are measured every other day to obtain the biological decay curve, Figure 1. These curves are very similar to the curves obtained by Bateman and by King after rather more complicated feeding procedures. On the basis of day-1 activities, the males are ranged into two groups with the same mean activities. Both groups of males are mated singly to several fresh virgin females daily. The number of females per male, necessary to insure complete exhaustion of the sperm supply, has been determined in separate experiments and has been found to vary between a maximum of 7·0 on day 2 and less than 1 on day 13.

The females mated to one group of males are allowed to lay eggs under optimal conditions, and the number of offspring sired each day is counted, giving a production curve with time, Figure 2. If it is agreed that the same degree of polyspermy obtains under most circumstances, this curve will give a measure of the relative sperm production on different days.

The offspring from this group of males are used for measuring the frequency of recessive sex-linked lethals by the M-5 technique. Thus a mutation curve with time is obtained.

The other group of males are used for the assay of ^{32}P incorporation during spermatogenesis. The females mated to this group may be regarded simply as collectors of newly ripe sperm. It is felt that this method of collection gives a more effective separation of the end point of spermatogenesis than would cytology, autoradiography or some physico-chemical approach. The method used so far is the following: The females are collected every day, the vaginæ with seminal receptacles are dissected out in Ringer's solution, fixed and extracted in alcohol-ether, hydrolysed in 1 N.HCl for 15 minutes at 60° C., washed in 3 washes of 2% Na_2HPO_{49}, and mounted for measurement of radioactivity. The curve obtained gives a measure of the radioactivity contained in the sperm and seminal fluid from ^{32}P incorporated in

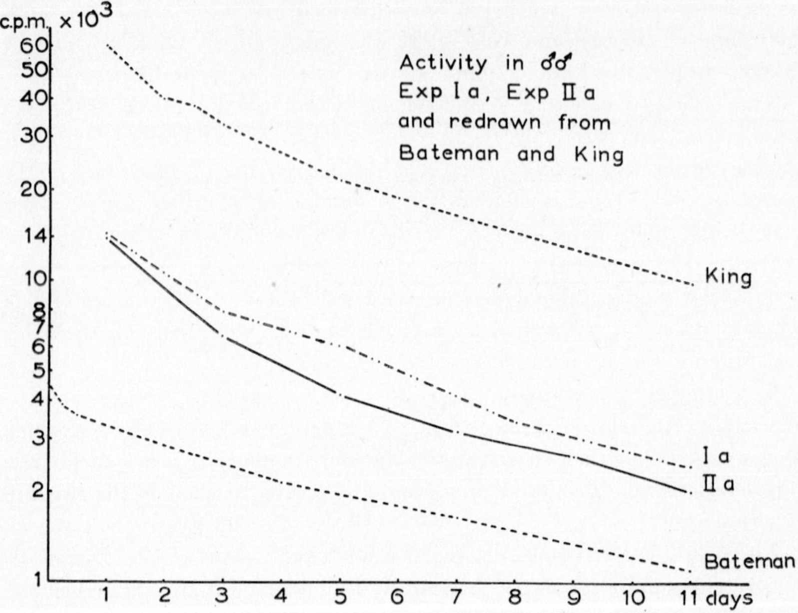

FIG. 1. ^{32}P radioactivities in *Drosophila* males after feeding for one hour.

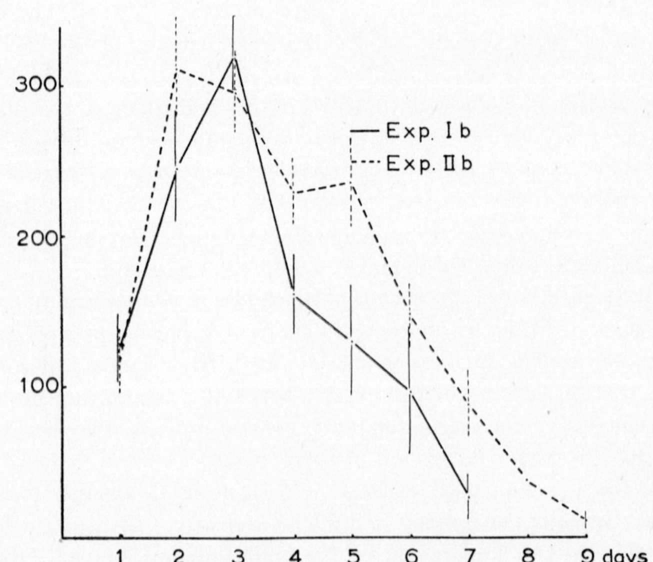

FIG. 2. Productivity of *Drosophila* offspring per male.

large non-extractable molecules, and chemically in fairly stable positions. According to Howard and Pelc (1951) this might be ^{32}P in DNA, but we do not consider this to be definitely established for our material.

The radioactivity curve of the females is corrected for two factors (in addition to radioactive decay and the number of males used): (1) For comparison between experiments, it is divided by the mean activity of the males on day 1, (2) it is divided by the number of offspring produced per male in the other mating group. If the assumption of a constant degree of polyspermy is accepted, this latter correction produces a measure of the activity per sperm. The curves obtained after these corrections are closely similar to the ^{32}P mutation curves found in the genetic part of the same experiments, Figures 3 and 4.

The complete procedure outlined above has been used in two experiments, see Table. In addition a number of part experiments have been performed to test one or another aspect of the methods employed. In a qualitative way, the latter findings tend to support the results reached in the two full experiments.

There are two obvious discrepancies between experiments I and II. Firstly, the mutation curve of I is not well reproduced in II. However, the same type of change is found in the incorporation curve, and this tends to support the interpretation that there is a real difference between the two experiments and not only a spurious variation. The difference in shape of the two productivity curves points in the same direction. The other difference between experiments is the lower incorporation figures on days 6–9 in II as compared to I. This necessitates a factor of 3·55 instead of 2·45 to bring the incorporation curve to coincide with the maximum of the mutation curve. The ^{32}P activity has been measured on different Geiger-Muller counters and scalers in the two experiments, but the close correlation of the incorporation curves on days 1–5 seems to indicate that the corrections introduced to compensate for differences in counter efficiency have been valid. Again, one is tempted to accept the differences as real.

If the difference found is real and biological, there is a possibility that the explanation may be found in the feeding period which in Experiment I was in mid-afternoon, in Experiment II was in the evening. The age of the males was the same in the two experiments, and a 6-hour time difference causing such a difference in mutation, etc. would indicate a greater diurnal variation than is usually recognised in *Drosophila*.

We are not as yet prepared to make a decision as to whether this close correlation between mutagenesis and ^{32}P incorporation should be taken as an indication of the importance of the transmutation of the ^{32}P-atoms incorporated into genetically important molecules, or whether the correlation simply shows a close relationship between synthesis and radiosensitivity.

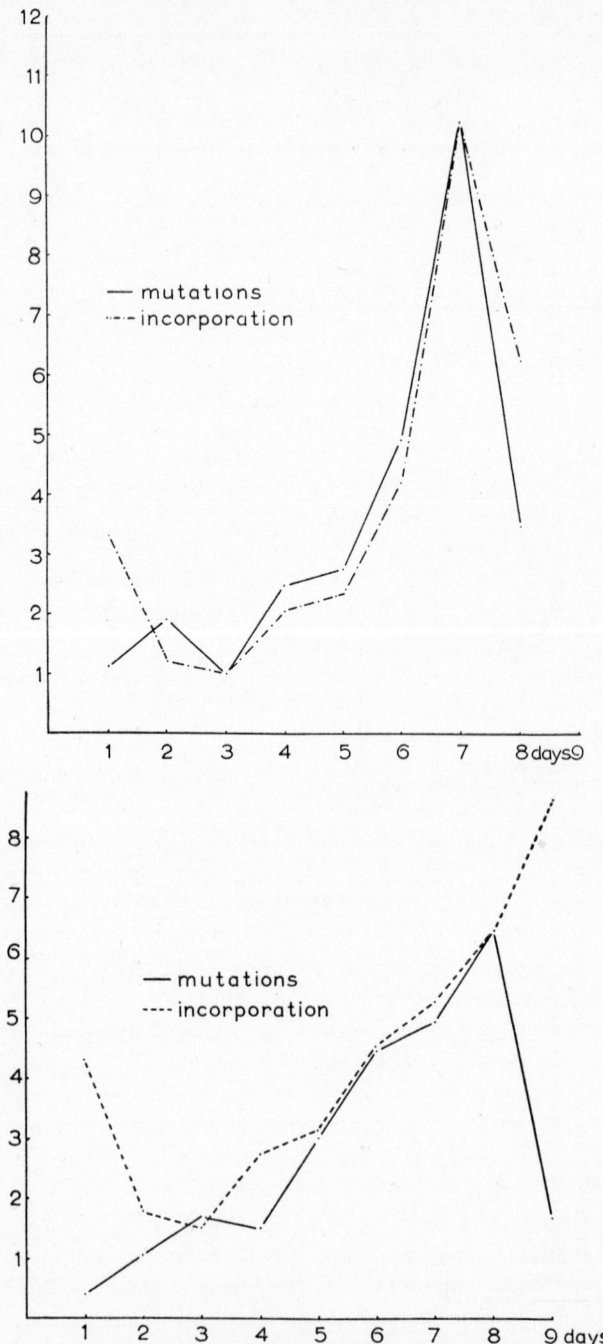

FIG. 3 and FIG. 4. Corrected ^{32}P incorporation and mutation curves. (See Table I for scale.)

TABLE

Days	Experiment I*a* and *b*							Experiment II*a* and *b*						
	No. of ♂♂ in *a* = 15. Day 1 activity per ♂ = 14200±750 No. of ♂♂ in *b* = 10. Day 1 activity per ♂ = 17100±1500							No. of ♂♂ in *a* = 20. Day 1 activity per ♂ = 12700±730 No. of ♂♂ in *b* = 20. Day 1 activity per ♂ = 12727±710						
	No. of offspring per ♂	No. of chrs. tested	(A) Sex-linked recessive lethals in %	Activity in ♂♂ in %	Cpm. in ♀♀ per ♂	(B) Cpm. in ♀♀ per ♂ per cpm. in ♂♂ per offspring	(A)/(B)	No. of offspring per ♂	No. of chrs. tested	(A) Sex-linked recessive lethals in %	Activity in ♂♂ in %	Cpm. in ♀♀ per ♂	(B) Cpm. in ♀♀ per ♂ per cpm. in ♂♂ per offspring	(A)/(B)
1	123·8	599	1·17	100·0	2·36	1·34·10^{-6}	0·87	116·3	758	0·40	100·0	1·79	1·21·10^{-6}	0·33
2	246·9	1,176	1·79	75·0	1·72	0·49	3·64	311·3	846	1·06	69·0	2·00	0·51	2·09
3	321·4	1,512	0·99	57·6	1·90	0·42	2·37	294·5	831	1·69	47·5	1·60	0·43	3·95
4	161·0	772	2·46	50·7	1·89	0·83	2·97	229·2	1,078	1·48	39·0	2·27	0·78	1·89
5	131·0	621	2·74	44·6	1·77	0·95	2·88	235·6	912	2·96	31·5	2·67	0·89	3·31
6	97·0	470	4·89	36·7	2·37	1·72	2·84	146·4	896	4·46	27·5	2·38	1·28	3·48
7	27·0	137	10·22	30·5	1·59	4·17	2·45	88·6	747	4·95	23·8	1·68	1·49	3·32
8	21·0	87	3·45	25·2	0·75	2·53	1·36	38·6	326	6·44	21·5	0·90	1·81	3·55
9	—	—	—	—	—	—	—	12·5	119	1·68	19·5	0·39	2·44	0·68
Total	1,129 ± 106							1,356 ±121						

It is hoped that investigations now in progress will substantiate one or the other of these alternatives.

Acknowledgments

This work has been supported by The Norwegian Cancer Society and The Norwegian Defence Research Establishment. The technical assistance of Miss Brit Falck Madsen and Mrs. Sonja Johansen is gratefully acknowledged.

REFERENCES

Bateman, A. J., 1955. *Heredity*, **9**, 187.

Howard, A., and Pelc, S. R., 1951. *Exp. Cell. Res.*, **2**, 178.

King, E., 1953. *J. Exp. Zool.*, **122**, 541.

Discussion

Alper. Does Dr. Oftedal have even a rough estimate as to the number of mutations per ^{32}P disintegration? Such data would be very interesting in view of Stent's finding that only one bacteriophage inactivation occurs for about twelve disintegrations of incorporated ^{32}P atoms.

Oftedal. At present we have not tried to estimate the dose to the gonads, nor the numbers of ^{32}P atoms remaining after hydrolysis, and it is impossible to say if we have effects even within the order of magnitude of Stent's, or if, in fact, radiation or transmutation constitutes the predominant mechanism in the production of the recessive lethals.

Auerbach. I notice that you found ^{32}P in spermatozoa which at the time of feeding had been mature or, in any case, postmeiotic. Do you think that there is an essential difference here between *Drosophila* and the mouse in which only stages up to meiosis become labelled?

Oftedal. This is one of the points which makes us feel uncertain that we are measuring only ^{32}P incorporated in DNA. The activity may reside in other components of the sperm, and even in the seminal fluid. This is suspected especially for the day-1 activities.

RECENT STUDIES ON CHROMOSOME BREAKAGE AND REJOINING

SHELDON WOLFF
Biology Division, Oak Ridge National Laboratory,
Oak Ridge, Tennessee, U.S.A.

The induction of chromosomal aberrations by ionising radiation is a phenomenon that has been extensively studied in the past, mainly because it gave a quantitative measure of the amount of radiation damage produced within the cell. The information obtained was particularly useful in the formulation of the target theory, which explained radiation effects in terms of a simple direct interaction of the radiation with certain specific loci in the cell. Recently, the study of aberration production has been approached with renewed vigour. In the main this seems to be due to the knowledge that, at low doses, radiation-induced cellular death is caused by damage within the nucleus of the cell. However, since much of the research in modern radiation biology is concerned with protection and recovery from radiation damage, the present trend in aberration studies is toward the utilisation of a chemical approach to elucidate the chemical nature of the induced damage.

Before proceeding, we should define some of the classic terms used to describe chromosome breakage and reunion. According to the target theory (Lea, 1955), either on or very close to the track of an ionising particle, an actual transverse rupture of the chromosome is produced, such that the two ends come apart. This is defined as a 'break' and it may now either not rejoin at all, or rejoin in one of two different ways, i.e. it may restitute (rejoin in its original condition, which is not detectable as an aberration), or may rejoin illegitimately with other broken ends. Lea (1955) reviewed and accumulated the extensive evidence indicating that the latter two processes, restitution and rejoining, occur for some considerable time (4 minutes average for *Tradescantia* chromatid aberrations) after the breaks have been formed. He also indicated the essential sameness of the two different types of rejoining. Studies on the seed of *Vicia faba* by Wolff and Atwood (1954) and on *Trillium* by Deschner and Sparrow (1955) have shown that radiation not only produces breaks but also has an independent effect on the rejoining of these breaks. The *Vicia* studies indicated that the aberration yield, with certain qualifications, was independent of the time that the breaks stayed open and that the decrease in the aberrations produced when the radiation was carried on under anoxic conditions was caused by less breakage and not by increased restitution.

The rationale used in the *Vicia* studies was based on Sax's earlier observation that if the radiation were either fractionated into several small doses with rest periods between, or if it were given at low intensities, the yield of two-hit aberrations decreased (Wolff, 1954). This was interpreted as indicating that at low intensities some of the breaks restituted before others were produced. Thus all the breaks were not open simultaneously and could not, therefore, give rise to as many two-hit aberrations. In *Vicia*

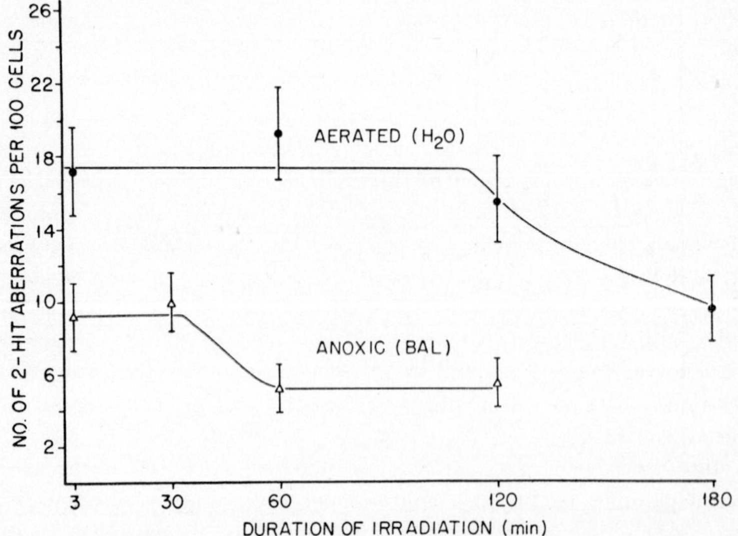

FIG. 1. Two-hit aberrations produced by radiation (600r) at intensities varying from 200 to 5r/minute. Modified from Wolff (1954).

(Fig. 1), it is found that in seed irradiated in air with 600r at intensities varying from 200 to 5r/min. and in those irradiated under anoxic conditions from 200 to 20r/min. there is no decrease in the number of two-hit aberrations (no intensity effect). This indicates that there is no rejoining of the breaks for at least 2 hours in those irradiated in air and for half an hour in those irradiated under anoxia (Wolff, 1954). These are the same breaks that have been found to require respiration and a source of energy before they can rejoin (Wolff and Luippold, 1955). Because of these two factors—(1) they stay open for a long period of time before rejoining, and (2) they require energy for the synthesis of the bonds formed in the rejoining process, it was postulated that these are breaks of covalent bonds (Wolff and Luippold, 1956).

However, the results of experiments with chelating agents by Mazia (1954) and with calcium deficiencies by Steffensen (1955) have led to reports that the chromosome consists of units held together by ionic bonds of calcium and/or magnesium ions and that perhaps radiation ruptures these bonds (Steffensen, 1955).

Since the restitution of breaks of ionic bonds would involve only electrical factors, these breaks would be expected to restitute very rapidly. Their existence would not have been observed in the previous *Vicia* experiments because the experimental design where the fastest dose took 3 minutes,

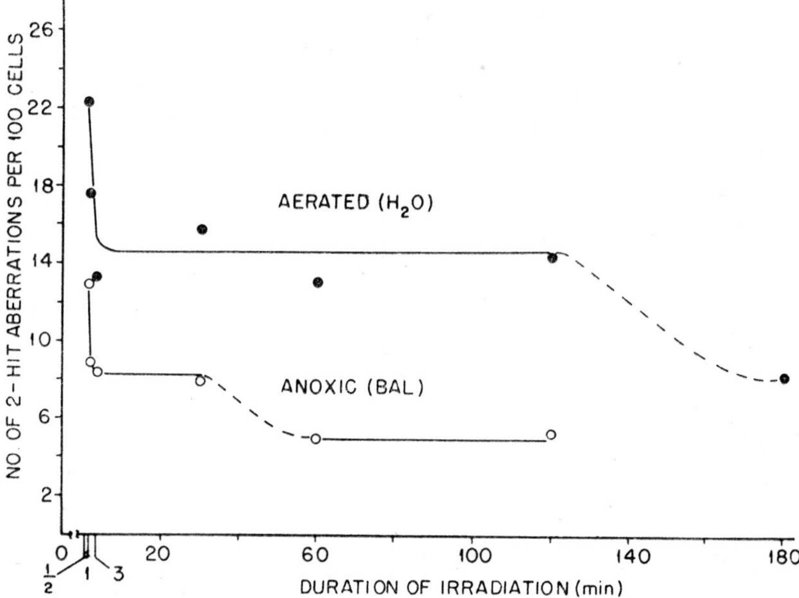

FIG. 2. Two-hit aberrations produced by radiation (600r) at intensities varying from 1,200 to 5r/minute. Data from Wolff and Luippold (1956*b*).

would not have allowed the separate detection of any class of break that restituted in less than 3 minutes. Therefore, the radiation was administered at the high intensities of 1,200 and 600r/min. in order to produce all the breaks within a shorter period of time. Figure 2 shows that at the very high intensities there is an increase in aberration yield above the plateaux that represent the covalent breaks. This indicates that, in addition to those breaks that stay open for a long period of time, there is another type that rejoins very rapidly. Only when the radiation is completed in less than a minute do these breaks coexist and thus contribute to the formation of two-hit aberrations. Therefore, it is seen that in *Vicia* there are two different types of breaks that can be classified by their time of rejoining—one is fast and the other slow. For the reasons previously mentioned, it is thought that the slow component consists of breaks of covalent bonds. The presence of a fast component is consistent with the idea that ionic bonds also are broken.

To test this concept further it was decided to treat the seeds with the chelating agent, versene (ethylene diamine tetra acetic acid) which had been found to be effective both in inducing chromosome breaks (Mazia, 1954) and in increasing crossing-over in *Drosophila* (Levine, 1955). In Table I it may be seen that versene breaks the chromosomes of *Vicia* too. It was then thought that after a pretreatment with versene, when irradiation at low intensities induced ionic breaks, chelation would occur and these breaks, instead of restituting immediately, would stay open and become available for the formation of aberrations. Table II contains the results of a typical

TABLE I

Aberrations produced by soaking Seed in Versene (0·001 M)

Total cells	Time in versene (hr.)	Number of aberrant cells	Aberrations (per 100 cells)
650	0	6	0·94 ± 0·38
750	3	27	3·60 ± 0·69
250	6	36	14·40 ± 2·4

Data from Wolff and Luippold (1956*b*).

experiment of this type. It may be seen that the effect of versene combined with radiation greatly exceeds the sum of the effects of either alone. This type of experiment gives extremely variable results, both between slides and between different experiments. Thus, although the increased aberration yield is qualitatively certain, the quantitative result may be poorly reproducible. These results, although they do not constitute a hard and fast proof, are consistent with the concept that breaks of ionic bonds are formed. Thus it appears that radiation produces at least two chemically different types of chromosome breaks.

It is too early to generalise from these results and to assume that the same phenomenon occurs in all organisms. However, it appears that an analogous effect is found in *Drosophila* sperm. It has been well known for many years that, in irradiation of *Drosophila* sperm, decreasing intensity of radiation does not cause a decrease in the number of two-hit aberrations. It is generally believed that this is because the broken chromosomes do not rejoin until the time of fertilisation. Here, then, is a type of break that remains open for long periods of time. It may be speculated that these breaks are analogous to those that require a source of energy for their rejoining and are represented by the plateaux of the *Vicia* curves. However, sperm respire very little (what little energy they do produce probably is necessary for motility) and the breaks cannot rejoin until they are supplied with the necessary energy from the fertilised egg. It must be noted, however, that in all the *Drosophila* intensity experiments, the fastest any dose was

TABLE II

Aberrations Produced by Combined Versene (0·001 M) and radiation (600r at 200r/min.). Seeds soaked for a Total of 5 hours

Slide no.	Treatment	Total cells	Normal cells (%)	Fragments (per 100 cells)	Dicentrics and rings (per 100 cells)
1	Versene, 3 hr.	50	96·0	2·0	2·0
2		50	100·0	0·0	0·0
3		50	98·0	2·0	0·0
Total ...		150	98·0	1·33	0·67
1	X-rays, 600r	50	66·0	24·0	16·0
2		50	80·0	12·0	10·0
3		50	80·0	8·0	12·0
4		50	78·0	12·0	10·0
5		44	72·7	9·1	13·6
6		56	71·5	16·6	17·85
Total ...		300	74·6	13·6	13·3
1	Versene, 3 hr. plus 600r	50	44·0	44·0	32·0
2		50	46·0	44·0	32·0
3		50	58·0	28·0	34·0
4		50	64·0	22·0	16·0
5		50	64·0	22·0	18·0
6		50	82·0	8·0	12·0
Total ...		300	60·3	28·0	24·0

Data from Wolff and Luippold (1956*b*).

administered was 8 minutes. Therefore these experiments, also, would not have demonstrated any rapidly closing breaks. In 1954, Haas *et al.* (1955), however, irradiated *Drosophila virilis* sperm at very high intensities so that they could compare doses administered in 1 minute with those in 20 minutes. They found an intensity effect. There were more aberrations at the very high intensity than at the low. So, in *Drosophila* too, it seems that there is a very rapidly closing break in addition to the one that does not rejoin for long periods of time. One might be ionic, the other covalent.

Attempts to modify the rejoining of the broken ends of *Vicia* chromosomes have dealt to date with only that type of break that stays open for long periods. By the application of enzyme inhibitors after irradiation, it was found that the rejoining of these breaks would be inhibited; that is, if cells containing breaks that ordinarily would restitute in 30 minutes were treated with certain respiratory inhibitors, the breaks would stay open and rejoin in a quantitatively determinable manner with breaks produced by subsequent radiation administered 75 minutes later (Wolff and Luippold, 1956). From experiments such as these, it was determined that, although the rate of restitution was dependent on oxidative processes, the amount of restitution was oxygen independent. In Table III is seen the result of a simple experiment which indicates that the inhibition of oxidative metabolism, and consequently the time that the breaks remain open, does not influence aberration

TABLE III

Independence of Aberration Yield on Time Breaks Stay Open

Post-irrad. (600r) treatment	Time open (min.)	Number of cells	Number of 2-hit aberrations (per 100 cells)
None	30	200	$11{\cdot}5 \pm 2{\cdot}4$
95% CO + 5% O_2 in dark ...	>75	200	$9{\cdot}5 \pm 2{\cdot}2$
$1{\cdot}5 \times 10^{-4}$ M DNP	>75	300	$10{\cdot}7 \pm 1{\cdot}9$
0° C.	>75	300	$11{\cdot}3 \pm 1{\cdot}9$

yield. This, of course, is true only if the radiation is administered at high enough intensity to produce all the breaks before any appreciable fraction of them can restitute. Another way of phrasing this is, that it is the numbers of breaks coexisting in the nucleus and not their time of coexistence that determines the yield of two-hit aberrations. By decreasing the time that breaks stay open before restituting one can decrease the numbers of breaks coexisting in the nucleus and thus can protect against some of the genetic damage that is manifested as two-hit aberrations.

So far, our work had indicated that the properties of rejoining, if not oxidative in themselves, require oxidative metabolism. Conger (1955), by means of an elegant fusion analysis for *Tradescantia* chromatid aberrations, has found that irradiation in nitrogen never results in more fusibility of the broken ends than would occur after irradiation in air. For these reasons, I have found untenable the position advocating that irradiation under anoxia produces the same number of breaks as irradiation in air but that more restitution occurs under anoxia (Schwartz, 1952; Baker and von Halle, 1953; Swanson and Schwartz, 1953; and Baker, 1956). Swanson, too, who was one of the supporters of the differential restitution hypothesis (Swanson and Schwartz, 1953), has reconsidered his views and now believes that the amount of restitution as defined by the advocates of this hypothesis is independent of oxygen (Swanson, 1955).

Recently Alper (1956) has produced a provocative paper in which she reinvokes the differential reunion hypothesis. However, without passing any judgment on the validity of her hypothesis, I wish to point out that she is not defining reunion in the same way that it is defined by most radiation cytologists. Actually she is describing a process by which a 'potential' chromosome break as described by Thoday (1953) and Swanson (1955) is transformed into an actual break which then is capable of entering into the reactions of rejoining. It then becomes obvious that her hypothesis can be fitted neatly into an explanation of the effects of radiation on chromosomes without necessarily implying an effect on classically defined rejoining. Both in her scheme and in Swanson's it is assumed that more actual chromosome breakage occurs if oxygen is present at the time of irradiation with X-rays. Whether this is caused by the repair of potential breaks by anoxia, as Swanson postulated, or by the transformation of potential to actual breaks by oxygen as Alper suggests, is still a moot point. It is hoped that her paper will induce more work to clarify the matter. At any rate it seems that anoxic irradiation does not cause more rejoining (either restitution or illegitimate rejoining) as defined by radiation cytologists, and that much of the oxygen controversy was caused by semantic difficulties that could have been avoided had the proponents of both sides been explicit in defining their terms.

In summary, it can be said that much of the experimental work on the induction of chromosome aberrations is now designed to disclose more about the chemical nature of the processes of chromosome breakage and reunion. Some of our own data are interpreted as indicating that there are at least two types of breaks: one that rejoins very rapidly and might be ionic in nature and another that remains open for long periods of time and then requires a source of energy for the synthesis of rejoining. These we think are the breaks of covalent bonds. With regard to chemical agents that modify the numbers of X-ray chromosome aberrations, it is concluded that, in all probability, none of them can affect the amount of rejoining, although some, i.e. metabolic inhibitors, affect the time that the breaks stay open. Other modifying agents, such as oxygen (or the lack of it), seem to affect only the amount of actual breakage and not rejoining as defined in the classical sense.

Acknowledgments

This work was performed under contract No. W-7405-eng-26 for the U.S. Atomic Energy Commission.

REFERENCES

ALPER, T., 1956. *Radiation Res.*, **5**, 573.
BAKER, W. K., 1956. *Mutation*, Brookhaven Symp. No. 8, 191.
BAKER, W. K., and VON HALLE, E. S., 1953. *Proc. nat. Acad. Sci. Wash.*, **39**, 152.
CONGER, A., 1955. Symp. on Genetic Recombination (discussion), *J. Cellular, Comp. Physiol.*, **45**, Suppl. 2, 309.
DESCHNER, E., and SPARROW, A. H., 1955. *Genetics*, **40**, 460.
HAAS, F. L., DUDGEON, E., CLAYTON, F. E., and STONE, W., 1955. *Genetics*, **39**, 453.
LEA, D. E., 1955. *Actions of Radiations on Living Cells*, 2nd ed. (Cambridge Press).
LEVINE, R. P., 1955. *Proc. nat. Acad. Sci. Wash.*, **41**, 727.
MAZIA, D., 1954. *Ibid.*, **40**, 521.
SAX, K., 1939. *Ibid.*, **25**, 225.
SCHWARTZ, D., 1952. *Ibid.*, **38**, 490.
STEFFENSEN, D., 1955. *Ibid.*, **41**, 155.
SWANSON, C., 1955. *Radiobiol. Symp.* 1954, p. 254 (London, Butterworth).
SWANSON, C., and SCHWARTZ, D., 1953. *Proc. nat. Acad. Sci. Wash.*, **39**, 1241.
THODAY, J. M., 1953. *Heredity*, **6**, suppl., p. 299.
WOLFF, S., 1954. *Nature, Lond.*, **173**, 501.
WOLFF, S., and ATWOOD, K. C., 1954. *Proc. nat. Acad. Sci. Wash.*, **40**, 187.
WOLFF, S., and LUIPPOLD, H. E., 1955. *Science*, **122**, 231.
Idem, 1955. *Progress in Radiobiol.*, 1956, p. 217 (Edinburgh, Oliver and Boyd).
Idem, 1956*b*. *Proc. nat. Acad. Sci. Wash.*, **42**, 510.

DISCUSSION

Fahmy. Our recent results on the effect of chelating agents on the genetic material *in vitro*, may have some bearing on Dr. Wolff's results. We applied various concentrations of some of these agents, viz. versene, sodium fluoride, and sodium salicylate to *Drosophila* salivary gland chromosomes previously squashed in isotonic saline. No marked effects on chromosome morphology were noticed until the concentrations of the chelating agents were definitely hypertonic, when granulation of the bands, and partial disintegration of the chromosomes took place. Isotonic solutions of the chelating agents in conjunction with 5% phenol, however, resulted in almost immediate disintegration of the chromosomes and in the liberation of DNA.

Pieces of rat liver and testis were also immersed in isotonic solutions of the chelating agents alone, in 5% phenol alone and in a mixture of the two. The tissues were then fixed and stained in Feulgen and the nuclei and chromosomes examined microscopically. Only the chelating agents in conjunction with phenol resulted in the liberation of the DNA. After this combined treatment the sperm nuclei and the chromosomes became Feulgen negative.

Our results suggest that in the genetic material the DNA is probably bound to the protein through metallic ions as well as by hydrogen bonding and other forces, since phenol is as important as the chelating agents for the breakdown of the protein-DNA complex.

In view of Dr. Wolff's results on the effect of versene and radiation on the yield of chromosome breaks, one wonders what would be the effect of versene and phenol in conjunction with radiation? Of course, the concentration of the phenol would have to be very low to break some protein-DNA linkages without complete liberation of the DNA.

Wolff. The results that you mention remind me very much of Mazia's results obtained by the use of chelating agents to break up chromosomes. I have purposely not tried to get the extremely drastic effects that Mazia and you have achieved, because once the chromosomes are disintegrated they are, of course, no longer useful for this type of cytological study on breakage and rejoining.

I would like to point out that all I can say about the experiments, is that the appearance of a fast-closing break and the results of the versene treatments are consistent with the concept that the chromosomes are made up of units bound together by metallic ions. However, I don't feel that it has been absolutely proved because it may be an over-simplification to think that all versene does is to cause chelation on or near the chromosome breaks. If versene binds divalent metal ions it could possibly inactivate those enzymes which require small amounts of metallic ions, or even possibly by the removal of calcium, effect membrane permeability.

Alexander. You have told us that when the dose is given in periods of less than a minute or so there is an increase in the number of chromosome aberrations scored by you. I would like to know how you can reconcile your observation with that of other workers who have failed to find such an effect. I refer in particular to the long series of experiments done at Oak Ridge and elsewhere, in which the biological effects produced by irradiation from an atomic explosion were compared with those obtained with conventional machines in the laboratory. The conclusion was that there was no difference in the number of chromosome breaks, even though the dose was given in one case in a micro-second, and in the others over several minutes.

Wolff. If I remember rightly, the experiments performed with atomic explosions are very difficult to compare with these experiments. At many of the bomb tests the irradiation consisted of mixed gamma and neutron radiation which complicates matters considerably. Also, I believe that even the control experiments performed in the laboratory were actually irradiated at a much higher intensity than usual. I think another difference that must be taken into account here is that the best of the cytological experiments at atomic tests were performed on *Tradescantia* microspores. Although we have observed this intensity effect on *Vicia* seed and barley seed and also think it has been observed in *Drosophila* sperm, we have not been able to find it in *Tradescantia.*

Alexander. Another question of mine concerns the greater effectiveness of pretreatment with versene for 6 hours rather than for 3 hours. The chelating action of versene is instantaneous and any chemical actions produced will therefore occur immediately, once the versene has reached a particular site. The implication would therefore be that it takes more than 3 hours for the versene to penetrate through material. On the face of it this seems to me unlikely.

Wolff. In regard to the time relations involved with versene treatments, I think this is due to the penetration and diffusion of versene into the tissues. Similar time-effect relations have been observed with sequestering agents used to soften peas before canning.

Ford. I should like to congratulate Dr. Wolff on a very clear presentation and ask whether he can exclude the possibility that the versene effect, in whole or in part, might be due to an alteration of the time-scale of the germination processes such that cell populations of different radiation sensitivity were compared? I have evidence that the radiation sensitivity of the *Vicia* seedling (measured as frequency of induced structural changes) increases fairly rapidly during germination.

Wolff. This is a problem that usually arises in aberration studies. However, we irradiated after only 5 hours of soaking. The first division does not occur until about 96 hours later. Since we are working with such a short portion of the mitotic cycle and also since we produce only chromosome aberrations which are induced in interphase, we do not think that we are sampling cells of different sensitivity.

Ford. A change of a few hours in the time interval between germination and irradiation does change quite seriously the yield of breaks.

Wolff. There is other evidence from our fractionated dose experiments on the seeds of *Vicia* which also indicate that the sensitivity does not change within the times utilised. We do not see any changes in radiation sensitivity until we observe chromatid aberrations, which indicates that the chromosomes were already doubled at the time of irradiation. As long as the radiations are performed at the early interphase where we don't see differences in sensitivity or chromatid aberrations, I do not believe that the results can be explained in terms of the differential sensitivities.

Auerbach. There was a paper by Clark in *Nature* quite recently, in which he showed that the frequency of a certain type of re-arrangement in *Drosophila melanogaster* is higher after very intense than after less intense X-radiation.

Wolff. Yes, this is another example of the intensity effect about which Dr. Alexander was wondering.

Muller. Would you please explain how you reconcile your finding that the breaks close earlier when anoxia is used, with the interpretation that the joining is a process that demands energy supplied by oxidation?

Wolff. My ideas on this subject were presented in a paper by myself and Luippold (1955) which was read by Dr. Gray at the conference in Cambridge last year. The experimental results can be summarised in a scheme represented as follows in which it can be seen that X-irradiation has two independent effects in regard to aberration production (Fig.). One is that it produces chromosome breaks and the other is that it affects

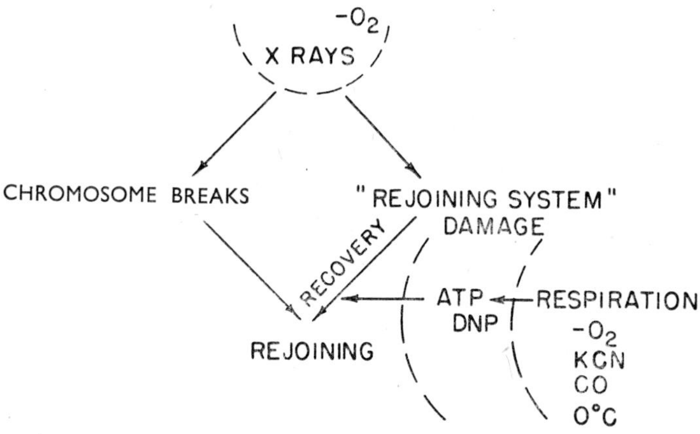

FIG. Schematic representation of the effects of irradiation on the production of two-hit chromosome aberrations.

the time that these breaks remain open. These independent effects were observed in *Vicia* by myself and Atwood and later confirmed in *Trillium* by Deschner and Sparrow. Since it was observed that under anoxic irradiation fewer breaks are obtained and that those breaks produced rejoin relatively quickly, it was postulated that a 'rejoining system' is damaged. The term 'rejoining system' is enclosed in quotation marks because we don't know just what it is. After the damage is produced, however, energy from ATP produced by cellular respiration is needed for recovery and chromosome break rejoining. We do not know whether the energy is utilised in the recovery of the 'rejoining system' or in the actual biosynthesis at the broken ends (or both). However, what the scheme does give, is a consistent explanation of the experimental results. I believe that under anoxic irradiation a typical oxygen effect of less damage is obtained. This is reflected both in fewer breaks and, because of less damage to the 'rejoining system', in earlier closing of these breaks. Conversely, by post-irradiation anoxia (actually inhibition of ATP formation) energy is not available and recovery cannot take place. Thus the breaks remain open. This later inhibition can be accomplished by low temperatures, enzyme inhibitors such as cyanide or carbon monoxide in the dark, by anoxia, or by dinitrophenol.

REFERENCE

Wolff, S., and Luippold, H. E., 1955. *Progress in Radiobiol.*, 1956, p. 217 (Edinburgh, Oliver and Boyd).

MODIFICATION OF X-RAY MUTAGENESIS IN *DROSOPHILA*

Relative Sensitivity of Spermatids and Mature Spermatozoa

I. I. Oster

Zoology Department, Indiana University, Bloomington, U.S.A.

For many years it had been thought that the chromosomes of *Drosophila* spermatozoa are both highly and uniformly sensitive to X-rays. However, more recently, several investigators working independently, have demonstrated that there are variations in the sensitivity of relatively mature germ cells. Lüning (1952*a*, *b*, *c* and *d*), Auerbach (1954), and Khishin (1955) have shown that spermatid stages of *Drosophila melanogaster* are much more susceptible to the damaging effects of X-rays than mature spermatozoa irradiated prior to ejaculation by the male; while the work of Bonnier and Lüning (1953), of Abrahamson and Telfer (1954) and of Muller, Herskowitz, and Oster (unpublished) has indicated that mature spermatozoa irradiated in the seminal receptacles and spermathecæ of females are also much more susceptible than mature spermatozoa treated in the male. A series of experiments was thereupon undertaken in order to determine whether these two most highly sensitive types of male germ cells are equally sensitive to X-rays and to see what factors may be involved in this heightened sensitivity.

Methods

For the irradiation of spermatids (series I) male prepupæ, characterised by a fully formed but colourless puparium which serves as an accurate time mark for determining their age, were collected and X-rayed when they were 48±2-hours-old. During this stage the pupal testis contains an accumulation of spermatids. A group of females (series II) which had been inseminated previously by males genotypically like those of series I were irradiated along with the 48-hour-old pupæ. Both groups of individuals received a dose of 2,000r (135 kV, dose delivered at a rate of 200r/minute through a 1 mm. thick aluminium filter).

The males carried a normal X chromosome marked by the gene for yellow body colour (*y*) and a Y chromosome (sc^8.Y) marked by its normal allele (y^+); the inseminated females and those to be inseminated, contained the markers and inversions designated y^{S1} sc^8 $B\,f\,In^{49}\,v$; *bw*; *e*. The males derived from irradiated pupæ were mated during the first three days after eclosion. The F_1 males were tested for translocations between autosomes II and III and between the Y chromosome and autosomes II and/or III by being backcrossed to y^{S1} sc^8 $B\,f\,In^{49}$ v; *bw*; *e* virgin females like their mothers. Their offspring (F_2) were examined for the absence of recombinants involving any of the three pairs of markers (*y* vs. y^+, *bw* vs. bw^+, and *e* vs. e^+) considered two pairs at a time. The F_1 females were tested for lethals by mating them to sc^8 . Y/y^{S1} sc^8 $B\,f\,In^{49}$ v; *bw*; *e* males and examining their offspring (F_2) for the absence of non-yellow males, a finding which would indicate that a lethal had been induced in the paternal X chromosome. Matings yielding insufficient flies for the determination of whether or not a lethal or translocation was present were retested by repeating with the F_2 flies crosses like those of the F_1.

Results and discussion

The results of the genetic tests were as follows: Here the percentages shown in parenthesis represent the induced sex-linked lethals, i.e. those remaining after the spontaneous frequency has been subtracted from the observed frequency. The absolute numbers are given as fractions, with the lethals or translocations in question forming the numerators and the total counts the denominators.

TABLE I

	Series I (spermatids)	Series II (spermatozoa)
Sex-linked lethal mutations ...	$\frac{98}{689}$ = 14·2% (14·1%)	$\frac{69}{1178}$ = 5·9% (5·8%)
Translocations involving autosomes II and III	$\frac{89}{419}$ = 21·2%	$\frac{87}{1322}$ = 6·6%
Translocations involving the Y chromosome and autosomes II and/or III	$\frac{46}{419}$ = 11·0%	$\frac{52}{1322}$ = 3·9%

These results indicate that spermatids represent the stage which is by far the most sensitive to the damaging effects of X-rays, being at least twice as sensitive as the mature spermatozoa carried in inseminated females.

In order to avoid ambiguity in the following discussion and others dealing with similar phenomena, we would like to introduce the word *reunion* as a general term for designating the rejoining of broken chromosomes, *restitution* for designating their rejoining in the same linear order as they had been in prior to treatment, and *disarrangement* for designating the rejoining of broken chromosomes in a new order. As it has been shown that the ends of chromosomes broken during the spermatid stages undergo reunion before fertilisation (Oster, 1955), unlike the breaks produced in mature spermatozoa, which remain open and reunite during fertilisation (Muller, 1938, 1939*a* and *b*, 1940), the possibility was considered whether this ability of the chromosomes to rejoin coupled with the drastic physical changes which

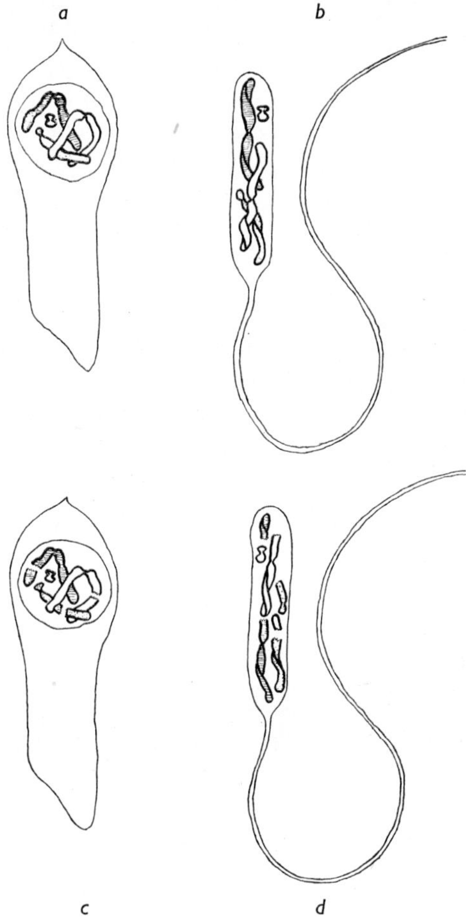

FIG. 1. (*a*) The chromosome complement of a spermatid.
(*b*) The chromosome complement of a spermatozoon.
(*c*) A spermatid containing chromosomes broken by X-rays.
(*d*) Separation of the broken ends of chromosomes following elongation of a spermatid to form a spermatozoon.

occur during spermiogenesis may be one of the main underlying causes of the heightened radiosensitivity of spermatids. That is, during the elongation of the spherical spermatid to form the conical spermatozoon and the consequent moving of the chromosomes and probable separation of their parts if broken, it might be expected that disarrangement rather than restitution would be favoured (Fig. 1). In order to test this hypothesis a series of experiments was undertaken in which 48-hour-old pupæ containing either a rod X-chromosome (X) or a ring-shaped X chromosome (X^{c2}) were irradiated with two doses of X-rays. An examination of their offspring yielded the following frequencies of sex-linked lethal mutations:

TABLE II

	X	X^{c2}
600r	$\frac{153}{2581} = 5{\cdot}9\%$	$\frac{60}{1041} = 5{\cdot}8\%$
2000r	$\frac{98}{689} = 14{\cdot}2\%$	$\frac{17}{101} = 16{\cdot}8\%$

The similarity of the mutation frequencies argues against the possibility that the sensitivity of the spermatid stage is related primarily to a moving apart of the broken chromosomes. For if this were the case, a ring chromosome would be expected to yield a lower mutation rate than an ordinary X-chromosome since those lethals associated with deficiencies or other

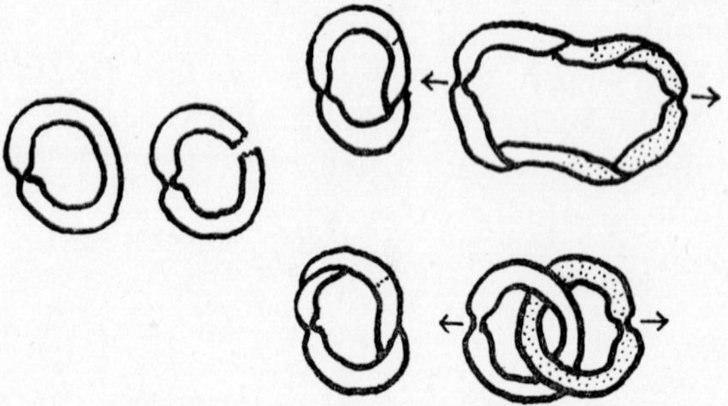

Fig. 2. Results of chromosome duplication after twisted restitution of a broken ring-shaped chromosome.

chromosomal disarrangements, would be expected often to be lost via twisted restitutions or opening out of the broken rings following the shuffling of the chromosomes during the transformation of the spermatids into spermatozoa (Fig. 2).

Although it has been shown in the case of many forms, including *Allium*, *Trillium*, *Ascaris*, female germ cells of *Drosophila*, and *Mus*, that the condensed state of the chromosomes (such as is found in late prophase, metaphase, anaphase, late oöcytes, and spermatozoa) is the most susceptible to damage by radiation (see, for instance, Oster, 1954 and Sparrow, 1951), the results of our comparison of the radiosensitivity of spermatids and spermatozoa, both of which have condensed chromosomes, have indicated that some other factor(s) may be responsible for the greater sensitivity of the former cells. It is of course possible that we have here a situation analogous to the one in *Trillium*. In this form, prophase, which is characterised by an extremely low rate of oxygen uptake, is the most radiosensitive stage of mitosis. Sparrow (1951) has interpreted this by proposing that during a period of lower consumption of oxygen by the cells a higher partial pressure would exist intracellularly, and that this would tend to increase sensitivity to X-ray damage, as had been shown by Thoday and Read (1947). Whether a similar explanation holds true for the 48-hour-old pupæ of *Drosophila* (containing a preponderance of the highly radiosensitive spermatids), which also undergoes a fall in oxygen uptake (Hadorn, 1951), may be investigated in further experiments.

Summary

Spermatids represent by far the most sensitive stage of gametogenesis studied, to the damaging effects of X-rays. This does not appear to be due merely to a separation of broken chromosomes favouring disarrangement of pieces during spermiogenesis.

Acknowledgment

This work was supported by a grant for work of Dr H. J. Muller and associates, from the Atomic Energy Commission (Contract AT (11–1)-195), and by a post-doctoral fellowship from the National Science Foundation of the U.S.A.

REFERENCES

ABRAHAMSON, S., and TELFER, J. D., 1954. *Rec. Genet. Soc. Amer.*, **33**, and *Genetics*, **39**, 955.

AUERBACH, C., 1954. *Z. ind. Abst. Ver.*, **86**, 113.

BONNIER, G., and LÜNING, K. G., 1953. *Hereditas*, **39**, 193.

HADORN, E., 1951. *Developmental Action of Lethal Factors in Drosophila in Advances in Genetics*, **4**, 53 (New York, Academic Press Inc.).

KHISHIN, A. F. E., 1955. *Z. ind. Abst. Ver.*, **87**, 97.

LÜNING, K. G., 1952*a*. *Hereditas*, **38**, 91.

Idem, 1952*b*. *Ibid.*, **38**, 321.

Idem, 1952*c*. *Acta Zool.*, **33**, 193.

Idem, 1952*d*. *Studies on X-ray Induced Mutations in Various Stages of Spermatogenesis in Drosophila melanogaster*, Doctoral Dissertation, University of Stockholm (Stockholm, Bonnier).

MULLER, H. J., 1938. *Collecting Net* (*Woods Hole, Mass.*), **13**, 181, 183, 198.

Idem, 1939*a*. *M.R.C. Special Report Ser.*, **236**, 14.

Idem, 1939*b*. (*Abst.*) *Proc. 7th Int. Genet. Cong., Edinburgh*, and 1941, *J. Genet. Suppl.*, p. 221.

Idem, 1940. *J. Genet.*, **40**, 1.

Oster, I. I., 1954. *Excerpta Medica, 8th Int. Cong. Cell Biology*, **8**, 406.
Idem, 1955. *Genetics*, **40**, 692.
Sparrow, A. H., 1951. *Ann. N.Y. Acad. Sci.*, **51**, 1508.
Thoday, J. M., and Read, J., 1947. *Nature, Lond.*, **160**, 608.

Discussion

Auerbach. Are you convinced that the excess of genetical effects during the sensitive stage refers only to breaks or restitutions? Could there not be a general increase in effectiveness which involves also gene mutations. The fact that the excess of lethals was almost as pronounced as that of translocations seems to be in favour of this interpretation.

Sobels. Since your experiments have ruled out one of the possible explanations for increased sensitivity in the spermatid stage, it would be interesting to hear your opinion on other possible interpretations of this phenomenon.

Oster. I think that it is premature to state definitely whether or not the heightened radiosensitivity of the spermatid stage is due mainly to an increase in the number of chromosome disarrangements, or to a general increase in the frequency of induced intra-genic changes as well as inter-genic aberrations. This problem may be attacked with methods which would discern between intra-genic and inter-genic changes in order to see whether or not they both increase after irradiation of spermatids. Such techniques might include studying the frequencies of visible and viable mutations induced in genes which when deficient are lethal, such as *cut*, *raspberry*, and *carnation*; or studying the frequencies of reversions of mutant alleles, for these would also be in the nature of intra-genic changes. However, the large scale on which experiments of this type would have to be carried out, because of the relative infrequency of such changes at specific loci, is against their being carried out at the present time.

RECOVERY OF CHROMOSOMES FROM X-RAY DAMAGE

K. NORDBACK AND C. AUERBACH
Institute of Animal Genetics, Edinburgh, Scotland

A number of workers (Abrahamson and Telfer, 1954; Baker and v. Halle, 1953, 1954; Lüning, 1954*a*, *b*; Telfer and Abrahamson, 1954*a*) have found that irradiated *Drosophila* ♂♂ carry more dominant lethals, translocations and other types of chromosome rearrangement in spermatozoa which are used on the first day after irradiation, than in spermatozoa which are used on the second day. These results raise two questions:

(1) Is the effect due to differences in radiosensitivity between chromosomes in fully mature and nearly mature spermatozoa, or is it due to recovery of chromosomes from radiation injury when irradiated spermatozoa are stored for one day in the testis?

(2) Does either of these mechanisms result in a general dose-reduction like the one found in photorecovery from ultraviolet radiation, or are only intergenic changes—breaks and rearrangements—reduced in second-day spermatozoa? Some evidence on these points is available in the quoted papers. We shall defer its discussion till after the presentation of our own results.

We planned our experiments with these two questions in mind. To tackle the first, we used the same plan which had been followed by Baker and v. Halle. Three series were run in each experiment. Series I consisted of progeny produced on the first day after irradiation of 3-4-day-old ♂♂. Series II consisted of progeny produced by the same ♂♂, but with new ♀♀, on the second day. Series III consisted of progeny produced on the second day by irradiated ♂♂ which had been kept unmated on the first day. For obtaining information on the second question we used a dual-purpose stock of tester ♀♀, so that we could score translocations and lethals in progenies from the same irradiated ♂♂. We were aware that this would probably not allow a decisive answer; for on either hypothesis a drastic reduction in breakage-frequency will be reflected in a reduction in lethal frequency. The difference is only one of degree, a more pronounced drop in lethal frequency being expected from a general dose-reduction than from a reduction in breakage-frequency. The numbers required for a clear decision between these alternatives are so high that we could not hope for more than suggestive evidence. The results of three experiments are presented in the Table.

Let us first consider translocations. In agreement with previous work, there is a marked drop in frequency from the first to the second day (Series I versus II). The combined χ^2 for the differences in the three experiments is about 10 for 1 D.F. In all experiments, translocation frequency in Series III was at least as low as in Series II, i.e. the drop in frequency occurred whether or not the ♂♂ had been with ♀♀ on the first day. It also was not dependent on the number of ♀♀; for in the third experiment the ♂♂ had been presented with more than twice as many ♀♀ as in the other two. This confirms Baker's results with dominant lethals and indicates that the observed difference is due to recovery of the chromosomes in stored

TABLE

Experiment	Series	*n*	Lethals %	*n*	Trans- locations
1	I	273	14·0	109	13·8
	II	413	9·5	236	8·1
	III	357	8·5	159	6·9
2	I	1022	10·0	735	9·5
	II	1060	8·5	917	5·8
	III	1056	8·8	820	6·1
3	I	278	6·9	283	6·0
	II	244	3·4	223	4·0
	III	634	6·7	260	3·5

spermatozoa and not to differential sensitivity of the chromosomes in germ cells furnishing successive batches of sperm. It is, of course, possible that the ♂♂ in Series III used or resorbed some of their spermatozoa during the first day or that their pool of mature sperm received an admixture of newly matured spermatozoa; but this would at best make translocation frequency in Series III intermediate between those in Series I and II. Moreover, dilution of stored sperm with newly matured spermatozoa should affect most of all very young ♂♂, in which the pool of mature sperm is low. In reality, however, the opposite was found in two experiments with newly hatched ♂♂. There was no difference between Series I and II, and the drop from Series I to III was only very slight. This is understandable on the recovery hypothesis: for in sexually active young ♂♂ there will be no opportunity for the storing of sperm, and even in sexually inactive ones stored spermatozoa may form too low a proportion of the germ cells to play a decisive role in determining the overall result for the second day.

Possibly, strains differ in the amount of mature sperm present in young ♂♂ for in one of Lüning's experiments (1954*b*) young ♂♂ gave essentially similar results to older ones in regard to the frequency of intergenic changes on the first and second day.

Like Baker, we found that recovery occurs no longer in spermatozoa which have been transferred to the ♀. In some series the day-to-day progeny from inseminated ♀♀ varied; but there was no consistent trend as with the ♂♂, and in other experiments there was no variation. One may speculate whether the absence of recovery in the ♀ is responsible for the fact that fewer genetical effects are produced when mature sperm is irradiated in the ♂ than when irradiation is given to inseminated ♀♀ (Bonnier, 1954; Bonnier and Lüning, 1953; Lüning, 1953; Lüning, 1953, 1954*b*). It would be interesting to see whether this difference persists when irradiation is given in nitrogen or by neutrons; for under these circumstances there is no difference in the frequency of genetical effect between first-day and second-day progeny (Baker and v. Halle, 1953, 1954; Telfer and Abrahamson, 1954*b*; see, however, Lüning, 1954*a*).

Turning now to the second question—whether there is recovery from intragenic changes as well as from intergenic ones—we look at the data on sex-linked lethals. In all three experiments, there was a marked drop in lethal frequency from the first to the second day. In two of them, lethal frequency in Series III was at least as low as in Series II, indicating again that recovery rather than differential sensitivity is responsible for the effect. Why Series III in the third experiment had the same mutation frequency as Series I, cannot be decided. It probably has to do with the fact that in this Series the ♀♀ were used for a longer time than in the others, so that some secondary effect of storing spermatozoa in the ♀ influenced the result. The difference between Series I and II is statistically well secured. The combined χ^2 is 7 for 1 D.F. The extent of the reduction in lethal frequency is not much less than that in translocation frequency, and the data are in good agreement with the assumption of a general dose-reduction for both intragenic and intergenic effects. But, as we had feared, they are not large enough to completely exclude the possibility that it is only the position-effect lethals which were reduced in frequency. Lüning, who supports the latter interpretation, could not completely exclude the possibility that there is also some reduction in the frequency of intragenic changes. Our interpretations thus do not differ radically, but only in emphasis. Unfortunately, it is difficult to devise a method which would allow a clear decision to be made.

It is certainly easier to visualise recovery from breaks than recovery from gene mutation. Actual or potential breaks may be repaired by restitution, but recovery from gene mutation implies reversion of the chemical change which constitutes the mutation. It is, however, not necessary to assume that what we call recovery is always actual repair of a completed injury. Alternatively, apparent recovery may occur when the chemical environment tends to counteract some intermediate step in mutagenesis. For photo-recovery from ultraviolet mutation, this appears the more acceptable explanation. It seems conceivable that among the chemical effects of X-radiation there are some which may fail to go to completion in the environment of the irradiated testis.

REFERENCES

ABRAHAMSON, S., and TELFER, J. D., 1954. *Genetics*, **39**, 955.
BAKER, W. K., and VON HALLE, E., 1952. *Ibid.*, **37**, 565.
Idem, 1953. *Proc. nat. Acad. Sci. Wash.*, **39**, 152.
Idem, 1954. *Science*, **119**, 46.
BONNIER, G., 1954. *Heredity*, **8**, 199.
BONNIER, G., and LÜNING, K. G., 1953. *Hereditas*, **39**, 193.
LÜNING, K. G., 1953. *Drosophila Inform. Serv.*, **27**, 99.
Idem, 1954*a*. *Hereditas*, **40**, 295.
Idem, 1954*b*. *Drosophila Inform Serv.*, **28**, 132.
Idem, 1954*c*. *Heredity*, **8**, 211.
TELFER, J. D., and ABRAHAMSON, S., 1954*a*. *Genetics*, **39**, 998.
Idem, 1954*b*. *Drosophila Inform. Serv.*, **28**, 161.

DISCUSSION

Lüning. In order to draw the conclusion that there is a recovery I think you will have to prove that the males really keep the sperm when not mated as I have shown the opposite for young males.

Auerbach. If I remember rightly, it was only the very young ♂♂ which in your experiments released their sperm irrespective of whether they were mated or not. Our experiments were done with 3-4-days-old ♂♂ which definitely regulate their sperm release by the opportunity for copulation. At least, I have always found this to be the case for the period from, say, the third to the 14th day, and I see little reason to suspect that they should behave differently on the first. Data from Professor Muller's laboratory also show that virgin ♂♂ store at least some of their spermatozoa, for there was a significant difference in the number of dominant lethals which a dose of X-rays produced on the first day in 3-4-day-old virgin ♂♂ and equally old ♂♂ which had been sexually active; the drop from the first to the second day was the same in both cases. We did several experiments with very young ♂♂ and got contradictory results, although in these ♂♂ the drop in the frequency of genetical effects from the first to the second day should be particularly clear, if it were due to sensitivity differences between fully mature and not quite mature sperm. In 3 out of 4 larger scale experiments with young ♂♂ there was no difference between the first and the second day, in the 4th the difference was as in older ♂♂.

Mossige. When males and females are kept separate for a while and then brought together, the males mate with a higher number of females per day—suggesting that they had more sperm stored.

Fahmy. The frequency of mating is not a measure of the amount of sperm ejaculated, since *Drosophila* males mate even when the testis and vesicula siminalis are almost empty, when they will only ejaculate seminal fluid with little or no sperm. The high frequency of mating after separation of the sexes, does not, therefore, necessarily mean a higher sperm reserve and cannot be used as evidence for the complete retention of sperm in unmated males.

Mossige. When I speak of matings I do not mean observed copulations but counts of either offspring or hatched eggs from females isolated after mating.

Hollaender. Have you found any effects of temperature on the recovery?

Auerbach. We did try the effect of temperature on recovery in one experiment, but unfortunately this was one of the experiments with young ♂♂, and we got no recovery at all.

Kanazir. We have tried to restore mutated gene-loci of *Salmonella* by using polymerised homologous DNA in a similar manner to Demerec and Lahr's recent work. Proline-less mutants which were obtained after UV or X-radiation are unable to grow on a 'minimum medium' (containing inorganic salts and glucose). They were treated with homologous DNA-extracts from a normal *Salmonella* strain and the number of auxotrophes was stored on minimum medium. The number of back mutations was found to be 10^{-5}–10^{-3} as compared to the frequency of spontaneous back mutations amounting to only 10^{-10}–10^{-9}. It is necessary for transformation that the mutants should pass through some divisions. It is difficult to establish beyond doubt that DNA is the active principle. Our preparations, however, lose their potency after DNA-se treatment and UV irradiation but not after trypsin treatment. 'Apurinic acid' and heterologous DNA was not effective. Transduction experiments using transformed mutants (treated with DNA) as donors for the reciprocal transductions, on the same 'proline-less' mutants as recipients, gave the same transduction frequency as that obtained by phage grown on wild type. In conclusion, our data indicate that changed gene loci (involved in proline synthesis) are restorable after treatment with polymerised homologous DNA.

LIST OF PARTICIPANTS

AUSTRALIA

Dr. R. D. BROCK, Commonwealth Scientific and Industrial Research Organisation, Canberra.

BELGIUM

Prof. Z. M. BACQ, Institut de Pathologie, 1 Rue des Bonnes Villes, Liège.

Prof. P. DESAIVE, Centre Anticancereux de l'Université, Liège.

Dr. A. DUNJIC, Institut du Cancer l'Université, Louvain.

CHINESE PEOPLE'S REPUBLIC

Prof. HU MAO-HUA, Dept. of Radiology, Peking Medical College, Peking.

Prof. WANG SHAO-HSUIN, Dept. of Radiology, Peking Medical College, Peking.

Prof. YUNG TU-SAN, Dept. of Radiology, First Medical College of Shanghai, Shanghai.

CZECHOSLOVAKIA

Dr. V. SLOUKA, Klinická nemocnice, Hradec Králové, and Biophysical Institute of the Czechoslovak Academy of Science, Brno.

Dr. J. SOŠKA, Brnenská, 1082, Blansko, and Biophysical Institute of the Czechoslovak Academy of Sciences, Brno.

Dr. J. SVEDA, Klinická nemocnice, Hradec Králové.

DENMARK

Dr. B. C. CHRISTENSEN, Finseninstitutet og Radiumstationen, Copenhagen.

Dr. M. FABER, Finsenlaboratoriet, Copenhagen.

Dr. P. B. HANSEN, Radiumstationen, Aarhus.

Dr. J. E. THYGESEN, Finseninstitutet og Radiumstationen Radiobiologiske Laboratorium, Strandboulevard 49, Copenhagen.

Prof. M. WESTERGAARD, Genetisk Institut, Universitetsparken 3, Copenhagen.

FRANCE

Prof. P. CHAMBON, Institut de Chimie biologique, Strasbourg.

Mlle. S. DAVYDOFF, Commissariat à l'Energie Atomique, Centre d'Etudes Nucleaires, Saclay.

Dr. J. F. DUPLAN, Laboratoire Pasteur de l'Institut du Radium, 26 Rue d'Ulm, Paris V[e].

Dr. M. Lefort, Laboratoire de Biologie Végétale, 12 Rue Cuvier, Paris.

Dr. H. Marcovich, Laboratoire Pasteur de l'Institut du Radium, 26 Rue d'Ulm, Paris V^{e}.

Mme. I. Pinset, Service de Biologie, Centre d'Etudes Nucleaires, Saclay.

Dr. J. Rodesch, Institut de Chimie biologique, Strasbourg.

GERMANY

Dr. J. Aurand, Max Planck-Institut für Biophysik, Forsthausstrasse 70, Frankfurt am Main.

Dr. R. Bauer, Goldammerweg 1, Köln-Vogelsang.

Dr. A. Catsch, Heiligenberg-Institut, Heiligenberg/Baden.

Dr. W. Dittrich, Universitäts-Frauenklinik, Hamburg 20.

Dr. U. Feine, Max Planck-Institut für Biophysik, Forsthausstrasse 70, Frankfurt am Main.

Dr. T. Fliedner, Czevny-Krankenhaus für Strahlenbehandlung der Universität, Heidelberg.

Dr. F. Gauwerky, Allg. Krankenhaus St. Georg, Hamburg.

Dr. G. Gerber, Max Planck-Institut für Biophysik, Forsthausstr. 70, Frankfurt am Main.

Dr. U. Hagen, Heiligenberg-Institut, Heiligenberg/Baden.

Dr. G. Höhne, Universitäts-Frauenklinik, Martinistr. 52, Hamburg/Eppendorf.

Dr. W. Jacobi, Max Planck-Institut für Biophysik, Forsthausstr. 70, Frankfurt am Main.

Dr. R. Koch, Heiligenberg-Institut, Heiligenberg/Baden.

Dr. H. A. Künkel, Universitäts-Frauenklinik, Hamburg/Eppendorf.

Dr. H. Maass, Universitäts-Frauenklinik, Hamburg/Eppendorf.

Dr. H.-J. Maurer, Strahleninstitut der Univ.-Frauenklinik, Erlangen.

Dr. H. Pauly, Max Planck-Institut für Biophysik, Forsthausstr. 70, Frankfurt am Main.

Prof. B. Rajewsky, Max Planck-Institut für Biophysik, Forsthausstr. 70, Frankfurt am Main.

Dr. I. Wolf, Max Planck-Institut für Biophysik, Forsthausstr. 70, Frankfurt am Main.

Dr. M. Zacharias, Institut für Kulturpflanzenforschung, Gatersleben Krs. Aschersleben.

GREAT BRITAIN

Dr. J. D. Abbatt, Dept. of Medicine, M.R.C. Cyclotron Building, Hammersmith Hospital, Ducane Road, London, W. 12. *Present address*: Medical Radiobiology, G.E.C. Atomic Energy Group, Erith, Kent.

Dr. P. Alexander, Chester Beatty Research Institute, Fulham Road, London, S.W.3.

Miss T. ALPER, M.R.C. Experimental Radiopathology Research Unit, Hammersmith Hospital, Ducane Road, London, W.12.

Dr. C. AUERBACH, Institute of Animal Genetics, The University, Edinburgh.

Dr. N. M. BLACKETT, Physics Dept., Institute of Cancer Research, Royal Cancer Hospital, Fulham Road, London, S.W.3.

Dr. W. M. C. BROWN, Institute of Radiotherapy, Western General Hospital, Edinburgh.

Dr. T. C. CARTER, M.R.C. Radiobiological Research Unit, A.E.R.E., Harwell, Berks.

Dr. A. R. CRATHORN, Pollards Wood Research Station of the Institute of Cancer Research, Royal Cancer Hospital, Chalfont St. Giles, Bucks.

Dr. W. M. DALE, Dept. of Biochemistry, Christie Hospital and Holt Radium Institute, Manchester 20.

Dr. M. I. DAVIS, Dept. of Radiotherapeutics, University of Cambridge, Downing Street, Cambridge.

Dr. D. L. DEWEY, Research Unit in Radiobiology, Mount Vernon Hospital, Northwood, Middx.

Dr. M. EBERT, M.R.C. Experimental Radiopathology Research Unit, Hammersmith Hospital, Ducane Road, London, W.12.

Dr. L. A. ELSON, Chester Beatty Research Institute, Royal Cancer Hospital, Fulham Road, London, S.W.3.

Dr. O. G. FAHMY, Pollards Wood Research Station, Chalfont St. Giles, Bucks.

Dr. C. E. FORD, M.R.C. Radiobiological Research Unit, A.E.R.E., Harwell, Berks.

Dr. C. W. GILBERT, Christie Hospital and Holt Radium Institute, Manchester 20.

Mr. J. GRANT, Orchard House, Caldecott Road, Abingdon-on-Thames, Berks.

Miss E. B. HARRISS, Physics Dept., Institute of Cancer Research, Royal Cancer Hospital, Fulham Road, London, S.W.3.

Dr. B. HOLMES, Dept. of Radiotherapeutics, University of Cambridge, Downing Street, Cambridge.

Mrs. S. HORNSEY, M.R.C. Experimental Radiopathology Research Unit, Hammersmith Hospital, London, W.12.

Dr. A. HOWARD, Research Unit in Radiobiology, Mount Vernon Hospital, Northwood, Middx.

Dr. P. HOWARD-FLANDERS, M.R.C. Experimental Radiopathology Research Unit, Hammersmith Hospital, London, W.12.

Dr. E. V. HULSE, M.R.C. Radiobiological Research Unit, A.E.R.E., Harwell, Berks.

Dr. L. G. LAJTHA, Dept. of Radiotherapy, Churchill Hospital, Oxford.

Dr. L. F. Lamerton, Physics Dept., Institute of Cancer Research, Royal Cancer Hospital, Fulham Road, London, S.W.3.

Dr. E. M. Ledlie, Dept. of Radiotherapy, Royal Marsden Hospital, Fulham Road, London, S.W.3.

Dr. J. F. Loutit, M.R.C. Radiobiological Research Unit, A.E.R.E., Harwell, Berks.

Dr. N. R. Mackay, Dept. of Radiotherapy, The Royal Marsden Hospital, Fulham Road, London, S.W.3.

Miss L. K. Mee, Dept. of Radiotherapeutics, University of Cambridge, Downing Street, Cambridge.

Prof. J. S. Mitchell, Dept. of Radiotherapeutics, University of Cambridge, Downing Street, Cambridge.

Dr. T. R. Munro, Tissue Culture Laboratory, Christie Hospital and Holt Radium Institute, Manchester 20.

Dr. R. J. Munson, M.R.C. Radiobiological Research Unit, A.E.R.E., Harwell, Berks.

Dr. G. J. Neary, M.R.C. Radiobiological Research Unit, A.E.R.E., Harwell, Berks.

Dr. M. G. Ord, Dept. of Biochemistry, The University, Oxford.

Dr. M. Owen, Nuffield Dept. of Medicine, Oxford.

Dr. S. H. Revell, Chester Beatty Research Institute, Royal Cancer Hospital, Fulham Road, London, S.W.3.

Dr. O. C. A. Scott, Research Unit in Radiobiology, Mount Vernon Hospital, Northwood, Middx.

Dr. C. L. Smith, Dept. of Radiotherapeutics, University of Cambridge, Downing Street, Cambridge.

Dr. K. A. Stacey, Chester Beatty Institute, Fulham Road, London, S.W.3.

Dr. L. A. Stocken, Dept. of Biochemistry, The University, Oxford.

Dr. A. J. Swallow, Tube Investments Research Laboratories, Hinxton Hall, near Saffron Walden, Essex.

Dr. D. Taylor, Radiobiological Research Unit, Radiotherapy Dept., Institute of Cancer Research, Royal Cancer Hospital, Fulham Road, London, S.W.3.

Dr. K. Williams, Medical Division, Building 354, A.E.R.E., Harwell, Berks.

INDIA

Dr. A. R. Gopal-Ayengar, Biology Division, Dept. of Atomic Energy, Indian Cancer Research Centre, Parel, Bombay-12.

ITALY

Dr. R. A. Silow, Food and Agriculture Organisation of the United Nations, Rome.

THE NETHERLANDS

Dr. D. W. van Bekkum, Medisch Biologisch Laboratorium R.V.O.-T.N.O., Rijswijk, Z.H.

Prof. J. A. Cohen, Medisch Biologisch Laboratorium R.V.O.-T.N.O., Rijswijk, Z.H.

Dr. H. M. Klouwen, Radiologische Werkgroept. n.o., Municipal Hospitak, Wagnerlaan 55, Arnhem.

Dr. F. H. Sobels, Genetisch Instituut, Stationsstraat 9, Utrecht.

Dr. H. J. Stuy, Natuurkundig Laboratorium, N.V. Philips, Eindhoven.

Dr. O. Vos, Medisch Biologisch Laboratorium R.V.O.-T.N.O., Rijswijk, Z.H.

NEW ZEALAND

Dr. H. D. Purves, Endocrinology Research, Medical School, Otago University, Dunedin.

NORWAY

Dr. T. Brustad, Norsk Hydros Institutt för Kreftforskning, Det Norske Radiumhospital, Oslo.

Dr. F. Devik, Institutt för Generell og Eksperimentell Patologi, Det Norske Radiumhospital, Oslo.

Dr. L. Eldjarn, Norsk Hydros Institutt för Kreftforskning, Det Norske Radiumhospital, Oslo.

Dr. K. Mikaelsen, Norges Landbrukshögskole, Vollebekk.

Dr. J. Mossige, Norsk Hydros Institutt för Kreftforskning, Det Norske Radiumhospital, Oslo.

Dr. P. Oftedal, Norsk Hydros Institutt för Kreftforskning, Det Norske Radiumhospital, Oslo.

Dr. A. Pihl, Norsk Hydros Institutt för Kreftforskning, Det Norske Radiumhospital, Oslo.

Dr. Ö. Strömnaes, Institutt för Genetikk, Universitetet, Karl Johansg. 47, Oslo.

SWEDEN

Prof. G. Bonnier, Genetiska Institutet, Stockholms Högskola, Teknologg. 8, Stockholm.

Dr. J. Braun, AB Atomenergi, Drottning Kristinas väg 47, Stockholm.

Dr. A. Brohult, Radiumhemmet, Karolinska sjukhuset, Stockholm 60.

Dr. S. Brohult, Forskningslaboratoriet LKB, Postfack 14, Bromma.

Dr. V. W. Burns, Gustaf Werners Institut för Kärnkemi, Uppsala.

Dr. L. Carlbom, AB Atomenergi, Drottning Kristinas väg 47, Stockholm.

Dr. B. Carlqvist, Försvarets forskningsanstalt, Sundbyberg 4.

Dr. C.-J. Clemedson, Försvarets forskningsanstalt, Sundbyberg 4.

Dr. L. Ehrenberg, Inst. för Organisk Kemi och Biokemi, Stockholms Högskola, Stockholm.

Dr. A. Forssberg, Radiofysiska institutionen, Stockholm 60.

Dr. S. Franzen, Radiumhemmet, Stockholm 60.

Prof. Å. Gustafsson, Statens skogsforskningsinstitut, Stockholm 51.

Dr. M. Helde, Radiofysiska institutionen, Stockholm 60.

Prof. G. Hevesy, Inst. för Organisk Kemi och Biokemi, Stockholms Högskola, Stockholm.

Dr. B. E. Holmberg, Försvarets forskningsanstalt, Sundbyberg 4.

Dr. M. Jaarma, Inst. för Organisk Kemi och Biokemi, Stockholms Högskola, Stockholm.

Dr. E. Klein, Inst. för Cellforskning, Karolinska institutet, Stockholm 60.

Dr. G. Klein, Inst. för Cellforskning, Karolinska institutet, Stockholm 60.

Dr. B. H. Larsson, Gustaf Werners Institut för Kärnkemi, Uppsala.

Dr. L.-G. Larsson, Radiumhemmet, Stockholm 60.

Dr. Ch. Lingen, Wenner-Grens Institut för Experimentell Biologi, Stockholms Högskola, Stockholm.

Dr. K. G. Lüning, Genetiska Institutet, Stockholms Högskola, Stockholm.

Dr. J. Mackey, Sveriges Utsädesförening, Svalöf.

Dr. A. Nelson, Försvarets Forskningsanstalt, Avd. 1, Sundbyberg 4.

Dr. B. A. Nohrman, Akademiska sjukhuset, Uppsala.

Dr. L. Révész, Inst. för Cellforskning, Karolinska institutet, Stockholm 60.

Dr. G. von Rosen, Box 82, Landskrona.

Dr. B. Swedin, Karolinska sjukhuset, Stockholm 60.

Dr. Th. Wahlberg, Radiofysiska institutionen, Stockholm 60.

Dr. D. von Wettstein, Statens skogsforskningsinstitut, Roslagsv., Stockholm 51.

Dr. J. Zajicek, Radiopatologiska institutionen Karolinska sjukhuset, Stockholm 60.

Dr. K. G. Zimmer, Statens skogsforskningsinstitut, Stockholm 51.

TURKEY

Dr. L. Tanberk, D.D. Hastanesi, Ankara.

U.S.A.

Dr. H. J. Curtis, Brookhaven National Laboratory, Upton, Long Island, New York.

Dr. S. Emerson, Biology Branch, Division of Biology and Medicine, U.S.A.E.C., Washington 25, D.C.

Captain J. A. English, U.S.N., Office of Naval Research, Keysign House, 429 Oxford Street, London, W.1, England.

Dr. C. ENTENMAN, U.S. Naval Radiological Defense Laboratory, San Francisco 24, California.

Dr. L. H. HEMPELMANN, Medical School, University of Rochester, Rochester, New York.

Dr. A. HOLLAENDER, Biology Division, National Laboratory, Oak Ridge, Tennessee.

Dr. F. HUTCHINSON, Gibbs Research Lab., Yale Univ., New Haven, Connecticut.

Dr. I. LEWIN, Chester Beatty Research Inst., Royal Cancer Hospital, Fulham Road, London, S.W.3, England, and The Division of Neoplastic Diseases, Montefiore Hospital, New York, N.Y.

Dr. R. LEWIN, Chester Beatty Research Inst., Royal Cancer Hospital, Fulham Road, London, S.W.3, England, and The Division of Neoplastic Diseases, Montefiore Hospital, New York, N.Y.

Dr. M. MENDELSOHN, Dept. of Radiotherapeutics, University of Cambridge, Cambridge, England, and the Sloan-Kettering Institute for Cancer Research, New York, N.Y.

Prof. H. J. MULLER, Indiana University, Jordan Hall 201 B, Bloomington, Indiana.

Dr. I. I. OSTER, Indiana University, Jordan Hall 201 B, Bloomington, Indiana.

Lt. Commander G. ROSENFELD, U.S.N., Naval Medical Research Institute, National Naval Medical Center, Bethesda, Maryland.

Prof. F. G. SHERMAN, Biology Dept., Brown University, 1 Providence, R.I.

Dr. E. L. SIMMONS, Argonne Cancer Research Hospital, The University of Chicago, Chicago 37, Illinois.

Dr. D. E. SMITH, Argonne National Laboratory, P.O. Box 299, Lemont, Illinois.

Prof. C. A. TOBIAS, Donner Laboratory, Berkeley 4, California.

Dr. S. WOLFF, Oak Ridge National Laboratory, Oak Ridge, Tennessee.

U.S.S.R.

Prof. O. A. BOGOMOLETS, Research Institute of Public Health, Kiev, Ukraine.

Prof. V. L. TROITSKY, Gamaleya Institute of Epidemiology and Microbiology of the Academy of Medical Sciences of the U.S.S.R., Moscow.

YUGOSLAVIA

Dr. D. KANAZIR, Institut "Boris Kidrik", P.O. Box 522, Belgrade.

SUBJECT INDEX

Acetylcholinesterase, activity in irradiated megakaryocytes, 380
Adenosine triphosphate and diphosphate
  concentration in irradiated ascites tumours, 46
  possible role in rejoining of chromosome breaks, 473
Adrenal response, in irradiation, 237
Aerosols, inhalation of Radon and Thoron, 310-18
Ageing, of mice after irradiation, 261
Albumins, changes in serum concentration in irradiated and cytotoxin treated dogs, 235
Alcohol dehydrogenase, radiation sensitivity of, wet and dry, 3
Aldehydes, interaction with cysteamine, 156
Alkoxyglycerols, in treatment of irradiated patients, 241-6
Alkylating agents
  as mutagens in *Drosophila*, 437; *see also under* Mustards, etc.
Alpha particles
  in irradiation of single cells and parts of the cell, 105-20
  in radiation chemistry, 17
Amino acids
  effects of irradiation on incorporation into proteins, 33-37; *see also under* Cysteine, etc.
Amino ethylisothiuronium (AET), as a protective compound
  in *E. coli*, 126
  in mammals, 135, 181-5
Aminopterin, effects of administration on iron turnover in bone marrow, 329
Aminothiols, as protective agents, 135
Anoxia, as modifying irradiation effects; *see under* Oxygen
Ascites tumour cells
  biochemical effects after irradiation, 43
Ascites tumour cells
  chromosome aberrations caused by X-rays and fast electrons, 86
  effect of irradiation damage on survivors, 80
Ascorbic acid, concentration changes in irradiated rats suprarenals, 238
Autoradiography
  in studies of ^{59}Fe-uptake by erythroid cells, 332, 333
  in studies of ^{131}I-labelled compounds from patients with thyrotoxicosis, 305
  ^{90}Sr distribution in bone, 287-97
Azides, as mutagens and effecting chromosome aberrations in *Drosophila*, 449

Bacteria
  α-particle irradiation of, 105
  growth and morphology after irradiation, 90
  in radiation protection and recovery studies, 126-8, 192-5
  induction in lysogenic-, 281
Bacteræmia, in irradiated animals, 221-4
Batylalcohol; *see* Alkoxyglycerols
Bone
  accumulation of ^{90}Sr in rabbits tibia, 287
  evaluation of dose distribution after ^{90}Sr uptake, 294
  radiosensitivity of growing bone at various oxygen pressures, 97
Bone marrow
  an assay of various types of erythroid cells, 349-54
  changes in nucleic acid concentration and synthesis, 54, 59
  iron uptake of erythroid cells, 333-8
  megakaryocytic changes in irradiation, 376

Bone marrow
radiomimetics and irradiation changes in white cell counts on, 372
transplant of bone marrow cells in radiation therapy, 128-44, 197-202, 214-19

Cancer cells; *see* Ascites tumour cells
Caronamide, combined effects of X-rays and — on kidney transport and respiration, 38
Catalase
effect of inhibition of, in radiation genetics, 449
inactivation of, in dry state, 9
iron incorporation after irradiation, 331
Cathepsin, activation with SH-compounds, 190
Cephalin; *see under* Phospholipids
Chelating agents
in isotope decontamination, 298
inducing chromosome aberrations, 446
Chimæras, produced by transplant of foreign tissue in irradiated animals, 197
Cholesterol, in irradiated rats, 238
Chromosomes, irradiation with alpha-particles, 108-17
Chromosome aberrations
effects of chelating agents, irradiation and oxygen on breakage and rejoining of chromosomes, 463-74
enhanced effects of X-rays and chemical mutagens, 451–3
induced by X-rays as compared to alkylating agents, 437-46
in irradiated *Yoshida* sarcoma cells, 86
produced in various stages of gametogenesis in *Drosophila*, 475, 481
Chromosome markers, in studies of radiation induced chimæras, 197
Clotting time of peritoneal fluid, impairment of, by radiation, 73
Coenzyme A, radiation sensitivity of, wet and dry, 3
Crosslinking of polymers, radiation effects on, 9
Cyanides, as mutagens in *Drosophila*, 449
Cysteamine-cystamine
mechanism of protective action, 136–7, 147–57
protective effect in *E. coli*, 128, in mammals, 135, 160–6, 181–5
Cysteine
mechanism of protective action, 136–7, 147, 192
protective effect on hibernating *loir*, 176, and rabbits, 192
Cytochrome C
binding to cysteamine, 155
inactivation in irradiated mouse liver slices, 25
iron incorporation after irradiation, 331
Cytochrome oxidase, radiation effects on, 25
Cytotoxins, producing physiological effects in irradiated animals, 231–40

Decontamination of radioactive isotopes, 298
Deoxyribonucleic acid (DNA)
irradiation and synthesis of, in:
bone marrow after whole body and partial irradiation of rats, 59
human bone marrow *in vitro*, 54
rat thymus, ^{32}P-studies of, 65
restoration of mutated *Salmonella* gene loci by homogenate of spleens of protected mice, 206
Deoxyribonucleinase, liberation of, from mitochondria and microsomes by irradiation, 69
Detrimentals
produced by X-rays in *Drosophila*, 425
rate of development of, under selection pressure, 433
Deuterons, in irradiation of enzymes, 3

Dihydroxyacetone phosphate, determinations in irradiated ascites cells, 44-5

Dihydroxydimethyl peroxide as mutagen in *Drosophila*, 449

Dimesyloxyalkanes, as mutagens in *Drosophila*, 437

Dinitrophenol, effect on kidney transport, 38

Direct and indirect effects
- on enzymes, 3–7
- on polymers and macromolecules, 8-15

Dithiocarbamate, protection by, 135, 162

Drosophila
- back mutations of, 407
- comparison of chemically and X-ray induced mutations, 437
- detrimental mutations, 425, 433
- enhancement of mutation rate after combined X-ray and chemical treatment, 449
- incorporation of ^{32}P and mutagenicity, 457
- recovery of chromosome damage, 481
- relative sensitivity of spermatids and mature spermatozoa, 475

E. coli
- growth and morphology after irradiation, 90
- protection and recovery of irradiated, 126–30, 193

Electrons, irradiation with, 10, 86

Energy transfer, in radiation chemistry, 8

Enzymes, effects of irradiation on
- acetylcholinesterase, 380
- alcohol dehydrogenase, 3
- catalase, 9, 331, 449
- Coenzyme A, 3
- cytochrome oxidase, 25
- deoxyribonuclease, 69
- invertase, 3
- succinic acid oxidase, 27

Epoxides, as mutagens in *Drosophila*, 438

Erhlich ascites tumour; *see under* Ascites tumours

Erythropoiesis, influence by irradiation
- in peripheral blood of patients given ^{32}P and X-rays, 382
- ^{59}Fe studies on iron uptake and turnover, 321–30, 333–8, 341–7, 349–54

Fertility, effects of irradiation
- on mice in air and in nitrogen, 253
- on mice with repeated small X-ray doses, 257
- restoration of rabbits primordial follicles, 274

Fibroblasts, α-particle irradiation, 114

Fluoroacetate and X-irradiation, effect on kidney transport and respiration, 38

Follicles, restoration after irradiation with single and fractionated doses, 274

Formaldehyde, as mutagen in *Drosophila*, 449

Fructose-1, 6-diphosphate, determination in irradiated ascites cells, 44-5

Globulins, changes in serum and concentration in irradiated and cytotoxin treated dogs, 235

Glutathione
- binding to cystamine, 149
- protective mechanism of, 137

Glycolysis
- inhibition by X-rays in ascites tumours, 43

Hæmoglobin synthesis in irradiated
- erythropoietic tissues; *see under* Erythropoiesis, iron turnover
- patients, changes in peripheral blood, 382

Histamine, protective effect of, 135

Hormonal influence in radiation protection, 161

Hydrogen bonds, possible role of, in radiobiology, 133

Hypophysectomised rats, radiation effects on, 237

Imines, as mutagens in *Drosophila*, 439
Immunity
  influence of radiation on natural, 221–4
  inactivation of immune response and growth of foreign tissue, 197–203, 204–10, 211–13, 214–220
Invertase, radiation sensitivity of, wet and dry, 3
Iodinated tyrosines, in patients with thyrotoxicosis, 305
Iodoacetic acid, reactivity with SH-compounds, 189
Isothiuronium ; *see under* Amino-ethylisothiuronium

Ketones, interaction with cysteamine, 156
Kidney
  action of X-rays and chemicals on transport, 38
  γ-activity after aerosol inhalation, 315

Lecithin ; *see under* Phospholipids
Leucopenia, effect of alkoxyglycerol esters on, 241
Leukæmia
  general implications of radiation induction, 388–96
  radiation induced in humans, 397–404
  stress with leukæmia as a test for recovery from radiation sickness, 217
Liver, liver cells, radiation effects on
  amino acid incorporation in proteins, 33
  enzyme inactivation, 26
  ^{32}P-incorporation in various cellular fractions, 49
Loir (*Glis Glis*), irradiation of hibernating as compared to non-hibernating, 176
Lymphocytes ; *see under* White cell counts
Lysogenic bacteria, induction of lysis by long term irradiation, 281

Macromolecules, radiation chemistry of, 3–7, 8–13
Megakaryocytes, in femurs of irradiated rats, 376
Mercury salts, combined effect of X-rays and, on kidney transport and respiration, 38
Metabolic effects on irradiated ascites cells, 43
Microsomes, irradiation effect on incorporation of amino acids into protein, 33
Mitochondria, radiation effects on
  enzymes, 25
  incorporation of amino acids into proteins, 33
Mitosis
  effects of α-particle beams, 108–20
  radiation effects at various stages on DNA-synthesis, 54
Modification of irradiation damage:
  1. Chromosome aberrations and mutations
    recovery of chromosomes from X-ray damage, 481
    recovery of mutants in bacteria, 128–9
    rejoining of chromosome breaks, 463
    relative sensitivity of spermatids and spermatozoa, 475
    restoration of mutated *Salmonella* loci with homologous DNA, 485
  2. Physiological effects
  Protective effect and mechanism of
    SH- and SS-compounds, 126, 134–8, 147–57, 160–6, 170–3, 176–9, 181–5, 187–90
    anoxia, 253
    changes in adrenal and pituitary functions, 237–9
    reduction of body temperature, 248–52
    shielding of minute areas of skin, 226–9
  Postirradiation therapy of
    leucopenia and thrombocytopenia with alkoxyglycerols, 241
    modifying effects of cytotoxins, 231–6

Modification of irradiation damage:
  Postirradiation therapy of
    whole body irradiated animals with transfer of bone marrow and spleen cells, 138–44, 197–202, 204–9, 211–13, 214–19
Muscle, γ-activity after inhalation of radioactive aerosols, 314
Mutations
  back mutations in *Drosophila* and other organisms, 407–15
  chemical mutagens, compared with radiation induced, 435–46
  detrimentals, induced by X-rays in *Drosophila*, 425–31, 433–6
  implications of irradiation in man, 416
  reversibility of radiation induced in *E. coli*, 128–30
  role of peroxides in mutagenesis, 449–54
  specific loci in the mouse, 421
Mustards
  as mutagens in *Drosophila*, 437, 453
  in stress of irradiated mice, 261
  pattern of blood response to, 372
Myleran, erythropoiesis after treatment with, 329
  patterns of blood response after treatment with, 372
Myoglobin, iron turnover after radiation, 331

Nervous system, response of peripheral, to irradiation, 76
Neuroblasts, irradiated, 125
Neurospora, back mutations, 407
Nitrogen mustard; *see under* Mustards
Nuclei, irradiation effects on incorporation of amino acids into proteins, 33

Organic peroxides
  participation of, in mutagenesis of *Drosophila*, 449–54
  role of, in radiobiology, 100
Ovary, restoration of follicles after irradiation, 274
Oxidation of SH-compounds, 189
Oxygen effect:
  in the irradiation of:
    anoxia as an explanation for radiation protection by SH-compounds, 163
    ascites tumour cells, chromosome aberrations of, 86
    bacteria, 90–6, 192–5
    mice, fertility of, 253
    polymers, 9
    tail bone of mice, 97
    Vicia, chromosome aberrations and rejoining, 463–74

Para-aminohippuric acid, transport in irradiated kidney, 38
Peritoneal fluid, impairment of clotting time, 73
Phages
  liberations from radiation lysed bacteria, 281
Phagocytosis and phagocytic index in irradiated animals
  after stress with dysentery, 221
  in serum of irradiated dogs, 233
Phenylalanine, incorporation into proteins, 33
Phospholipids, incorporation of ^{32}P, 49
Pituitary response to irradiation, 237
Polonium, as α-particle source, 105–17
Polyethylene, radiation and cross-linking, 10
Polymers, radiation chemistry, 8–15
Polymethacrylate, radiation chemistry, 8
Polystyrene, oxygen effect on irradiation, 11
Probability diagram, in hæmatological test, 361
Protection, mechanism of and protective agents; *see under* Modification
Proteins
  amino acid incorporation *in vivo* of various cell fractions after irradiation, 33
  concentration changes in serum after irradiation, 235

Proteins
interactions of, with:
SH- and SS-compounds, 153
radioactive yttrium (^{90}Y), 298
radiochemical changes, 11-12
Pyruvic acid, concentrations in irradiated ascites tumour cells, 45

Radicals
formation by X-rays and α-particles and implications for radiobiology, 16–20, 99
mean lifetime in cells, 7
participation in degradation of polymers, 9–15
Radon, inhalation of aerosols, 310
Recovery from irradiation effects; *see under* Modifications
Redox potential of SH-compounds, 188
Relative biological efficiency of different ionising radiations, 16–21
Respiration of irradiated
ascites tumour cells, 43
rabbit kidney slices, 38
various fractions of mouse liver tissue slices, 26
Respiratory tract, concentration of inhaled aerosols, 312
Rhesus monkeys, life span of red cells, 357
Ribonucleic acid (RNA)
changes in concentration and ^{32}P-incorporation of irradiated bone marrow, 59
^{32}P-incorporation in irradiated rats' thymus, 65
Ribomononucleotides, irradiation effects on ^{32}P incorporation, 49

Salmonella, restoration of mitotic activity with homologous DNA, 99
Selection pressure, on induced detrimental mutations, 433
Sensitive volume, of irradiated enzymes, 3–7, 31
Sensitising agents, constitution and effects of, 170
Shielding, partial of irradiated animals
effect on iron turnover, 341
modification of skin reaction through grids, 226
DNA-synthesis in bone marrow, 59
Skin, modification of irradiation damage by shielding with wires, 226
Spermatids, and spermatozoa relative sensitivity in *Drosophila*, 475
Sphingomyelin; *see under* Phospholipids
Spleen
grafts of cells in irradiation therapy, 204–9, 211–13
relative DNA turnover after transfer of homogenates, 206
Strontium (^{90}Sr)
evaluation of dose distribution in bone, 294
pattern of uptake, 287
Succinic acid oxidase and -dehydrogenase, radiation effects in tissue slices, 25
Sulphydryl compounds
chemistry of, 187
in protection, 126–9, 135, 147–57, 160–6, 170–3, 176–9, 181–5, 192–5
Suprarenals, radiation changes in, 237
Synkavit, mechanism of radiosensitisation, 174

Tail bone of mouse, oxygen dependence of radiosensitivity, 97
Temperature, influence on radiation effects in
bacterial recovery, 126–7
degradation of polymers, 9
hibernators, 176
protection of mammals, 248
Thoron, inhalation of radioactive aerosols, 310
Thrombocytopenia, alcoxyglycerol ester treatment, 241
Thymus, irradiation effects on DNA and RNA, 65
Thyrotoxicosis, studies with radioiodine, 305

Time factor in radiobiology
  induction of lysogenic bacteria at low dose rates, 281
  influence on rejoining of breaks in *Vicia*, 463
  life span of long term irradiated and stressed mice, 261
  repeated small doses on fertility of mice, 257
  restoration of follicles, single and fractionated doses, 274
  survival of long continued and fractionated irradiated mice, 267
Tissue cultures, in α-particle irradiation, 108–20
Transformation of induced mutants, 485
Trypan blue, test on skin reactions, 234
Tryptamine, mechanism of protection, 191
Typhoid toxins, in stress of irradiated mice, 261
Tyrosines, assay in thyrotoxic patients, 305

Versene; *see under* Chelating agents
Vicia, chromosome breaks and rejoining, 463
Viscosity changes of irradiated polymers, 11

Water, irradiation effects on, 16–18
Water content, effect on radiation sensitivity of enzymes, 3
White cell counts in
  irradiated animals, 231–6, 372–5
  patients, 243–6, 382–7
  persons employed in radiological work 361–71
Whole body irradiation
  versus partial; *see under* Shielding
  and studies of
    biochemical and physiological effects, 49–52, 65–8, 73–5
    erythropoiesis, 321–32, 333–8, 341–8, 349–54
    fertility, 257–60
    nerve reactions, 76–9
    protection, recovery and immunity, 134–44, 160–6, 176–80, 181–6, 197–202, 204–9, 211–13, 214–19, 221–4, 231–5, 237–40
    survival and weight changes, 248–52, 261–5, 267–72

Yoshida, ascites tumours, biochemical irradiation effects, 43
Yttrium, studies of ^{90}Y-equilibrium with proteins, 298

INDEX TO CONTRIBUTORS

Bold figures refer to papers read; ordinary figures indicate contribution to discussion

Abbatt, J. D., 293, **305**, 309
Alexander, P., **8**, 14, 99, 101, 132, 158, 159, 186, 230, 265, 471, 472
Almeida, A. B., **49**
Alper, T., **90**, 100, 131, 169, 173, 455, 462
Auerbach, C., 132, 168, 284, 413, 423, 436, 448, 454, 455, 462, 472, 480, **481**, 484, 485
Aurand, K., **267**, **310**

Bacq, Z. M., 21, 69, 79, 159, **160**, 168, 169, 174, 191, 236, **237**, 239, 355
Barnes, D. W. H., **197**, **211**
Belcher, E. H., **321**
Björnerstedt, R., **294**
Bogomolets, O., **231**, 236
Boico, A., **231**
Bonnier, G., **433**
Brohult, A., **241**, 247
Burns, V. W., 117
Butler, J. A. V., **33**

Carlqvist, B., 303
Carter, T. C., 230, 253, 414, **416**, 423, 424
Catsch, A., **181**, 185, 186
Chambon, P., **59**, 64
Christensen, B. C., 403
Clemedson, C.-J., **294**
Cohen, J. A., **134**, 145, 179, 185, 213, 380, 381, 395
Cohn, P., **33**
Crathorn, A. R., **33**, 37, 117
Curtis, H. J., 239, **261**, 265, 266

Dale, W. M., 132, 159
Davis, M. I., **114**, 118, 119, 120
Denko, J., **214**
Desaive, P., **274**, 280
Devik, F., **226**, 230
Diadiucha, G., **231**
Dittrich, W., **86**, 89
Drášil, V., **204**
Dunjic, A., **341**, 348
Duplan, J. F., **192**

Ebert, M., 84, 280
Ehrenberg, L., 456
Eldjarn, L., **147**, 159, 174, 339
Ellis, F., **54**
Elson, L. A., 145, 247, 266, 303, 354, **372**
English, J. A., 293
Engström, A., **294**
Esnouf, M. P., **211**, 213

Faber, M., 340, **397**, 404
Fahmy, M. J., **437**, 447
Fahmy, O. G., 203, **437**, 447, 448, 470, 485
Farran, H. E. A., **305**
Fischer, P., **237**
Fliedner, T., 247, 355, 360
Ford, C. E., 145, **197**, 202, 203, 414, 448, 472
Forssberg, A., 37, 47, 48
Friedenstein, A. J., **221**

Gerber, G., **25**
Gilbert, C. W., **357**, 360
Gopal-Ayengar, A. R., 120
Gros, C. M., 59

Hagen, U., 32, **187**, 191
Haigh, M. V., **357**
Hamerton, J. L., **197**

Harriss, E. B., **333**, 338, 339, 340
Hart, H. E., **298**
Healey, R., **261**
Helde, M., **361**, 371
Hevesy, G. C. de, 57, 63, 68, 118, 331, 332, 339, 340
Höhne, G., **43**, 48, 174, **176**
Hollaender, A., **123**, 131, 132, 133, 145, 169, 203, 371, 381, 395, 485
Holmes, B., 53, 57
Hornsey, S., **248**, 252, 253, 254
Howard, A., 117, 414
Howard-Flanders, P., 84, 97, 101, 203, 230
Hulse, E. V., 247, 266, 339, **349**, 354, 355
Hutchinson, F., **3**, 7, 14, 32, 75, 84, 118, 120, 283

Jacobi, W., **310**
Jacobson, L. O., **214**
Jaudel, C., **59**
Jonsson, S., **425**

Kanazir, D., 99, 145, 236, 239, 280, 415, 485
Karpfel, Z., **204**
Klein, G., 220
Koch, R., **170**, 174, 175, 179, 191
Künkel, H. A., **43**, **176**, 179

Lajtha, L. G., **54**, 57, 58, 68, 84, 118, 225, 266, 292, 332, 338, 340, 395
Lamerton, L. F., 303, **321**, 332, 348, 354, 387
Langendorff, H., **257**
Langendorff, M., **257**
Larsson, L.-G., 309
Laszlo, D., **298**
Latarjet, R., **281**
Lavrick, V., **231**
Ledlie, E. M., **382**, 387
Levtchouk, G., **231**
Lewin, R., **298**, 303, 304
Lewis, Y. S., **73**
Lindell, B., **376**
Loutit, J. F., 144, **197**, 203, 213, 247, 387, **388**, 395, 396, 403
Lüning, K. G., **425**, 484

Maas, H., **43**, **176**
Mackay, N. R., 293
Maisin, H., **341**
Maisin, J., **341**
Maldague, P., **341**
Mandel, P., **59**
Marcovich, H., 79, 84, 132, 168, **192**, 196, **281**, 283, 284
Martinovic, P., **237**
Mendelsohn, M. L., **38**, 42
Mitchell, J., 174
Mossige, J. C., **457**, 485
Muller, H. J., 145, **407**, 413, 414, 415, 423, 432, 436, 473
Munro, R., **108**, 118, 120
Munson, R. J., **105**, 117, 118

Nelson, A., **294**
Nordback, K., **481**

Oftedal, P., 168, **457**, 462
Oliver, R., **54**
Ord, M. G., **65**
Oster, I. I., **407**, 414, 447, **475**, 480
Owen, M., **287**, 292, 293

Paterson, E., **357**
Pauly, H., **25**
Pavlović, M., **237**
Pihl, A., **147**, 158, 159

Rajewsky, B., 7, **25**, 32, 89, 119, 120, 265, **267**
Rathgen, G. H., **43**
Révész, L., **80**, 84, 85
Rodesch, J., **59**
Rosoff, B., **298**

Schraub, A., **310**
Scott, O. C. A., 85, 169
Sherman, F. G., **49**, 53, 117
Simmons, E. L., **214**, 220, 396
Simon-Reuss, I., **114**
Sladić, G., **237**

Slouka, V., **76**, 79
Smith, C. L., **114**, 119, 292
Smith, D. E., **73**, 75, 79, 168
Sobels, F. H., 100, 145, 284, **449**, 455, 456, 480
Soška, J., **204**, 210
Stein, G., **16**
Stern, K. G., **298**
Stocken, L. A., **65**, 68, 69, **211**
Swallow, A. J., 14, **16**, 99, 100, 133, 168

Taylor, D., 339
Tobias, C. A., 101, 424
Troitsky, V. L., **221**, 225, 284
Tumanjan, M. A., **221**

van Bekkum, D. W., 132, **134**, 196, 203, 210, 253, 332
Vaughan, J., **287**
von Wettstein, 414
Vos, O., **134**, 144, 145, 146, 210

Westergaard, M., 414, 447, 455
Williams, K., 292, 304
Wolf, I., **267**
Wolff, S., 202, 252, 253, 455, **463**, 471, 472, 473

Zajicek, J., **376**, 380, 381
Zekhova, Z., **231**
Zimmer, K. G., 13

PRINTED IN GREAT BRITAIN BY
OLIVER AND BOYD LTD.
EDINBURGH